The Future of Life (2002)

Pheidole in the New World: A Hyperdiverse Ant Genus (2003)

From So Simple a Beginning: The Four Great Books of Darwin, edited with introductions (2005)

Nature Revealed: Selected Writings, 1949–2006 (2006)

The Creation: An Appeal to Save Life on Earth (2006)

THE SUPERORGANISM

The Beauty, Elegance, and Strangeness of Insect Societies

BERT HÖLLDOBLER

AND

EDWARD O. WILSON

LINE DRAWINGS BY MARGARET C. NELSON

W. W. NORTON & COMPANY

NEW YORK • LONDON

Frontispiece: Plate 1. Weaver ants (*Oecophylla smaragdina*) cooperate in arranging leaves during the construction of their leaf tent nests.

Printed in the United States of America
First Edition

For information about special discounts for bulk purchases, please contact W. W. Norton Special Sales at specialsales@wwnorton.com or 800-233-4830

Manufacturing by RR Donnelley, Willard, OH
Book design by Abbate Design
Production manager: Anna Oler

Library of Congress Cataloging-in-Publication Data

Hölldobler, Bert, 1936–
The superorganism : the beauty, elegance, and strangeness of insect societies / Bert Hölldobler and Edward O. Wilson ; line drawings by Margaret C. Nelson. — 1st ed.
p. cm.
Includes bibliographical references and index.
ISBN 978-0-393-06704-0 (hardcover)
1. Insect societies. I. Wilson, Edward O. II. Title.
QL496.H65 2009
595.7'1782—dc22

2008038547

W. W. Norton & Company, Inc.
500 Fifth Avenue, New York, N.Y. 10110
www.wwnorton.com

W. W. Norton & Company Ltd.
Castle House, 75/76 Wells Street, London W1T 3QT

1 2 3 4 5 6 7 8 9 0

FOR MARTIN LINDAUER, *our colleague and friend, whose pioneering work and inspiration in experimental sociobiology contributed greatly to the conception of an insect society as a functional superorganism*

Let him who boasts the knowledge of actually existing things, first tell us of the nature of the ant.

—ST. BASIL

CONTENTS

NOTE TO THE GENERAL READER

Imagine that 1 million years ago, long before the origin of humanity, a team of alien scientists landed on Earth to study its life-forms. Their first report would surely include something like the following: *This planet is teeming with more than 1,000 trillion highly social creatures, representing at least 20,000 species!* Their final report would surely contain the following key points:

- Most of the highly social forms are insects (six legs, two antennae on the head, three body parts). All live on the land, none in the sea.
- At maturity, each colony contains as few as 10 members to as many as 20 million members, according to species.
- The members of each colony are divided into two basic castes: one or at most a small number of reproductives and a larger number of workers who conduct the labor in an altruistic manner and do not, as a rule, attempt to breed.
- In the great majority of the colonial species—namely, those belonging to the order Hymenoptera (ants, bees, wasps)—the colony members are all female. They produce and care for males during short periods of time prior to the mating season. The males do no work. After the mating season, any of these drones that remain in the nest are expelled or killed by their worker sisters.
- On the other hand, in a minority of the highly social species, belonging to the order Isoptera (termites), a king typically lives with the queen, the reproductive female. Unlike hymenopteran workers, those of termites often belong to both sexes, and in some species, labor is divided to some degree between the sexes.
- More than 90 percent of the signals used in communication by these strange colonial creatures are chemical. The substances, the pheromones, are released

from exocrine glands located in various parts of the body. When smelled or tasted by other colony members, they evoke a particular response, such as alarm, attraction, assembly, or recruitment. Sound or substrate-borne vibrations and touch are also used by many species in communication, but ordinarily just to augment the effects of pheromones. Some signals are complex, combining smell, taste, vibration (sound), and touch. Notable examples are the waggle dance of honeybees, the recruitment trails of fire ants, and the multimodal communication of weaver ants.

- The social insects distinguish their own nestmates from members of other colonies by using receptors on their antennae to smell the hydrocarbons in the outer layer of their hard-shelled cuticles. They use different blends of these chemicals to identify different castes, life stages, and ages among their nestmates.

- Each colony is integrated tightly enough by its communication system and caste-based division of labor to be called a superorganism. The social organizations, however, vary greatly among the social insect species, and we can recognize different evolutionary grades of superorganismic organization. A "primitive" (less derived) grade is represented by several ponerine species, where members of the colony have full reproductive potential and there is considerable interindividual reproductive competition within each colony. Highly advanced grades are represented, for example, by the leafcutter ant genera *Atta* and *Acromyrmex* and the *Oecophylla* weaver ants, where the queen caste is the sole reproductive, and the hundreds of thousands of sterile workers occur as morphological subcastes that are tightly integrated in division of labor systems. These societies exhibit the ultimate superorganism states, where interindividual conflict within the colony is minimal or nonexistent.

- The superorganism exists at a level of biological organization between the organisms that form its units and the ecosystems, such as a forest patch, of which it is a unit. This is why the social insects are important to the general study of biology.

Such is the array of phenomena on which we two Earth-born biologists will now expand. The ants, bees, wasps, and termites are among the most socially advanced nonhuman organisms of which we have knowledge. In biomass and impact on ecosystems, their colonies have been dominant elements of most of the land habitats for at least 50 million years. Social insect species existed for more than an equivalent span of time previously, but were relatively much less common. Some of the ants, in particular, were similar to those living today. It gives pleasure to think that they stung or sprayed formic acid on many a dinosaur that carelessly trampled their nests.

The modern insect societies have a vast amount to teach us today. They show how it is possible to "speak" in complex messages with pheromones. And they illustrate, through thousands of examples, how the division of labor can be crafted with flexible behavior programs to achieve an optimal efficiency of a working group. Their networks of cooperating individuals have suggested new designs in computers and shed light on how neurons of the brain might interact in the creation of mind. They are in many ways an inspiration. The study of ants, President Lowell, of Harvard University, said when he bestowed an honorary degree on the great myrmecologist William Morton Wheeler in the 1920s, has demonstrated that these insects, "like human beings, can create civilizations without the use of reason."

The superorganisms are the clearest window through which scientists can witness the emergence of one level of biological organization from another. This is important, because almost all of modern biology consists of a process of reduction of complex systems followed by synthesis. During reductive research, the system is broken down into its constituent parts and processes. When they are well enough known, the parts and processes can be pieced back together and their newly understood properties used to explain the emergent properties of the complex system. Synthesis is in most cases far harder than reduction. For example, biologists have come far in defining and describing the molecules and organelles that compose the foundation of life. At the next higher major level of biological organization, biologists have further described in precise detail many of the emergent structures and properties of cells. But this achievement is still a long way from understanding fully how molecules and organelles are assembled, arranged, and activated to create a complete living cell. Similarly, biologists have learned the properties of many of the species that compose the living parts of a few ecosystems—for example, ponds and forest patches. They have worked out large-scale processes, including material and energy cycles. But they are far from mastering the many complex ways in which species interact to create the higher-level patterns.

Social insects, in contrast, offer a far more accessible connection between two levels of biological organization. The lower-level units in this case, the organisms, are relatively simple in the way they interact to create colonies, and thus, the colonies themselves are not nearly so complex in structure and operation as cells and ecosystems. Both of these levels, organism and colony, can be easily viewed and experimentally manipulated. As we will show in the chapters that follow, it is now possible to press far ahead in this fundamental enterprise of biology.

We will conclude this introduction with a guess. If alien scientists had landed to study Earth's prehuman biosphere, one of their first projects would have been to set up beehives and ant farms. This is our biased guess, because we have been fascinated by the social insects, and in particular the ants, during our entire scientific lives. The reader will find this slant throughout this book. We have chosen examples mainly from the ants and focused on those with which we are most familiar, but we repeatedly "glance over the fence," and especially to the honeybees, the best studied of the social insect species. This book is not intended to be as comprehensive a monograph as *The Ants* (1990). Rather, our intention here is to present the rich and diverse natural history facts that illustrate superorganismic traits in insect societies and to trace the evolutionary pathways to the most advanced stages of eusociality. Our intent in doing so is to revive the superorganism concept, with emphasis on colony-level adaptive traits, such as division of labor and communication. Finally, in presenting the subject this way, we visualize the colony as a self-organized entity and a target of natural selection.

In this book, we view the insect colony as the equivalent of an organism, the unit that must be examined in order to understand the biology of colonial species. Consider one of the most organism-like of all insect societies, the great colonies of the African driver ants. Viewed from afar, the huge raiding column of a driver ant colony seems like a single living entity. It spreads like the pseudopodium of a giant amoeba across 70 meters or so of ground. A closer look reveals it to comprise a mass of several million workers running in concert from the subterranean nest, an irregular network of tunnels and chambers dug into the soil. As the column emerges, it first resembles an expanding sheet and then metamorphoses into a treelike formation, with the trunk growing from the nest, the crown an advancing front the width of a small house, and numerous branches connecting the two. The swarm is leaderless. The workers rush back and forth near the front. Those in the vanguard press forward for a short distance and then turn back into the tumbling mass to give way to other advancing runners. These predatory feeder columns are rivers of ants coming and going. The frontal swarm, advancing at 20 meters an hour, engulfs all the ground and low vegetation in its path, gathering and killing all the insects and even snakes and other larger animals unable to escape. After a few hours, the direction of the flow is reversed, and the column drains backward into the nest holes.

To speak of the colonies of driver ants—or other social insects, such as the gigantic colonies of leafcutter ants (described in Chapter 9), the honeybee societies, or the termite colonies—as more than just tight aggregations of individuals is to conceive of superorganisms and invite a detailed comparison between the society and a conventional organism.

In the 18 years since we wrote *The Ants*, an astounding wealth of information has been revealed from the phylogenetic primitive (ancestral) ant species belonging to the poneromorph group, the subject to which Chapter 8 is devoted. Although some species of this group exhibit all the key superorganismic traits, such as castes, division of labor, and sophisticated communication (topics treated in Chapters 5 and 6), the societies of many other poneromorph species are characterized by intense competition among nestmates for reproductive privileges. Group members are organized in dominance hierarchies, which, from time to time, are challenged and overthrown by members of the society ready to take the top position. Although the division of labor and communication in these societies is quite primitive, the behavioral interactions among nestmates are complex, with dominance displays and submissive behaviors, chemical signaling of reproductive status, and even individual recognition. These societies exhibit superorganismic traits, but are far from possessing the ultimate superorganismic organization exhibited by the driver ants and the leafcutter ants. ■

THE SUPERORGANISM

|| Plate 2. Two-meter-high mound of the nest of a colony of *Formica polyctena* in a forest in southern Finland.

1

THE CONSTRUCTION OF A SUPERORGANISM

Consider a honeybee gathering nectar from a flower bed. Although simple in appearance, the act is a performance of high virtuosity. The forager was guided to this spot by dances of her nestmates that contained symbolic information about the direction, distance, and quality of the nectar source. To reach her destination, she traveled the bee equivalent of hundreds of human miles at bee-equivalent supersonic speed. She has arrived at an hour when the flowers are most likely to be richly productive. Now she closely inspects the willing blossoms by touch and smell and extracts the nectar with intricate movements of her legs and proboscis. Then she flies home in a straight line. All this she accomplishes with a brain the size of a grain of sand and with little or no prior experience.

Our forager is part of a superorganism, a colony with many of the attributes of an organism but one step up from organisms in the hierarchy of biological organization. The basic elements of the superorganism are not cells and tissues but closely cooperating animals. To follow one bee home, to peer into the hive she enters, to observe the mass of nest inhabitants in their full organized frenzy is to understand why social insects—the colonial bees, wasps, ants, and termites—are species for species the most abundant of land-dwelling arthropods. Although they represent only 2 percent of the approximately 900,000 known insect species in the world, they likely compose more than half the biomass. In a patch of Amazonian rain forest near Manaus, where a measurement was actually made, social insects composed 80 percent. Ants and termites alone composed nearly 30 percent of the entire animal biomass in this same sample, and ants alone weighed four times as much as the combined mammals, birds, reptiles, and amphibians.[1] Social insects

1 | E. J. Fittkau and H. Klinge, "On biomass and trophic structure of the central Amazonian rain forest ecosystem," *Biotropica* 5(1): 2–14 (1973).

prevail at every level in all forests around the world except the coldest and wettest. In one sample from the canopy of Peruvian rain forest, ants made up 69 percent of all the individual insects.[2] In this specialized environment, they function not only as predators and scavengers but also as cryptic herbivores, collecting the rich sugary excrement of aphids, treehoppers, and other sap-feeding homopterous insects they tend like cattle.[3]

An odd parity exists between the social insects and humanity. About 6.6 billion individuals compose *Homo sapiens*, the most social and ecologically successful species in vertebrate history. And the number of ants alive at any given time has been estimated conservatively at 1 million billion to 10 million billion. If this latter estimate is correct, and given that each human weighs on average very roughly 1 or 2 million times as much as a typical ant, then ants and people have (again, very roughly) the same global biomass.[4]

WHY COLONIES ARE SUPERIOR

Environmental domination by ants and other social insects is the result of cooperative group behavior. When multiple workers address the same tasks, they use "series-parallel" operations: each worker can switch from one task to another as need demands, so that no task goes unattended for long and each step in the task is soon completed. Workers are also more inclined than solitary insects to be aggressive, even suicidal. There is little Darwinian loss in their bravery: individual casualties incurred during foraging and nest defense leave unharmed the rest of the colony, especially the all-important reproductive caste, and lost workers are soon replaced. In addition to this fighting edge, the enlarged insect power and coordinated actions

---

2 | T. L. Erwin, "Canopy arthropod biodiversity: a chronology of sampling techniques and results," *Revista Peruana de Entomologia* 32: 71–77 (1989).

3 | J. H. Hunt, "Cryptic herbivores of the rainforest canopy," *Science* 300: 916–917 (2003).

4 | The number of insects alive on Earth at any given time has been calculated by ecologist Carrington Bonner Williams to be, to the nearest order of magnitude, 1 billion billion, or 10^{18}; see C. B. Williams, *Patterns in the Balance of Nature and Related Problems in Quantitative Ecology* (New York: Academic Press, 1964). We suppose that ants make up 10 percent, order of magnitude, of the individual living insects worldwide, with tropical forests and all other terrestrial and aquatic habitats considered collectively and the large numbers of tiny collembolans and comparably small insects included. We also suppose the average ant to have a dry weight of about 0.5 to 1.0 milligrams and humans to have an average dry weight of about 10 kilograms.

enable members of colonies to construct complex nests with superior defensive ramparts and interior microclimate control.

Endowed with the advantages of colonial life, the social insects have managed to displace solitary insects, such as cockroaches, grasshoppers, and beetles, from the most favored nest sites and defensible foraging ranges. In the most general terms, social insects control the center of the land environment, while solitary insects predominate in the margins. Where social insects take territorial possession of the larger and more enduring spaces of the vegetation and ground, the solitary forms occupy the peripheral twigs, leaf surfaces, mudflats, and wet or very dry and crumbling portions of dead wood. In short, solitary forms tend to prevail over social insects only in the more remote and transient of living spaces.[5]

THE CONSTRUCTION OF SUPERORGANISMS

Reflection on the success of social life allows us to address a classic question of biology: *How does a superorganism arise from the combined operation of tiny and short-lived minds?* The answer is relevant to studies of lower levels of biological organization and the related question that also presents itself: *How does an organism arise from the combined operation of tiny and short-lived cells?*

The object of most research conducted on social insects during the past half century can be expressed in a single phrase: *the construction of superorganisms*. The first level of construction is sociogenesis, the growth of the colony by the creation of specialized castes that act together as a functional whole. Castes are created by algorithms of development, the sequential decision rules that guide the body growth of each colony member step by step until the insect reaches its final, adult stage. In the social hymenopterans (ants, social bees, and social wasps), the sequence is roughly as follows. At the first decision point, depending on its physiological condition, the developing female egg or larva is shunted onto one or the other of two paths of physical development. If the immature insect takes the path leading to more extended growth and development, it will turn into a queen upon reaching

---

5 | General accounts of the dominance of social insects and the reasons for it are given in E. O. Wilson, *Success and Dominance in Ecosystems: The Case of the Social Insects* (Oldendorf/Luhe, Germany: Ecology Institute, 1990); and B. Hölldobler and E. O. Wilson, *The Ants* (Cambridge, MA: The Belknap Press of Harvard University Press, 1990).

the adult stage. If it takes the other path, it will curtail growth and development and end up a worker. In some species of ants, the worker-bound larva encounters a second decision point on the road to adulthood, from which one path leads it to maturity as a major worker ("soldier") and the other to maturity as a minor worker.

These specialists, working together as a functional unit, are guided by sets of behavioral rules that operate in the following manner. If in a given context the worker encounters a certain stimulus, it predictably performs one act, and if the same stimulus is received in a different context, the worker performs a different act. For example, if a hungry larva is encountered in the brood chamber, the worker offers it food; if a larva is found elsewhere, the worker carries it, whether hungry or not, to the brood chamber and places it with other larvae. And so on through a repertory of a few dozen acts. The totality of these relatively sparse and simple responses defines the social behavior of the colony.

Nothing in the brain of a worker ant represents a blueprint of the social order. There is no overseer or "brain caste" who carries such a master plan in its head. Instead, colony life is the product of self-organization. The superorganism exists in the separate programmed responses of the organisms that compose it. The assembly instructions the organisms follow are the developmental algorithms, which create the castes, together with the behavioral algorithms, which are responsible for moment-to-moment behavior of the caste members.

The algorithms of caste development and behavior are the first level in the construction of a superorganism. The second level of construction is the genetic evolution of the algorithms themselves. Out of all possible algorithms, generating the astronomically numerous social patterns they might produce, at least in theory, only an infinitesimal fraction have in fact evolved. The sets of algorithms actually realized, each of which is unique in some respect to a living species, are the winners in the arena of natural selection. They exist in the world as a select group that emerged in response to pressures imposed by the environment during the evolutionary history of the respective species.

THE LEVELS OF ORGANIZATION

Life is a self-replicating hierarchy of levels. Biology is the study of the levels that compose the hierarchy. No phenomenon at any level can be wholly characterized without incorporating other phenomena that arise at all levels. Genes prescribe proteins, proteins self-assemble into cells, cells multiply and aggregate to form organs, organs arise as parts of organisms, and organisms gather sequentially into societies, populations, and ecosystems. Natural selection that targets a trait at any of these levels ripples in effect across all the others. All levels of organization are primary or secondary targets of natural selection. For example, the genes that distinguish the Africanized honeybee (or "killer bee"), which was accidentally introduced into Brazil in the 1950s, include induction of restless and aggressive behavior in workers. Under free-living conditions, Africanized colonies outcompete those of other strains. To some extent, they also penetrate and alter wild environments, including especially the canopies of tropical forests.

As ecosystems change by biological invasions, such as those of the Africanized honeybees, or by shifts in climate or by any other means, the relative abundances of the species composing the ecosystems also change. Some species are likely to drop out and new ones invade. As a consequence, the selection pressures on the individuals and societies are altered, with eventual consequences for the inherited traits of at least some of the species.

The dynamism of ecosystems is consequently eternal. Biological hierarchies are reverberating systems within which, depending on the histories of the species and the environmental niches they occupy, social order may or may not evolve.

The principal target of natural selection in the social evolution of insects is the colony, while the unit of selection is the gene. Because the traits of the colony are summed products of the traits of the colony members and those traits differ genetically among the members, as well as from one colony to the next, the evolution of the social insects is grounded in the flux of changing gene frequencies across generations. That flux in turn reflects the complex interplay of behavior both by colonies and the individual members that compose them.

EUSOCIALITY AND THE SUPERORGANISM

The sociobiology of insects is most effectively constructed with the concept of the superorganism, with reference to both its origin and evolution. Which of the insect societies deserve to be designated a superorganism? In the broadest sense, the term *superorganism* is appropriate for any insect colony that is eusocial, or "truly social," and that means combining three traits: first, its adult members are divided into reproductive castes and partially or wholly nonreproductive workers; second, the adults of two or more generations coexist in the same nests; and third, nonreproductive or less reproductive workers care for the young. For those who prefer a stricter definition, the term *superorganism* may be applied only to colonies of an advanced state of eusociality, in which interindividual conflict for reproductive privilege is diminished and the worker caste is selected to maximize colony efficiency in intercolony competition.[6]

In the chapters that follow, we will draw examples from insects and arthropods at all levels of social evolution, with emphasis on the eusocial species. The ants, for example, are all eusocial,[7] and they also vary enormously, according to species, in the complexity of their social organization. Specifically, they differ widely in mature colony size and the degree to which the workers are specialized for particular tasks. They further vary substantially in the rate of information exchange among colony members, the number of kinds of behavioral acts performed by the colony as a whole, and the amount of collaboration by workers in the performance of these acts, as well as in the architecture of their nests, including the homopterous "cattle" sheds and other physical structures they build.

At one end of the spectrum in the social evolution of ants are the anatomically primitive *Prionomyrmex* (formerly *Nothomyrmecia*) *macrops* "dawn ants" of Australia and species of the cosmopolitan genus *Amblyopone*. Their colonies, with fewer than 100 workers, employ only elementary communication signals. They engage in little or no division of labor other than that distinguishing queens and workers, and they build simple nests. At the opposite extreme of the spectrum are the mighty *Atta* leafcutters, doryline driver ants, ecitonine army ants, and *Oecophylla* weaver ants, whose colonies of hundreds of thousands to millions of workers contain advanced caste systems. These "civilized" species employ complex division of labor and communication systems, and they build elaborate nests, such as the silken pavilions of

---

6 | H. K. Reeve and B. Hölldobler, "The emergence of a superorganism through intergroup competition," *Proceedings of the National Academy of Sciences USA* 104(23): 9736–9740 (2007).

7 | Among the ant species that have fewer than the full complement of eusocial traits are the workerless parasites. In addition, several ponerine species, such as *Pachycondyla* (formerly *Ophthalmopone*) *berthoudi*, whose mated sister workers reproduce (that is, they are "gamergates"), there is an absence of strict overlap between generations and no castes; see C. Peeters and R. Crewe, "Worker reproduction in the ponerine ant *Ophthalmopone berthoudi*: an alternative form of eusocial organization," *Behavioral Ecology and Sociobiology* 18(1): 29–37 (1985). Also, a few ant species reproduce by parthenogenesis; hence, there are no castes, and all members of each colony are clones; see K. Tsuji, "Obligate parthenogenesis and reproductive division of labor in the Japanese queenless ant *Pristomyrmex pungens*: comparison of intranidal and extranidal workers," *Behavioral Ecology and Sociobiology* 23(4): 247–255 (1988).

the *Oecophylla* weaver ants or, in the case of the ecitonine army ants, shelters created by an acrobatic interlocking of their own bodies.

Between the two extremes, occupying almost every conceivable point in the gradient of social complexity, are thousands of other ant species. Together they provide a clear view of the likely evolutionary origins of the intermediate and advanced grades of superorganisms.

A BRIEF HISTORY OF INSECT SOCIOBIOLOGY

The concept of the superorganism is venerable in its own evolution. It arose during a period of intense interest in evolutionary philosophy in the late nineteenth and early twentieth centuries. A succession of prominent thinkers, including Ernst Haeckel, Herbert Spencer, and Giti Fechner, wrote of the hierarchical structure underlying order in the entire universe, and they expatiated on the unique properties that emerge within each level of the grand order of creation. William Morton Wheeler, in his famous 1911 essay "The Ant-Colony as an Organism," brought the concept explicitly into sociobiology. "The ant-colony is an organism," he wrote, "and not merely the analogue of the person."[8] The colony, Wheeler pointed out, has several diagnostic qualities of this status:

1 | It behaves as a unit.
2 | It shows some idiosyncrasies in behavior, size, and structure, some of which are peculiar to the species and others of which distinguish individual colonies belonging to the same species.
3 | It undergoes a cycle of growth and reproduction that is clearly adaptive.
4 | It is differentiated into "germ plasm" (queens and males) and "soma" (workers).

Wheeler, in his later summary work *The Social Insects, Their Origin and Evolution* (New York: Harcourt Brace, 1928), was also the first to call the social insect colony a superorganism. He reinforced the concept of social homeostasis, consisting of the physiological and behavioral processes by which the colony keeps itself

8 | W. M. Wheeler, "The ant-colony as an organism," *The Journal of Morphology* 22(2): 307–325 (1911).

in optimal condition for growth and reproduction. "We have seen," he explained, "that the insect colony or society may be regarded as a super-organism and hence as a living whole bent on preserving its moving equilibrium and integrity."

The history of insect sociobiology can be fruitfully viewed as the evolution of the superorganism concept as it has waxed and waned and waxed again.[9] Of all the species whose colonies rank as advanced superorganisms, the best known, and indeed one of the best-known animal species of any kind, is the honeybee *Apis mellifera*.[10] Advanced superorganisms also exist among the termites, reaching an apogee among the mound-building macrotermitines of the African tropics. But the social insects that boast the greatest number of such extreme species, that embrace the largest number of evolutionary lines, and that have been studied across the most species, especially during the last several decades, are the ants. These insects also happen to have been the focus of our own personal research and will be the principal subject of the accounts to follow.

In general, bees and wasps offer the important advantage to scientists of many living species still at the earliest and intermediate stages of colonial evolution; thus, they display most clearly the likely evolutionary origins of social life itself. Ants and termites, on the other hand, reveal little of the first stages of colonial evolution, because all their species are eusocial; but in compensation, they have the most to tell us concerning the evolution of superorganisms. Of these two hegemonic groups, ants are by far the more diverse—more than 14,000 species of ants are known versus about 2,000 species of termites—and their biology has been better studied.

---

9 | Because in addition to cooperation, conflict also occurs or at least has the potential to occur due to some forms of opposing interests among genetically differing members of the same colony, F. L. W. Ratnieks and H. K. Reeve have suggested that *superorganism* as a unit term is problematic and might better be replaced by reference to "community of interests," with some behaviors having superorganismic qualities and others not; see F. L. W. Ratnieks and H. K. Reeve, "Conflict in single-queen hymenopteran societies: the structure of conflict and processes that reduce conflict in advanced eusocial species," *Journal of Theoretical Biology* 158(1): 33–65 (1992). We disagree, holding that while ambiguities do exist, the term *superorganism* is sufficiently clear-cut and more than sufficiently heuristic to justify using it to denote a fundamental unit of biological organization. See also T. D. Seeley, "The honey bee colony as a superorganism," *American Scientist* 77(6): 546–553 (1989).

10 | R. F. A. Moritz and E. E. Southwick, *Bees As Superorganisms: An Evolutionary Reality* (New York: Springer-Verlag, 1992); T. D. Seeley, *The Wisdom of the Hive: The Social Physiology of Honey Bee Colonies* (Cambridge, MA: Harvard University Press, 1995); The Honeybee Genome Sequencing Consortium (C. W. Whitfield, G. E. Robinson, et al.), "Insights into social insects from the genome of the honeybee *Apis mellifera*," *Nature* 443: 931–949 (2006); and R. E. Page Jr. and G. V. Amdam, "The making of a social insect: developomental architectures in social design," *BioEssays* 29(4): 334–343 (2007).

To gain perspective at this point from our narrative, the following thumbnail sketch of the history of myrmecology (the scientific study of ants) might be helpful.

Leaving aside pioneering but minimally influential works of René Antoine Ferchault de Réaumur (*Mémoires pour Servir à l'Histoire des Insectes* [Amsterdam: Pierre Motier, 1737]) and the Reverend William Gould (*An Account of English Ants* [London: A. Millar, 1747]), the modern scientific study of ants can fairly be said to have been launched in 1810 by Pierre Huber's *Recherches sur les Moeurs des Fourmis Indigènes* (Paris: Chez J. J. Paschoud). For the next 150 years, myrmecology consisted largely of taxonomy and natural history. This foundational descriptive work was rich and productive, and it continues unabated today: possibly half the species remain undiscovered, and of those given a scientific name, only a tiny fraction—1 percent or fewer—have been examined intensively in the field or laboratory. Much of the pleasure in the study of ants still consists of discovering new forms of social behavior and ecological adaptations in little-known groups, nowadays mostly in the tropics, and applying that knowledge to improve reconstructions of ant evolution.

Since around 1950, the number of researchers and publications in myrmecology has grown exponentially, while the range of topics addressed has expanded at nearly equal pace. At the risk of oversimplification, the history of this past six decades can be encapsulated as follows.

From the 1950s through the 1970s, researchers worked out much of the basic plans of chemical communication, the evolution of caste systems, and many of the physiological factors that determine caste in a diversity of ant species. This work played a key role in the foundation of sociobiology.

In the 1970s and 1980s, sociobiology was established as a new discipline built on physiology, ecology, and evolutionary theory. In this synthesis, the social insects were given a central role. Toward the end of this period, attention was focused especially on the forces of selection that shape colony structure and life cycles. But ants in particular came to play an important role in general population and community ecology, particularly in studies of communal foraging and competition.

The 1990s and early 2000s saw important advances in analyzing the self-organization of colonies based on simple rules of individual worker behavior. A productive branch of population genetics also emerged: sociogenetics, the analysis of genetic relatedness among colonies and members as well as the hereditary basis of some forms of social behavior. It was quickly followed by a first effort at sociogenomics, the decoding of entire genomes in the search for the genes critical for social evolution. In 2006, the complete sequencing of the honeybee genome was announced.

Although each decade saw the appearance of newly popular topics, with small research industries growing around them, studies on those enjoying earlier favor continued without interruption. The two oldest subjects, systematics and scientific natural history, actually experienced a renaissance during the 1980s and 1990s and accelerated during the 2000s.

All this research has created a linkage, albeit still tenuous in places, across all levels of biological organization, from molecule to ecosystem. We think it timely to attempt the revised synthesis based on that perception. In upgrading the concept of the superorganism as an organizing theme, we will address not just the analogies frequently cited for a century between organisms and superorganisms but also, and at increasing depth, the principles by which the two entities are built and maintained.

The chapters that follow expand on the tentative conclusions just cited. We trust that our exposition will make clear the substantial importance of social insects for general biology. The content is also arranged to present sociobiology as it truly is, in its full range, aligning cause-and-effect explanations from genetics to behavioral science and ecology. By shifting emphasis back toward empirical studies of colony-level selection, the prime mover of social evolution, we aim also to promote a more fruitful union between sociobiology and behavioral ecology. Finally, by stressing the algorithms that direct the self-construction of colonies, we hope to assist in establishing more clearly the relevance of sociobiology to the general principles of developmental biology and systems theory.

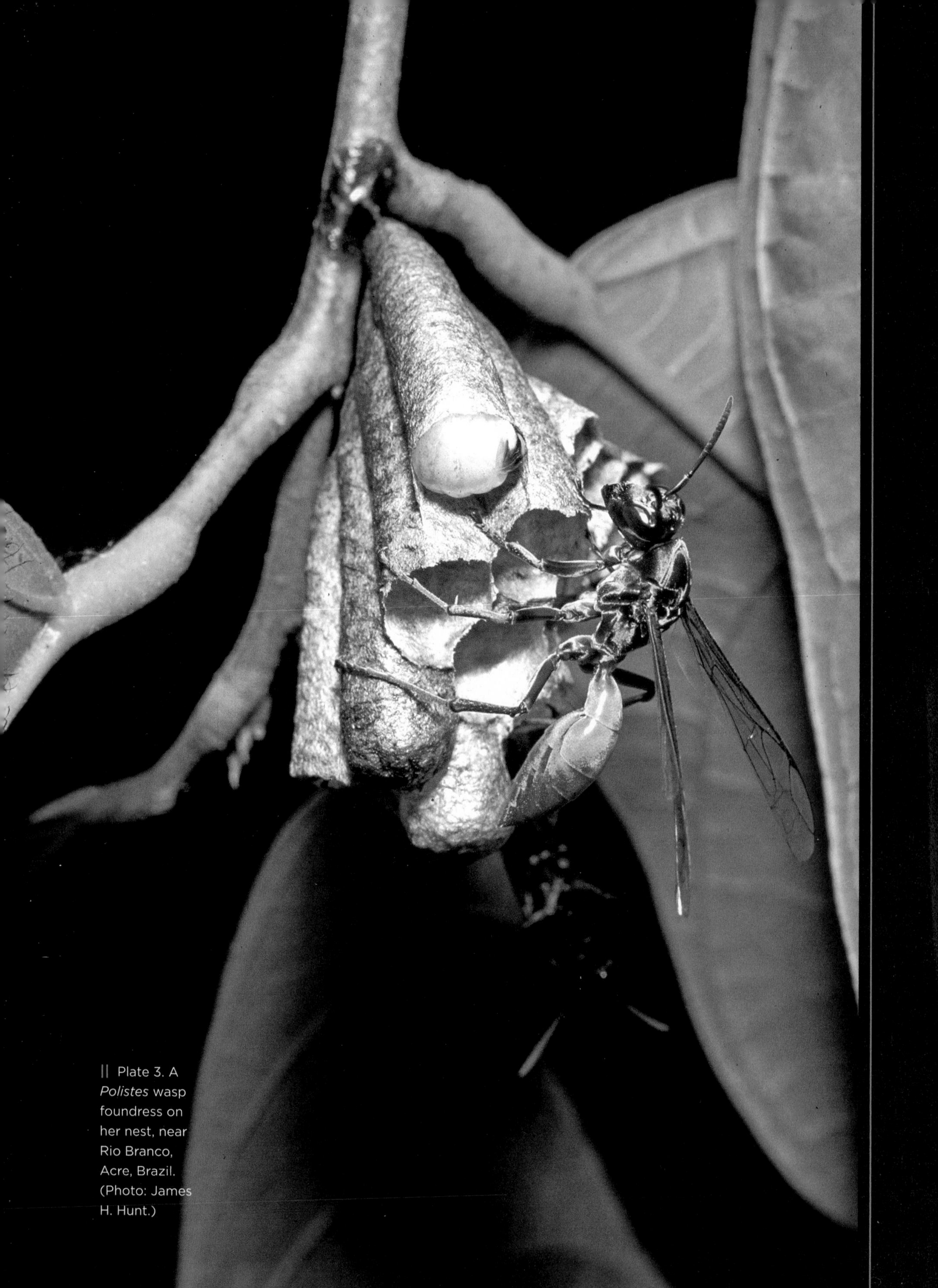

|| Plate 3. A *Polistes* wasp foundress on her nest, near Rio Branco, Acre, Brazil. (Photo: James H. Hunt.)

2

GENETIC SOCIAL EVOLUTION

Through geological time, organisms of many species, from bacteria to human beings, have evolved to live at least part of their lives in groups. A very few of these, distributed sparsely across the multitude of animal phylogenetic lines, have divided their societies into specialized reproductive and nonreproductive castes to create superorganisms. Much of what is known of this second advance comes from the more than 16,000 known species of highly social insects and other arthropods.[1]

AN ABRIDGED HISTORY OF THE GENETIC THEORY OF SOCIAL EVOLUTION[2]

The story of the genetic theory begins in 1859 with Charles Darwin's *On the Origin of Species.* In constructing the theory of evolution by natural selection, the master naturalist encountered in the colonies of social insects the "one special difficulty, which first appeared to me insuperable, and actually fatal to my theory." How, he asked, could the worker castes of ants and other social insects have evolved if they are sterile and leave no offspring? Moreover, how could sterile lines of descent differentiate into different castes, so strikingly exemplified by the highly specialized soldiers and minor workers of many ant and termite species? The answer is simple

1 | This estimate of the number of species applies to the advanced, "eusocial" grade of evolution (with reproductive and nonreproductive castes, overlapping adult generations, and brood care). It is based on data in E. O. Wilson, *The Insect Societies* (Cambridge, MA: The Belknap Press of Harvard University Press, 1971), updated to 2007.

2 | A second history, based on some important differences in perspective, is presented in D. S. Wilson and E. O. Wilson, "Rethinking the theoretical foundation of sociobiology," *The Quarterly Review of Biology* 82(4): 327–348 (2007); and in D. S. Wilson and E. O. Wilson, "Survival of the selfless," *New Scientist* 196: 42–46 (2007).

and was no doubt easily grasped by Darwin: the target of selection in these insects is the family, not the individual. Suppose that the same elements of heredity—today one would say the genes—are plastic in effect, able to produce within the same family both the royal castes, responsible for reproduction, and the sterile worker castes, responsible for labor. Suppose further that families able to generate sterile worker castes are able to survive and reproduce better than families unable to generate such castes. Then the hereditary elements responsible for the caste systems would spread through the population of competing families.

For Darwin, the evolution of insect colonies resembled the creation of domestic vegetables through artificial selection. By saving and planting only the seeds of those plants whose nonreproductive parts grow largest and taste best, the domestic species is forced to evolve in a direction favored by the gardener. Of the theoretically troublesome ants he wrote:

> I believe that natural selection, by acting on the fertile parents, could form a species which regularly produce neuters, either all of a large size with one form of jaw, or all of a small size with jaws having a widely different structure; or lastly, and this is the climax of our difficulty, one set of workers of one size and structure, and simultaneously another set of workers of different size and structure. (*The Origin of Species*, p. 241)

Darwin has been proven essentially correct in the idea of colony-level selection, but he overlooked an important detail. By picturing the insect colony as genetically uniform, with the sterile workers mere hereditary duplicates of the queen and her mate, he oversimplified social evolution. He did not know about the scrambling actions of meiosis and recombination in Mendelian heredity, causing the workers of each colony to vary considerably among themselves in their hereditary makeup. He was also unaware that in most species of ants, social bees, and social wasps, the workers (all of which are female) have ovaries, and many are capable of reproduction on their own. In other words, workers are often potential reproductive rivals of their mother and of each other.

After particulate-and-recombinant heredity was established as a fact, in the early twentieth century, it was the geneticist Alfred H. Sturtevant in 1938 who first realized that evolution in the social insects is driven not just by one but by three levels of selection, namely, those forces operating between members of the same colony, those between colonies, and those between entire populations of colonies.[3] Furthermore, the selection forces at the different levels can either work in concert, rendering cooperation and social evolution more likely, or work against one another, thus slowing, stalling, or even reversing social evolution.

Hence, the problem of sterile workers had not entirely gone away as a result of Darwin's insight concerning the well-flavored vegetable. Altruism, which is a contributive if not essential condition for the creation of a superorganism, still begged for a full explanation.

In 1945, Sewall Wright, one of the architects of the Modern Synthesis of evolutionary theory, addressed multilevel selection in an attempt to solve the altruism problem.[4] Wright combined selection with a boost in frequency of the altruism gene by genetic drift in small populations. The model was, however, notably incomplete in its explanation of the spread of the gene through populations as a whole. A further advance, addressing both group selection and the role of close hereditary kinship, was made by G. C. Williams and D. C. Williams in 1957.[5]

In 1932, however, J. B. S. Haldane, another of the architects of the Modern Synthesis, appears in a much overlooked paper to have already partly solved the final conundrum. Altruism, he observed, can evolve if by the hereditary acts of sacrifice the altruist increases the Darwinian success of relatives. "A consideration of these [altruistic] traits," he wrote, "involves the consideration of small groups. For a character of this type can only spread through the population if the genes determining it are borne by a group of related individuals whose chances of leaving offspring are increased by the presence of these genes in an individual member of the group whose own private viability they lower."[6]

In 1955, as part of an introduction to population genetics, Haldane provided

---

3 | A. H. Sturtevant, "Essays on evolution, II: on the effects of selection on social insects," *The Quarterly Review of Biology* 13(1): 74–76 (1938).

4 | S. Wright, "Tempo and mode in evolution: a critical review (a review of *Tempo and Mode in Evolution*, by George G. Simpson)," *Ecology* 26(4): 415–419 (1945). A superb review of the history of the development of the genetical theory of altruism, including accounts of the Wright and other key models, has been provided in E. Sober and D. S. Wilson, *Unto Others: The Evolution and Psychology of Unselfish Behavior* (Cambridge, MA: Harvard University Press, 1998).

5 | G. C. Williams and D. C. Williams, "Natural selection of individually harmful social adaptations among sibs with special reference to social insects," *Evolution* 11(1): 32–39 (1957).

6 | J. B. S. Haldane, *The Causes of Evolution* (London: Longmans, Green, 1932; paperback reprint, Ithaca, NY: Cornell University Press, 1966).

an exact solution to the altruism problem. He applied the problem to the behavior of human beings and social insects with a precision and clarity of language difficult to improve:

> Let us suppose that you carry a rare gene which affects your behaviour so that you jump into a flooded river and save a child, but you have one chance in ten of being drowned, while I do not possess the gene, and stand on the bank and watch the child drown. If the child is your own child or your brother or sister, there is an even chance that the child will also have the gene, so five such genes will be saved for one lost in an adult. If you save a grandchild or nephew the advantage is only two and a half to one. If you only save a first cousin, the effect is very slight. If you try to save your first cousin once removed the population is more likely to lose this valuable gene than to gain it. But on the two occasions when I have pulled possibly drowning people out of the water (at an infinitesimal risk to myself) I had no time to make such calculations. Paleolithic men did not make them. It is clear that genes making for conduct of this kind would only have a chance of spreading in rather small populations where most of the children were fairly near relatives of the man who risked his life. It is not easy to see how, except in small populations, such genes could have been established. Of course the conditions are even better in a community such as a beehive or ants' nest, whose members are all literally brothers and sisters."[7]

For another decade, Haldane's principle, later called kin selection,[8] languished unattended. In 1964, William D. Hamilton (who cited Haldane) published his highly influential genetic theory of social evolution. Hamilton built the solid mathematical foundation for kin selection and expressed it in the more general theory of inclusive fitness. Consistent with Haldane's inequality principle and extending it beyond survival to include reproductive success, Hamilton's rule, as it is usually called, states very simply that a hereditary altruistic trait will spread through a population if rb exceeds (>) c, where b is the increase in units of offspring produced by the recipient of the altruistic behavior, r is the fraction of genes that the altruist

---

7 | J. B. S. Haldane, "Population genetics," *New Biology* 18: 34–51 (1955). For an excellent account of the history of these developments in evolutionary biology, see L. A. Dugatkin, *The Altruism Equation: Seven Scientists Search for the Origins of Goodness* (Princeton, NJ: Princeton University Press, 2006).

8 | The term *kin selection* was first used by J. Maynard Smith, "Group selection and kin selection," *Nature* 201: 1145–1147 (1964).

shares with the recipient by common descent, and *c* is the cost to the altruist, also measured in units of offspring.[9]

Hamilton expanded this rule to the whole society, comprising the altruist plus all its beneficiaries within the whole society conceived as a network of interacting organisms. He called the total effect of all the interactions with the altruist its "inclusive fitness." Let us suppose, Hamilton said,

> that the genotype of *A* [the altruist] simply gives rise to a fixed pattern of social behavior and that this has fixed average effects on *A* and on relative *B*, and possibly on many other individuals as well. All the effects which *A* causes may be weighted by their approximate *b*s and collected together in a quantity which may be named the *inclusive fitness effect* of *A*. . . . .
>
> If an altruistic act by *A* greatly increases the fitness of *B*, *A*'s inclusive fitness may be increased in spite of decrease in *A*'s individual fitness.[10]

Inclusive fitness might have gone largely unnoticed for a long time, like Haldane's insight that preceded it, except for a brilliant observation that Hamilton added in his introductory paper of 1964.[11] From his experience as a student of social insects, he proposed what was later to be called the haplodiploid hypothesis.[12] All known species of the insect order Hymenoptera, which includes the ants, bees, and wasps, determine sex by haploid-diploidy or, more concisely expressed, haplodiploidy. Eggs that are haploid, that is, unfertilized and hence whose genes consist entirely of half the genes of the mother, become males. In contrast, eggs that are diploid, having been fertilized and whose genes come half from the mother and half from the father, become females.

As a consequence of haplodiploidy, a strange asymmetrical network of genetic relationships is set up among close relatives. All of a male's genes are identical to half the genes of his mother; the degree of genetic relationship of son and mother

---

9 | W. D. Hamilton, "The genetical evolution of social behaviour, I, II," *Journal of Theoretical Biology* 7(1): 1–52 (1964).

10 | W. D. Hamilton, "Altruism and related phenomena, mainly in social insects," *Annual Review of Ecology and Systematics* 3: 193–232 (1972).

11 | W. D. Hamilton, "The genetical evolution of social behaviour, I, II," *Journal of Theoretical Biology* 7(1): 1–52 (1964).

12 | The expression *haplodiploid hypothesis* was first used by M. J. West Eberhard in "The evolution of social behavior by kin selection," *The Quarterly Review of Biology* 50(1): 1–33 (1975).

is thus one-half. Each female shares half her genes with her mother (degree of relationship one-half) but only one-fourth with each brother and (this is the key to the haplodiploid hypothesis) three-fourths with each sister (degrees of relationship one-fourth and three-fourths, respectively).[13]

Because sisters share three-fourths of one another's genes instead of the usual one-half delegated by the more common diplodiploid (XX/XY) form of heredity, it can be reasonably supposed that sisters will favor each other over their own daughters. If this is true, Hamilton reasoned, then it should be easier for societies consisting of siblings—in this case, sisters—to arise in Hymenoptera species than in other insects.

Hamilton suggested that the haplodiploid biasing effect can explain two peculiarities he (and others) thought to exist in the pattern of advanced social behavior among the insects. First, at least when he wrote in the 1960s (this early date is important), colonies with nonreproducing workers appeared to be almost limited to the order Hymenoptera. The one known exception was the order Isoptera, or termites, which have an ordinary, diplodiploid (XX/XY) mode of sex determination. Second, as expected from the haplodiploid hypothesis, male social hymenopterans almost never cooperate with their sisters, limiting their activity to reproduction.

This application of Mendelian heredity became an early stanchion of the new discipline of sociobiology, assembled by one of us (Wilson) in *The Insect Societies* (1971) and *Sociobiology: The New Synthesis* (1975).[14] (The other key elements and the earlier derived principles of division of labor and communication are reviewed in Chapters 5 and 6 of the present work.) The evolutionary principle of Hamilton's inclusive fitness theory had far-reaching importance. It became the foundation of gene selectionism. The result was to bring down group selection (what drives social

---

13 | The reasoning concerning the degree of genetic relations of a female with a haplodiploid sex-determining system is as follows. *Female to brother*: half her genes come from her father, of which the brother has none (being parthenogenetic in origin) and half from the mother; half of the half from the mother, or one-fourth, is therefore what she shares with her brother. *Female to sister*: all the genes they receive from the father (half of their own genes) are identical because the father is haploid and has always only the same set to give his daughters; and half the genes a female and her sister received from their diploid mother are identical, so that overall half of half, or one-fourth, the genes they possess from the mother are identical; one-half plus one-fourth is three-fourths. (A male shares all of his genes with one-fourth his sister's genes.)

14 | Edward O. Wilson, in *The Insect Societies* (Cambridge, MA: The Belknap Press of Harvard University Press, 1971), was the first to synthesize sociobiology on the groundwork of population biology and to suggest the name of the new discipline, while *Sociobiology: The New Synthesis* (Cambridge, MA: The Belknap Press of Harvard University Press, 1975) extended the idea to include all organisms, including human beings.

evolution is the success of the group) and to catapult kin selection into the mainstream of discussion on genetic social evolution. Richard Dawkins's pellucid and best-selling *The Selfish Gene* simultaneously and independently helped bring the subject to a broad public audience.[15]

Unfortunately, the haplodiploid hypothesis, which proposes a genetic bias toward the evolution of eusociality in the order Hymenoptera, has been shown to be incorrect. Many haplodiploid species that form aggregations did not evolve eusociality, and a growing number of diplodiploid species, in addition to the termites, have been discovered that are eusocial.

The hypothesis of haplodiploid bias nonetheless survived in a spin-off role proposed in 1976 by Robert Trivers and Hope Hare.[16] For the haplodiploid bias to work, and for sisters to prefer to make sisters instead of daughters, they reasoned, it is necessary for them to invest more in new female reproductives than in males. The mother queen, in contrast, should prefer an equal investment in the two sexes, since her relationship to them is not female biased as a result of haplodiploid sex inheritance. The difference should therefore result in conflict between the mother queen and her worker daughters, a dispute that tips to one or the other contender according to the way the colony is organized and hence to the contender who can exercise control. This prediction proved correct, at least in part.[17]

The haplodiploidy hypothesis led to the early but erroneous assumption that a high degree of relatedness is needed for kin selection to be possible. This was a misconception of the essence of Hamilton's rule, which is the mathematical representation of kin selection theory. According to these principles, social evolution is effected by both genetic and ecological factors. Hamilton's rule shows that

---

15 | R. Dawkins, *The Selfish Gene* (New York: Oxford University Press, 1976).

16 | R. L. Trivers and H. Hare, "Haplodiploidy and the evolution of the social insects," *Science* 191: 249–263 (1976).

17 | J. J. Boomsma and A. Grafen, "Intra-specific variation in ant sex ratios and the Trivers-Hare hypothesis," *Evolution* 44(4): 1026–1034 (1990); J. J. Boomsma and A. Grafen, "Colony-level sex ratio selection in the eusocial Hymenoptera," *Journal of Evolutionary Biology* 4(3): 383–407 (1991); L. Sundström, "Sex ratio bias, relatedness asymmetry and queen mating frequency in ants," *Nature* 367: 266–268 (1994); A. F. G. Bourke and N. R. Franks, *Social Evolution in Ants* (Princeton, NJ: Princeton University Press, 1995); and N. J. Mehdiabadi, H. K. Reeve, and U. G. Mueller, "Queens versus workers: sex-ratio conflict in eusocial Hymenoptera," *Trends in Ecology and Evolution* 18(2): 88–93 (2003). The countervailing, or dissolutive effect of queen-worker conflict over the sex investment ratio is inferred, but not yet tested. If the queen lays eggs with a 1:1 sex ratio and workers destroy immatures to attain a 3:1 ratio in favor of new queens over males, there is at least a loss in colony-level fitness due to the conflict, albeit relatively low.

altruism can evolve when relatedness between altruist and beneficiary (r) and cost to the altruist (c) is relatively low, but benefit (b) to the recipient is very high. Several authors, including Hamilton himself, have criticized the preoccupation with genetic factors and neglect of ecological parameters in the application of kin selection theory.[18]

In essence, a gene for altruism that encodes its bearer to act altruistically toward individuals having above-average probability of sharing the same gene will spread through the population, provided ecological pressure favors such altruistic interactions. Taking a gene-selectionist perspective, Andrew Bourke and Nigel Franks expressed it in the following way: "What the gene loses in the sacrifice of the altruistic body it occupies, it can redeem many times over in the enhanced survival or reproduction of the beneficiaries."[19] Although the propagation of such a gene for altruism will be greatly facilitated by a high degree of relatedness between the interacting individuals, one has always to take into account the cost-benefit ratio in Hamilton's rule when considering the importance of genetic relatedness. In conclusion, kin selection theory explains the evolution of altruism toward kin, and this includes the care for direct offspring, as well as altruism toward nondescendant kin (also called collateral kin).[20]

This brings us to another misunderstanding of the term *kin selection*. John Maynard Smith, who coined the term *kin selection*, defined it as follows: "By kin selection I mean the evolution of characteristics which favour the survival of close relatives of the affected individual, by processes which do not require any discontinuities in population breeding structure. In this sense, the evolution of placental and of parental care (including 'self-sacrificing' behaviour such as the injury-feigning) are due to kin selection, the favoured relatives being the children of the affected individual."[21]

Indeed, kin selection theory is identical to Hamilton's inclusive fitness theory; therefore, when taking a gene-selectionist perspective (which is the perspective of

18 | For an excellent account of these misunderstandings, see A. F. G. Bourke and N. R. Franks, *Social Evolution in Ants* (Princeton, NJ: Princeton University Press, 1995).

19 | A. F. G. Bourke and N. R. Franks, *Social Evolution in Ants* (Princeton, NJ: Princeton University Press, 1995).

20 | Strictly speaking, nondescendant kin is a more precise expression, because if a nondescendant kin is a lineal relative such as parent or grandparent being helped, those nondescendant kin are not *collateral* kin, at least in the srictest sense of the word. See J. L. Brown, *Helping and Communal Breeding in Birds: Ecology and Evolution* (Princeton, NJ: Princeton University Press, 1987).

21 | J. Maynard Smith, "Group selection and kin selection," *Nature* 201: 1145–1147 (1964).

kin selection theory), it does not make sense to exclude direct offspring. Thus, kin selection theory is meant to explain the evolution of altruism toward nondescendant kin as well as the evolution of parental care.

MULTILEVEL NATURAL SELECTION

The problem with a strictly gene-selectionist perspective is that it does not explicitly address the targets of selection.[22] The gene-selectionist approach accounts for the spread of genes and for the change of frequencies of genes in the population. The gene-selectionist view focuses more on "replicators" (genes) and less on the "vehicles," the phenotypes of the bearers of genes.[23] However, selection targets the totality of traits of a "vehicle" (individual or group of individuals) that exhibit variation. Selection is a sorting process based on causal interaction of traits with the biotic and abiotic environment. Traits that warrant a superior adaptation to these environmental parameters will be favored by selection, while other traits, less efficiently adapted, will be disfavored. However, only those phenotypic positive characteristics that have a genetic basis will be evolutionarily selected, that is, passed on to the next generation. Ultimately, it is the change in the frequency of the trait-prescribing genes in the population that indicates that evolution is taking place. As long as the traits are adaptive, genes encoding such traits will be passed to other bearers in ever-changing combinations of alleles in subsequent generations of individuals or groups of individuals. If we are satisfied with a gene-selectionist perspective, the dynamics of change in the frequencies of alleles encoding for certain traits should be sufficient. However, if we want to understand the sorting process, a key element in behavioral ecology and sociobiology, a gene-selectionist perspective is not the solution.

All selection is multilevel. Elements that vary genetically at each level of biological organization serve as the *targets* of selection—whether genes, organelles, cells, organisms, or superorganisms. The ultimate *unit* of evolution, however, is the gene,

---

22 | E. O. Wilson and B. Hölldobler, "Eusociality: origin and consequences," *Proceedings of the National Academy of Sciences USA* 102(38): 13367–13371 (2005).

23 | Replicator-vehicle dichotomy has been elaborated by R. Dawkins in his book *The Extended Phenotype: The Gene as the Unit of Selection* (San Francisco: W. H. Freeman, 1982) and in "Replicators and vehicles," in King's College Sociobiology Group, eds., *Current Problems in Sociobiology* (New York: Cambridge University Press, 1982), pp. 45–64.

or ensemble of alleles of interacting genes, by which the varying traits of the higher units are encoded.

Three forces of natural selection can be distinguished at work on the superorganism according to the target of selection:[24] *group selection* (between groups), the differential survival and reproduction of entire cooperative groups as a result of the frequency and kind of alleles encoding social actions in each; *individual direct selection* (within groups), accruing from the differential personal survival and reproduction of each of the colony members; and *nondescendant (collateral) kin selection*, differential fitness of colony members due to their favoring or disfavoring by collateral and other nondescendant relatives, that is, relatives other than personal offspring. Nondescendant kin selection is a true force, comprising actions by one colony member directed to others according to the degree of pedigree relatedness. Between-group selection tends to be *binding* in its phenotypic effects; it tends to be opposed by individual direct selection (within-group selection), which is generally *dissolutive*. The *inclusive fitness* of the prescribing genotype, of individual colony members and hence statistically the colonies they compose, is the nonadditive product of the three forces.

A subprinciple of multilevel selection is that if the proportions of variants are changed at any level, from genes to superorganismic colonies, the effects can reverberate up and down through all the levels and, upon traveling down to the proportions of competing alleles, result in genetic evolution.[25]

---

24 | E. O. Wilson and B. Hölldobler, "Eusociality: origin and consequences," *Proceedings of the National Academy of Sciences USA* 102(38): 13367–13371 (2005).

25 | Different aspects of multilevel selection, and of the history of research on it, are provided by E. O. Wilson, *The Insect Societies* (Cambridge, MA: The Belknap Press of Harvard University Press, 1971); W. D. Hamilton, "Innate social aptitudes of man: an approach from evolutionary genetics," in R. Fox, ed., *Biosocial Anthropology* (London: Malaby Press, 1975), pp. 133–155; M. J. Wade, "Soft selection, hard selection, kin selection, and group selection," *American Naturalist* 125(1): 61–73 (1985); B. Hölldobler and E. O. Wilson, *The Ants* (Cambridge, MA: The Belknap Press of Harvard University Press, 1990); D. C. Queller, "Quantitative genetics, inclusive fitness, and group selection," *American Naturalist* 139(3): 540–558 (1992); L. A. Dugatkin and H. K. Reeve, "Behavioral ecology and levels of selection: dissolving the group selection controversy," *Advances in the Study of Behavior* 23: 101–133 (1994); J. Maynard Smith and E. Szathmáry, *The Major Transitions in Evolution* (New York: W. H. Freeman, 1995); A. F. G. Bourke and N. R. Franks, *Social Evolution in Ants* (Princeton, NJ: Princeton University Press, 1995); W. D. Hamilton, *Narrow Roads of Gene Land, Volume 1: Evolution of Social Behaviour* (New York: W. H. Freeman, 1996); R. H. Crozier and P. Pamilo, *Evolution of Social Insect Colonies: Sex Allocation and Kin Selection* (New York: Oxford University Press, 1996); T. D. Seeley, "Honey bee colonies are group-level adaptive units," *American Naturalist* 150(Supplement): S22–S41 (1997); S. A. Frank, *Foundations of Social Evolution* (Princeton, NJ: Princeton University Press, 1998); E. Sober and D. S. Wilson, *Unto Others: The Evolution and Psychology of Unselfish Behavior* (Cambridge, MA: Harvard University Press, 1998);

In 1966, George C. Williams elaborated the principle of parsimony,[26] consistent with similar reasoning by William D. Hamilton,[27] which holds that adaptations at all levels of biological organization can be explained by natural selection operating between individuals and hence the genes they carry. This work helped stimulate a substantial body of new theory of kin networks and the strategies of interaction among individuals within societies. But it also unjustifiably turned attention away from group selection (intrademic or trait group selection[28] and interdemic selection[29]). It

---

multiple authors, L. Keller, ed., *Levels of Selection in Evolution* (Princeton, NJ: Princeton University Press, 1999); R. E. Page Jr. and J. Erber, "Levels of behavioral organization and the evolution of division of labor," *Naturwissenschaften* 89(3): 91–106 (2002); W. J. Alonso and C. Schuck-Paim, "Sex-ratio conflicts, kin selection, and the evolution of altruism," *Proceedings of the National Academy of Sciences USA* 99(10): 6843–6847 (2002); L. Avilés, J. A. Fletcher, and A. D. Cutter, "The kin composition of social groups: trading group size for degree of altruism," *American Naturalist* 164(2): 132–144 (2004); R. Axelrod, R. A. Hammond, and A. Grafen, "Altruism via kin-selection strategies that rely on arbitrary tags with which they coevolve," *Evolution* 58(8): 1833–1838 (2004); J. Korb and J. Heinze, "Multilevel selection and social evolution of insect societies," *Naturwissenschaften* 91(6): 291–304 (2004); A. F. G. Bourke, "Genetics, relatedness and social behaviour in insect societies," in M. D. E. Fellowes, G. J. Holloway, and J. Rolff, eds., *Insect Evolutionary Ecology* (Proceedings of the 22nd Symposium of the Royal Entomological Society, University of Reading, UK, 2004) (Cambridge, MA: CABI Pub., 2005), pp. 1–30; F. L. W. Ratnieks, K. R. Foster, and T. Wenseleers, "Conflict resolution in insect societies," *Annual Review of Entomology* 51: 581–608 (2006); M. A. Novak, "Five rules for the evolution of cooperation," *Science* 314: 1560–1563 (2006); S. A. West, A. S. Griffin, and A. Gardner, "Social semantics: altruism, cooperation, mutualism, strong reciprocity and group selection," *Journal of Evolutionary Biology* 20(2): 415–432 (2007); S. A. West, A. S. Griffin, and A. Gardner, "Evolutionary explanations for cooperation," *Current Biology* 17(16): R661–R672 (2007); D. S. Wilson and E. O. Wilson, "Rethinking the theoretical foundation of sociobiology," *The Quarterly Review of Biology* 82(4): 327–348 (2007); D. S. Wilson and E. O. Wilson, "Survival of the selfless," *New Scientist* 196: 42–46 (2007); and E. O. Wilson, "One giant leap: how insects achieved altruism and colonial life," *BioScience* 58(1): 17–25 (2008).

26 | G. C. Williams, *Adaptation and Natural Selection* (Princeton, NJ: Princeton University Press, 1966). The basic argument was repeated and expressed by Richard Dawkins in *The Selfish Gene* (New York: Oxford University Press, 1976).

27 | W. D. Hamilton, "The genetical evolution of social behaviour, I, II," *Journal of Theoretical Biology* 7: 1–52 (1964).

28 | D. S. Wilson, "A theory of group selection," *Proceedings of the National Academy of Sciences USA* 72(1): 143–146 (1975); M. J. Wade, "The evolution of social interactions by family selection," *American Naturalist* 113(3): 399–417 (1979); and M. J. Wade, "Evolution of interference competition by individual, family, and group selection," *Proceedings of the National Academy of Sciences USA* 79(11): 3575–3578 (1982).

29 | E. O. Wilson, in *Sociobiology: The New Synthesis* (Cambridge, MA: The Belknap Press of Harvard University Press, 1975), reviews and evaluates the interdemic models of Richard Levins ("Extinction," in M. Gerstenhaber, ed., *Some Mathematical Questions in Biology* [Lectures on Mathematics in the Life Sciences, vol. 2] [Providence, RI: American Mathematical Society, 1970], pp. 77–107) and S. A. Boorman and P. R. Levitt ("Group selection on the boundary of a stable population," *Proceedings of the National Academy of Sciences USA* 69[9]: 2711–2713, 1972), which define between them two dynamic conditions under which altruistic alleles can spread by rendering demes (local populations) carrying them superior in survival than demes that do not. For

sometimes proved unnecessarily confusing and even ineffective in evaluating the operation of selection against the primary targets of selection (as opposed to genes, the units of selection). In particular, it neglected targets above the level of the individual, which are based on the interaction of groups and the dynamics of metapopulations.

The major result derived from these between-group selection models was that altruism can evolve, provided the interacting group members exhibit greater than average genetic similarity. However, as Andrew Bourke and Nigel Franks point out, "The crucial similarity required by the models is at the locus determining selfish or altruistic behavior. And kinship, as well as being a universal factor generating genetic similarity, is about the only one that can form the basis for the preferential allocation of social behavior without provoking within-genome conflict between the locus for social behavior and other loci."[30] So intrademic group selection models are not necessarily in conflict with kin selection models, and the controversy over these alternative models is mainly one of semantics.[31]

To summarize briefly to this point, a gene for altruism will decrease within groups under individual-level selection (within-group selection) because altruistic individuals suffer fitness costs, whereas the personal fitness of selfish individuals is enhanced because they do not suffer the cost but instead gain the benefit from their group mates' altruistic acts. The only way an altruism gene can be propagated by group selection is if groups with more altruists are more productive than those with fewer altruists. Thus, positive between-group selection must exceed negative within-group selection; and this is why group selection entails two major selection components: within-group selection, acting against the spread of altruism genes, and between-group selection, acting in favor of the spread of altruism genes. This concept does not contradict kin selection and is in full agreement with the "component of selection" concept proposed by Michael Wade: "Groups with altruists must be more productive than those without."[32]

---

example, in the Levins model, which introduced the concept of the metapopulation, altruistic alleles can increase when the differential survival rate demes favored by altruists exceed the loss in personal fitness of the altruists caused by their altruisms.

30 | A. F. G. Bourke and N. R. Franks, *Social Evolution in Ants* (Princeton, NJ: Princeton University Press, 1995).

31 | D. S. Wilson, "The group selection controversy: history and current status," *Annual Review of Ecology and Systematics* 14: 159–187 (1983).

32 | M. J. Wade, "Kin selection: its components," *Science* 210: 665–667 (1980).

Finally, David S. Wilson argued that the evolution of so-called weak altruism does not require relatedness, but could arise in genetically random groups of individuals. In "weak altruism," an individual's own personal fitness is increased by the altruistic act, but that of the group members even more. This is very different from what D. S. Wilson calls "strong altruism," which refers to personal fitness relative to *all* the individuals in the population.[33] Most theoreticians agree that the evolution of such "strong altruism" requires relatedness or genetic similarity. Although there are some cases in social insects that fit the definition of weak altruism—for example, pleometrotic colony founding, where several unrelated queens (foundresses) cooperate—such behavior can also be explained by individual selection. For instance, if there is little or no chance to found a new colony alone, individuals should be selected to join other foundresses, even if they will probably not end up being the queen that heads the mature colony.[34]

The ultimate agent of natural selection at all levels is always the environment. The cell is the environment for the genes and organelles, while the world outside the organism is the environment for the organism and colony. The environment does more than just select variants at each of these levels. It also affects the expression of the genes all across the hierarchy of levels. The traits of the varying units at each level result from the interactions of alleles and environment. The phenotypic expression of each allele (or ensemble of alleles) varies more or less predictably as a

---

33 | D. S. Wilson, "Structured demes and trait-group variation," *American Naturalist* 113(4): 606–610 (1979); D. S. Wilson, "The group selection controversy: history and current studies," *Annual Review of Ecology and Systematics* 14: 159–187 (1983); and D. S. Wilson, "Weak altruism, strong group selection," *Oikos* 59(1): 135–140 (1990); for a detailed and balanced discussion of these issues, see A. F. G. Bourke and N. R. Franks, *Social Evolution in Ants* (Princeton, NJ: Princeton University Press, 1995).

34 | See for example discussions in: S. H. Bartz and B. Hölldobler, "Colony founding in *Myrmecocystus mimicus* Wheeler (Hymenoptera: Formicidae) and the evolution of foundress associations," *Behavioral Ecology and Sociobiology* 10(2): 137–147 (1982); S. W. Rissing, G. B. Pollock, M. R. Higgins, R. H. Hagen, and D. R. Smith, "Foraging specialization without relatedness or dominance among co-founding ant queens," *Nature* 338: 420–422 (1989); D. C. Queller, "The evolution of eusociality: reproductive head starts of workers," *Proceedings of the National Academy of Sciences USA* 86(9): 3224–3226 (1989); H. K. Reeve and F. L. W. Ratnieks, "Queen-queen conflict in polygynous societies: mutual tolerance and reproductive skew," in L. Keller, ed., *Queen Number and Sociality in Insects* (New York: Oxford University Press, 1993), pp. 45–85; D. C. Queller, F. Zacchi, R. Cervo, S. Turillazzi, M. T. Henshaw, L. A. Santorelli, and J. E. Strassmann, "Unrelated helpers in social insects," *Science* 405: 784–787 (2000); J. Field, G. Shreeves, S. Sumner, and M. Casiraghi, "Insurance-based advantage to helpers in a tropical hover wasp," *Nature* 404: 869–871 (2000); G. Shreeves and J. Field, "Group size and direct fitness in social queues," *American Naturalist* 159(1): 81–95 (2002).

response to different environments. This "norm of reaction"—the pattern of phenotypic variability in response to the array of particular states of the environment—is itself subject to control by regulatory genes plus the interactions of the ensembles they compose. Hence, the norm of reaction evolves in response to selection. By such means, social behavior is rendered both programmed and flexible, not just in a general sense but also in a way critical for social evolution, in patterns that are adaptive to the organism and group to which the organism belongs.

Finally, the environment itself is always altered to some degree by the actions of organisms. In extreme cases among the social insects, colonies change their environment radically through the construction of nests that provide microclimate control. A few enrich their food supply by growing fungi or herding sugar-producing insect symbionts on vegetation near or within the nest.

What, then, are the forces that bring organisms together and subsequently transform them by altruism into superorganisms? Consider first the context of the origin of a superorganism. Each species of solitary beings evolves in a crucible of intense and ceaseless pressures from the environment. Sometimes the pressures nudge the members of the species into extended parental care. Less commonly, they lead to cooperative aggregates, and then, at the extreme, they produce superorganisms possessing castes and reproductive division of labor. Predators hunt the evolving colonies, parasites invade their social networks, and competitors vie for their food and nest sites. The colonies or their foundresses must accommodate temperature and humidity changes encountered when they disperse from their birth site. As they search for the right microhabitat, and even after settling there, they must survive changes from one locality to the next and from day to day. Either the populations of colonies adapt to this unforgiving kaleidoscopic world by genetic evolution, and at every level of biological organization, or else they and their prescriptive alleles die away and are replaced by others better prepared.

THE EVOLUTION OF EUSOCIALITY

Eusociality, the care of the offspring of a reproductive caste by a worker caste, is the most advanced level of social life in the insects. Although rare in evolution, the condition once attained has often been spectacularly successful. The eusocial insects, and in particular the ants and termites, overall tend to dominate the more persistent and defensible parts of the environments they inhabit.

Why has eusociality been so successful? The well-documented answer is that organized groups beat solitaires in competition for resources, and large, organized groups beat smaller ones of the same species.[35] Then why has eusociality been so rare? The answer is that it requires nondescendant altruism, which is behavior benefiting others at the cost of the lifetime production of offspring by the altruist.

An examination of the early evolutionary pathways to eusociality provides the key to the evolution of a superorganism. In all the known clades whose extant species display the earliest stages of eusociality, their behavior protects a persistent, defensible resource from predators, parasites, or competitors. The resource is invariably a nest plus dependable food within foraging range of the nest inhabitants. This holds for both invertebrate and vertebrate eusocial species. The latter includes the well-studied naked mole rat.[36]

The females of many species of aculeate wasps, for example, construct nests, then provision them with paralyzed prey for the larvae to consume. Among the 50,000 to 60,000 known aculeate species, at least seven independent lines have reached the eusocial condition.[37] In contrast, of the more than 70,000 parasitoid and other apocritan hymenopteran species, whose females travel from prey to prey to lay their eggs, none is known to be eusocial. Nor is any one of the hugely diverse 5,000 described species of sawflies and horntails. Larvae of some sawfly species form aggregations, but not eusocial colonies, and the adults lead solitary lives.[38]

Almost all of the thousands of known species of bark and ambrosia beetles, which compose the families Scolytidae and Platypodidae, depend on ephemeral dead wood for shelter and food. Many also care for their young in burrows they dig. An extremely few of the latter are able to cut and sustain burrows in living wood, allowing the coexistence of numerous generations. Among these latter few a single one, the Australian eucalyptus-boring *Platypus* (formerly *Austroplatypus*)

---

35 | B. Hölldobler and E. O. Wilson, *The Ants* (Cambridge, MA: The Belknap Press of Harvard University Press, 1990); W. R. Tschinkel, *The Fire Ants* (Cambridge, MA: The Belknap Press of Harvard University Press, 2006).

36 | P. W. Sherman, J. U. M. Jarvis, and R. D. Alexander, eds., *The Biology of the Naked Mole-Rat* (Princeton, NJ: Princeton University Press, 1991).

37 | E. O. Wilson and B. Hölldobler, "Eusociality: origin and consequence," *Proceedings of the National Academy of Sciences USA* 102(38): 13367–13371 (2005).

38 | J. T. Costa, *The Other Insect Societies* (Cambridge, MA: The Belknap Press of Harvard University Press, 2006).

incompertus, is known to have developed eusociality. Because of the persistence of the habitat of this species, tunnel systems housing presumably the same families are estimated to have survived for up to 37 years.[39]

In parallel manner, the handful of known eusocial aphids and thrips are gall inducers, enjoying a rich food supply in a secure, defensible home of their own making.[40] The vast majority of other known aphid and adelgid species, roughly 4,000 in number, and thrips species, about 5,000 strong, often form aggregations, but do not form galls or divide labor. Similarly, several snapping shrimp species of the genus *Synalpheus*, out of roughly 10,000 known decapod crustacean species, have reached the eusocial level. *Synalpheus* is highly unusual among decapods in constructing and defending nests in sponges.[41]

CROSSING THE EUSOCIALITY THRESHOLD

The key preadaptation for eusociality in social hymenopterans is progressive provisioning, a behavior that in solitary species is by individual direct selection. Although experimental field studies of the ecological pressures on pre-eusocial species have scarcely begun, one published example is especially instructive. Females of the sphecid wasp *Ammophila pubescens* provision their soil burrows with caterpillars. Because they are forced to open and close their nests to keep the larvae inside fed, they lose many of their eggs to cuckoo flies.[42] It is entirely reasonable to suppose that if a second *Ammophila* female were available to serve as a guard, the loss of eggs would be considerably reduced.

Simultaneous progressive provisioning, by which multiple larvae are reared at the same time,[43] is especially potent as a preadaptation in the order Hymenoptera. From this wholly solitary adaptation, it is but one short step in evolution for adult

39 | D. S. Kent and J. A. Simpson, "Eusociality in the beetle *Austroplatypus incompertus* (Coleoptera: Curculionidae)," *Naturwissenschaften* 79(2): 86–87 (1992).

40 | B. J. Crespi, "Eusociality in Australian gall thrips," *Nature* 359: 724–726 (1992); and D. L. Stern and W. A. Foster, "The evolution of soldiers in aphids," *Biological Reviews of the Cambridge Philosophical Society* 71(1): 27–79 (1996).

41 | J. E. Duffy, C. L. Morrison, and R. Ríos, "Multiple origins of eusociality among sponge-dwelling shrimps (*Synalpheus*)," *Evolution* 54(2): 503–516 (2000).

42 | J. Field and S. Brace, "Pre-social benefits of extended parental care," *Nature* 428: 650–652 (2004).

43 | J. Field, "The evolution of progressive provisioning," *Behavioral Ecology* 16(3): 770–778 (2005).

offspring to remain at the nest and help their mother raise siblings instead of dispersing to rear brood of their own.[44]

Let us first consider whether the evolution of cooperation among individuals that live in groups can be a preadaptation for the evolution of eusociality, which is the evolution toward permanent reproductive altruism. There are many species of bees and wasps in which foundresses associate, each fully endowed to reproduce on her own, and a reproductive division of labor is thereafter established through dominance interactions.[45] Such groups are called semisocial. Although division of labor is exhibited in such semisocial groups, we cannot call these groups eusocial, because the division of labor occurs among individuals of the same generation. Further, in most cases, the condition is only temporary and does not lead to permanent reproductive altruism. It may become a primitively eusocial system, however, when offspring of the dominant foundress and perhaps also of the subdominant individuals remain in the nest as helpers, rearing siblings produced by their mothers. The colonies then display reproductive division of labor with an overlap of generations, the two main criteria for eusociality.

Foundress association and the semisociality stemming from it, however, are not prerequisites. More likely, eusociality, at least in Hymenoptera, is derived from parental care by a single reproductive female, the condition called subsocial behavior.

So let us consider the evolution of parental care and the subsequent evolution of sib-social care (whereby siblings rear younger siblings in the natal nest). A step-by-step account of the selection processes could be the following:

- At first, alleles originate that program organisms in such a way that they survive and reproduce better than other competing organisms in the population possessing different sets of alleles. The favored alleles increase in frequency in the gene pool because identical copies of them will be present in larger numbers of offspring. Such is simple Darwinian fitness and the absolute prerequisite condition for evolution whether the allele bearers engage in parental care or not.

---

44 | E. O. Wilson, *The Insect Societies* (Cambridge MA: The Belknap Press of Harvard University Press, 1971); and C. D. Michener, *The Social Behavior of the Bees: A Comparative Study* (Cambridge, MA: The Belknap Press of Harvard University Press, 1974).

45 | C. D. Michener, *The Bees of the World* (Baltimore, MD: Johns Hopkins University Press, 2000); S. Turillazzi and M. J. West-Eberhard, eds., *Natural History and Evolution of Paper-Wasps* (New York: Oxford University Press, 1996); and J. H. Hunt, *The Evolution of Social Wasps* (New York: Oxford University Press, 2007).

- Next, ecological conditions favor selection for parental care, such that the parental individuals will invest energy and take survival risks to protect and raise healthy offspring. Then the alleles that program such behavior in parent organisms will increase in frequency across generations relative to alleles that do not program parental behavior. In other words, individuals practicing parental care suffer costs: they risk their own survival and they invest energy, which could otherwise produce a greater number of offspring. But the parental caregivers more than make up the difference by helping a larger proportion of their offspring reach healthy maturity.

- If ecological conditions favor cooperation in groups, alleles that program the expression of cooperation in individuals (for example, joint defense of nests and brood) will increase in frequency relative to alleles that do not program such behavior. The reason is that cooperative individuals will raise, on average, more and better-endowed offspring than solitary individuals. Thus, *individuals* will be selected to seek cooperation. Such cooperation can evolve even among genealogically unrelated individuals (if they are cocarriers of such cooperation genes). Possible mutations of genes that program selfish cheating within the group might be kept in check by between-group competition; this is natural selection acting at the group (colony) level.

- When we speak of between-group selection, we should keep in mind that the group is just one of the targets of multilevel selection. It is important to consider the effects of selection that targets different phenotypic levels. This distinction, however, does not inform us much about the population genetic substrate that is affected by between-group selection. In the gene view approach, the outcome is explored by focusing on the inclusive fitness of the members of the group. When we do this with the aid of simple mathematical models, we learn that cooperation is most likely to evolve faster in groups of individuals more closely related by pedigree because the inclusive fitness of each individual is higher than in groups of less pedigree-related individuals. Cooperative individuals also suffer costs from giving aid, because they invest to some degree in the survival of foreign offspring, but they gain cooperation from other breeders in nest and brood defense (mutualism).[46] Although they might raise fewer of their own

46 | M. J. West-Eberhard, "Polygyny and the evolution of social behavior in wasps," *Journal of the Kansas Entomological Society* 51(4): 832–856 (1978); and J. Seger, "Cooperation and conflict in social insects," in J. R. Krebs

offspring, they obtain greater "assurance" that the offspring will survive due to collective protection. Groups of consistently cooperative breeders will collectively raise more offspring than groups with many cheaters.

- It is difficult, however, to devise aprioristic scenarios and models that explain how eusociality evolves from such cooperative groups consisting of wholly unrelated individuals. Further, all known examples of such cooperative groups fall into the category "weak altruism" (sensu David S. Wilson).
- On the basis of mathematical theory, it would seem that the most reasonable hypothesis for the evolution of eusocial behavior is that it originates from parental care behavior. In Hymenoptera, it is exclusively the mother (as opposed to the father) that cares for the young. Alleles that prescribe maternal care behavior were favored during presocial evolution by ecologically generated selection forces.
- Further ecological pressure may favor cooperative defense of the nest, cooperative foraging, and cooperative rearing of the young. In advanced subsocial systems, young adults could remain for some time in the natal nest and help to rear siblings. Such is the transition from a subsocial to a primitively eusocial organization.[47] In more advanced eusocial states, sib-social helpers remain with the natal nest more or less permanently. The sib-social helpers have evolved to workers that feed and protect the mother (queen). Thus, they ensure a significant increase in the queen's reproduction, and they protect and raise their younger siblings at the cost of their own personal reproduction.
- Groups (colonies) exhibiting such reproductive division of labor survive better and are more productive under certain ecological conditions than groups not exhibiting it. Thus, between-group selection shapes features that enhance colony productivity relative to other competing colonies. A striking example is provided in a study by Sam Beshers and James Traniello of the fungus-growing ant

---

and N. B. Davies, eds., *Behavioural Ecology: An Evolutionary Approach,* 3rd ed. (Boston: Blackwell Scientific, 1991), pp. 338–373.

47 | D. C. Queller, "The origin and maintenance of eusociality: the advantage of extended parental care," in S. Turillazzi and M. J. West-Eberhard, eds., *Natural History and Evolution of Paper-Wasps* (New York: Oxford University Press, 1996), pp. 218–234; R. Gadagkar, *The Social Biology of* Ropalidia*: Toward Understanding the Evolution of Eusociality* (Cambridge, MA: Harvard University Press, 2001); J. Field and S. Brace, "Pre-social benefits and extended parental care," *Nature* 428: 650–652 (2004); and E. O. Wilson, "One giant leap: how insects achieved altruism and colonial life," *BioScience* 58(1): 17–25 (2008).

Trachymyrmex septentrionalis. They discovered that the size frequency distribution of workers affects colony sexual reproduction in populations of this species.[48]

- "Ultimately, the unfolding of the evolutionary process entails selection on the genotypes of the founding female of the colony and her mates, operating through colony traits determined by the genotypes of the worker offspring they produce."[49]
- Thus, in the conventional scenario of the evolution of eusociality presented to this point (the total forgoing of personal reproduction for the sake of enhanced sib-social care and queen reproduction), groups of related individuals are needed. Kinship does not merely accelerate the propagation of a "eusociality gene"; it makes such a gene more likely to spread (versus be eliminated) in the first place. Hamilton's rule specifies, albeit very abstractly and without reference to the targets of multilevel selection, the minimum cost-benefit threshold at which the gene starts spreading. Higher-pedigree relatedness, it would appear to follow, drops this threshold.

It is important to keep in mind that mathematical gene-selectionist (inclusive fitness) models can be translated into multilevel selection models and vice versa. As Lee Dugatkin, Kern Reeve, and several others have demonstrated, the underlying mathematics is exactly the same; it merely takes the same cake and cuts it at different angles. Personal and kin components are distinguished in inclusive fitness theory; within-group and between-group components are distinguished in group selection theory. One can travel back and forth between these theories with the point of entry chosen according to the problem being addressed.[50]

To summarize the standard model to this point, a reproductive division of labor implies that incipient workers care for offspring other than their own, meaning that in effect they sacrifice some production of their own offspring. In semisocial

48 | S. N. Beshers and J. F. A. Traniello, "The adaptiveness of worker demography in the attine ant *Trachymyrmex septentrionalis*," *Ecology* 75(3): 763–775 (1994).

49 | R. E. Owen, "The genetics of colony-level selection," in M. D. Breed, and R. E. Page Jr., eds., *The Genetics of Social Evolution* (Boulder, CO: Westview Press, 1989), pp. 31–59; A. F. G. Bourke and N. R. Franks, *Social Evolution in Ants* (Princeton, NJ: Princeton University Press, 1995); and E. O. Wilson, "One giant leap: how insects achieved altruism and colonial life," *BioScience* 58(1): 17–25 (2008).

50 | L. A. Dugatkin and H. K. Reeve, "Behavioral ecology and levels of selection: dissolving the group selection controversy," *Advances in the Study of Behavior* 23: 101–133 (1994).

systems (foundress associations), individuals can reverse this sacrifice either by producing more brood that would receive care formerly given to the partner's offspring or by increasing the amount of care given to their own existing offspring. However, the almost complete sacrifice of personal reproduction in eusocial systems, the so-called strong altruism, cannot evolve among nonrelatives in either inclusive fitness models or multilevel selection models.

Kern Reeve has made the mathematical argument for this conclusion as follows:[51]

> Hamilton's inequality, $rB - r'C > 0$, where r is the altruist's relatedness to its own offspring and r' is the altruist's relatedness to the recipient's offspring, is the key prediction of both inclusive fitness theory *and* multilevel selection theory, as we would expect by the principle of equivalence. If this rule is not satisfied, then altruism cannot evolve either in inclusive fitness models *or* in multilevel selection models, so it is the *only possible* starting point if selection is involved in the evolution of altruism.
>
> B and C above refer to the number of offspring that the altruism helps to create for the beneficiary and the number of offspring that the altruist gives up because of its altruism, respectively. Larger B is equivalent to stronger positive between-group selection, and larger C is equivalent to stronger negative within-group selection, showing Hamilton's rule is firmly connected to multilevel selection theory. Now what if altruist and recipient are unrelated? Then we have $-r'C > 0$, which cannot be satisfied if there is a positive C (i.e., a sacrifice for being altruistic). Now you might think: But can't the altruism still evolve because of group selection in this case? No. In every trait-group selection model of altruism, altruism evolves *only* if altruists have more offspring than do nonaltruists across the entire population, i.e., $C < 0$, in accordance with Hamilton's rule.
>
> So why do people have the impression that group selection can favor altruism among nonrelatives? The answer: Because group selectionists have changed the meaning of altruism to mean reduced reproduction relative to one's group members ("weak altruism"). But this is not the same kind of altruism that enters into C—the latter refers to the number of offspring given up *on average across the entire population* ("strong altruism"), not to the number given up relative to other group members. Weak altruism among nonrelatives evolves in group selection models *only* when C

51 | K. Reeve, personal communication (2007).

< 0, i.e., there is no strong altruism. And precisely the same is true in inclusive fitness models. Hamilton's rule is the condition for the evolution of strong altruism in either inclusive fitness or multilevel selection models.

To see this mathematically, let the number of offspring produced by an altruist in a group of nonrelatives be equal to pG, where p is the fraction of group offspring produced by the altruist and G is the total number of offspring produced by the group. A nonaltruist produces a fraction p' of the offspring and its group produces G' total offspring. Let $p < p'$ (negative within-group selection) and $G > G'$ (positive between-group selection). The altruism allele spreads in both inclusive fitness and group selection models when $pG > p'G'$, i.e., when altruists produce on average more offspring than do nonaltruists across the entire population. Weak altruism can evolve among nonrelatives in both models, but strong altruism can evolve in neither.

We now ask: What is the origin of the hypothetical "eusociality gene"?

An important clue, offered by several authors, is that helper behavior (or sib-social care behavior) is evolutionarily derived from maternal care behavior.[52] In early writings, Mary Jane West-Eberhard offered an "ovarian groundplan" scenario in which queen and worker phenotypes, even though they carry identical caste-determining alleles, diverge from each other by environmentally induced developmental pathways built from the same ancestral program. Working off this conjecture, Timothy Linksvayer and Michael Wade introduced a model of the evolution of sib-social care from maternal care with heterochrony.[53] They first note that ancestral maternal care genes are expressed only after mating and the completion

52 | R. Dawkins, "Twelve misunderstandings of kin selection," *Zeitschrift für Tierpsychologie* 51(2): 184–200 (1979); M. J. West-Eberhard, "The epigenetical origins of insect sociality," in J. Eder and H. Rembold, eds., *Chemistry and Biology of Social Insects* (Proceedings of the Tenth Congress of the International Union of Social Insects, Munich, 18–22 August 1986) (Munich: Verlag J. Peperny, 1987), pp. 369–372; M. J. West-Eberhard, "Flexible strategy and social evolution," in Y. Ito, J. L. Brown, and J. Kikkawa, eds., *Animal Societies: Theories and Facts* (Tokyo: Japan Scientific Societies Press, 1987), pp. 35–51; R. D. Alexander, K. M. Noonan, and B. J. Crespi, "The evolution of eusociality," in P. W. Sherman, J. U. M. Jarvis, and R. D. Alexander, *The Biology of the Naked Mole-Rat* (Princeton, NJ: Princeton University Press, 1991), pp. 3–44; and M. J. West-Eberhard, "Wasp societies as microcosms for the study of development and evolution," in S. Turillazzi and M. J. West-Eberhard, eds., *Natural History and Evolution of Paper-Wasps* (New York: Oxford University Press, 1996), pp. 290–317.

53 | T. A. Linksvayer, and M. J. Wade, "The evolutionary origin and elaboration of sociality in the aculeate Hymenoptera: maternal effects, sib-social effects, and heterochrony," *The Quarterly Review of Biology* 80(3): 317–336 (2005); and T. A. Linksvayer, "Direct, maternal, and sibsocial genetic effects on individual and colony traits in an ant," *Evolution* 60(12): 2552–2561 (2006).

of reproductive development. In the evolutionarily derived condition, aspects of the reproductive developmental program are co-opted so that maternal care behaviors are expressed before reproduction and toward siblings instead of their own offspring. This modification of the expression of genes that regulate the timing of the expression of behaviors, called "behavioral heterochrony," has been implicated in the evolution of eusociality in termites as well as the evolution of helping behavior in other animal groups.[54, 55] In the Linksvayer-Wade model, at least some genetic variation underlies variation in the timing of the expression of maternal care behavior, as it does in physiological and behavioral traits generally. The authors suggest "that there may often be a small number of genes underlying this behavioral heterochrony, permitting rapid social evolution once the appropriate mutations arise." Queens and workers are increasingly divergent in these and other behavioral traits, but at certain times of the colony cycle, each colony member must perform similar brood care behavior. Therefore, Linksvayer and Wade propose, alleles affecting brood care have pleiotropic (phenotypically flexible) effects on both maternal care and sib-social care, and a certain genetic correlation between maternal and sib-social care is maintained. In addition, the social environment is an important factor in the expression of individual and colony phenotypes.[56] Genes expressed in social partners (indirect genetic effects), together with zygotic genes (direct genetic effects), act on the phenotype. Linksvayer and Wade have suggested various experimental scenarios that could disentangle direct and indirect genetic effects likely to play a major role in the evolutionary dynamics and expression of social traits.[57]

In fact, the responsiveness to indirect genetic effects may be another preadaptation that favors the transition to semisociality and eusociality.[58] Several natural history examples support this hypothesis. One such case is the Japanese stem-nesting xylocopine bee

---

54 | C. A. Nalepa and C. Bandi, "Characterizing the ancestors: paedomorphosis and termite evolution," in T. Abe, D. E. Bignell, and M. Higashi, eds., *Termites: Evolution, Sociality, Symbiosis, Ecology* (Boston: Kluwer Academic Publishers, 2000), pp. 53–75; and M. J. West-Eberhard, *Developmental Plasticity and Evolution* (New York: Oxford University Press, 2003).

55 | I. G. Jamieson, "Behavioral heterochrony and the evolution of birds' helping at the nest: an unselected consequence of communal breeding?" *American Naturalist* 133(3): 394–406 (1981).

56 | M. J. West-Eberhard, *Developmental Plasticity and Evolution* (New York: Oxford University Press, 2003).

57 | T. A. Linksvayer and M. J. Wade, "The evolutionary origin and elaboration of sociality in the aculeate Hymenoptera: maternal effects, sib-social effects, and heterochrony," *The Quarterly Review of Biology* 80(3): 317–336 (2005).

58 | E. O. Wilson, "One giant leap: how insects achieved altruism and colonial life," *BioScience* 58(1): 17–25 (2008).

Ceratina flavipes. The vast majority of the females of this species provision their nests with pollen and nectar as solitary foundresses, but in slightly more than 0.1 percent of the cases, two individuals cooperate. When this happens, the pair divide the labor: one lays the eggs and guards the nest entrance while the other forages.[59]

Even more striking are solitary bees that behave like semisocial species when forced together experimentally. In *Ceratina* and *Lasioglossum*, the coerced partners proceed variously to divide labor in foraging, tunneling, and guarding.[60] Furthermore, in at least two species of *Lasioglossum*, females engage in back following, with one bee leading another to the nest—a characteristic of primitively eusocial bees. The division of labor appears to be the result of a preexisting behavioral groundplan in which solitary individuals tend to move from one job to another after the first is completed. In eusocial species, the algorithm is transferred to the avoidance of a job already being filled by another nestmate. It is evident that progressively provisioning bees and wasps are already "spring-loaded" for a rapid shift to eusociality once ecological factors favor the change.

The results of the forced-group experiments fit the fixed threshold model of the origin of labor division proposed for the emergence of the phenomenon in established insect societies.[61] The model posits that variation, sometimes genetic in origin and sometimes purely phenotypic, exists in the response thresholds associated with various tasks. When two or more individuals interact, those with the lowest threshold are the first to begin the task. The activity inhibits their partners, who are then more likely to move on to whatever other tasks are available. Thus, once again, the group impact of a single phenotypically flexible allelic change inhibiting

---

59 | S. F. Sakagami and Y. Maeta, "Sociality, induced and/or natural, in the basically solitary small carpenter bees (*Ceratina*)," in Y. Itô, J. L. Brown, and J. Kikkawa, eds., *Animal Societies: Theories and Facts* (Tokyo: Japan Scientific Societies Press, 1987), pp. 1–16.

60 | S. F. Sakagami and Y. Maeta, "Sociality, induced and/or natural, in the basically solitary small carpenter bees (*Ceratina*)," in Y. Itô, J. L. Brown, and J. Kikkawa, eds., *Animal Societies: Theories and Facts* (Tokyo: Japan Scientific Societies Press, 1987), pp. 1–16; W. T. Wcislo, "Social interactions and behavioral context in a largely solitary bee, *Lasioglossum* (*Dialictus*) *figueresi* (Hymenoptera, Halictidae)," *Insectes Sociaux* 44(3): 199–208 (1997); and R. Jeanson, P. F. Kukuk, and J. H. Fewell, "Emergence of division of labour in halictine bees: contributions of social interactions and behavioural variance," *Animal Behaviour* 70(5): 1183–1193 (2005).

61 | G. E. Robinson and R. E. Page Jr., "Genetic basis for division of labor in an insect society," in M. D. Breed and R. E. Page Jr., eds., *The Genetics of Social Evolution* (Boulder, CO: Westview Press, 1989), pp. 61–80; E. Bonabeau, G. Theraulaz, and J.-L. Deneubourg, "Quantitative study of the fixed threshold model for the regulation of division of labour in insect societies," *Proceedings of the Royal Society of London B* 263: 1565–1569 (1996); and S. N. Beshers and J. H. Fewell, "Models of division of labor in social insects," *Annual Review of Entomology* 46: 413–440 (2001).

dispersal from the natal nest would seem to be enough to carry preadapted species across the eusocial threshold.

The difference in roles between a mother and her nonreproductive offspring is not genetic in nature. Rather, as shown by evidence from the primitively eusocial species, the two castes represent different phenotypes of the same evolutionarily modified genome.

Altruism and eusociality are thus evidently born from the appearance of a phenotypically flexible eusocial allele (or ensemble of such alleles) in a progressively provisioning mother and between-group selection acting on emergent group traits, socially binding in nature and sufficiently powerful to overbalance the dissolutive effects of individual direct (within-group) selection.

One small step, so to speak, for a newly created worker caste, one giant leap for Hymenoptera.[62]

Another example of flexibility, genetic at the base and just at the eusociality threshold, is provided by the ground-nesting halictid sweat bee *Halictus sexcinctus*. The species appears to be genetically polymorphic at one locality within its range in southern Greece, with colonies of one strain founded by cooperating females and colonies of a second strain founded by a single territorial female whose offspring serve as workers.[63]

Considerable progress has been made in the study of the genetic architecture and developmental physiology underlying division of labor and eusociality.[64] In fact, several of the recent findings support the prediction made by West-Eberhard's "ovarian groundplan model" and the Linksvayer-Wade heterochrony model for the evolution of eusociality. The suggestion that many of the same genes will be expressed in adults performing sib-social care behaviors as in adults performing maternal care behaviors finds increasing support from experimental studies. Robert Page and Gro Amdam have concluded from analytical studies in selective breeding, genetic mapping, functional genomics, and endocrinology and physiology that complex social behavior can evolve from tentatively simple heterochronic changes in reproductive signaling

---

62 | E. O. Wilson, "One giant leap: how insects achieved altruism and colonial life," *BioScience* 58(1): 17–25 (2008).

63 | M. H. Richards, E. J. von Wettberg, and A. C. Rutgers, "A novel social polymorphism in a primitively eusocial bee," *Proceedings of the National Academy of Sciences USA* 100(12): 7175–7180 (2003).

64 | G. E. Robinson, "Genomics and integrated analysis of division of labor in honeybee colonies," *American Naturalist* 160(6, Supplement): S160–S172 (2002).

systems. "The origins of complex social behavior," they conclude, "from which insect societies emerge, are derived from ancestral developmental programs. These programs originated in ancient solitary insects and required little evolutionary remodeling."[65]

Exactly what kind of selection drives the species across the eusociality threshold? Concrete examples of this adaptation and the transition it affords are provided by halictid sweat bees and polistine wasps. In one recently documented case, two species of sweat bees that switched from collecting the pollen of many plant species to collecting pollen from only a few plant species also reverted from a primitively eusocial life back to solitary life. Specialization on a limited array of plants as a source of food is advantageous in the environment in which the reverted species live. Such genetic change also shrinks the length of the harvesting season and removes the possibility of overlapping generations (and hence the formation of a eusocial colony) and with it the advantage that might accrue from the presence of guard bees. Evolution in the reverse direction is easily conceivable and very likely has occurred: adaptation to a broader array of food plants sets the stage for multiple generations, with overlapping generations in the same nest.[66] Similar evidence has been adduced for primitively eusocial wasps.[67] In crossing the line to eusociality, a single allele (heterochrony gene) disposing daughters to stay could be fixed in the population at large if the advantage given it by the little group over solitaires sufficiently outweighed the advantage given the worker if it were to leave and try on its own.

It seems to follow as an overarching principle that a change on the order of the final step to eusociality can occur with the substitution of only one or a small set of alleles. There are precedents in other studies of insect social genetics. Throughout the great diversity of living ant species, for example, the coexistence of winged reproductive females and wingless worker females is a basic trait of colonial life. Judging from the phylogenetically well-separated flies (order Diptera) and butterflies (order Lepidoptera), wing development throughout the winged insects is directed by an unchanged regulatory gene network. Over 110 million years ago, the earliest ants (or else their immediate ancestors) altered the regulatory network so that some of

---

65 | R. E. Page Jr. and G. V. Amdam, "The making of a social insect: developmental architectures in social design," *BioEssays* 29(4): 334–343 (2007).

66 | B. Danforth, "Evolution of sociality in a primitively eusocial lineage of bees," *Proceedings of the National Academy of Sciences USA* 99(1): 286–290 (2002).

67 | J. H. Hunt and G. V. Amdam, "Bivoltinism as an antecedent to eusociality in the paper wasp genus *Polistes*," *Science* 308: 264–267 (2005).

the genes could be shut down under the influence of diet or some other environmental factor. Thus was produced a wingless worker caste.[68]

COUNTERVAILING FORCES OF SELECTION

While some individual direct selection may play an auxiliary role in the origin of eusociality, the force that targets the maintenance and elaboration of eusociality is by necessity environmentally based between-group selection, which acts on the emergent traits of the group as a whole. An examination of the behavior of the most primitively eusocial ants, bees, and wasps shows that the emergent traits are initially dominance behavior, reproductive division of labor, and very likely some form of alarm communication mediated by pheromones. A species in the earliest stage of eusociality can be expected to be a kind of neurogenetic hybrid, at least in the following sense: on the one hand, the newly emergent traits favor the group; on the other hand, much of the rest of the genome, having been the target of millions of years of individual direct selection, favors personal dispersal and reproduction.

For the binding effects of group selection (between-group selection) to outweigh the dissolutive effects of individual direct selection (within-group selection), the candidate insect species evidently must have only a very short evolutionary distance to travel, such that no more than a very small number of emergent traits are needed to form a eusocial colony. The reduction of that distance is achieved by a particular set of preadaptations. The rarity of these preadaptations in just the right combination, when added to the high bar to eusociality set by countervailing individual direct selection (within-group selection), may be enough to explain the general phylogenetic rarity of eusociality.

PASSING THE POINT OF NO RETURN

In the earliest stage of eusociality, the offspring remaining in the nest would be expected to assume the worker role simply in conformity to the preexisting

---

68 | E. Abouheif and G. A. Wray, "Evolution of the gene network underlying wing polyphenism in ants," *Science* 297: 249–252 (2002).

behavioral ground rule inherited from the pre-eusocial ancestor. Subsequently, a morphological worker caste can emerge by a further genetic change in which the expression of maternal care genes is rerouted to precede foraging, thus reversing the normal sequence in the adult developmental groundplan of the ancestor.[69] The rerouting is programmed to remain part of the phenotypic plasticity of the alleles that prescribe the overall groundplan. This origin of an anatomically distinct worker caste appears to mark the "point of no return" in evolution at which eusocial life becomes irreversible.[70]

With the passing of the point of no return, a distinction has to be made between the evolutionary origin of eusociality and its maintenance.[71] Ecological selection forces and kin structures of advanced societies are often very different from those of ancestral eusocial groups. This is illustrated with an individual selection (inclusive fitness) model developed by Kern Reeve and one of us (Hölldobler) to explore the selection options under certain environmental conditions in the early evolutionary state of eusociality versus in the advanced state of a superorganism.[72]

Individuals are faced with the problem of how to divide their energy between intercolony competition and intracolony competition—that is, within-group tugs-of-war over shares of resources gained through intergroup competition (between-group tugs-of-war). The model clearly demonstrates that when ecological and genetic factors advance a society to near the upper extreme of the superorganism continuum, subsequent selection may result in the complete loss of costly physiological structures involved in within-group competition. Thus, this model exemplifies

---

69 | G. V. Amdam, K. Norberg, M. K. Fondrk, and R. E. Page Jr., "Reproductive ground plan may mediate colony-level selection effects on individual foraging behavior in honey bees," *Proceedings of the National Academy of Sciences USA* 101(31): 11350–11355 (2004); G. V. Amdam, A. Csondes, M. K. Fondrk, and R. E. Page Jr., "Complex social behaviour from maternal reproductive traits," *Nature* 439: 76–78 (2006); and R. E. Page Jr. and G. V. Amdam, "The making of a social insect: developmental architectures of social design," *BioEssays* 29(4): 334–343 (2007).

70 | E. O. Wilson, *The Insect Societies* (Cambridge, MA: The Belknap Press of Harvard University Press, 1971); and E. O. Wilson and B. Hölldobler, "Eusociality: origin and consequence," *Proceedings of the National Academy of Sciences USA* 102(38): 13367–13371 (2005).

71 | A. F. G. Bourke and N. R. Franks, *Social Evolution in Ants* (Princeton, NJ: Princeton University Press, 1995); B. J. Crespi, "Comparative analysis of the origins and losses of eusociality: causal mosaics and historical uniqueness," in E. P. Martins, ed., *Phylogenies and the Comparative Method in Animal Behavior* (New York: Oxford University Press, 1996), pp. 253–287; and C. D. Michener, *The Bees of the World* (Baltimore, MD: Johns Hopkins University Press, 2000).

72 | H. K. Reeve and B. Hölldobler, "The emergence of a superorganism through intergroup competition," *Proceedings of the National Academy of Sciences USA* 104(23): 9736–9740 (2007).

how to specify the conditions leading to a point of no return in eusocial evolution, that is, a point at which the capacity for "selfishness" has become insignificant because the underlying organs (for example, ovaries and spermatheca) important for within-group competition degenerate or become completely lost. Once such organs become obsolete through progressive selective elimination, they are unlikely to be restored in a single mutational step.

Intergroup competition facilitated by ecological factors of resource patchiness is associated with the most elaborate cooperation and communication systems. The nested tugs-of-war model by Reeve and Hölldobler predicts that within-group cooperation increases as the number of competing groups (and between-group competition intensity) increases, because greater cooperation improves group competitiveness and a greater number of competing groups increases the pressure of between-group competition. At this point, members invest all of their energy in within-group cooperation to outcompete other groups, and the society can be regarded as a "superorganism" in the fullest sense. High within-group relatedness becomes insignificant as long as within-group relatedness remains higher than between-group relatedness. In such advanced social organizations, the colony effectively becomes the major target of selection; that is, it is a coherent "extended phenotype" of the genes within colony members. Selection therefore optimizes caste demography, patterns of division of labor, and communication systems at the colony level. For example, colonies that employ the most effective recruitment system to retrieve food or that exhibit the most powerful colony defense against enemies and predators will be able to raise the largest number of reproductive females and males every year and thus will have the greatest fitness within the population of colonies. We have recently argued that the extreme levels of cooperation exhibited by the advanced eusocial insects ultimately must be explained by invoking the *binding* force of between-group selection rather than by appealing to genetic relatedness, which only amplifies ecologically driven selective forces for cooperation without causing them.[73] According to this view, between-group selection must be invoked

73 | E. O. Wilson, and B. Hölldobler, "Eusociality: origin and consequences," *Proceedings of the National Academy of Sciences USA* 102(38): 13367–13371 (2005); D. S. Wilson and E. O. Wilson, "Rethinking the theoretical foundations of sociobiology," *The Quarterly Review of Biology* 82(4): 327–348 (2007); D. S. Wilson and E. O. Wilson, "Survival of the selfless," *New Scientist* 196: 42–46 (2007); and E. O. Wilson, "One giant leap: how insects achieved altruism and colonial life," *BioScience* 58(1): 17–25 (2008).

to understand cooperation in insects. However, it is easy to conduct a general, purely individual (inclusive fitness) model of cooperation mediated by between-group competition. This is not surprising, as it is now well established that trait group selection models can be mathematically translated into individual (inclusive fitness) models and vice versa, so the two classes of models cannot be considered alternatives to each other.[74] The truly interesting problem is to determine (with either inclusive fitness or equivalent trait group selection models) how intergroup competition can increase the extent to which social groups can be viewed as coherent vehicles for gene propagation—that is, superorganisms.

Although the nested tugs-of-war model just described is an individual selection model (inclusive fitness model), there is no disagreement when the evolutionary process is expressed from a group selection perspective, whereby the force targeting the maintenance and elaboration of eusociality is by necessity environmentally based group selection (between-group selection), which acts on the emergent traits of the group as a whole.[75]

On occasion, it has been claimed that the often low degree of within-colony relatedness in advanced eusocial insect colonies is evidence for the insignificance of pedigree relatedness in social evolution. As just pointed out, however, a high degree of relatedness is not needed for the maintenance of advanced eusocial societies, and greater genetic variability (hence lower relatedness) within such colonies should therefore not be used as evidence for the insignificance of relatedness in the evolution of eusociality, at least in its more evolutionarily advanced stages.[76] Most

---

74 | L. A. Dugatkin and H. K. Reeve, "Behavioral ecology and levels of selection: dissolving the group selection controversy," *Advances in the Study of Behavior* 23: 100–133 (1994); A. Traulsen and M. A. Nowak, "Evolution of cooperation by multilevel selection," *Proceedings of the National Academy of Sciences USA* 103(29): 10952–10955 (2006); and L. Lehmann, L. Keller, S. West, and D. Roze, "Group selection and kin selection: two concepts but one process," *Proceedings of the National Academy of Sciences USA* 104(16): 6736–6739 (2007).

75 | E. O. Wilson, "One giant leap: how insects achieved altruism and colonial life," *BioScience* 58(1): 17–25 (2008); see also L. A. Dugatkin and H. K. Reeve, "Behavioral ecology and levels of selection: dissolving the group selection controversy," *Advances in the Study of Behavior* 23: 100–133 (1994).

76 | For a further debate of these issues, see E. O. Wilson and B. Hölldobler, "Eusociality: origin and consequences," *Proceedings of the National Academy of Sciences USA* 102(38): 13367–13371 (2005); K. R. Foster, T. Wenseleers, and F. L. W. Ratnieks, "Kin selection is the key to altruism," *Trends in Ecology and Evolution* 21(2): 57–60 (2006); T. Wenseleers and F. L. W. Ratnieks, "Comparative analysis of worker reproduction and policing in eusocial Hymenoptera supports relatedness theory," *American Naturalist* 168(6): E163–E179 (2006); L. Lehmann and L. Keller, "The evolution of cooperation and altruism—a general framework and a classification of models," *Journal of Evolutionary Biology* 19(5): 1365–1376 (2006), with discussions in the same issue of this

entomologists agree that multiple mating in social Hymenoptera species is a derived trait. Multiple mating in advanced eusocial insects could evolve for many reasons.[77] After the point of no return, colonies can afford a lower average within-colony relatedness. In fact, at this advanced stage of eusociality, which we regard as the ultimate superorganismic grade, greater genetic variability is most likely adaptive for the colony as a whole. For example, it has been argued that multiple mating by the queen lowers the queen-worker conflict over sex allocation and worker reproduction, thus increasing overall colony productivity.[78]

In addition, increased genetic diversity among workers might easily arise as a means of improving overall resistance to disease.[79] Just this correlation has been

---

journal; H. K. Reeve, and B. Hölldobler, "The emergence of a superorganism through intergroup competition," *Proceedings of the National Academy of Sciences USA* 104(23): 9736–9740 (2007); D. S. Wilson and E. O. Wilson, "Rethinking the theoretical foundation of sociobiology," *The Quarterly Review of Biology* 82(4): 327–348 (2007); D. S. Wilson and E. O. Wilson, "Survival of the selfless," *New Scientist* 196: 42–46 (2007); and E. O. Wilson, "One giant leap: how insects achieved altruism and colonial life," *BioScience* 58(1): 17–25 (2008).

77 | R. E. Page Jr., "The evolution of multiple mating behavior by honey bee queens (*Apis mellifera* L.)," *Genetics* 96(1): 263–273 (1980); R. H. Crozier and R. E. Page Jr., "On being the right size: male contributions and multiple mating in social Hymenoptera," *Behavioral Ecology and Sociobiology* 18(2): 105–115 (1985); F. L. W. Ratnieks and H. K. Reeve, "Conflict in single-queen hymenopteran societies: the structure of conflict and processes that reduce conflict in advanced eusocial species," *Journal of Theoretical Biology* 158(1): 33–65 (1992); M. J. F. Brown and P. Schmid-Hempel, "The evolution of female multiple mating in social Hymenoptera," *Evolution* 57(9): 2067–2081 (2003); and H. Schlüns, R. F. A. Moritz, P. Neumann, P. Kryger, and G. Koeniger, "Multiple nuptial flights, sperm transfer and the evolution of extreme polyandry in honeybee queens," *Animal Behaviour* 70 (1): 125–131 (2005).

78 | C. K. Starr, "Sperm competition, kinship, and sociality in the aculeate Hymenoptera," in R. L. Smith, ed., *Sperm Competition and the Evolution of Animal Mating Systems* (New York: Academic Press, 1984), pp. 427–464; R. F. A. Moritz, "The effects of multiple mating on the worker-queen conflict in *Apis mellifera* L.," *Behavioral Ecology and Sociobiology* 16(4): 375–377 (1985); M. Woyciechowski and A. Lomnicki, "Multiple mating of queens and the sterility of workers among eusocial Hymenoptera," *Journal of Theoretical Biology* 128(3): 317–327 (1987); P. Pamilo, "Evolution of colony characteristics in social insects, II: number of reproductive individuals," *American Naturalist* 138(2): 412–433 (1991); D. C. Queller, "Worker control of sex ratios and selection for extreme multiple mating by queens," *American Naturalist* 142(2): 346–351 (1993); and F. L. W. Ratnieks and J. J. Boomsma, "Facultative sex allocation by workers and the evolution of polyandry by queens in social Hymenoptera," *American Naturalist* 145(6): 969–993 (1995).

79 | P. Schmid-Hempel, *Parasites in Social Insects* (Princeton, NJ: Princeton University Press, 1998); P. W. Sherman, T. D. Seeley, and H. K. Reeve, "Parasites, pathogens, and polyandry in social Hymenoptera," *American Naturalist* 131(4): 602–610 (1988); J. F. A. Traniello, R. B. Rosengaus, and K. Savoie, "The development of immunity in a social insect: evidence for the group facilitation of disease resistance," *Proceedings of the National Academy of Sciences USA* 99(10): 6838–6842 (2002); and A. Stow, D. Briscoe, M. Gillings, M. Holley, S. Smith, R. Leys, T. Silberbauer, C. Turnbull, and A. Beattie, "Antimicrobial defences increase with sociality in bees," *Biology Letters* 3(4): 422–424 (2007).

found in colonies of the leafcutter *Acromyrmex echinatior* in the control of a virulent soil fungus and in honeybees.[80,81] Further correlative evidence favoring the hypothesis of disease resistance has been obtained in ant and other social insect species in which the queen increases the genetic diversity of her worker progeny by mating with multiple males.[82] (However, among the fungus-growing ants as a whole, including many species with singly mated queens and regardless of the aforementioned case of *Acromyrmex echinatior*, the evidence for the enhancement of disease resistance by genetic diversity remains ambiguous.[83]) Favoring the disease hypothesis is the recent discovery that the potency of antimicrobial defenses in bee populations rises steeply from solitary species to semisocial species and beyond to advanced eusocial species.[84]

In another functional category, an increase in the genetic diversity of honeybee nestmates is positively correlated with increased stability in hive temperatures and productivity and fitness.[85] This homeostatic effect appears to arise from the enhanced flexibility of colonies that harbor bees with innately different patterns of response. A similar conclusion has been tentatively advanced for genetic variation in worker specialization, as documented in the ant *Formica selysi*.[86] Furthermore, greater genetic variability in the work force can be favored by colony-level selection (between-group selection), as suggested in the harvester ant *Pogonomyrmex occidentalis*. Colonies with greater genetic variation have overwhelmingly higher growth

---

80 | W. O. H. Hughes and J. J. Boomsma, "Genetic diversity and disease resistance in leaf-cutting ant societies," *Evolution* 58(6): 1251–1260 (2004).

81 | D. R. Tarpy and T. D. Seeley, "Lower disease infections in honeybee (*Apis mellifera*) colonies headed by polyandrous vs monandrous queens," *Naturwissenschaften* 93(4): 195–199 (2006); and T. D. Seeley and D. R. Tarpy, "Queen promiscuity lowers disease within honeybee colonies," *Proceedings of the Royal Society of London B* 274: 67–72 (2007).

82 | R. H. Crozier and E. J. Fjerdingstad, "Polyandry in social Hymenoptera—disunity in diversity?" *Annales Zoologici Fennici* 38: 267–285 (2001); and A. J. Denny, N. R. Franks, S. Powell, and K. J. Edwards, "Exceptionally high levels of multiple mating in an army ant," *Naturwissenschaften* 91(8): 396–399 (2004).

83 | T. Murakami, S. Higashi, and D. Windsor, "Mating frequency, colony size, polyethism and sex ratio in fungus-growing ants (Attini)," *Behavioral Ecology and Sociobiology* 48(4): 276–284 (2000).

84 | A. Stow, D. Briscoe, M. Gillings, M. Holley, S. Smith, R. Leys, T. Silberbauer, C. Turnbull, and A. Beattie, "Antimicrobial defences increase with sociality in bees," *Biology Letters* 3(4): 422–424 (2007).

85 | J. C. Jones, M. R. Myerscough, S. Graham, and B. P. Oldroyd, "Honey bee nest thermoregulation: diversity promotes stability," *Science* 305: 402–404 (2004); and H. R. Mattila and T. D. Seeley, "Genetic diversity in honey bee colonies enhances productivity and fitness," *Science* 317: 362–364 (2007).

86 | T. Schwander, H. Rosset, and M. Chapuisat, "Division of labour and worker size polymorphism in ant colonies: the impact of social and genetic factors," *Behavioral Ecology and Sociobiology* 59(2): 215–221 (2005).

and reproduction rates than those with less variation.[87] This rise in fitness may be due to the enhancement of division of labor by a spread among workers of genetic predispositions to specialization. Such a disposition has been discovered in the polymorphic worker caste of the Florida harvester, *Pogonomyrmex badius*: some heritability occurs in adult worker size, which, with allometric growth of imaginal disks during larval development in the final larval instar, differentiates colony members into small-headed minors and large-headed majors.[88] On the other hand, no correlations of colony efficiency with degree of relatedness were detected in experiments on the Argentine ant *Linepithema humile*.[89]

---

87 | B. J. Cole and D. C. Wiernasz, "The selective advantage of low relatedness," *Science* 285: 891–893 (1999).
88 | F. E. Rheindt, C. P. Strehl, and J. Gadau, "A genetic component in the determination of worker polymorphism in the Florida harvester ant *Pogonomyrmex badius*," *Insectes Sociaux* 52(2): 163–168 (2005).
89 | H. Rosset, L. Keller, and M. Chapuisat, "Experimental manipulation of colony genetic diversity had no effect on short-term task efficiency in the Argentine ant *Linepithema humile*," *Behavioral Ecology and Sociobiology* 58(1): 87–98 (2005).

3

SOCIOGENESIS

All on its own, a major advance or regression in evolutionary grade has a special significance for the understanding of evolution itself. The creation of superorganisms is such an event. Of the 2,600 or so living families of insects and other arthropods, for example, only 15 are known to contain eusocial species, that is, possess colonies with low-reproductive or nonreproductive castes.[1] And of the 740 nonhuman vertebrate families, only one, the naked mole rats (Bathyergidae), is known to have reached the same grade of social organization.[2]

Among the most important questions in sociobiology is the nature and number of genetic and physiological steps required for a species to make the extraordinary advance to eusociality. The answer will be expressed as the rules of construction of the physiology and behavior of colony members. These rules will prove to be multilevel and hierarchical, operating through the life cycle of the members across the three levels of their biological organization—genomic, organismic, and societal. Further, to reveal the independent pathways to eusociality followed in evolution, it will be necessary to address the life cycles of many species. The same is true for identification of the ecological pressures and opportunities that have drawn different phyletic lines across the eusociality threshold.

Bear in mind that each eusocial species occupies a particular niche in the

---

1 | The arthropod families known to have eusocial species are Apidae (honeybees, bumblebees, and stingless bees), Halictidae (sweat bees), Formicidae (ants), Vespidae (vespid wasps), Sphecidae (sphecid wasps), six families of termites (composing the order Isoptera), Platypodidae (ambrosia beetles), Thripidae (gall thrips), Aphididae (aphids), and Alphaeidae (snapping shrimps). The number of arthropod families is based on comprehensive listings by specialists in S. P. Parker, ed., *Synopsis and Classification of Living Organisms*, Vol. 2 (New York: McGraw-Hill, 1982).

2 | The number of vertebrate families is also based on S. P. Parker, ed., *Synopsis and Classification of Living Organisms*, Vol. 2 (New York: McGraw-Hill, 1982).

environment, comprising a unique envelope of habitat, nest site, food, and enemies on which social grouping confers competitive superiority. The life cycle of the species has evolved across thousands or millions of years down to the eyeblink of time when at last it has caught the attention of human intelligence.

THE COLONY LIFE CYCLE

For purposes of analysis, the life cycle of a colony can be efficiently thought to begin as an egg destined to produce a virgin queen. For most kinds of social insects, the life cycle spirals out from that starting point in the following steps: following her development from egg through larva and pupa, the newly adult queen leaves her birth nest; she mates with one or more males from other nests; she searches for a nest site; she lays eggs that produce workers; the workers divide into castes; they perform cooperative labor; they grow in number; they communicate by an ever-expanding repertoire of pheromones; and in the improbable event that every step of the sequence to this point has gone well, the worker population reaches mature size, whereupon the queen lays fertilized eggs that give rise to the next generation of virgin queens, along with unfertilized eggs that yield males. Both reproductives are destined to leave the nest and mate with reproductives of the opposite sex from other nests. An example of such a life cycle is found in the European ant species *Myrmica ruginodis*, illustrated in Figure 3-1.

Viewed as a product of evolution by natural selection, the colony life cycle is the genes' roundabout way of putting as many of their copies as possible into the next generation.

SOCIAL ALGORITHMS

What the genes prescribe is not the life cycle in any literal sense but rather its epigenetic program, a kind of molecular and organismic operating manual by which the colony assembles itself. In describing this process, many biologists have found it helpful to use the language of the physical and computer sciences. For their part, researchers in the physical and computer sciences have done the reverse by employing ants, honeybees, and other social insects as models of self-organizing systems.[3] The steps of the program, in insects and machines, are envisioned as sequences of *decision rules*, or, phrased more biologically, *epigenetic rules*. The programs unfold in linear manner. As each successive binary *decision point* is reached, the individual colony member proceeds down one pathway or another until it comes either to the next decision point or to the end of the sequence. A particular program may guide the gradual anatomical and physiological development of individual colony members into one caste or another, or it may cause changes in a member's behavior within the ambit of its caste repertoire. The passage from one decision point to the next may last for weeks, as sometimes occurs in caste specialization, or it may be as brief as seconds, as in the recognition of nestmates. A complete sequence of decision points that produces a caste, product, or full behavioral response is called an *algorithm*. The progress from one decision point to another is the analog of linear programming of computer software design. A simple example is given in Figure 3-2, depicting the first steps in the identification of colony odors.

Overall, then, two broad classes of algorithms work simultaneously to guide the autoconstruction of an insect colony. While a colony member is still an egg or larva, certain events occur that channel its development into either an adult reproductive or an adult worker. In some species of ants and termites, a larva already launched on the worker pathway encounters a second decision point, at which it proceeds in final anatomical development to maturity as either a minor worker or a major worker. Minor workers become labor generalists, while major workers usually become soldiers or storage castes.[4]

---

3 | A pioneering effort to draw the parallels of insect societies and machine computation can be found in D. R. Hofstadter, *Gödel, Escher, Bach: An Eternal Golden Braid* (New York: Basic Books, 1979; 1999, 20th anniversary edition with a new preface by the author). Recent contributions are summarized in G. Weiss, ed., *Multiagent Systems, A Modern Approach to Distributed Artificial Intelligence* (Cambridge, MA: MIT Press, 1999); S. Camazine, J.-L. Deneubourg, N. R. Franks, J. Sneyd, G. Theraulaz, and E. Bonabeau, *Self-Organization in Biological Systems* (Princeton, NJ: Princeton University Press, 2001); F. Klügl, *Multiagentensimulation: Konzepte, Werkzeuge, Anwendungen* (Munich: Addison-Wesley, 2001); M. Dorigo and T. Stützle, *Ant Colony Optimization* (Cambridge, MA: MIT Press, 2004); and essays by E. O. Wilson and B. Hölldobler, "Dense heterarchies and mass communication as the basis of organization in ant colonies," *Trends in Ecology and Evolution* 3(3): 65–68 (1988); J. W. Pepper and G. Hoelzer, "Unveiling mechanisms of collective behavior," *Science* 294: 1466–1467 (2001); B. Schouse, "Getting the behavior of social insects to compute," *Science* 295: 2357 (2002); and J. H. Fewell, "Social insect networks," *Science* 301: 1867–1870 (2003).

4 | These developmental pathways have been authoritatively described by Diana E. Wheeler in "Developmental and physiological determinants of caste in social Hymenoptera: evolutionary implications," *American Naturalist*

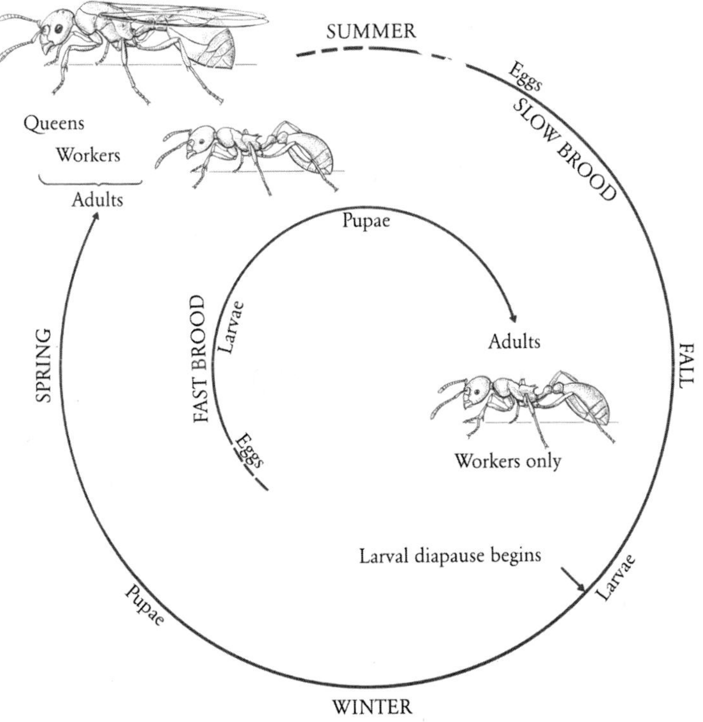

FIGURE 3-1. The annual cycle of brood development in a mature colony of *Myrmica ruginodis*. The mother queen continues to lay eggs intermittently through the spring and summer. Many of the larvae that hatch early in the season are able to complete development by the end of the summer and become workers (fast brood). Others persist as larvae through the winter and can become workers or queens the following spring (slow brood). The full development of fast brood requires about three months, that of slow brood almost a year. From E. O. Wilson, *The Insect Societies* (Cambridge, MA: The Belknap Press of Harvard University Press, 1971).

Once it has reached the adult stage, the social insect behaves by algorithms of the second major kind, performing labor appropriate to its physical caste and adult age. Decision by decision, the insect responds to those stimuli to which its sensory and nervous systems are programmed to respond. These stimuli compose the highly filtered sensory world of the caste to which it belongs. A caste member performs specialized tasks because it has lower response thresholds to stimuli linked to those tasks. Each caste, in a phrase, is characterized by a unique lowered-threshold profile.

A caste member, however, is far from a simple automaton. As a result of its

128(1): 13–34 (1986), and "The developmental basis of worker caste polymorphism in ants," *American Naturalist* 138(5): 1218–1238 (1991).

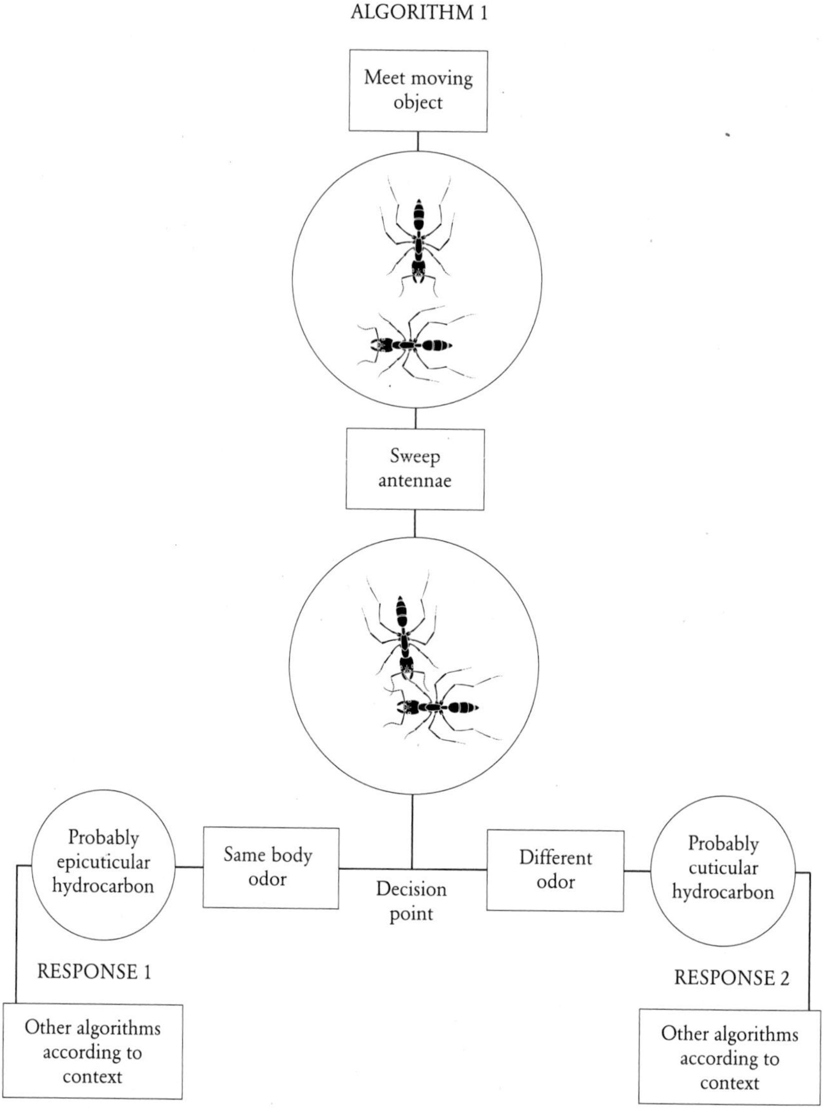

FIGURE 3-2. Diagram of a simple algorithm guiding the social behavior of an ant. In this case, a worker discriminates between fellow colony members and aliens and responds accordingly.

individual experience, the insect modifies its responses to some kinds of the stimuli and not others. The learning is nevertheless limited and statistically predictable. It is also biased: the insect is innately prepared to make certain kinds of responses and counterprepared (hence predisposed to resist) making other kinds of responses. The ensemble of biases, prepared and counterprepared, appears adaptive; there is every reason to believe (although the principle is not yet proved) that the pattern of flexibility in social interaction created by the combination has been fashioned by natural selection.

The social insect is also programmed to switch algorithms with a readiness appropriate to its caste and personal experience. If, for example, an ant is repairing a breach in the nest wall and encounters a misplaced larva, it automatically picks up the larva and returns it to the brood chamber. We therefore speak of behavior of social insects as context specific. There is no reason to suppose that the insect is thinking in human manner about the reasons for its actions or about their possible consequences. Rather, it has simply switched from one algorithm to another. The ability of most adult colony members to move from one task to another is a well-documented universal property of social insects, and the flexibility the capacity provides is generally regarded to be a prime cause of their ecological success.

Hundreds or thousands of such colony members operating autonomously might seem at first to be a recipe for chaos. Yet somehow their simultaneously operating algorithms combine to guide each member through the confusion of colony life. And together the mob of algorithm-guided individuals manages to form the colony, a higher unit integrated in its actions into patterns that allow it to survive and reproduce as a colony.

How did such order come into existence? The best short answer available is that natural selection at the colony level creates algorithms that maximize efficient order. Genes prescribe algorithms, which guide colony members by means of the nuances of sensory thresholds, context, and innate flexibility in a manner that draws the appropriate response from the colony as a whole. It is the integration of these modules of individual behavior that determines the fate of the colonies and hence of the genes that prescribed their construction.

The algorithms discovered thus far are surprisingly simple. At most decision points, they permit the colony member a binary choice. In contrast, the final products of the algorithms, by the patterns they form at the colony level in communication and division of labor, are relatively complex. In theory, the power of simple algorithms to produce elaborate yet precise patterns is virtually unlimited. It has the decisive advantage of allowing creatures with tiny brains to create complex societies. As a mathematical principle, even the simplest decision rules can produce enormously variable games of choice. A 3-by-3 tic-tac-toe game has 50,000 configurations, while a 50-move chess game can generate more pathways than there are atoms in the visible universe.[5] As rules are multiplied, the potential outcomes increase at a superexponential rate. In an insect colony, one algorithm with three successive decision points has only eight outcomes. But seven such algorithms in combination have over 2 million outcomes. From such a vast pool of potential outcomes, natural selection has chosen a microscopically small set of algorithms and algorithmic combinations.

By obeying algorithms, by using them as rules of thumb, each worker is able to make quick instinctive decisions in the midst of seeming chaos. Conditioned by the ongoing simultaneous decisions of other colony members, the colony as a whole creates emergent patterns of adaptive response that are difficult and perhaps even impossible to predict from the observed behavior of individuals alone. An evolutionary trade-off exists in the properties of the algorithms between simplicity and quickness on the one side and calculation and delay on the other side. Either of these two opposing properties, simplicity or calculation, can produce emergent patterns, but the limitation imposed on computing capacity by small brain size has tipped the social insects to the simple and quick.

SELF-ORGANIZATION AND EMERGENCE

To add one last concept from computer science, social insect workers are cellular automata, defined as agents programmed to function interactively as a higher-level system. They have this trait because their colony as a whole lacks command and control by a still higher-level system. It therefore must be self-organized. Through the combined senses and brains of its members, the colony operates as an information-processing system. The environment challenges it with problems: the workers must locate an adequate nest site, find the right food items and bring them home, establish home ranges and territories, defend against enemies, and care for the helpless young. These disjunct problems press on the colony at almost all times. The algorithms of individual development and behavior contain the solutions to all of them. With algorithms, the colony masters the problems natural selection has designed it to solve. The requisite information is distributed among the colony members. Thus, a distributed colony intelligence is created greater than the intelligence of

5 | J. H. Holland, *Emergence: From Chaos to Order* (Reading, MA: Addison-Wesley, 1998).

any one of the members, sustained by the incessant pooling of information through communication.[6]

The following are two famous examples of emergent properties that arise from the actions of individuals but are displayed only by the colony as a whole:

- When resting, colonies of New World tropical army ants (genus *Eciton*) form shelters using their own massed bodies. By choosing the right location to construct the living bivouac and then adjusting its position and shape, the colony as a whole regulates the temperature and humidity inside the bivouac, while also protecting the queen and growing brood from outside intruders. To forage for food, hundreds of thousands of the workers form tight columns or, in one species (*E. burchelli*), fan-shaped swarms that spread over the ground like a single organism. They gather large quantities of prey, then stream in the reverse direction to rejoin the rest of the colony in the bivouac.[7]

- Fungus-growing termites (*Macrotermes natalensis*) of Africa build immense and elaborately designed nests whose architecture precisely regulates the temperature and composition of the interior air. As air warms in the central core of the nest from the metabolic heat of the termites living there, it rises by convection to a large chamber in the upper part of the mound and then out to a flat, capillary-like network next to the outer nest wall. In these latter chambers, the air is cooled and refreshed. As

---

6 | The idea of self-organization based on distributed intelligence has deep roots. It is true that the earliest writers, relying on what must have seemed common sense, assumed that the insect societies are governed by some sort of central command. Charles Butler, speaking of honeybees in *The Feminine Monarchie* (Oxford: J. Barnes, 1609), with a deep bow to the English ruler, declared that "For the Bees aborre as well polyarchie, as anarchie, God having shewed in the unto mean express patterne of a perfect monarchie, the most natural and absolute forme of government." Perhaps the first perception of self-organization was that of Pierre Huber in his *Recherches sur les Moeurs des Fourmis Indigènes* (Paris: J. J. Paschoud, 1810). Writing of nest construction by the ant *Formica fusca*, he stated: "I am convinced that each ant acts independently of its companions. The first that hits upon an easy plan of execution immediately produces the outline of it; others only have to continue along these same lines guided by an inspection of the first efforts." A host of authors from the 1930s onward, from J. Freisling on the social wasps, T. C. Schneirla on army ants, P.-P. Grassé on termites, and C. D. Michener and S. W. T. Batra on halictid bees, came closer to formulating self-organization as the governing principle. Finally, it was made an explicit idea by one of the present authors—E. O. Wilson, in the 1971 synthesis *The Insect Societies* (Cambridge, MA: The Belknap Press of Harvard University Press)—and has been the subject of a small research industry by researchers whose discoveries are chronicled in this and succeeding chapters.

7 | Reviews of army ant behavior are given in T. C. Schneirla, *Army Ants: A Study in Social Organization*, H. R. Topoff, ed. (San Francisco: W. H. Freeman, 1971); and W. H. Gotwald, *Army Ants: The Biology of Social Predation* (Ithaca, NY: Comstock Publishing Associate of Cornell University Press, 1995).

this occurs, the air sinks to the lower, inhabited chambers, and the cyclical journey begins anew. The entire nest thus operates as an air-conditioning system, thereby keeping, in *Macrotermes natalensis* at least, the core living area within a degree of 30°C and carbon dioxide concentration between 2.6 and 2.8 percent.[8]

While such emergent properties are marvelous to behold, their engineering is not intrinsically mysterious. The extremes of higher-level traits may at first appear to have a life of their own, one too complex or fragile to be reduced to their basic elements and processes by deductive reasoning and experiment. But such separatist holism is in our opinion a delusion, the result of still insufficient knowledge about the working parts and processes. An important merit of insect sociobiology, as opposed to vertebrate and especially human sociobiology, is that the colony organizations it addresses contain a large array of emergent phenomena simple enough to be explained by scaling up from the behavior of the constituent elements. This is the advantage provided us by the small brains of the social insects and the general quick and simple decisions they must make with limited algorithms.

PHYLOGENETIC INERTIA AND DYNAMIC SELECTION

The evolution of superorganisms proceeds by a clash of force and inertia. The environment presents to the species particular problems and opportunities. But given the severe restrictions in the species' potential caused by adaptations previously made, only a small array of possible solutions are available for its further evolution. As a result of this phylogenetic inertia, the same problems and opportunities, expressed as forces of natural selection, can result in radically different solutions among species. The visible result is a variety of anatomical structures and patterns of behavior among species, each of which serves approximately the same function.

The balance of inertia and force in evolution is vividly illustrated by the differences in the recruitment systems of the wingless, earthbound ants and their equivalent in complexity of social organization, the winged and airborne honeybees.

---

8 | M. Lüscher, "Air-conditioned termite nests," *Scientific American* 205(1): 138–145 (1961); and C. Noirot and J. P. E. C. Darlington, "Termite nests: architecture, regulation and defence," in T. Abe, D. E. Bignell, and M. Higashi, eds., *Termites: Evolution, Sociality, Symbioses, Ecology* (Norwell, MA: Kluwer Academic Publishers, 2000), pp. 121–139.

The principal instruments of the ants are chemical trails. The most thoroughly researched such system is that of the red imported fire ant, *Solenopsis invicta*, a native of South America that has been introduced into the United States and is now a major pest. When a solitary worker foraging away from the nest encounters a sizable dead insect or some other food object too large to carry, she returns to the nest at a slower, more deliberate pace. At frequent intervals during the trip, she extrudes and drags her sting along the ground, much like a pen tip drawn over paper to deposit an inked line. As the sting touches the surface, a mix of pheromones flows down and out of the Dufour's gland, a small finger-shaped gland located at the rear of the abdomen. This material, weighing only about a billionth of a gram, consists of a blend of compounds, each fulfilling a special function in the complex trail signal. The principal compounds for trail orientation are two α-farnesenes and two homofarnesenes, accompanied by a still unidentified component that serves as an attractant. Oddly, these substances remain inactive unless the ants have been induced by yet another, still unidentified component in the glandular secretions. The two substances responsible for attraction and induction require about 250 times the relative concentration of the orientation pheromones. This proportionality explains previous observations that the more desirable the food find, the more intense the trail laying by the recruiter ants. Relatively large quantities of discharged trail pheromone provide a sufficient amount of initial attraction and induction to get the recruitment process started. Once turned on by these relatively short-lived signals, the ants also follow trails consisting of small amounts of orientation pheromones. The ants are able to detect the complex by smelling the vapor it emits over distances as great as a centimeter. When activated, they do not, however, simply track the liquid trace itself. Instead, they move through the vapor created by diffusion of the pheromone into the air. All around the liquid deposit and within the vapor cloud is an "active space," semiellipsoidal in shape, within which the pheromones are at high enough concentration to be sensed by the ants. As the ants travel through this vapor tunnel back out toward the food site, they continuously sweep the air with their antennae, the principal organs of smell, testing the air for odorant molecules and keeping themselves within the active space around the trail[9] (see Chapter 6).

9 | The glandular origin of the trail pheromones was discovered by E. O. Wilson in 1959 ("Source and possible nature of the odor trail of fire ants," *Science* 129: 643–644), the first of what was to be a long list of such exocrine signals identified in ant communication. The chemistry and roles of the first ant trail components were worked out by R. K. Vander Meer and his collaborators in the 1980s; see R. K. Vander Meer, F. M. Alvarez, and C. S.

In the course of their travel through the active space, the individual ants make decisions whether to go the distance to the food site, and if that is the case whether to lay reinforcing trails of their own. It is as though they were saying, pheromonally, "Yes, there *is* food here, and it's good enough for more of you to follow the trail too." As a result of many such decisions, based on the circumstances under which they are made, the aggregate of responders create mass communication, by which multiple workers communicate a compound of information about the quantity and quality of the food source. The information controls the number going to the food site. It is an emergent phenomenon that can only be generated by groups of workers and acted on by other groups of workers.

The mass communication works as follows. The number of workers drawn out of the nest increases with the amount of pheromones in the trail, as well as the vigor with which the recruitment is conducted. As additional workers reach the food and are satisfied, more trails are laid and still more workers emerge from the nest and run to the food. At first the buildup is exponential, but it decelerates toward a limit as workers pile up on the food mass, making it difficult for newcomers to get through. Frustrated workers wander about, and most soon return home. The backflow combined with the short life of the active space from the pheromone contributed by each worker, lasting no more than a few minutes, causes the number of ants at the food to equilibrate not far from the optimum needed. As the food is eaten or carried off in fragments, its diminishing amount is matched by the decline in number of workers attending.[10]

The mass communication is more finely calibrated to equilibrium by additional adjustments in the responses of individual workers. The more desirable the food find, the higher the percentage of positive responses, the greater the trail-laying effort, the more the trail pheromone presented to the colony, and the more the

---

Lofgren, "Isolation of the trail recruitment pheromone of *Solenopsis invicta*," *Journal of Chemical Ecology* 14(3): 825–838 (1988); and R. K. Vander Meer, C. S. Lofgren, and F. M. Alvarez, "The orientation inducer pheromone of the fire ant *Solenopsis invicta*," *Physiological Entomology* 15(4): 483–488 (1990). The fundamentals of ant trail and other forms of chemical communication generally are reviewed in B. Hölldobler and E. O. Wilson, *The Ants* (Cambridge, MA: The Belknap Press of Harvard University Press, 1990), and in R. K. Vander Meer and L. E. Alonso, "Pheromone directed behavior in ants," in R. K. Vander Meer, M. D. Breed, K. E. Espelie, and M. L. Winston, eds., *Pheromone Communication in Social Insects: Ants, Wasps, Bees, and Termites* (Boulder, CO: Westview Press, 1998), pp. 159–192.

10 | E. O. Wilson, "Chemical communication among workers of the fire ant *Solenopsis saevissima* (Fr. Smith)," *Animal Behaviour* 10(1–2): 134–147, 148–158, 159–164 (1962).

newcomer ants that emerge from the colony. The same effect occurs when the food is placed close to the nest.

A close examination of the vivacity of trail laying has revealed still more sophistication in adjustment of the mass response. Six elements of "salesmanship" can be added to the transmission of the pheromonal communication by individuals to strengthen the overall recruitment signal:

1 | A trail is laid by the recruiter while returning to the food site.
2 | The recruiter runs faster.
3 | The recruiter waggles her head next to nestmates.
4 | The recruiter brushes nestmates with her antennae.
5 | The recruiter, approaching nestmates, displays regurgitated food held between her mandibles.
6 | The recruiter leads nestmates with an outgoing trail.

When motivation of the colony is high, at least under laboratory conditions when the foragers are presented with a concentrated sugar solution, only three of the six signals are sufficient to elicit following. When motivation is low, as when a weak concentration of the sugar solution is offered, all six of the signals are needed. Gradations in the number of workers receiving these combinations accounts for variation in the number responding[11] (see Chapter 6).

The domesticated honeybee (*Apis mellifera*) lives in a wholly different sensory world. The foragers must fly far from the nest in search of flower patches sufficient to feed their large and metabolically very active colonies with pollen and nectar. The collective reach of the foragers is enormous. If each bee were the length of a human and the hive were Austin, Texas, the searchers from a single colony could patrol the entire state, making round-trips as far as the borders in as little as half an hour.[12]

The colony members speed the collection of nectar by a division of labor. Foragers by and large continue to forage, while back at the nest, the nectar loads are

---

11 | D. Cassill, "Rules of supply and demand regulate recruitment to food in an ant society," *Behavioral Ecology and Sociobiology* 54(5): 441–450 (2003).

12 | Worker bees can travel 6 kilometers or more from the hive and back in a single trip at 25 kilometers per hour; see T. D. Seeley, *The Wisdom of the Hive: The Social Physiology of Honey Bee Colonies* (Cambridge, MA: Harvard University Press, 1995, p. 47). When bee length is scaled up to human length by about 200, the round-trip would be the equivalent of 2,400 kilometers or more traveled at a speed of roughly 5,000 kilometers per hour.

received by younger workers, who either feed the nectar to hungry nestmates or store it in the combs for future consumption. This specialization increases the efficiency of the colony as a whole, but it also creates two potential bottlenecks. First, discovery of a rich new flower patch can flood the hive with more nectar than the active processors can manage. Second, and conversely, slow periods in the field can result in unemployed processors.

The honeybee colony employs three interacting signals to open the bottlenecks and adjust the collection and flow of nectar to colony needs.[13]

The *waggle dance*, a vibrating figure-eight run conducted on the vertical comb surfaces by successful foragers, alerts other colony members to the existence of nectar sources outside the nest. The direction and duration of the middle segment of the figure eight (during which the bee waggles its body laterally) corresponds respectively to the direction of the food site relative to the sun as the bee leaves the hive and the distance of the site from the hive. The time spent dancing by the returning bees and the liveliness of their dances suggest the value of the food discovery, in particular how rich it is, how far away, and how much it is needed by the colony.

When foragers have discovered a nectar source needed by the colony and there are not enough foragers to exploit it, bees alert one another by the *shaking signal*. The active signaler climbs on top of a nestmate, grips her with its legs, then vibrates its whole body for 1 or 2 seconds before moving on to another bee. The result is a movement of bees to the dance floor and then a dispatch of more foragers to the food site.

On the other hand, when more nectar is coming into the nest than can be handled by active processors, more processors are recruited by means of the *tremble dance*. The signaling bees run irregularly about the combs, shaking their bodies back and forth, right and left, with their forelegs held aloft, like Saint Vitus' dancers. Many bees contacting the dancers convert to nectar processing.

In conducting the intricate maneuvers of nestmate recruitment, fire ants and honeybees make decisions based on context and their own experience. They have minds of a primitive sort. This is not to say they possess humanlike reflective consciousness, with competing scenarios unwinding back and forth in time through vast banks of language-coded memory and weighted with self-awareness and

---

13 | This account is drawn from the magisterial account of social behavior in honeybees in T. D. Seeley, *The Wisdom of the Hive: The Social Physiology of Honey Bee Colonies* (Cambridge, MA: Harvard University Press, 1995).

meaning. Rather, their minds consist of a far simpler form of perceptual consciousness, in which fragments of memory are joined to immediate-time perception to create versatile representations and communication.[14] In the judgment of Thomas D. Seeley, after extended studies under natural field conditions, "It is now clear that we cannot explain the behavior of a bee producing communication signals in terms of simple responses to immediate stimuli. Instead, to understand the signal production behavior of a worker bee, we must view her as a sophisticated decision maker, one capable of integrating numerous pieces of information (both current perceptions and stored representations) as she chooses the general type and specific form of signal that is appropriate for a particular situation."[15]

The bee, in short, apparently "thinks" like a motorist driving homeward along a familiar route, integrating automatically from multiple subconscious sources the inner road map and his destination, yet pondering neither the trip nor the operation of the automobile.

While all this behavior is very sophisticated by insect standards, with the exceptional flexibility that perceptual consciousness allows, the worker honeybee still consistently follows the decision rules of her species and hence her genome. She assesses the labor needs of the colony and acts accordingly in a predictable manner. Her decision rules can be stated as follows:

1 | Not enough nectar collectors in the field? If yes, and if you also have immediate knowledge of a producing flower patch, perform the waggle dance.

2 | Is the flower patch rich or the weather fine or the day early or does the colony need substantially more food? Perform the dance with appropriately greater vivacity and persistence.

3 | Not enough active foragers to send into the field? Perform the shaking maneuver.

4 | Not enough nectar processors in the hive to handle the nectar inflow? Perform the tremble dance.

---

14 | This distinction between reflective and perceptual consciousness was suggested in D. R. Griffin, *Animal Minds: Beyond Cognition to Consciousness* (Chicago: University of Chicago Press, 2001).

15 | T. D. Seeley, "What studies of communication have revealed about the minds of worker honey bees," in T. Kikuchi, N. Azuma, and S. Higashi, eds., *Genes, Behavior and Evolution of Social Insects* (Sapporo, Japan: Hokkaido University Press, 2003), pp. 21–33.

Hundreds of bees making such decisions more or less simultaneously yield the overall response of the superorganism. As the colony need grows, communication spreads and more workers respond. As the need subsides, the number of engaged workers tapers off. By the law of large numbers, the workers' personal idiosyncrasies, mistakes, and lucky guesses are summed. When added to deviations either up or down through inappropriate vivacity or tepidness, they tend to cancel out and hold the colony response hour by hour close to its optimum while narrowing fluctuation around that level.

|| Plate 5. The small bee in the photograph with the honeybee is a primitively eusocial bee of the genus *Halictus*. (Photo: Gro Amdam.)

4

THE GENETIC EVOLUTION OF DECISION RULES

We now turn to the great evolutionary transition from organism to superorganism. Solitary bees and wasps practicing maternal care by progressive provisioning are only a short evolutionary step from the creation of a eusocial society. At this stage, each female builds a nest of her own, then supplies it at intervals with pollen or arthropod prey to nourish her young. All that is needed to attain colonial life is for some of the adult offspring to forgo reproduction in order to help their mother rear more of her progeny. Alternatively, colonial behavior might arise when adults leave home and go to other nests where they serve as nonreproductive helpers.

THE GENETIC ORIGIN AND FURTHER EVOLUTION OF EUSOCIALITY

As simple as these crucial steps to eusociality may seem, they have occurred only rarely in evolution. The crucial preadaptations and sufficiently potent selection forces needed to drive such a presocial-to-eusocial transition have been considered in previous chapters. Now we turn to the substance of the transition itself in an effort to answer the following two questions:

1 | How many decision rules, and of what kind, are needed to create a eusocial species?
2 | What kind of genetic change, and at what magnitude, is required to prescribe the decision rules?

In addition to the origin of eusociality, a second transition of colonial existence is possible, from the primitive superorganism to the advanced superorganism—for example, from the simple colonies of the halictid bees to the highly complex organizations of the honeybees or from the Australian *Prionomyrmex* (formerly *Nothomyrmecia*) dawn ants to the elaborate societies of leafcutter ants. This final advance requires a great many sequential steps in evolution. Like the origin of eusociality itself, it has occurred only rarely. Its existence poses a pair of questions parallel to the first just offered:

3 | Once the eusocial threshold has been crossed, how many decision rules, and of what kind, are needed to transform the primitively eusocial species thus created into an advanced eusocial species?

4 | What kind of genetic change, and of what magnitude, is required to prescribe this amount of social change?

Forty years of research devoted to the communication and caste systems of social insects, addressing both the origin of eusociality and the subsequent rise to advanced eusociality, has yielded considerable progress in identifying the decision rules. Only since the mid-1980s, however, has research reached down to the next level of biological organization to address the genetic changes that prescribe the decision rules. Because that field of investigation is in its infancy and therefore less complicated—so far!—we will begin there.[1]

SOCIOGENETICS AND SOCIOGENOMICS

The most advanced research on the genetics of insect social behavior to date has been that conducted on the domestic honeybee *Apis mellifera*. More recently, a flurry of important discoveries have also been made with ants. The steps in the brief history of these endeavors have recapitulated those of genetics as a whole. First came *sociogenetics*, the discovery and mapping of variations in single genes or small

1 | The history of hymenopteran and, in particular, honeybee sociogenetics is reviewed in R. E. Page Jr., J. Gadau, and M. Beye, "The emergence of hymenopteran genetics," *Genetics* 160(2): 375–379 (2002); R. E. Page Jr. and J. Erber, "Levels of behavioral organization and the evolution of division of labor," *Naturwissenschaften* 89(3): 91–96 (2002); G. E. Robinson, "Sociogenomics takes flight," *Science* 297: 204–205 (2002); G. E. Robinson, C. M. Grozinger, and C. W. Whitfield, "Sociogenomics: social life in molecular terms," *Nature Reviews / Genetics* 6: 257–271 (2005); and R. E. Page Jr. and G. V. Amdam, "The making of a social insect: developmental architectures of social design," *BioEssays* 29(4): 334–343 (2007).

ensembles of genes that affect particular traits in behavior or caste anatomy, along with the analysis of their patterns of distribution among kin within and among colonies. Once identified, these alleles can be mapped to chromosome position, and the proteins they transcribe can be identified. Next, the provenance of each phenotype is tracked one by one, from gene to cellular component, next to neuron structure or mediating hormone, then to a pheromone produced or a response threshold modified. Finally, the learning process selectively biased to produce adaptive behavior is open to analysis.

In theory at least, sociogenetics, by proceeding from one trait to the next, can map virtually all the genes responsible for the behavioral decision rules as well as the molecular and cellular developmental processes that manufacture the rules. Nonetheless, while very effective as the opening wedge, sociogenetics need not be relied on for the complete mapping of the genome. It has already been supplemented and in time is likely to be replaced by *sociogenomics*,[2] the top-down method that starts with all or most of the genes as a whole and then sorts out those responsible for social behavior. The method proceeds by the complete or near-complete sequencing of the species' base pairs (the genetic "letters"), followed by the mapping of the genes, each of which comprises large numbers of base pairs. Using resources such as expressed sequence tags and microarrays, researchers can obtain snapshots of all the genes active in the formation and function of individual cells and tissues. They can follow similar gene expression profiles to track the patterns and timing of expression of the genes that affect social behavior. Finally, they can identify those genes activated in individual colony members as a result of changes in their environment.

At the present time, the bottom-up approach of sociogenetics is well along in the honeybees and ants, while the top-down approach of sociogenomics is in its infancy. Now, however, progress in sociogenomics is accelerating due to the full sequencing in 2006 of the honeybee genome.[3]

The merit of sociogenomics emerges at different levels. It is the most effective tool for the rapid discovery of genes that prescribe the decision rules of development and behavior. Unlike classical Mendelian genetics, it is not dependent on the discovery of multiple alleles, particularly those that are pathological and therefore most

2 | The term *sociogenomics* was first used by Gene E. Robinson in "Integrative animal behaviour and sociogenomics," *Trends in Ecology and Evolution* 14(5): 202–205 (1999).

3 | The Honeybee Genome Sequencing Consortium, "Insights into social insects from the genome of the honeybee *Apis mellifera*," *Nature* 443: 931–949 (2006).

easily perceived and mapped by intergenerational analysis. Coupled to microarray technology, it is a gateway to the study of metabolic events. With this information, research can proceed more rapidly into proteinomics, the tracking of protein change during cell assembly.

Finally, genomics in general (with sociogenomics following suit) speeds the discovery of genetically complex traits, those controlled by multiple genes at different chromosome sites—as opposed to multiple alleles, which are distributed through the population at single sites. Because classical genetics so commonly reveals the presence of "background" genes that modify the expression of single major genes, it seems likely that most or even all traits will eventually be found to be prescribed by multiple genes. The number of identified human polygenic traits is far fewer than the more than 1,600 single-gene traits known. Yet researchers expect that as genomics matures, the number will come to exceed that of single-gene traits.[4] The same turnaround is likely to occur in the genetic analysis of insect social behavior.

HONEYBEE SOCIOGENOMICS

The domestic bee *Apis mellifera* has always been and continues to be the premier model species in the study of advanced eusocial behavior. Since the first publication on honeybees of the modern era, Charles Butler's *Feminine Monarchie* (1609), discoveries have flowed at the organism and colony levels, justifying Karl von Frisch's famous remark about his own research in the 1930s and 1940s that, to scientists, "the life of bees is like a magic well. The more you draw from it, the more there is to draw."

The most celebrated characteristic of *Apis mellifera*, next to its honey and pollination services, is the waggle dance. Foraging workers, on returning to the hive after successful searches for food sources or new nest sites, run figure eights on the vertical comb surfaces, with the middle segment of their body symbolically representing the flight to be taken outward. This latter "waggle run" contains information about the direction of the target with reference to the sun, as well as conveying distance from the hive. Timed buzzing and odor secretion enhances the message. Circular

---

4 | A. M. Glazier, J. H. Nadeau, and T. J. Aitman, "Finding genes that underlie complex traits," *Science* 298: 2345–2349 (2002).

"round dances" supplant the full waggle dance to inform nestmates that the target is close to the nest.

Research in recent years has revealed, as noted in Chapter 3, other performances on the honeybee dance card. If the returning foragers discover many food handlers unemployed as they unload their harvest, they perform the "shaking dance" to bring more workers onto the dance floor and thence out to the field. If the reverse occurs—in other words, if the foragers have trouble passing on their load—they engage in "tremble dances" to draft more bees as food handlers. (For a detailed description, see Chapter 6.)

In addition to their terpsichorean program, honeybees employ pheromones. These substances, secreted from glands distributed over the body, variously alarm or recruit nestmates, distinguish them from alien bees, and classify them according to gender, caste, and age. As workers pass through their natural adult life span of about 40 days, their socially active glands variously grow and shrink in programmed sequences in concert with the labor roles they assume. The progression can be speeded up or reversed according to the needs of the colony. And as the workers shift among specialties, their receptiveness to particular signals rises or falls (see Chapter 5 for details).

Finally, worker bees possess an extraordinary memory capacity. They learn the odor of the colony to which they belong. On foraging flights, they use landmarks as well as the instructions given by their dancing nestmates. They can recall the location of up to five flower beds or other food sites, together with the approximate time of day when each is most productive.

The decoders of the honeybee genome, led by G. M. Weinstock and G. E. Robinson,[5] recognized early on that virtually every biological system of the species has been altered to some degree during evolution. Thus, they were not surprised to find widespread corresponding alterations among the 10,157 genes they identified—yet at sites and in ways that could not have been predicted.

Some of the honeybee genes are modified from recognizable ancient precursors. For example, the yellow protein pigmentation gene has multiplied and assumed mediation in the production of royal jelly, the special food used to raise new queen bees. Other genes, however, including those that prescribe the complex processes

5 | The Honeybee Genome Sequencing Consortium, "Insights into social insects from the genome of the honeybee *Apis mellifera*," *Nature* 443: 931–949 (2006).

of smell and taste together with others that encode food processing and storage, have evidently evolved since the honeybees (and possibly their ancestors among the solitary bees as well) split from other insect lineages. Oddly, the genome of honeybees and that of the one other hymenopteran thus far sequenced, a species of parasitic wasp, suggest that the line leading to the order Hymenoptera (ants, bees, and wasps) diverged before that leading to the origin of the other major holometabolous orders—Coleoptera (beetles), Diptera (flies), and Lepidoptera (moths and butterflies).

Using microarrays, Whitfield, Robinson, and colleagues then began the next step in the sociogenomic program, which was to identify the genes that program the division of labor.[6] They found that extensive, age-related changes in the microarrays coincide with changes in the brain and behavior as young bees shift from activities within the hive to foraging outside. This switch in lifestyle is usually complete at about 8 days following eclosion by the young adult bees from the pupal stage. Later shifts back and forth between hive and outdoor duties, to be described later in more detail (see Chapter 5), are due both to the genetic programs of aging and to multiple environmental effects, such as shortages of hive or foraging specialists.

SOCIOGENOMIC CONSERVATION

One principle emerging from genetic analyses of social insects, albeit tenuously, is that much of colonial behavior is prescribed by genes conserved from solitary behavior and modified in expression to produce a social phenotype. In other words, there may be relatively few truly novel "social genes." This principle, if true, could explain why the honeybee, with about a million neurons and highly complex social behavior, has genes comparable in number to those of the decidedly solitary fruit fly *Drosophila melanogaster*, possessing only about a quarter as many neurons and far simpler adult behavior.[7]

---

6 | C. W. Whitfield, Y. Ben-Shahar, C. Brillet, I. Leoncini, D. Crauser, Y. LeConte, S. Rodriguez-Zas, and G. E. Robinson, "Genomic dissection of behavioral maturation in the honey bee," *Proceedings of the National Academy of Sciences USA* 103(44): 16068–16075 (2006).

7 | G. L. G. Miklos and R. Maleszka, "Deus ex genomix," *Nature Neuroscience* 3(5): 424–425 (2000). The same disparity exists between *Drosophila* and the nematode worm *Caenorhabditis elegans*; the latter has a comparably complex genome but only 302 neurons.

Genomic conservation is nicely illustrated by the foraging (*for*) gene of insects. In *Drosophila melanogaster*, *for*^{r} (the superscript *r* is for rover) flies have higher levels of *for* mRNA than do *for*^{s} (sitter) flies and hence more genetic transmission capacity and more kinase, the protein produced that promotes foraging activity. Individuals with rover alleles are evidently better adapted to patchy, more widely distributed food sources. The same forager gene also occurs in honeybees, but is expressed in a very different, social function. As worker bees age or are stressed by an insufficient number of foragers among their nestmates, their forager genes increase in expression, evidently inducing them to shift from work within the hive to searching for food outside.[8] This change is apparent in the alteration of the microarray pattern that occurs at the same time.[9]

A second gene retained from solitary ancestry and reassigned to a social function is the period gene. In *Drosophila melanogaster*, the gene affects various temporal phenomena, including circadian (approximately 24-hour) locomotor rhythms, mating behavior, and developmental time from egg to adult. In honeybee workers, the period gene is expressed, as evidenced by its raised mRNA levels, when the bees change from work inside the hive to foraging outside. Hive workers perform their tasks without circadian rhythms, whereas the trips of foraging bees are strongly circadian. Hence, the gene for solitary behavior in this case has been utilized to aid the division of labor in honeybee colonies.[10]

A different kind of genomic conservation is sustained in the basic caste system of ants. Throughout the ants, the coexistence of winged reproductive females with wingless workers is a basic trait. The origin of wings dates back within the class Insecta some 300 million years, and all of the winged species existing today evidently originated from a single ancestor at least that remote in time. Judging from two phylogenetically well-separated winged groups analyzed to date—the flies (order Diptera) and butterflies (order Lepidoptera)—there exists throughout the insects a common regulatory gene network directing wing development. The ants, members

---

8 | Y. Ben-Shahar, A. Robichon, M. B. Sokolowski, and G. E. Robinson, "Influence of gene action across different time scales on behavior," *Science* 296: 741–744 (2002).

9 | C. W. Whitfield, Y. Ben-Shahar, C. Brillet, I. Leoncini, D. Crauser, Y. LeConte, S. Rodriguez-Zas, and G. E. Robinson, "Genomic dissection of behavioral maturation in the honey bee," *Proceedings of the National Academy of Sciences USA* 103(44): 16068–16075 (2006).

10 | D. P. Toma, G. Bloch, D. Moore, and G. E. Robinson, "Changes in *period* mRNA levels in the brain and division of labor in honey bee colonies," *Proceedings of the National Academy of Sciences USA* 97(12): 6914–6919 (2000).

of the order Hymenoptera, possess the same network. Over 110 million years ago, their ancestral species, evidently descended from solitary wasps, altered the gene regulatory network so that some of the genes could be shut down under the influence of certain environmental stimuli to produce a wingless worker caste. It is reasonable to suppose that the same genes within the network would thus be curtailed throughout the ants, but such turns out not to be the case. When researchers examined a variety of ants, they found that different species shut down different genes. The original genome has remained the same, but somehow the gene expression pattern producing wingless workers has shifted around during the evolution of the ants.[11] Either that or the more than 14,000 known species of ants as a whole are polyphyletic, with two or more lines having evolved independently from the alate wasp ancestors.

THE FIRE ANT CASE

A corollary of genetic conservatism is the large amount of change that can occur by the modification of a very few genes within the genome. As a consequence, evolution in the social organization of a species can occur within a few generations. The effect is promoted by amplification, the magnification of a small change in a gene to a greater change in downstream behavior of individual colony members to a still greater transformation in the emergent properties of the colony as a whole. Because social organization is the phenotype farthest removed in cause and effect from the gene, it is the one also most subject to ultimate magnification.

A textbook example of amplification is the shift that has occurred in the red imported fire ant *Solenopsis invicta* in the southern United States. A native of southern Brazil and northern Argentina, this small ant was introduced around the mid-1930s into the port of Mobile, Alabama, almost certainly by stowaway colonies in seaborne cargo. By the early 1940s, it was spreading outward from Mobile at the rate of 8 kilometers (5 miles) a year and then faster by leapfrogging in nursery stock and other ground cargo. By the 1970s, it occupied most of the southern tier of states, from eastern Texas to the Carolinas. During the 1990s, it reached southern

11 | E. Abouheif and G. A. Wray, "Evolution of the gene network underlying wing polyphenism in ants," *Science* 297: 249–252 (2002). The species analyzed for the network belonged, one each, to the genus *Formica* (subfamily Formicinae) and genera *Crematogaster*, *Myrmica*, and *Pheidole* (all subfamily Myrmicinae).

California and parts of the West Indies and about then or soon after Hong Kong and southern China. The early U.S. population, like that in the South American homeland, was monogyne or oligogyne, that is, containing either one or a small number of closely related functioning queens. Its colonies were also clearly demarcated by territorial behavior, which spreads nests out. Sometime during the 1970s, the monogyne and oligogyne fire ants began to be replaced by a polygyne (many queens) variant, whose "unicolonial" populations contain numerous small queens and defend no territorial boundaries. The genes for polygyny may have been in the U.S. immigrant all along, only to find expression by recombination and natural selection. Both monogyne-oligogyne and polygyne variants have been found in at least the Argentine portion of the native range.[12]

It turns out that the difference between the monogyne-oligogyne and polygyne forms stems from variation in a single major gene, *Gp-9*.[13] When workers carry the polygyne allele, normally in the heterozygous condition, they kill all queens that are homozygous for the monogyne condition. As a result, the colony is changed to the polygyne state. Because of the intensity of this selective assassination, colonies are converted when only as few as 15 percent of the workers carry the polygyne allele of *Gp-9*.[14]

The two *Gp-9* alleles of the introduced U.S. population have now been sequenced. Their product appears to be a key molecular component in the recognition of the odor of nestmates. The effect of the polygyne allele is evidently to reduce or knock out entirely the ability to discriminate among workers and also among potential egg-laying queens of different fire ant colonies. The latter ability is also an important means of regulating queen number.[15]

The alteration of just this one gene has thus worked a profound change throughout much of the U.S. population of the red imported fire ant. With territorial boundaries erased, local populations now coalesce into a single sheet of

---

12 | K. G. Ross, E. L. Vargo, and L. Keller, "Social evolution in a new environment: the case of introduced fire ants," *Proceedings of the National Academy of Sciences USA* 93(7): 3021–3025 (1996).

13 | K. G. Ross and L. Keller, "Genetic control of social organization in an ant," *Proceedings of the National Academy of Sciences USA* 95(24): 14232–14237 (1998).

14 | L. Keller and K. G. Ross, "Selfish genes: a green beard in the red fire ant," *Nature* 394: 573–575 (1998); and K. G. Ross and L. Keller, "Experimental conversion of colony social organization by manipulation of worker genotype composition in fire ants (*Solenopsis invicta*)," *Behavioral Ecology and Sociobiology* 51(3): 287–295 (2002).

15 | M. J. B. Krieger and K. G. Ross, "Identification of a major gene regulating complex social behavior," *Science* 295: 328–332 (2002).

intercompatible ants spread across the inhabited landscape. Furthermore, colonies are no longer founded by virgin queens that mate and disperse far from the mother nest. The bodies of the queens are smaller, their fat bodies reduced, and their initial egg production lower. New colonies or, more accurately put, new extensions of the supercolony, are instead created by fissioning, the emigration from the mother colony of inseminated queens accompanied by part of the worker force.

There is still more to the amplification of the gene-to-phenotype sequence in the polygyne strain of the fire ant. A higher proportion of the queens remain unmated, skewing the sex ratio within the supercolony. Finally, because of the sprawl and fluidity of the polygyne system, the workers and queens are no longer related genetically beyond the level of random mating.[16]

GENETIC VARIATION AND PHENOTYPIC PLASTICITY

It is probable that every behavioral trait contributing to division of labor and other basic phenomena of insect social organization varies genetically among colony members and hence among colonies. To take one instructive example, allele differences among honeybee colonies have been estimated to contribute 59 percent of the variance in quantities of pollen stored in the hive.[17] R. E. Page Jr. and his coworkers, using carefully controlled insemination to produce genetically homogeneous colonies of known parentage, have narrowed variation in pollen storage by honeybee colonies to allele substitutions at three quantitative trait loci (QTLs), which are genes or small ensembles of linked genes. QTLs have other diverse effects: they induce quantitative variation in foraging behavior by individual bees, associative learning performance, and response thresholds to sucrose feeding.[18] Still other QTLs have been located in honeybees for behavior during colony defense as well as body size.[19]

---

16 | For a comprehensive review of the literature on fire ants, see W. R. Tschinkel, *The Fire Ants* (Cambridge, MA: Harvard University Press, 2006).

17 | G. J. Hunt, R. E. Page Jr., M. K. Fondrk, and C. J. Dullum, "Major quantitative trait loci affecting honey bee foraging behavior," *Genetics* 141(4): 1537–1545 (1995).

18 | R. E. Page Jr. and J. Erber, "Levels of behavioral organization and the evolution of division of labor," *Naturwissenschaften* 89(3): 91–106 (2002).

19 | G. J. Hunt, R. E. Page Jr., M. K. Fondrk, and C. J. Dullum, "Major quantitative trait loci affecting honey bee foraging behavior," *Genetics* 141(4): 1537–1545 (1995); and R. E. Page Jr., J. Gadau, and M. Beye, "The emergence of hymenopteran genetics," *Genetics* 160(2): 375–379 (2002).

The first insights are also being achieved concerning the external regulation of gene activity. To a substantial degree, queen honeybees regulate the activities of workers by the "queen substance," a primer pheromone released from the queen's mandibular gland and distributed throughout the colony by grooming and the exchange of regurgitated food among the workers. The pheromone transiently regulates expression in the workers of several hundred genes and persistently regulates the expression of 19 genes. One of the functions in this feedback loop is to delay the shift as workers age from nursing tasks in the hive to foraging for food. It evidently works by activating nursing genes and repressing foraging genes.[20]

The picture emerging from these various early advances in ant and honeybee genetics is that genes, expressed through the mediation of sensory input and hormones, affect the threshold of task-specific stimuli to which individual colony members respond. Coupled with the prior lifetime experience, they create variations in the thresholds of response among the individual insects. The array of responses result in a division of labor favorable to the survival and reproduction of the colony as a whole. They set the stage for further ecologically guided evolution through the phylogenetic labyrinth.[21]

---

20 | C. M. Grozinger, N. M. Sharabash, C. W. Whitfield, and G. E. Robinson, "Pheromone-mediated gene expression in the honey bee brain," *Proceedings of the National Academy of Sciences USA* 100(Supplement 2): 14519–14525 (2003).

21 | As reviewed by C. Detrain, J. L. Deneubourg, and J. M. Pasteels, eds. (with multiple authors), *Information Processing in Social Insects* (Basel: Birkhäuser Verlag, 1999); and S. N. Beshers and J. H. Fewell, "Models of division of labor in social insects," *Annual Review of Entomology* 46: 413–440 (2001).

|| Plate 6. An *Oecophylla longinoda* (weaver ant) major worker carries a minor worker to a place where the minor is needed for special work, such as attending honeydew-secreting homopterans or nursing small larvae.

5

THE DIVISION OF LABOR

A superorganism is a colony of individuals self-organized by division of labor and united by a closed system of communication. Its members choose their labor roles by a small number of relatively simple algorithms that evolved by natural selection at the colony level.

Each insect colony, as we have stressed, can be envisaged as a factory inside a fortress. The factory is the egg-laying queen, in company with the nurse workers, who rear her progeny, and the foragers, who supply food for all. The fortress is the nest, the workers that build it, and the members of the soldier caste that defend it. In species with the simplest organization, the factory and fortress roles are interchangeable: the workers can switch from one to another on short notice. At the opposite extreme, in species with the most complex colonies, the roles are less easily switched; in many cases, the roles are permanently fixed in castes physically specialized to perform them.

PARALLELS: ORGANISM AND SUPERORGANISM

Caste and division of labor of social insect colonies display many traits similar in function to the cells and organs of organisms. The parallels between these two levels of organization, between superorganism and organism, were first suggested by William Morton Wheeler in 1910.[1] Elaborated by many biologists subsequently, they can be summarized in modern terms as shown in Table 5-1.

The evolution of systems at the two levels has also proceeded on parallel tracks but with one key difference: where organisms are constructed to replicate as many of their

1 | W. M. Wheeler, "The ant colony as an organism" (lecture delivered at the Marine Biological Laboratory, Woods Hole, MA, 2 August 1910), *Journal of Morphology* 22(2): 307–325 (1911).

TABLE 5-1. *Functional parallels between an organism and a superorganism*

ORGANISM	SUPERORGANISM
Cells	Colony members
Organs	Castes
Gonads	Reproductive castes
Somatic organs	Worker castes
Immune system	Defensive castes; alarm-defense communication; colony recognition labels
Circulatory system	Food distribution, including regurgitation between nestmates (trophallaxis), distribution of pheromones,and chemical cues
Sensory organs	Combined sensory apparatus of colony members
Nervous system	Communication and interactions among colony members
Skin, skeleton	Nest
Organogenesis: growth and development of the embryo	Sociogenesis: growth and development of the colony

personal hereditary traits as possible, colonies are constructed to replicate as many as possible of the hereditary superorganismic traits. The link between the two is causal: the superorganismic properties emerge solely from the summed genetically guided actions of the colony members, which are shaped in turn by colony-level selection. Foremost among the superorganismic properties are caste and division of labor.

THE ECOLOGY OF CASTE SYSTEMS

Each species has adapted its social organization to the complex environmental pressures and opportunities unique to its history. The most important single trait is colony size. Following on colony size, and intimately dependent on it, are the degrees of complexity: first, of the communication codes, second, of the caste systems, and, third, of division of labor.

A striking example of these relationships, including the constraining role of colony size, is provided by the attine ants. This unusually abundant and diverse group, which range from the southern United States to Argentina, are the only ants known to cultivate fungi for food. As depicted in Figure 5-1, modern species closest in

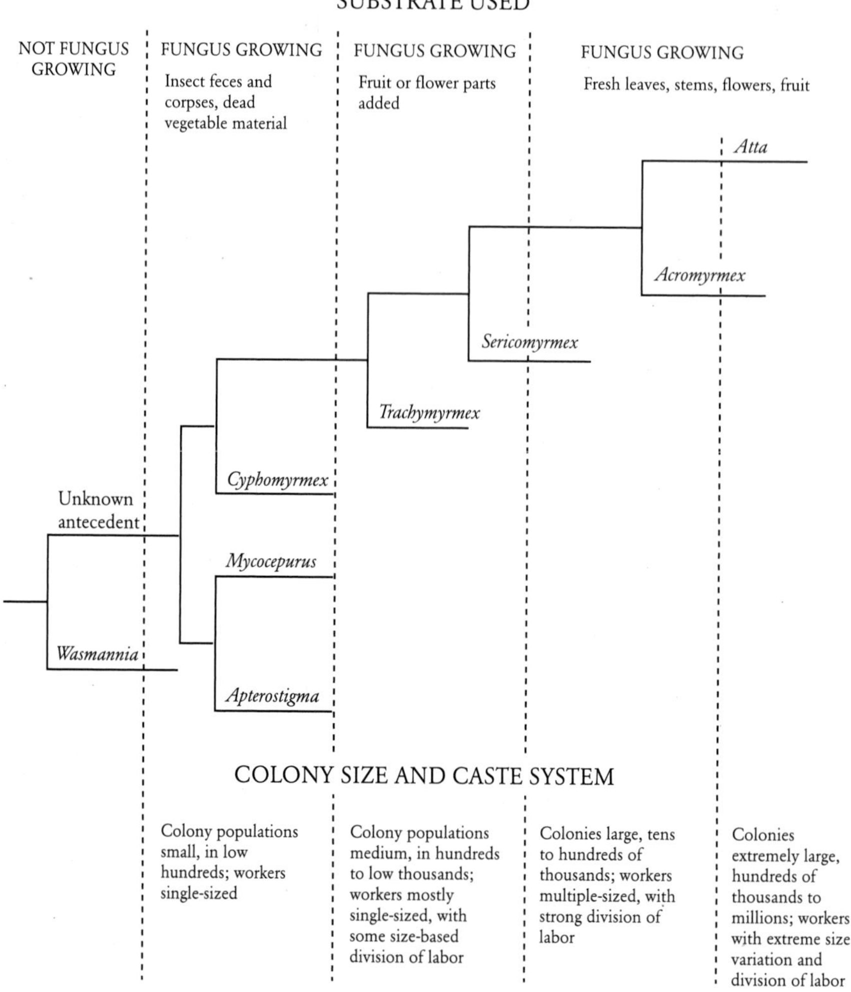

FIGURE 5-1. The evolution of division of labor and social complexity in the attine fungus-growing ants, interpreted as a function of environmental adaptation. With utilization of different substrates during their adaptive radiation, the proliferating clades of attines also settled on increasing mature colony sizes and corresponding increases in specialized worker subcastes and division of labor. The phylogeny is taken from J. K. Wetterer, T. R. Schultz, and R. Meier, "Phylogeny of fungus-growing ants (tribe Attini) based on mtDNA sequence and morphology," *Molecular Phylogenetics and Evolution* 9(1): 42–47 (1998); the other data are condensed from multiple sources, as reviewed by B. Hölldobler and E. O. Wilson, *The Ants* (Cambridge, MA: The Belknap Press of Harvard University Press, 1990).

genetic relationship to the earliest lineages grow the fungi on insect feces and corpses, combined with bits of decaying vegetation. Living representatives of late evolutionary lineages add bits of fruit. Finally, a major breakthrough was achieved by a still more advanced lineage represented by the modern genera *Acromyrmex* and *Atta*. These "leafcutters" use fresh leaves, twigs, and flower parts as the primary fungus substrate.

Given the vastly increased food supply made available by the shift to fresh vegetation, leafcutters are today among the most abundant of all insects, as well as the dominant herbivores, species for species, of the American tropical forests.

As the preferred garden substrate in the evolution of fungus-growing ants increased in availability, mature colony size also grew, ranging among species from thousands in the middle-level genera *Trachymyrmex* and *Sericomyrmex* to hundreds of thousands or even millions in *Acromyrmex* and *Atta*. Accompanying these trends was an increase in subdivision of the worker caste, yielding well-defined minors, medias, majors, and then, in *Atta*, supermajors, along with a correspondingly complex division of labor (Figure 5-2; also see Chapter 9).

During the evolution of insect societies, from over 100 million years ago forward, division of labor appears to have arisen very easily. In the case of ant species, it is all but automatic, even in the otherwise most elementary of associations. When newly inseminated queens of the harvester ant *Pogonomyrmex californicus* are placed together in the laboratory, duplicating a common occurrence in nature, they readily cooperate in colony founding. They also divide labor, with some specializing in nest excavation while others stand by. Which individuals undertake this task, at least under experimental conditions, can be predicted by the individual initiative they display in nest excavation when previously kept alone.[2] It is

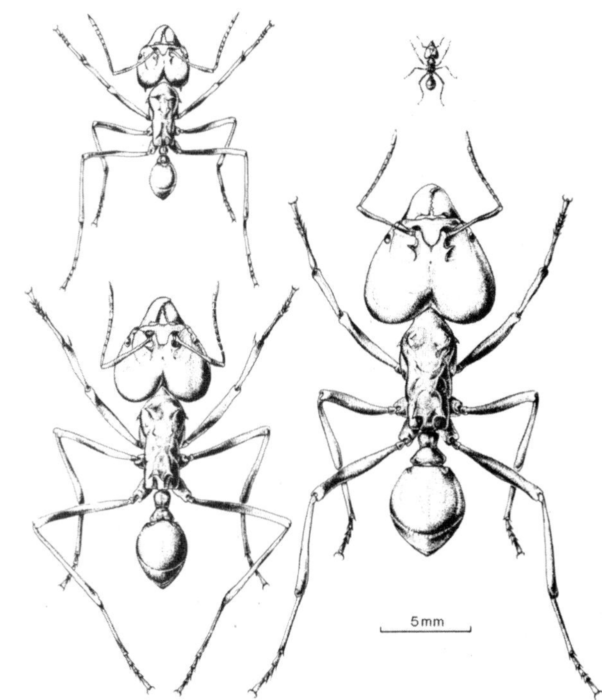

FIGURE 5-2. The worker subcastes of the South American leafcutter ant *Atta laevigata*. Species of the genus *Atta* have the most complex systems of division of labor known in the ants. Original drawing by Turid Hölldobler-Forsyth from G. F. Oster and E. O. Wilson, *Caste and Ecology in the Social Insects* (Princeton, NJ: Princeton University Press, 1978).

2 | S. H. Cahan and J. H. Fewell, "Division of labor and the evolution of task sharing in queen associations of the harvester ant *Pogonomyrmex californicus*," *Behavioral Ecology and Sociobiology* 56(1): 9–17 (2004).

a remarkable fact that the same predictability occurs in the normally solitary foundress queens of a similar species, *Pogonomyrmex barbatus*, when forced together in laboratory arenas.[3] The clear inference is that variation in behavior among the queens, whether genetic in origin or not, constitutionally predisposes them to divide labor when they find themselves part of a group. Females of xylocopine bees are usually solitary, but when several females congregate in one of the scarce nest sites or one female usurps the site from another, a dominance order is favored and the subordinates either leave or stand guard.[4]

Thereafter in the course of evolution, the smallest of differences in predisposition and experience can be amplified into a nonreproductive division of labor. Workers of the North American fungus-growing ant *Trachymyrmex septentrionalis* are monomorphic, varying in size only slightly and with little change in body proportions. Yet they participate differently in regard to five principal roles: small workers lean toward brood care and gardening, while large workers statistically prefer nest maintenance, foraging, and preparation of the substrate.[5] Workers of the monomorphic European ant *Temnothorax* (formerly *Leptothorax*) *albipennis*, whose small colonies usually nest in rock crevices, diversify among themselves into nest workers grown fat with lipid stores and lean workers that conduct most of the foraging. Those differences presumably reverberate into other categories of behavior.[6]

In a further example, Deborah Gordon and her collaborators report that in the American harvester ant *Pogonomyrmex barbatus*, "the extent to which one worker group actively performs a task affects the activity of other worker groups. Brief antennal contacts with workers engaged in the same or different tasks may influence

---

3 | J. H. Fewell and R. E. Page Jr., "The emergence of division of labour in forced associations of normally solitary ant queens," *Evolutionary Ecology Research* 1(5): 537–548 (1999).

4 | C. D. Michener, *The Social Behavior of the Bees: A Comparative Study* (Cambridge, MA: The Belknap Press of Harvard University Press, 1974); D. Gerling, H. H. W. Velthuis, and A. Hefetz, "Bionomics of the large carpenter bees of the genus *Xylocopa*," *Annual Review of Entomology* 34: 163–190 (1989); and T. Dunn and M. H. Richards, "When to bee social: interactions among environmental constraints, incentives, guarding, and relatedness in a facultatively social carpenter bee," *Behavioral Ecology* 14(3): 417–424 (2003). Similar results were obtained with a solitary halictine bee *Lasioglossum* species and the communal *Lasioglossum hemichalceum* by R. Jeanson, P. F. Kukuk, and J. H. Fewell, "Emergence of division of labour in halictine bees: contributions of social interactions and behavioural variance," *Animal Behaviour* 70(5): 1183–1193 (2005).

5 | S. N. Beshers and J. F. A. Traniello, "Polyethism and the adaptiveness of worker size variation in the attine ant *Trachymyrmex septentrionalis*," *Journal of Insect Behavior* 9(1): 61–83 (1996).

6 | G. B. Blanchard, G. M. Orledge, S. E. Reynolds, and N. R. Franks, "Division of labour and seasonality in the ant *Leptothorax albipennis:* worker corpulence and its influence on behaviour," *Animal Behaviour* 59(4): 723–738 (2000).

an ant's subsequent behavior."[7] In the search for task-related cues that might be used by the ants for identifying task groups, Gordon and colleagues investigated the cuticular hydrocarbon blends and found clear differences, for example, between patrolling (scouting) ants, foragers, and nest maintenance workers (Plate 7). In a subsequent study, Gordon and her coworkers provided experimental evidence that these hydrocarbon cues may indeed function as recognition labels and affect the behavioral activity of the recipient ants.[8] Differences in cuticular hydrocarbon profiles between brood tender and field workers had been demonstrated earlier in the European carpenter ant species *Camponotus vagus*[9] and the Asian ponerine species *Harpegnathos saltator*[10] and more recently in the African myrmicine *Myrmicaria eumenoides*,[11] suggesting that the phenomenon may be general in ants.

THE EVOLUTION OF CASTE: PRINCIPLES

Among a large sample of better-studied species of ants, several consistent trends in the evolution of division of labor have become evident:

- In at least a few of the anatomically and socially most primitive species and in the early stages of colony growth of a few other species, division of nonreproductive labor among workers is very weak or absent altogether. Their labor in this lower

---

7 | D. M. Gordon and N. J. Mehdiabadi, "Encounter rate and task allocation in harvester ants," *Behavioral Ecology and Sociobiology* 45(5): 370–377 (1999).

8 | D. M. Gordon, "Group-level dynamics in harvester ants: young colonies and the role of patrolling," *Animal Behaviour* 35(3): 833–843 (1987); D. Wagner, M. J. F. Brown, P. Broun, W. Cuevas, L. E. Moses, D. L. Chao, and D. M. Gordon, "Task-related differences in the cuticular hydrocarbon composition of harvester ants, *Pogonomyrmex barbatus*," *Journal of Chemical Ecology* 24(12): 2021–2037 (1998); D. M. Gordon and N. J. Mehdiabadi, "Encounter rate and task allocation in harvester ants," *Behavioral Ecology and Sociobiology* 45(5): 370–377 (1999); and M. J. Greene and D. M. Gordon, "Cuticular hydrocarbons inform task decisions," *Nature* 423: 32 (2003).

9 | A. Bonavita-Cougourdan, J. L. Clément, and C. Lange, "Functional subcaste discrimination (foragers and brood-tenders) in the ant *Camponotus vagus* Scop.: polymorphism of cuticular hydrocarbon patterns," *Journal of Chemical Ecology* 19(7): 1461–1477 (1993).

10 | J. Liebig, C. Peeters, N. J. Oldham, C. Markstädter, and B. Hölldobler, "Are variations in cuticular hydrocarbons of queens and workers a reliable signal of fertility in the ant *Harpegnathos saltator*?" *Proceedings of the National Academy of Sciences USA* 97(8): 4124–4131 (2000).

11 | F. Lengyel, S. A. Westerlund, and M. Kaib, "Juvenile hormone III influences task-specific cuticular hydrocarbon profile changes in the ant *Myrmicaria eumenoides*," *Journal of Chemical Ecology* 33(1): 167–181 (2007); the authors present evidence that recognition cues are influenced by juvenile hormone.

PLATE 7. The Western red harvester ant *Pogonomyrmex barbatus*. *Above*: Nest maintenance workers. *Below*: Forager workers.

evolutionary grade, and also to a large extent in higher grades, is divided according to need as perceived by individual colony members. Workers unoccupied, hence available for labor, patrol inside the nest as well as outside while foraging for work—or else they are recruited by nestmates to sites short on labor.

- When species have evolved more complex societies, true castes have made their appearance, comprising individuals that devote long periods of their lives to certain tasks, such as care of the young, nest construction, or foraging. In a majority of such species, the castes are primarily physiological, displaying at most weak differences among nestmates in size and anatomy. The caste systems are the outcome, variously according to species, either of dominance relations or age differences or differences in experience among nestmates.

- The trend toward still larger, more complex societies has been accompanied by a hardening of the mechanisms that differentiate worker subcastes and labor roles. The most extreme such diversification is achieved in the minority of species that form physical castes. Within the physical castes are folded more finely differentiated physiological castes.

A theoretically perfect labor system would consist of a specialist for each labor role, with the role performed in proportions appropriate to the colony's needs. But ant castes fall far short of that level.[12] The diversity of castes is severely constrained by moment-by-moment, unpredictable shifts in the environment and the exigencies they impose on the colony. To approach maximum efficiency, the workers must be able to change from one role to another, often within a few minutes.

Do any ant species (or termite species) exist that lack worker subcastes altogether? The condition would consist of a statistical uniformity of role selection by workers on the basis of labor availability only, regardless of the size, age, or position in the colony dominance hierarchy of individual workers. Such labor neutrality is evidently very rare; at least, extremely few cases have been reported in the literature. Among those known to us include the primitive Australian dawn ant *Prionomyrmex* (formerly *Nothomyrmecia*) *macrops*[13] (Plate 8) and the relatively primitive

---

12 | E. O. Wilson, "The ergonomics of caste in the social insects," *American Naturalist* 102: 41–66 (1968); and G. F. Oster and E. O. Wilson, *Caste and Ecology in the Social Insects* (Princeton, NJ: Princeton University Press, 1978).

13 | P. Jaisson, D. Fresneau, R. W. Taylor, and A. Lenoir, "Social organization in some primitive ants, I: *Nothomyrmecia macrops* Clark," *Insectes Sociaux* 39(4): 425–438 (1992).

PLATE 8. The Australian dawn ant *Prionomyrmex* (formerly *Nothomyrmecia*) *macrops. Above*: Queen and worker castes are identical, except for the queen's slightly heavier thorax. *Below*: A worker feeds larvae with the remains of a captured fly.

American dampwood termite *Zootermopsis angusticollis*.[14] It is also known to exist in immature colonies of the Australian ectatommine ant *Rhytidoponera metallica*, but in this case is replaced in older, mature colonies by an age-correlated division of labor.[15] Age-correlated castes may be absent in the primitive amblyoponine ant species *Amblyopone pallipes* or, if present, are at most very weak.[16] In contrast, there is strong evidence for temporal division of labor in the related species *Amblyopone silvestrii*[17] and for the amblyoponine species *Prionopelta amabilis*.[18] Young workers attend the brood and queen, while older workers are more active in foraging. Young workers of *Prionopelta amabilis* moreover have better developed ovaries than do foragers, and their eggs are fed to the queen. Only a trace of labor division in concert with age is present in the small mature colonies of the myrmicine *Temnothorax* (formerly *Leptothorax*) *unifasciatus*, whose workers tend to specialize not on role but on "spatial fidelity zones," within which they perform tasks in a readily interchangeable way.[19] Finally, nonreproductive subcastes appear to be absent in primitively eusocial bees and wasps.

DOMINANCE ORDERS IN CASTE DETERMINATION

Thus, some division of labor among nonreproductive workers, one of the diagnostic traits of superorganisms, is close to universal in the eusocial insects. The most elementary way that labor can be divided is by dominance hierarchies arising from competition for reproductive rights. The precedent for this role of dominance in

14 | J. F. A. Traniello and R. B. Rosengaus, "Ecology, evolution and division of labour in social insects," *Animal Behaviour* 53(1): 209–213 (1997).

15 | M. L. Thomas and M. A. Elgar, "Colony size affects division of labour in the ponerine ant *Rhytidoponera metallica*," *Naturwissenschaften* 90(2): 88–92 (2003).

16 | Age-based castes were not detectable in studies by James F. A. Traniello, as reported in "Caste in a primitive ant: absence of age polyethism in *Amblyopone*," *Science* 202: 770–772 (1978), but they were described in the same species by J. P. Lachaud, D. Fresneau, and B. Corbara, "Mise en évidence de sous-castes comportementales chez *Amblyopone pallipes*," *Actes des Colloques Insectes Sociaux* 4: 141–147 (1988).

17 | K. Masuko, "Temporal division of labor among workers in the ponerine ant, *Amblyopone silvestrii* (Hymenoptera: Formicidae)," *Sociobiology* 28(1): 131–151 (1996).

18 | B. Hölldobler and E. O. Wilson, "Ecology and behavior of the primitive cryptobiotic ant *Prionopelta amabilis* (Hymenoptera: Formicidae)," *Insectes Sociaux* 33(1): 45–58 (1986).

19 | A. B. Sendova-Franks and N. R. Franks, "Spatial relationships within nests of the ant *Leptothorax unifasciatus* (Latr.) and their implications for the division of labour," *Animal Behaviour* 50(1): 121–136 (1995).

evolution is the separation of adult members of colonies into reproductive and non-reproductive castes. The phenomenon is especially well marked during the founding of colonies by multiple queens of bees, wasps, and ants.[20] In the South American ponerine ant *Pachycondyla inversa*, for example, unrelated queens often found colonies cooperatively. They then form dominance hierarchies based on overtly aggressive interactions. A single one of the subordinates ends up devoting herself to foraging for food, while the others remain in the nest as reproductrices.[21] Similar patterns have been documented in detail in the harvester ant *Pogonomyrmex californicus* and the fungus-growing ant *Acromyrmex versicolor*, although in these cases dominant orders are not known.[22]

The same dominance patterns then spread with evident ease from division of labor among queens to division of labor among workers. The phenomenon was first observed by biologists in colonies of social wasps, where subordinate workers not only refrain from laying eggs but also tend to spend more time foraging.[23] Other examples have come to light in studies of ponerine ants. In the Australian *Pachycondyla sublaevis*, there is no anatomically distinct queen caste. Instead, the top-ranked worker mates and lays eggs (thus becomes a "gamergate") and joins those

---

20 | E. O. Wilson, *The Insect Societies* (Cambridge, MA: The Belknap Press of Harvard University Press, 1971); C. D. Michener, *The Social Behavior of the Bees: A Comparative Study* (Cambridge, MA: The Belknap Press of Harvard University Press, 1974); B. Hölldobler and E. O. Wilson, *The Ants* (Cambridge, MA: The Belknap Press of Harvard University Press, 1990); K. G. Ross and R. W. Matthews, eds., *The Social Biology of Wasps* (Ithaca, NY: Comstock Publishing Associates of Cornell University Press, 1991).

21 | K. Kolmer and J. Heinze, "Rank orders and division of labour among unrelated cofounding ant queens," *Proceedings of the Royal Society of London B* 267: 1729–1734 (2000). High-ranking queens of this species are distinguished from subordinates by their cuticular hydrocarbon patterns, which presumably imparts an odor to which nestmates respond. In particular, the hydrocarbons of the dominants include considerable amounts of pentadecane and heptadecane (J. Tentschert, K. Kolmer, B. Hölldobler, H.-J. Bestmann, J. H. C. Delabie, and J. Heinze, "Chemical profiles, division of labor and social status in *Pachycondyla* queens (Hymenoptera: Formicidae)," *Naturwissenschaften* 88(4): 175–178, 2001).

22 | S. H. Cahan and J. H. Fewell, "Division of labor and the evolution of task sharing in queen associations of the harvester ant *Pogonomyrmex californicus*," *Behavioral Ecology and Sociobiology* 56(1): 9–17 (2004); and S. W. Rissing, G. B. Pollock, M. R. Higgins, R. H. Hagen, and D. R. Smith, "Foraging specialization without relatedness or dominance among co-founding ant queens," *Nature* 338: 420–422 (1989).

23 | The pattern was first suggested as part of an optimum fitness model by M. J. West-Eberhard, "Intragroup selection and the evolution of insect societies," in R. D. Alexander and D. W. Tinkle, eds., *Natural Selection and Social Behavior: Recent Research and New Theory* (New York: Chiron Press, 1981), pp. 3–17; and evidence for it adduced in the review by R. L. Jeanne, "Polyethism," in K. G. Ross and R. W. Matthews, eds., *The Social Biology of Wasps* (Ithaca, NY: Comstock Publishing Associates of Cornell University Press, 1991), pp. 389–425.

just beneath her in caring for the brood, while the workers of lower rank forage outside the nest for food. The youngest workers in this species take the top position.[24] A similar pattern of dominance-based labor allocation, which may be unrelated to age, occurs in the ponerine ant *Odontomachus brunneus*. In this subtropical American species, dominant workers drive subordinates away from the brood piles by aggressive posturing and grooming. Subordinates respond by lowering their bodies while rapidly "shivering" their antennae and walking in the direction indicated by the dominant. High-ranking workers remain in the brood zone, medium-level workers stay in the nest but away from the brood, and the lowest-ranking individuals, which are also the oldest and hence marked by withered ovaries, evidently conduct the foraging[25] (Figure 5-3). For other examples from the highly diverse poneromorph ants, see Chapter 8.

A remarkable case of dominance behavior among a highly specialized worker subcaste has been discovered in the myrmicine ant species *Acanthomyrmex ferox*. This species possesses a distinct worker subcaste system consisting of gigantic majors and small minors. Strangely, the majors, which usually function as soldiers, possess large ovaries with the same number of ovarioles as those of the queens. In contrast, the minor workers have only one-third the number of ovarioles. In the presence of the queen, majors lay only unviable trophic eggs. Majors (but not minors) exhibit occasional aggressive interactions that feature antennal boxing and spectacular shaking contests (Figure 5-4). This agonistic behavior leads to a ranking of the soldiers, and after the queen's removal, the contests intensify but the established ranking remains stable. All majors then lay viable, male-destined haploid eggs. However, the highest-ranked individual is the most productive one. She also patrols the egg-pile, frequently stopping to cannibalize eggs laid by subordinates.[26]

Finally, the aggressive interactions of ants may reinforce division of labor in

---

24 | F. Ito and S. Higashi, "A linear dominance hierarchy regulating reproduction and polyethism of the queenless ant *Pachycondyla sublaevis*," *Naturwissenschaften* 78(2): 80–82 (1991); S. Higashi, F. Ito, N. Sugiura, and K. Ohkawara, "Worker's age regulates the linear dominance hierarchy in the queenless ponerine ant, *Pachycondyla sublaevis* (Hymenoptera: Formicidae)," *Animal Behaviour* 47(1): 179–184 (1994).

25 | S. Powell and W. R. Tschinkel, "Ritualized conflict in *Odontomachus brunneus* and the generation of interaction-based task allocation: a new organizational mechanism in ants," *Animal Behaviour* 58(5): 965–972 (1999).

26 | B. Gobin and F. Ito, "Sumo wrestling in ants: major workers fight over male production in *Acanthomyrmex ferox*," *Naturwissenschaften* 90(7): 318–321 (2003).

FIGURE 5-3. Dominance interactions control division of labor in the ponerine ant *Odontomachus.* As depicted here, a dominant worker directs a subordinate away from the brood area and into the role of forager. Based on S. Powell and W. R. Tschinkel, "Ritualized conflict in *Odontomachus brunneus* and the generation of interaction-based task allocation: a new organizational mechanism in ants," *Animal Behaviour* 58(5): 965–972 (1999).

unexpected and inadvertent ways. Garbage dump workers of the leafcutter *Atta cephalotes*, for example, tend to be confined to this role by the aggressive behavior of nestmates, who respond to the garbage odors clinging to the bodies of the garbage dump workers.[27]

27 | A. G. Hart and F. L. W. Ratnieks, "Task partitioning, division of labour and nest compartmentalisation collectively isolate hazardous waste in the leafcutting ant *Atta cephalotes*," *Behavioral Ecology and Sociobiology* 49(5): 387–392 (2001).

FIGURE 5-4. In *Acanthomyrmex ferox*, gigantic soldiers lay trophic eggs; when the queen is absent, they switch to laying viable eggs. Their individual roles are based on dominance hierarchies, sustained in turn by the aggressive display illustrated here. Based on B. Gobin and F. Ito, "Sumo wrestling in ants: major workers fight over male production in *Acanthomyrmex ferox*," *Naturwissenschaften* 90(7): 318–321 (2003).

TEMPORAL CASTES

With or without aggressive interactions among nestmates, which may be a phenomenon mostly limited to ponerines, division of labor among workers is nearly universal in ants. It arises as a statistical change in behavior that occurs with aging of the adult colony members. The shift is almost always from work inside the nest, including care of the brood and reproductive castes, to work outside the nest, made up largely of defense of the nest and foraging for food.

Because of the looseness of the correlation between labor and age, and the ease with which workers can switch from inside work to outside work and back again, this pattern is often called temporal division of labor, and the different classes are called temporal castes. What is not immediately clear is the extent to which temporal division of labor is caused by the aging process, in which case temporal castes might equally well be called age castes. Alternatively, perhaps the aging pattern is just an incidental effect of nest geography. Possibly it is the result of workers wandering farther and farther from the center of their birth in the brood chambers in search of work, so that those picking up tasks in the outer nest chambers and outside the nest happen to be older.

This second hypothesis was proposed for ants in 1993 by Ana Sendova-Franks and her husband Nigel Franks, of Bristol, along with Christopher Tofts:[28] "The model assumes that tasks are ordered—this arises naturally from the structure of nests, and simplistically is just the distance at which the task can be performed from the center of the brood pile. It also assumes that ants are born into the first task, which is located on the brood pile. Given these assumptions, ants using the 'forage for work' strategy will show temporal polyethism, that is, the age of ants performing the later tasks is usually greater than that of the ants performing the earlier tasks. This arises as a result of the new young ants increasing the number of individuals available to work on the brood pile and displacing older ants on average."[29]

The three researchers based their ideas on the reasonable assumption that in the course of evolution, ant colonies tend to adopt the simplest algorithm that yields a favorable result. The hypothesis, including its mathematical development, has come to be called the "foraging for work model." That label, however, is unintentionally a bit misleading and might be more aptly named the "centrifugal model." Entomologists have long known that idle worker ants wander about; they "patrol" the colony's nest and home range, in the course of which they pick up available tasks in an opportunistic fashion. Patrollers are also available for recruitment by nestmates to tasks where there is a labor shortage. An equally venerable principle is that ants switch readily from one task to another and at a rate that varies according to context and priority of urgency. The overall result is an efficient allocation of labor by the colony as a whole.[30]

---

28 | C. Tofts, "Algorithms for task allocation in ants (a study of temporal polyethism: theory)," *Bulletin of Mathematical Biology* 55(5): 891–918 (1993).

29 | A. Sendova-Franks and N. R. Franks, "Task allocation in ant colonies within variable environments (a study of temporal polyethism: experimental)," *Bulletin of Mathematical Biology* 55(1): 75–96 (1993).

30 | For the extensive literature of patrolling, opportunism, and division of labor, the following reviews are in varying degrees comprehensive: E. O. Wilson, *The Insect Societies* (Cambridge, MA: The Belknap Press of Harvard University Press, 1971); G. F. Oster and E. O. Wilson, *Caste and Ecology in the Social Insects* (Princeton, NJ: Princeton University Press, 1978); B. Hölldobler and E. O. Wilson, *The Ants* (Cambridge, MA: The Belknap Press of Harvard University Press, 1990); G. E. Robinson, "Regulation of division of labor in insect societies," *Annual Review of Entomology* 37: 637–665 (1992); A. F. G. Bourke and N. R. Franks, *Social Evolution in Ants* (Princeton, NJ: Princeton University Press, 1995); T. D. Seeley, *The Wisdom of the Hive: The Social Physiology of Honey Bee Colonies* (Cambridge, MA: Harvard University Press, 1995); D. M. Gordon, "The organization of work in social insect colonies," *Nature* 380: 121–124 (1996); D. M. Gordon, "Interaction patterns and task allocation in ant

Two questions of interest remain, however. First, is temporal division of labor, which so widely occurs in the social hymenopterans as well as in some termites, biased by the innate physiological changes of aging? Second, if such is the case, in what manner and to what degree does the bias occur? Insofar as it addresses these questions, the centrifugal model, while elegant and heuristically valuable, is not favored by the empirical evidence.[31] This evidence, drawn from multiple sources, can be summarized as follows:

- For most species of ants, nest architecture militates against pure centrifugal emigration as the basis of labor division. A very short distance usually exists from the brood piles to the nest walls and beyond to the outside world, often only a few ant lengths (or tarsal steps, so to speak), yet rarely do the youngest workers take this trip of a minute or so (or less) and thence devote themselves to nest building or foraging.
- In many species of ants, the pupae are sequestered from the rest of the brood, usually in a drier part of the nest. The new workers emerging from them tend to find and assume the care of the eggs and larvae, in spite of the fact that they start their postpupal, adult life well removed from those two life stages.
- Temporal castes in honeybees occur in tasks for which physiological specialization is required, as in nectar processing and brood care, but do not occur in tasks such as nest defense, for which physiological specialization is unnecessary.[32]

colonies," in C. Detrain, J. L. Deneubourg, and J. M. Pasteels, eds., *Information Processing in Social Insects* (Basel: Birkhäuser Verlag, 1999), pp. 51–67; S. N. Beshers, G. E. Robinson, and J. E. Mittenthal, "Response thresholds and division of labor in insect colonies," in C. Detrain, J. L. Deneubourg, and J. M. Pasteels, eds., *Information Processing in Social Insects* (Basel: Birkhäuser Verlag, 1999), pp. 115–139; S. N. Beshers and J. H. Fewell, "Models of division of labor in social insects," *Annual Review of Entomology* 46: 413–440 (2001); and W. R. Tschinkel, *The Fire Ants* (Cambridge, MA: The Belknap Press of Harvard University Press, 2006).

31 | An early review of some of the countervailing evidence, as well as a general critique of the use of minimalist modeling like that employed, was made in G. E. Robinson, R. E. Page Jr., and Z. Y. Huang, "Temporal polyethism in social insects is a developmental process," *Animal Behaviour* 48(2): 467–469 (1994); J. F. A. Traniello and R. B. Rosengaus, "Ecology, evolution and division of labour in social insects," *Animal Behaviour* 53(1): 209–213 (1997); and S. K. Robson and S. N. Beshers, "Division of labour and 'foraging for work': simulating reality versus the reality of simulations," *Animal Behaviour* 53(1): 214–218 (1997). A defense of the model was provided in N. R. Franks, C. Tofts, and A. B. Sendova-Franks, "Studies of the division of labour: neither physics nor stamp collecting," *Animal Behaviour* 53(2): 219–224 (1997).

32 | B. R. Johnson, "Organization of work in the honeybee: a compromise between division of labour and behavioural flexibility," *Proceedings of the Royal Society of London B* 270: 147–152 (2003).

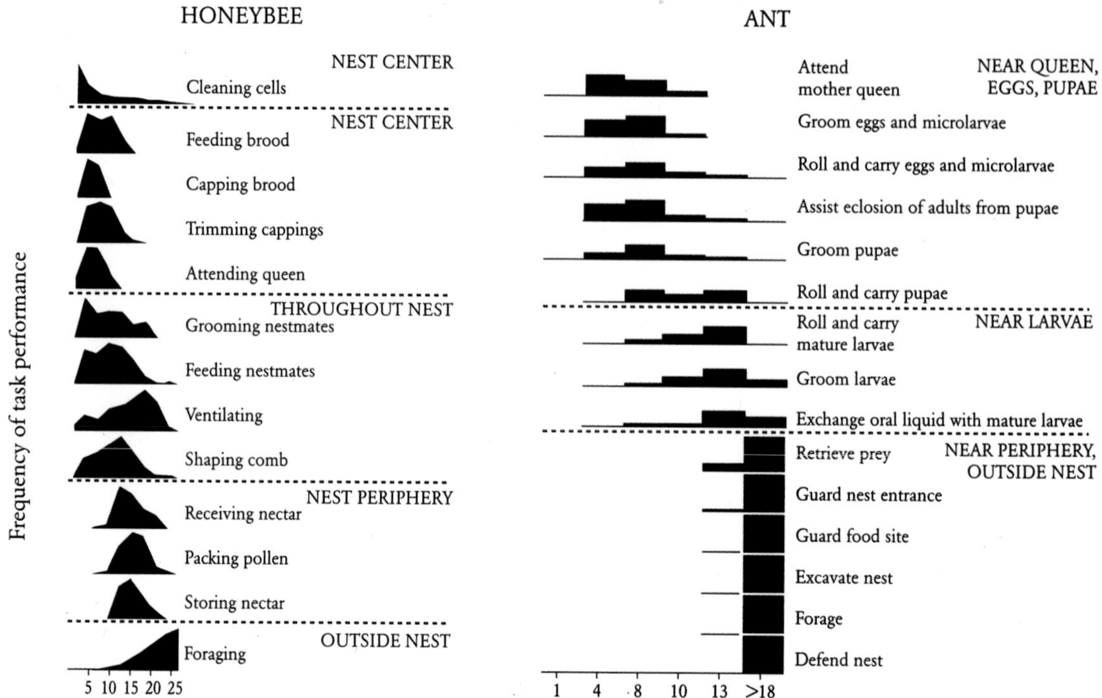

FIGURE 5-5. The temporal division of labor, based on changes in behavior in adult workers with aging, is shown for a domestic colony of the honeybee (left), compared here with comparable data for minor workers of a laboratory colony of the North American ant *Pheidole dentata*. The insects shift in central tendency from one linked set of tasks to another as they move their activities outward from the nest center. The similarities of the two species is due to evolutionary convergence. The sum of frequencies in each histogram is 1.0. From B. Hölldobler and E. O. Wilson, *The Ants* (Cambridge, MA: The Belknap Press of Harvard University Press, 1990), based on data published by T. D. Seeley (bees) and E. O. Wilson (ants).

- The ant *Prionomyrmex* (formerly *Nothomyrmecia*) *macrops* and termite *Zootermopsis angusticollis* have central nest sites, with tasks distributed centrifugally, yet there is no centrifugal age distribution as predicted by the Tofts-Franks model. This bit of phylogenetic evidence suggests that the centrifugal pattern of temporal division of labor evolves as a biological trait and is not just a geometric epiphenomenon.

- In further contrast to the purely geometric centrifugal model, there is abundant evidence, at least in ants and honeybees, of innate developmental programs in physiology and behavior that guide the observed age-related trajectories in division of labor. The widespread occurrence of the pattern in the social hymenopterans may reflect what Mary Jane West-Eberhard has called the "ovaries groundplan" in solitary aculeate wasps that provision their nests: a cycle of egg laying and foraging

> that is repeated. Preserved in primitively eusocial wasps of the genera *Aulopus* and *Zethrus*, the groundplan emerges as a three-phased temporal development in young adults of more advanced species. In subordinates, the ovaries first develop, then decline. Next, there are queen-like details, such as nest construction, and finally queen-like oviposition in dominants and foraging in subordinates.[33]

To press on beyond the last point, the innate driving forces include dominance hierarchies, such as those headed by the egg-laying workers, which, for example, are the youngest individuals in queenless colonies of *Pachycondyla sublaevis* and also among workers of queenright colonies of *Odontomachus brunneus*. The effects redound through all the stages of temporal polyethism. Not enough is known to determine whether this newly discovered form of overt control occurs in only a few species of ants or is more widespread.

A rough correspondence thoroughly documented in honeybees and ants occurs between age, physiological condition, and labor. The pattern of change through the life of the honeybee has in fact been known since antiquity. Aristotle, in his *History of Animals*, observed that field bees have lost hair from their bodies and therefore must be older.[34] In the modern era, the phenomenon has been recorded in ever-growing detail for nearly 400 years, starting with Charles Butler's *Feminine Monarchie* in 1609.[35] In briefest terms, it unfolds as follows (Figure 5-5). Each worker bee has a maximum natural longevity of 65 to 80 days. The first 21 days are spent in developing through the egg, larval, and pupal stages. Then, very broadly, the workers, upon emerging as adults, spend the first one or two weeks as nurses, caring for the queen and brood. Next, they continue working inside the hive but now as food storers, converting nectar to honey and storing the honey in the beeswax combs. Finally, when 20 to 30 days old, they turn into field bees,

---

33 | M. J. West-Eberhard, "Wasp societies as microcosms for the study of development and evolution," in S. Turillazzi and M. J. West-Eberhard, eds., *Natural History and Evolution of Paper Wasps* (New York: Oxford University Press, 1996), pp. 290–317.

34 | Cited by R. E. Page Jr. and S. D. Mitchell, "Self organization and adaptation in insect societies," in Philosophy of Science Association, *Proceedings of the Biennial Meeting of the Philosophy of Science Association in 1990*, Vol. 2: *Symposia and Invited Papers* (Chicago: University of Chicago Press, 1991), pp. 289–298

35 | Charles Butler, in *The Feminine Monarchie: On a Treatise Concerning the True Ordering of Them* (Oxford: Joseph Barnes, 1609), notes, "The young Bees as best able, beare the greatest burdens: for they not only worke abroad but also watch and ward at home both early & late. . . . But the labour of the old ones is only in gathering, which they wil never give over, while their wings can bear them."

foraging more or less exclusively on behalf of the colony for food until they die, more often of external "natural" causes than old age.[36]

During early adulthood, the worker secretes a proteinaceous glandular food for the larvae. Then, in middle age, she produces enzymes that convert nectar to honey. The larval food of the early stage is manufactured by the hypopharyngeal gland, an organ whose growth and activity are mediated by juvenile hormone. An increase in juvenile hormone reduces the hypopharyngeal gland and prepares the bee for later age-related roles. At first, the diminished gland shifts its production from the proteinaceous food for larvae to invertase, the enzyme that converts nectar to honey. The increase of the hormone then reduces the sensitivity of the bee to mechanical disturbance of the nest as well as to the alarm pheromone released by her nestmates when such intrusions occur.[37] The result is that the bee is more "focused" on the single task of foraging, the task for which older individuals are generally specialized.[38]

The sequence of internal changes in the honeybee are programmed, but the timing of the stages can be altered by certain social stimuli or the absence of these stimuli. When Zhi-Yong Huang and Gene E. Robinson reared honeybees in isolation, the bees displayed prematurely elevated rates of juvenile hormone synthesis and a corresponding reduction of the hypopharyngeal glands, comparable to that exhibited by aging bees reared in a normal social environment. Thus deprived of the ordinary inhibitory stimuli of demands from the brood and older workers, the bees moved quickly and prematurely toward the final role of foraging.[39]

---

36 | Recent reviews of the honeybee temporal caste pattern are provided in G. E. Robinson, "Regulation of division of labor in insect societies," *Annual Review of Entomology* 37: 637–665 (1992); T. D. Seeley, *The Wisdom of the Hive: The Social Physiology of Honey Bee Colonies* (Cambridge, MA: Harvard University Press, 1995); Z.-Y. Huang and G. E. Robinson, "Social control of division of labor in honey bee colonies," in C. Detrain, J. L. Deneubourg, and J. M. Pasteels, eds., *Information Processing in Social Insects* (Basel: Birkhäuser Verlag, 1999), pp. 165–186; and R. E. Page Jr. and G. V. Amdam, "The making of a social insect: developmental architectures of social design," *BioEssays* 29(4): 334–343 (2007).

37 | G. E. Robinson, "Modulation of alarm pheromone perception in the honey bee: evidence for division of labor based on hormonally regulated response thresholds," *Journal of Comparative Physiology* 160(5): 613–619 (1987).

38 | Reviews of these and similar phenomena involving changes in sensory thresholds and their effects on division of labor and other social phenomena are provided in chapters by S. N. Beshers, E. Bonabeau, C. Dreller, Z.-Y. Huang, J. Mittenthal, R. F. A. Moritz, R. E. Page Jr., G. E. Robinson, and G. Theraulaz in C. Detrain, J. L. Deneubourg, and J. M. Pasteels, eds., *Information Processing in Social Insects* (Basel: Birkhäuser Verlag, 1999).

39 | Z.-Y. Huang and G. E. Robinson, "Social control of division of labor in honey bee colonies," in C. Detrain, J. L. Deneubourg, and J. M. Pasteels, eds., *Information Processing in Social Insects* (Basel: Birkhäuser Verlag, 1999), pp. 165–186.

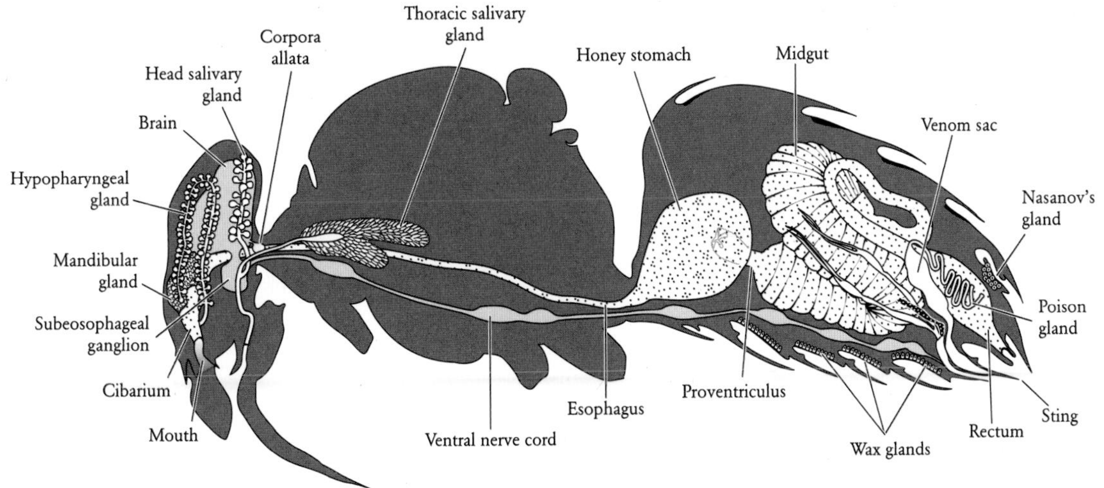

FIGURE 5-6. The major internal organs of a honeybee worker. Modified from C. D. Michener, *The Social Behavior of the Bees: A Comparative Study* (Cambridge, MA: The Belknap Press of Harvard University Press, 1974).

THE PHYSIOLOGY OF TEMPORAL CASTES

The following sequence of control in a normal colony envisioned by Huang and Robinson consists of these steps: The *colony* provides signals to the sensory system of the *individual bee*, altering *brain activity* in a way that sends inhibitory or excitatory signals to the corpora allata (a pair of endocrine glands located in the head). The corpora allata alter their production of *juvenile hormone*, and the juvenile hormone titer affects the size and activity of the *hypopharyngeal gland* as well as the *behavior* of the bee (Figure 5-6).

Huang and Robinson call this envisioned causal chain the activator-inhibitor model of the temporal division of labor and express it as follows: "Workers emerge with low levels of hormone, which are programmed to increase. Once the hormone reaches a critical level in a worker, the worker becomes a forager. The level of an inhibitor also becomes high in foragers, and unlike the activator, the inhibitor is transferred from bee to bee so that other individuals are inhibited from becoming foragers."[40]

---

40 | Z.-Y. Huang and G. E. Robinson, "Social control of division of labor in honey bee colonies," in C. Detrain, J. L. Deneubourg, and J. M. Pasteels, eds., *Information Processing in Social Insects* (Basel: Birkhäuser Verlag, 1999), pp. 165–186.

A balanced role allocation would thus be maintained in the colony by the interplay of activation and inhibition circulated through the classical endocrine and behavioral cycle of individual insects. The cycle would be programmed to proceed in a certain sequence, but its timing could be speeded or slowed according to circumstance. The sequence could even be reversed when commanded by extreme or otherwise unusual conditions. Experiments extending back to the 1920s have demonstrated that when a shortage of wax-producing bees arises, some of the older workers redevelop their wax glands and recommence comb building.[41] This result has been augmented by experiments demonstrating that the principal excitatory stimulus is the combination of a high influx of nectar into the hive and a lack of empty combs for storing the incoming nectar.[42] Further, when a honeybee colony's forager force is decimated by a sudden rainstorm, leaving only younger bees that have not previously flown, the hypopharyngeal glands of many of the hive bees quickly regress. As a result, these bees no longer produce high-grade food for the larvae and instead speed up their maturation into field bees. In the reverse direction to that driven by the aging process, the original development of the hypopharyngeal glands is prolonged beyond the normal time of regression—provided the workers are forced to continue on duty as nurse bees. It is even possible that if the population of nurse workers in the hive has been decimated, older bees that have previously flown may regenerate their hypopharyngeal glands and become nurses again. However, the recovery of the hypopharyngeal glands requires the presence of larvae. Tanya Pankiw and Robert Page have in fact identified a pheromone emanating from larvae that stimulates pollen foraging.[43] This signal might also affect the activation of the hypopharyngeal glands. In addition, a primer pheromone has recently been identified, ethyloleate, that inhibits the transition from nurse to forager. Its action mediates a

---

41 | In particular, the earlier research (in chronological order) by W. J. Nolan, G. A. Rösch, Z. Orösi-Pál, B. D. Milojevic, V. C. Moskovljevic-Filipovic, and J. B. Free, extending to growth and regression of exocrine glands, has been reviewed, for example, in E. O. Wilson, *The Insect Societies* (Cambridge, MA: The Belknap Press of Harvard University Press, 1971), on which part of our present account is based.

42 | S. C. Pratt, "Optimal timing of comb construction by honeybee (*Apis mellifera*) colonies: a dynamic programming model and experimental tests," *Behavioral Ecology and Sociobiology* 46(1): 30–42 (1999); S. C. Pratt, "Condition-dependent timing of comb construction by honeybee colonies: how do workers know when to start building?" *Animal Behaviour* 56(3): 603–610 (1998); and S. C. Pratt, "Collective control of the timing and type of comb construction by honey bees (*Apis mellifera*)," *Apidologie* 35(2): 193–205 (2004).

43 | T. Pankiw, R. E. Page Jr., and M. K. Fondrk, "Brood pheromone stimulates pollen foraging in honey bees (*Apis mellifera*)," *Behavioral Ecology and Sociobiology* 44(3): 193–198 (1998).

negative-feedback loop in the colony's division of labor: when ethyloleate is produced by foraging workers and then transmitted in regurgitated food, it inhibits physiological and behavioral maturation in the nurse workers.[44] Interestingly, this substance may also be a component of the honeybee brood pheromone.[45] The activator-inhibitor model is still mostly in the hypothetical stage, but it is at least apparent that several behavioral and physiological feedback loops affect the division of labor in honeybee societies.[46]

Another approach to understanding division of labor, taken by Robert Page and his coworkers, has been to analyze the differences between foragers who are more likely to collect pollen, a source of protein, and foragers who are more likely to collect nectar, a source of carbohydrate. The study has revealed a complex set of behavioral, physiological, and anatomical traits that vary concordantly with the bees' foraging inclination.[47] Of course, that inclination depends fundamentally on the colony's moment-by-moment need for water, nectar, or pollen, but all things being equal, individual bees vary in their propensities, which are already evident very early in their adult life. For example, the divergence in workers' response to sucrose solutions can be measured within several hours after the bees have eclosed from their pupae.

Responsiveness to sugar was tested quite simply by touching sugar solutions in a series of increasing concentration to the antennae. If the concentration is above threshold, the bee extends her proboscis. Bees differ in their response to sugar within the first 4 hours after emerging as adults. Those that respond to the lowest concentration of sucrose solution tend subsequently to initiate foraging at an earlier

---

44 | I. Leoncini, Y. Le Conte, G. Gostagliola, E. Plettner, A. L. Toth, M. Wang, Z. Huang, J.-M. Bécard, D. Crauser, K. N. Slessor, and G. E. Robinson, "Regulation of behavioral maturation by a primer pheromone produced by adult worker honey bees," *Proceedings of the National Academy of Sciences USA* 101(50): 17559–17564 (2004).

45 | R. E. Page Jr. and T. Pankiw, "Synthetic bee pollen foraging pheromones and uses thereof," United States Patent No. US 6,535,828 B2 (2003).

46 | Z.-Y. Huang and G. E. Robinson, "Social control of division of labor in honey bee colonies," in C. Detrain, J. L. Deneubourg, and J. M. Pasteels, eds., *Information Processing in Social Insects* (Basel: Birkhäuser Verlag, 1999), pp. 165–186.

47 | R. E. Page Jr. and J. Erber, "Levels of behavioral organization and the evolution of division of labor," *Naturwissenschaften* 89(3): 91–106 (2002); G. V. Amdam, K. Norbert, A. Hagen, and S. W. Omholt, "Social exploitation of vitellogenin," *Proceedings of the National Academy of Sciences USA* 100(4): 1799–1802 (2003); G. V. Amdam and S. W. Omholt, "The hive bee to forager transition in honey bee colonies: the double repressor hypothesis," *Journal of Theoretical Biology* 223(4): 451–464 (2003); G. V. Amdam, A. Csondes, M. K. Fondrk, and R. E. Page Jr., "Complex social behaviour derived from maternal reproductive traits," *Nature* 439: 76–78 (2006); and C. M. Nelson, K. E. Ihle, M. K. Fondrk, R. E. Page Jr., and G. V. Amdam, "The gene vitellogenin has multiple coordinating effects on social organization," *PLoS Biology* 5(3): 673–677 (2007).

age and are more inclined to collect water and pollen. Those less responsive show a higher propensity to become nectar foragers.[48]

At first it may seem strange that nectar foragers are less responsive to sucrose than pollen foragers. However, the nectar foragers are more discriminating in their responses over the range of naturally occurring nectar, making them more critical while evaluating the floral "market." Bees more inclined to collect pollen are less discriminating and collect, on average, lower concentrations of sugar. Thus, although forager bees collect both pollen and nectar, and according to colony needs readily switch from pollen to nectar or vice versa, there is a tendency to specialize by collecting disproportionate loads of either nectar or pollen.

Detailed genetic analysis conducted by Robert Page and his colleagues have further disclosed that sucrose sensitivity and the age of foraging are controlled by a network of interacting genes.[49] These behavioral and physiological traits are probably linked through common neurobiochemical pathways under the regulatory control of at least two hormones involved in reproductive signaling in insects: vitellogenin (Vg) and juvenile hormone (JH). Gene Robinson, Gro Amdam, and their coworkers discovered that foraging honeybees have elevated levels of JH and decreased levels of Vg in their blood (hemolymph) relative to bees that perform tasks inside the nest.[50] A number of studies indicate that increasing the blood titer of JH or adding methoprene, an analog of the hormone, results in the early onset of foraging.[51] Juvenile hormone is produced in the corpora allata, which are paired glands associated

---

48 | T. Pankiw and R. E. Page Jr., "Response thresholds to sucrose predict foraging division of labor in honeybees," *Behavioral Ecology and Sociobiology* 47(4): 265–267 (2000).

49 | R. E. Page Jr., M. K. Fondrk, G. J. Hunt, E. Guzmán-Novoa, M. A. Humphries, K. Nguyen, and A. S. Greene, "Genetic dissection of honeybee (*Apis mellifera* L.) foraging behavior," *Journal of Heredity* 91(6): 474–479 (2000).

50 | For reviews, see G. Bloch, D. E. Wheeler, and G. E. Robinson, "Endocrine influences on the organization of insect societies," in D. W. Pfaff, A. P. Arnold, A. M. Etgen, S. E. Fahrbach, and R. T. Rubin, eds., *Hormones, Brain and Behavior*, Vol. 3 (New York: Academic Press, 2002), pp. 195–235; and R. E. Page Jr., R. Scheiner, J. Erber, and G. V. Amdam, "The development and evolution of division of labor and foraging specialization in a social insect (*Apis mellifera* L.)," *Current Topics in Developmental Biology* 74: 253–286 (2006).

51 | While juvenile hormone (JH) modulates age polyethism with foragers exhibiting high JH titers in honeybees (and some ants), in primitively eusocial species and in wasps of the genera *Polistes* and *Ropalidia*, JH accelerates ovarian development and does not promote foraging behavior. See P. I. Röseler, "Reproductive competition during colony esablishment," in K. G. Ross and R. W. Matthews, eds., *The Social Biology of Wasps* (Ithaca, NY: Comstock Publishing Associates of Cornell University Press, 1998), pp. 309–335; and M. Agrahari and R. Gadagkar, "Juvenile hormone accelerates ovarian development and does not affect age polyethism in the primitive eusocial wasp, *Ropalidia marginata*," *Journal of Insect Physiology* 49(3): 217–222 (2003).

with the brains of insects (see Figure 5-6). However, removal of these glands and hence curtailment of JH production does not affect the age at which the bees begin foraging. JH, in other words, is not necessary to initiate foraging, bringing into question some of the assumptions proposed in the activator-inhibitor model.

Further, detailed analyses have revealed that juvenile hormone and vitellogenin regulate each other's level. High Vg suppresses JH, and high JH suppresses Vg, a reciprocity that results in a rapid change in both hormones just prior to the onset of foraging. It is important to note that vitellogenin is not only a hormone but also a precursor of egg yolk proteins. Worker bees begin synthesizing Vg soon after they emerge as adults. When the Vg blood titer increases, JH production is suppressed, resulting in a low JH blood titer. Further studies by Gro Amdam and her coworkers (as well as other analyses) have revealed that Vg proteins are incorporated into "brood food." This vital substance comprises glandular secretions produced by the hypopharyngeal gland of nurse bees and fed to developing larvae. As the larvae are fed, the nurse bee Vg titers fall until the protein no longer suppresses JH production. Then, as JH titers rise, they suppress Vg production, rapidly driving Vg down and resulting in bees with low Vg and higher JH. With this combination, the bees initiate foraging. Although it is apparent that high JH is not an essential player, it is not known what actually triggers the initiation of foraging. Perhaps it is the low Vg titer in the blood, with an elevated JH titer serving to prepare the bees for the strenuous, energy-consuming foraging activities about to be assumed outside the nest.

Bees genetically selected for foraging behavior differ in their Vg blood titers soon after they emerge as adults. Those more prone to forage for pollen have higher Vg titers than those likely to collect nectar. The system can be experimentally manipulated by injecting newly emerged bees with double-stranded RNA (dsRNA) encoding a short segment of the vitellogenin gene. The dsRNA interferes with Vg production, resulting in greatly reduced Vg blood titers in young bees. Such "Vg knockdown bees" were found to forage earlier in life than control bees. Also in response to vitellogenin knockdown, juvenile hormone activity increases rapidly, which is correlated with rapid social behavioral ontogeny and a short life span of control bees.[52] Thus, vitellogenin appears to affect an entire suite of traits that differ in bees more inclined to collect pollen versus those

52 | G. V. Amdam, K.-A. Nilsen, K. Norberg, M. K. Fondrk, and K. Hartfelder, "Variation in endocrine signaling underlies variation in social life history," *American Naturalist* 170(1): 37–46 (2007).

inclined toward nectar. High titers of vitellogenin in young bees results in a faster decline of Vg later. Vg, in short, primes the bees to be more disposed to pollen foraging than to nectar collecting.

These discoveries have led to a new perspective of the evolution of the physiological mechanisms of division of labor in honeybees. Vg and JH are part of the reproductive regulatory hormonal network of insects. The finding that these substances are involved in the onset of foraging, as well as choice of food harvested, is a remarkable case of parsimony in evolution. Using this discovery as a clue, Gro Amdam, Robert Page, and their colleagues have pressed on to investigate the relationships between ovary development and foraging behavior. Worker honeybees in fact possess functional ovaries that are reduced in size compared with those of queens; worker ovaries also exist in various stages of reproductive activation. In the absence of the queen, the ovaries of young workers are activated, so that at least some workers can lay unfertilized eggs that develop into males.[53]

A further link has been found in the course of this remarkable spate of research: bees with larger ovaries have higher levels of ovary activation, forage earlier in life, and are more likely to collect pollen and nectar with a lower concentration of sugar. Young workers, in particular, possessing larger ovaries also have higher titers of Vg, linking ovary size, hormone titer, and behavior.

From all these correlations, researchers conclude that foraging, division of labor, and a tendency in honeybees for specialization in food harvesting most likely evolved from the ordinary reproductive cycle of the solitary ancestors. Such is the conception of Mary Jane West-Eberhard's "groundplan hypothesis," discussed previously in this chapter and in Chapter 2. The first step envisioned was a shift in the timing of reproductive hormonal signaling events, involving JH and another class of hormones (ecdysteroids), from the late pupal stages to the mature adult stage. In young bees, this shift turns on the production of vitellogenin and behavior related to reproductive maturity. Thus, in one simple change in timing, vitellogenin-producing females bypassed the phases of mating, disposal, diapause, and estivation that characterize the ancestral prereproductive period. They substituted instead the maternal reproductive behaviors of larval care, nest defense, and foraging. The second step was the evolution of a feedback interaction between vitellogenin and JH, resulting in a regulatory

53 | Because honeybee workers are not mated, worker-laid eggs have a single (haploid) set of chromosomes. Haploid eggs produce males.

mechanism that enabled vitellogenin to become a pacemaker for division of labor. Higher blood titers of vitellogenin keep bees in the nest performing maternal, non-foraging tasks. These include brood care, processing and storage of newly collected food, provisioning the brood cells, cleaning the cells, and other housekeeping work.

When their vitellogenin supply is exhausted, the bees automatically commence foraging. Following this transition to outside work, bees with larger, active ovaries preferentially forage for pollen and the protein it contains, as did their reproductively activated solitary ancestors while provisioning their brood. Those with smaller, inactive ovaries forage primarily for nectar, as do nonreproductive solitary insect females.[54]

Parallel physiological and behavioral changes most likely occur during the aging process of ants, although this supposition has not yet been investigated in any depth. The result is a programmed sequence of role allocations comparable to that of honeybees, as illustrated by the European mound-building species *Formica polyctena* (Figure 5-7). Just as in bees, the sequence of roles can be reversed when events in the environment cause a severe perturbation of ratios in the age cohorts. In one simple but elegant demonstration, Sophie Ehrhardt transferred workers of the European ant *Myrmica rubra* that had been engaged in nursing to isolated cells containing only soil, and, conversely, other workers exclusively engaged in excavation and foraging to cells containing brood but no soil.[55] The erstwhile nurses soon devoted themselves to soil excavation, while those previously working as foragers turned into nurses. Of even greater interest is that both groups, when returned to their entire assembly of nestmates in the home nest, reverted to their original functions.

The condition of temporal castes in the social insects, at least in the honeybee *Apis mellifera* and species of ants where the phenomenon has been studied best, conform to the following broad pattern. As workers age, they pass through a series of loosely defined labor roles that entail care of the queen and brood at first, then labor elsewhere in the nest, and finally foraging outside the nest.

---

54 | G. V. Amdam, K. Norbert, M. K. Fondrk, and R. E. Page Jr., "Reproductive ground plan may mediate colony-level selection effects on individual foraging behavior in honey bees," *Proceedings of the National Academy of Sciences USA* 101(31): 11350–11355 (2004); and G. V. Amdam, A. Csondes, M. K. Fondrk, and R. E. Page Jr., "Complex social behaviour derived from maternal reproductive traits," *Nature* 439: 76–78 (2006).

55 | S. Ehrhardt, "Über Arbeitsteilung bei *Myrmica-* und *Messor*-Arten," *Zeitschrift für Morphologie und Ökologie der Tiere* 20(4): 755–812 (1931).

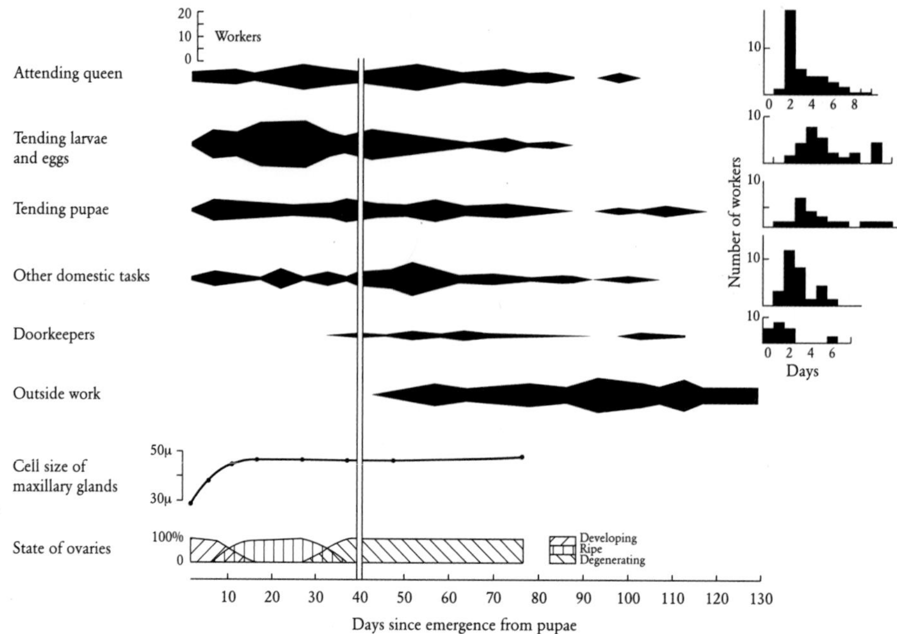

FIGURE 5-7. Temporal division of labor in ants in finer detail. As shown, workers of *Formica polyctena* change through time in both behavior and glandular development. The histograms on the right show the number of days that workers persisted in the given task. These data reveal, for example, that few workers specialize in attending the queen, while a comparatively large number specialize in caring for the brood. The double vertical line roughly separates the period of nest work from that of foraging service. Redrawn from J. H. Sudd, *An Introduction to the Behaviour of Ants* (London: Arnold, 1967), based on the data of D. Otto, in E. O. Wilson, *The Insect Societies* (Cambridge, MA: The Belknap Press of Harvard University Press, 1971).

In most ant species, as in honeybees, the ontogenetic transitions in task activities unfold with change in the workers' ovary activity. Young interior (nest) workers have better-developed ovaries than exterior (field) workers. They often produce trophic eggs, laid not for reproduction but for nutrition; the eggs are fed to the queens and to the larvae. In some species, trophic eggs produced by nurse workers are a major nutritional source for immature nestmates and queens. In these species, foraging workers carry well-developed oocytes in their ovaries. They also contain many yellow bodies, which are residues from oocyte source cells and indicate a history of egg-laying activity. Egg feeding is common across a phylogenetically diverse array of ant species.[56]

---

56 | For review, see B. Hölldobler and E. O. Wilson, *The Ants* (Cambridge, MA: The Belknap Press of Harvard University Press, 1990).

For example, young nurse workers of *Aphaenogaster* (formerly *Novomessor*) *cockerelli* and *Aphaenogaster albisetosus*, two common species in the southwestern United States, have developed ovaries and appear to feed trophic eggs to the larvae and queen (Plates 9 and 10). Some forager ants of these species readily revert to interior worker tasks during severe shortages of brood tenders, provided their ovaries can still be reactivated. Once the ovaries have completely degenerated, however, these ants do not revert to interior tasks.[57] Similar observations have been made with harvester ant species of the genus *Pogonomyrmex* and weaver ants of the genus *Oecophylla.* A few studies indicate that foraging ants, like foraging honeybees, have an elevated juvenile hormone blood titer.[58] It remains to be determined whether ants engaged in brood and queen tending have higher vitellogenin titers than do exterior workers.

Karl Heinz Bier, a pioneer in the study of physiological mechanisms that regulate division of labor and reproduction in ants, postulated that the postpharyngeal gland, a finger-shaped organ in the head, is the source of "profertile substances" that the nurse ants feed to the queens and to larvae destined to develop into reproductive females. In his studies of mound-building wood ants (*Formica polyctena*), he observed that brood-tending workers isolated from queens and larvae commence to lay viable eggs. He reasoned that such nurse workers prevented from feeding their profertile substances to larvae and queens use these high-quality nutrients for the development of their own ovaries and eggs.[59] Subsequent research confirmed that young nurse workers of *Formica* exhibit a high secretory activity of the postpharyngeal gland. And further studies, employing radioactive tracers, have revealed that the postpharyngeal glands of these brood (and queen) tenders have a significant nutritive role. Parallel studies with different species of ants support the nutritive function of the postpharyngeal gland

---

57 | B. Hölldobler and N. F. Carlin, "Colony founding, queen control and worker reproduction in the ant *Aphaenogaster* (= *Novomessor*) *cockerelli* (Hymenoptera: Formicidae)," *Psyche* (Cambridge, MA) 96: 131–151 (1989).

58 | K. Sommer, B. Hölldobler, and H. Rembold, "Behavioral and physiological aspects of reproductive control in a *Diacamma* species from Malaysia (Formicidae, Ponerinae)," *Ethology* 94(2): 162–170 (1993); and C. Brent, C. Peeters, V. Dietemann, R. Crewe, and E. Vargo, "Hormonal correlates of reproductive status in the queenless ponerine ant *Streblognathus peetersi*," *Journal of Comparative Physiology A* 192(3): 315–320 (2006).

59 | K. Bier, "Die Bedeutung der Jungarbeiterinnen für die Geschlechtstieraufzucht im Ameisenstaat," *Biologisches Zentralblatt* 77(3): 257–265 (1958).

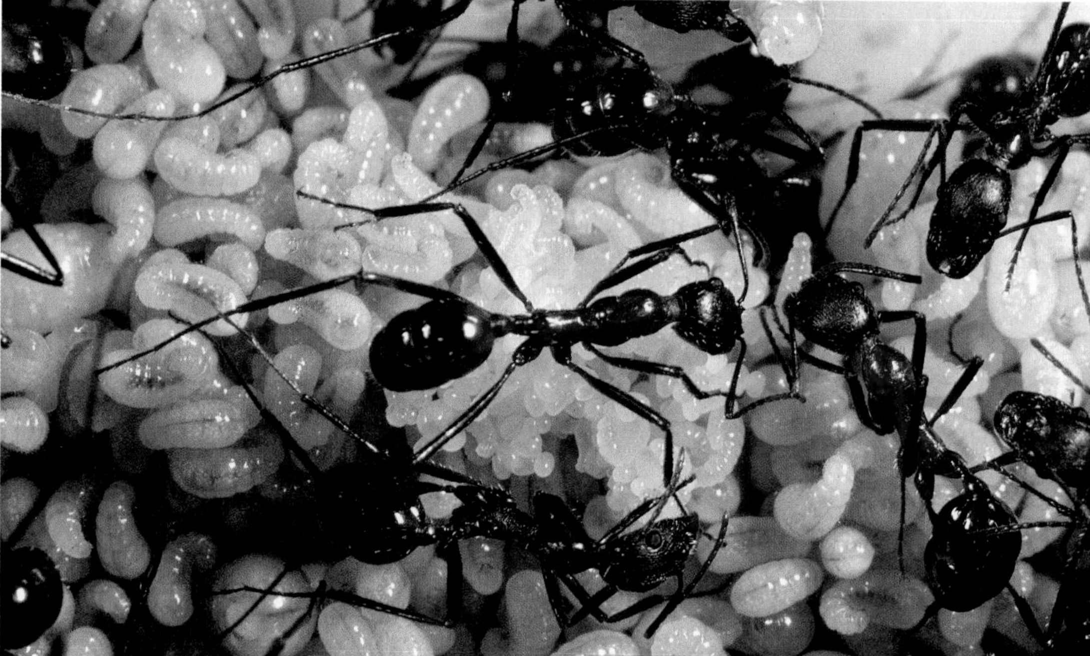

PLATE 9. *Aphaenogaster* (formerly *Novomessor*) *cockerelli. Above*: A queen with brood and workers. *Below*: In the absence of the queen, workers produce viable eggs (instead of trophic eggs). These are not fertilized and therefore develop into males.

PLATE 10. *Aphaenogaster* (formerly *Novomessor*) *cockerelli. Above*: Workers isolated from the queenright colony begin to produce viable eggs. *Below*: When these workers are reunited with their colony, nestmates aggressively police them to prevent them from reproducing.

in brood-tending ants.[60] It remains to be seen whether the profertile substances postulated by Bier are related in function to vitellogenin, functioning in ants in the same way this substance functions in honeybees—as "brood food." Certainly, in ants, as in all hymenopterans, viable eggs are loaded with vitellogenin. Moreover, eggs of certain formicine and myrmicine ant species (*Camponotus festinatus, Solenopsis invicta, Monomorium pharaonis*) contain the same kind of vitellogenin as honeybee eggs.[61] The same is true of species in the relatively primitive poneromorph genera *Amblyopone*, *Odontomachus*, and *Platythyreus*. Species in the poneromorph subfamilies Ponerinae and Ectatomminae, on the other hand, have a wide diversity of vitellogenin structures.[62]

During the growth period of a social insect colony, prior to production of the next generation of reproductive forms, the worker population operates as a growth-maximizing machine, and for a time at least during this period, it approaches a stable age distribution.[63] The major roles can be expected, under these circumstances, to follow the aging process, at least as a statistical central tendency, with the bulk of the workers passing without interruption from one temporal caste to the next. The phenomenon has been observed in controlled laboratory and field populations maintained in constant, relatively safe conditions and fed at will.

The programming of temporal castes in honeybees and ants is consistent with what is currently the most widely accepted genetic theory of aging. Under natural conditions, foraging is by far the most dangerous activity for the insect, due to greater exposure to enemies, sudden changes in weather, and simple loss of way. Judging from still limited data, the loss of all workers, depending on species, is

---

60 | See review in B. Hölldobler and E. O. Wilson, *The Ants* (Cambridge, MA: The Belknap Press of Harvard University Press, 1990).

61 | T. Martinez and D. Wheeler, "Identification of vitellogenin in the ant, *Camponotus festinatus*: changes in hemolymph proteins and fat body development in workers," *Archives of Insect Biochemistry and Physiology* 17: 143–155 (1991); E. L. Vargo and M. Laurel, "Studies on the mode of action of a queen primer pheromone of the fire ant *Solenopsis invicta*," *Journal of Insect Physiology* 40(7): 601–610 (1994); and P. V. Jensen and L. W. Børgesen, "Yolk protein in the pharaoh's ant: influence of larvae and workers on vitellogenin and vitellin content in queens," *Insectes Sociaux* 42(4): 397–409 (1995).

62 | D. Wheeler, J. Liebig, and B. Hölldobler, "Atypical vitellins in ponerine ants (Formicidae: Hymenoptera)," *Journal of Insect Physiology* 45(3): 287–293 (1999).

63 | M. V. Brian, *Social Insect Populations* (New York: Academic Press, 1965).

between 1 and 10 percent of the entire population per day.[64] It has been commonly observed that this loss, as expected, is concentrated in the foragers. During the first 14 days of adult life monitored in a colony of the Middle Eastern hornet *Vespa orientalis*, for example, workers suffered mortality of 8.8 percent when confined to the nest and 42.5 percent when allowed to forage outside.[65] Comparable observations have confirmed the same pattern in ants.[66]

With mortality much higher outside the nest, it should be to the advantage of the colony for workers to delay foraging to a later period of their lives.[67] Looked at another way, we can expect aging itself to follow a timetable set by the seasonal availability of resources, which in turn sets the best times to forage. Guided by natural selection, the species programs the rest of the stages of temporal polyethism in a manner optimal for survival and reproduction of the colony as a whole; in essence, mortality abroad pulls the temporal life stages behind it.

The overall pattern by which honeybees and ants shift their age-based division of labor reflects the time scale of important variations in the environment of the colony. When changes occur on a short-term basis—say, the discovery of a rich food source or the mass emergence of reproductive adults from their pupae—workers readily switch tasks within their age-based block of tasks. Only a few may shift to a different age-based block of tasks to assist the roles that have opened up. When, on the other hand, changes occur on a long-term basis—such as response to a prolonged food shortage, the decimation of foragers by severe weather, or the die-off of brood by disease—workers can accelerate or slow down their exocrine and behavioral programs to fill the gaps more fully and precisely.

---

64 | E. O. Wilson, *The Insect Societies* (Cambridge, MA: The Belknap Press of Harvard University Press, 1971); S. O'Donnell and R. L. Jeanne, "Implications of senescence patterns for the evolution of age polyethism in eusocial insects," *Behavioral Ecology* 6(3): 269–273 (1995).

65 | J. Ishay, H. Bytinski-Salz, and A. Shulov, "Contributions to the bionomics of the Oriental hornet (*Vespa orientalis* Fab.)," *Israel Journal of Entomology* 2: 45–106 (1967).

66 | See B. Hölldobler and E. O. Wilson, *The Ants* (Cambridge, MA: The Belknap Press of Harvard University Press, 1990).

67 | A. Tofilski, "Influence of age polyethism on longevity of workers in social insects," *Behavioral Ecology and Sociobiology* 51(3): 234–237 (2002).

GENETIC VARIABILITY IN CASTE DIFFERENTIATION

As we have stressed, temporal castes are never rigidly programmed. They are instead expressed as central tendencies. A wide variance occurs in behavioral traits within each age cohort and for a variety of reasons. Among the potentially important factors forcing plasticity is genetic variation that causes workers at the outset to choose preferentially certain tasks over others. A few cases of such genetic influence on task choice have been reported, most notably in the inclination of honeybee foragers to forage pollen versus nectar, but research has been too limited to judge whether the phenomenon is widespread.

In the South American social wasp *Polybia occidentalis*, the honeybee *Apis mellifera*, and the American leafcutter ant *Acromyrmex versicolor*, some workers are consistent corpse removal specialists.[68] However, this could also be due to learning—that is, repeated performance of the same task. Other evidence is more supportive of genetic variance. Glennis Julian and Jennifer Fewell, for example, found that *Acromyrmex versicolor* workers of different matrilines vary in the age of transition from inside the nest to foraging, although not in task performance rates. In this species, moreover, division of labor appears to be affected by genetic dispositions of individuals for task preference and age-related task choice.[69] In the case of honeybees, variation in predisposition of individual bees to assume nest guarding and foraging also have a partial genetic base.[70] In this species at least, such variation is favored by the multiple inseminations of the queens by different males. Its expression is likely mediated by differences in sensory and neural thresholds, a property already experimentally demonstrated for specialization on different food items during foraging.[71]

There are other predispositions in which genetic influence might be profitably sought. Among the specialists and their more flexibly oriented nestmates, there exist

---

68 | S. O'Donnell and R. L. Jeanne, "Forager specialization and the control of nest repair in *Polybia occidentalis* Olivier (Hymenoptera: Vespidae)," *Behavioral Ecology and Sociobiology* 27(5): 359–364 (1990); S. T. Trumbo and G. E. Robinson, "Learning and task interference by corpse-removal specialists in honey bee colonies," *Ethology* 103(11): 966–975 (1997); and G. E. Julian and S. Cahan, "Undertaking specialization in the desert leaf-cutter ant *Acromyrmex versicolor*," *Animal Behaviour* 58(2): 437–442 (1999).

69 | G. E. Julian and J. H. Fewell, "Genetic variation and task specialization in the desert leaf-cutter ant, *Acromyrmex versicolor*," *Animal Behaviour* 68(1): 1–8 (2004).

70 | R. E. Page Jr. and G. E. Robinson, "The genetics of division of labour in honey bee colonies," *Advances in Insect Physiology* 23: 117–169 (1991).

71 | T. Pankiw and R. E. Page Jr., "Response thresholds to sucrose predict foraging division of labor in honeybees," *Behavioral Ecology and Sociobiology* 47(4): 265–267 (2000).

in the honeybee and in at least some ant species a statistically distinguishable group sometimes referred to as elites. They are persistently well above average in the tempo of their activities, their personal productivity, and the degree to which they stimulate and help organize nestmates.[72]

MEMORY IN DIVISION OF LABOR

Watched for only a few hours, a colony of social insects might be interpreted as consisting of automata driven with the same uniform set of decision rules. But that is far from the case. Each member of the colony is distinct in some manner or other that affects its behavior. Each has a mind of its own. By mind we do not mean a reflective, self-aware, wide-roaming consciousness of the human kind, but rather a cognitive consciousness built with a relatively complex brain that can store information from all its sensory modalities (taste, smell, touch, sight, and sound) as well as some memory of the events it has experienced during its short life. In fact, it is now known that heritable variation in learning performance affects preference in the workers.[73] In many cases, moreover, life is not short: while honeybee workers die of old age within a few weeks, workers of some ant species can live several years. Labor is driven by priorities influenced at least in part by prior cognition: search for this task, or finish the task in progress, or patrol the surround in search for whatever needs doing, or pause to stand guard, or simply rest.[74]

In certain categories, social insects have a memory capacity impressive even by

---

72 | M. Möglich and B. Hölldobler, "Social carrying behavior and division of labor during nest moving in ants," *Psyche* (Cambridge, MA) 81(2): 219–236 (1974) (this was one of the first papers identifying elites); G. F. Oster and E. O. Wilson, *Caste and Ecology in the Social Insects* (Princeton, NJ: Princeton University Press, 1978); and S. K. Robson and J. F. A. Traniello, "Key individuals and the organization of labor in ants," in C. Detrain, J. L. Deneubourg, and J. M. Pasteels, eds., *Information Processing in Social Insects* (Basel: Birkhäuser Verlag, 1999), pp. 239–259.

73 | J. S. Latshaw and B. H. Smith, "Heritable variation in learning performance affects foraging preferences in the honey bee (*Apis mellifera*)," *Behavioral Ecology and Sociobiology* 58(2): 200–207 (2005).

74 | The conception of a rich cognitive consciousness of honeybee workers is explored in T. D. Seeley, "What studies of communication have revealed about the minds of worker honey bees," in T. Kikuchi, N. Azuma, and S. Higashi, eds., *Genes, Behavior and Evolution of Social Insects* (Sapporo, Japan: Hokkaido University Press, 2003), pp. 21–33. The general subject of subhuman consciousness is addressed in D. R. Griffin, *Animal Minds: Beyond Cognition to Consciousness* (Chicago: University of Chicago Press, 2001). For a most recent review see R. Menzel, B. Brembs, and M. Giurfa, "Cognition in invertebrates," in J. H. Kaas, *Evolution of Nervous Systems*, Vol. 2: *Evolution of Nervous Systems in Invertebrates* (New York: Academic Press, 2006), pp. 403–422.

human standards. Most species of ants learn their colony odor, a complex bouquet of hydrocarbons resident in the outer cuticular layer of the exoskeleton. Foraging workers of honeybees and some species of ants learn the terrain of their home range based on visual landmarks, and they employ a celestial compass system using spectral and polarization channels and vector integration when homing from a foraging excursion. This has been studied in great detail by Rüdiger Wehner and his coworkers in the formicine desert ant genus *Cataglyphis*. The foragers meander on the outward travel of a foraging trip, but take a direct route home. This requires the ability to integrate the many turning angles and to measure the distance traveled. How ants measure distance has remained obscure until recently. Mathias Wittlinger and his collaborators addressed the problem by lengthening the legs with stilts or shortening them by cutting off the tarsi. In this ingenious way, they were able to demonstrate that the ants measure distance by counting steps.[75] At least one forest-dwelling ant species, the African ponerine *Pachycondyla tarsata* (formerly *Paltothyreus tarsatus*), memorizes the detailed outlines of tree crowns beneath which it passes on outgoing trips, then integrates and reverses the information upon its return to run a straight line home.[76] Honeybees searching for flower beds or new nest sites perform essentially the same operation as the canopy "scanners" by using features of the landscape they pass. Similarly, nestmates who follow their waggle dances back in the hive keep in mind the instructions of direction and distance as they fly outward in search of the same sites. In addition, bees associate color, shape, and scent with the rewarding nectar they collect in the flowers, and they consistently return to the same kind of flowers on subsequent trips.[77]

Training experiments have shown that worker bees are able to learn signals in

---

75 | R. Wehner, "The ants' celestial compass system: spectral and polarization channels," in M. Lehrer, ed., *Orientation and Communication in Arthropods* (Basel: Birkhäuser Verlag, 1997), pp. 145–185; R. Wehner, "Desert ant navigation: how miniature brains solve complex tasks," *Journal of Comparative Physiology A* 189(7): 579–588 (2003); M. Wittlinger, R. Wehner, and H. Wolf, "The ant odometer: stepping on stilts and stumps," *Science* 312: 1965–1967 (2006); and M. Knaden and R. Wehner, "Ant navigation: resetting the path integrator," *Journal of Experimental Biology* 209(1): 26–31 (2006).

76 | B. Hölldobler, "Canopy orientation: a new kind of orientation in ants," *Science* 210: 86–88 (1980); P. S. Oliveira and B. Hölldobler, "Orientation and communication in the Neotropical ant *Odontomachus bauri* Emery (Hymentoptera, Formicidae, Ponerinae)," *Ethology* 83: 154–166 (1989); and A. P. Baader, "The significance of visual landmarks for navigation of the giant tropical ant, *Parponera clavata* (Formicidae, Ponerinae)," *Insectes Sociaux* 43(4): 435–450 (1996).

77 | B. H. Smith, G. A. Wright, and K. C. Daly, "Learning-based recognition and discrimination of floral odors," in N. Dudareva and E. Pichersky, eds., *Biology of Floral Scent* (Boca Raton, FL: Taylor & Francis, 2006), pp. 263–295.

every known sensory modality. They can learn them quickly in most cases, and they can simultaneously master multiple tasks, each dependent on several modalities. Tasks can be memorized and performed in a sequence, as in the programs of visits to different flowers at specific times of the day. Isolated worker bees can be trained to walk through relatively complex mazes requiring as many as five turns in sequence in response to such clues as the distance between two spots, the color of a marker, and the angle of a turn in the maze. After associating a given color once with a reward of a 2-molar sucrose solution, they can remember it for as long as 6 days. If given the experience three times in a row, they remember the color for at least 2 weeks. The location of a food site in the field can be remembered for a period of 6 to 8 days after last visiting the location; on one occasion, worker bees were observed dancing out the location of a site after 2 months of winter confinement. Ants can perform comparable feats. Workers of *Formica pallidefulva* are able to learn a six-point maze with relative ease at a rate only two to three times slower than that achieved by laboratory rats. Workers of *Formica polyctena* can remember their way through mazes for periods of up to 4 days, while those of *Formica rufa*, operating under more natural circumstances, can simultaneously memorize the position of four separate landmarks and recall them well enough for use in orientation as much as a week later.[78]

Thus, while a brief surveillance of an insect colony may seem to disclose a confusing kaleidoscope of activity, longer periods of observation reveal that patterns are built from many quite individual minds linked by a high degree of organization. That amount of order is central to the colony's survival and reproduction.

---

78 | The learning capacities of ants and honeybees were a favored subject from the 1930s through the 1960s; these studies attained a high level of sophistication, especially in experiments by R. Jander, M. Lindauer, R. Menzel, T. C. Schneirla, R. Wehner, and K. Weiss. This large body of work is reviewed in E. O. Wilson, *The Insect Societies* (Cambridge, MA: The Belknap Press of Harvard University Press, 1971), pp. 210–218. For later reviews see R. Menzel, "Learning, memory and 'cognition' in honey bees," in R. P. Kesner and D. S. Olton, eds., *Neurobiology of Comparative Cognition* (Hillsdale, NJ: Erlbaum, 1990), pp. 237–292; R. Menzel, R. J. De Marco, and U. Greggers, "Spatial memory, navigation and dance behaviour in *Apis mellifera*," *Journal of Comparative Physiology A* 192: 889–903 (2006); R. Menzel, B. Brembs, and M. Giurfa, "Cognition in invertebrates," in J. H. Kaas, *Evolution of Nervous Systems*, Vol. 2: *Evolution of Nervous Systems in Invertebrates* (New York: Academic Press, 2006), pp. 403–422; R. Menzel and M. Giurfa, "Dimensions of cognition in an insect, the honeybee," *Behavioral and Cognitive Neuroscience Reviews* 5(1): 24–40 (2006); B. H. Smith, G. A. Wright, and K. C. Daly, "Learning-based recognition and discrimination of floral odors," in N. Dudareva and E. Pichersky, eds., *Biology of Floral Scent* (Boca Raton, FL: Taylor & Francis, 2006), pp. 263–295; and R. Wehner, "The ants' celestial compass systems: spectral and polarization channels," in M. Lehrer, ed., *Orientation and Communication in Arthropods* (Basel: Birkhäuser Verlag, 1997), pp. 145–185.

TASK SWITCHING AND BEHAVIOR PLASTICITY

Part of the Darwinian prerequisites of colony success is the ability of the workers to switch tasks quickly and precisely. The prerequisite can be understood in theory by approaching it as a problem in labor optimization.[79] A nonsocial insect, such as a solitary wasp, has no choice when addressing a task but to perform it as an *unbroken series* of steps:

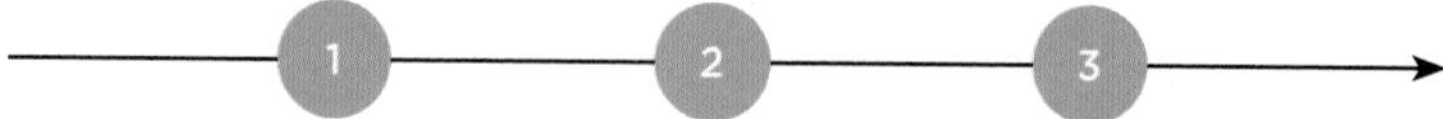

A colony can perform many such tasks simultaneously in *parallel series*:

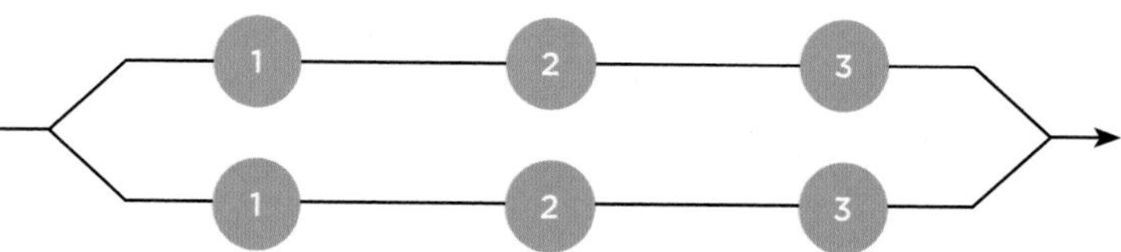

The whole process can be speeded up if the workers switch opportunistically to perform whatever task is closest to hand or fill in for another worker that has halted, and so forth. Such is the *series-parallel* process, the one actually observed in social insects:

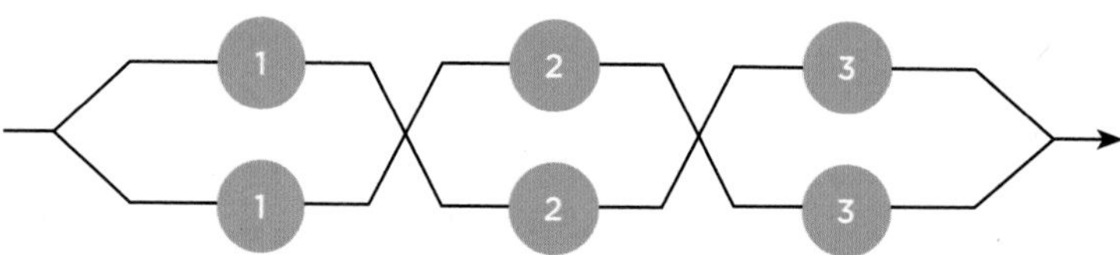

The efficiency of the colony is enhanced still more if workers are able, when the need of the colony is great, to switch not just from one step to another in a task sequence but also to move to an entirely different task sequence, as from nest construction to the retrieval of a misplaced larva. This plasticity, too, seems to be a universal property of insect colonies.

---

79 | The theory is developed at some length in G. F. Oster and E. O. Wilson, *Caste and Ecology in the Social Insects* (Princeton, NJ: Princeton University Press, 1978).

Ants and other social insects are good at what they do, and they get better by means of cooperative labor. Their behavior fulfills principles of ergonomic efficiency embodied in the Barlow-Proschan theorems.[80] When individual competence is low, the first theorem says, the reliability of a system of individuals acting together is lower than the summed competence of the individuals acting singly; but when individual competence is high, above a certain threshold level, the reliability of the system based on cooperation is greater. According to the second theorem, one redundant system, whose parts that can be switched back and forth (as in colony members), is more reliable than two otherwise identical systems with no such backup parts.

Efficiency of the system—the insect colony—can be moved up another notch if groups of workers are specialized in size, anatomical proportions, physiology, and behavioral competence to perform certain roles. The most immediate and efficient way to divide labor is by physiology and behavior, rather than by size and anatomy, since these traits can be changed more readily to achieve a shift in major labor roles—as from nurse to forager or the reverse. That, too, is an observed principle of division of labor in the social insects.

To conclude that the central tendencies of role change with aging have a genetic basis is not to imply that it is rigidly determined. Instead, the ensemble of genes that program the sequence of role changes have, like all such hereditary units, a norm of reaction, the array of possible outcomes in physiology and behavior determined by the interaction of the genes and the particular environment in which the development occurs. An extremely narrow norm of reaction means that only one outcome occurs, regardless of the environment. An extremely wide norm of reaction means that a different outcome occurs in each one of the environments in which the development can occur.

The hereditary control of change in temporal division of labor falls between the two extremes and to different levels of intermediacy according to the trait affected. That is, while it is overall plastic, it is far from unlimited. Another way of putting the matter is that the temporal division of labor is plastic to a degree that is limited and genetically programmed according to the trait affected.

The genetics of temporal division of labor is still in the earliest stage of

---

80 | Applied in G. F. Oster and E. O. Wilson, *Caste and Ecology in the Social Insects* (Princeton, NJ: Princeton University Press, 1978).

exploration. The timing of the transition from nest work to guarding and foraging has been shown to be influenced by gene differences in honeybees, and a genetic component is indicated indirectly for *Temnothorax* ants.[81, 82] Furthermore, some circumstantial evidence suggests that task allocation in the polygynous ectatommine ant *Gnamptogenys striatula* depends on matrilines.[83] Finally, in *Formica selysi*, a formicine species that varies in the number of inseminating males per queen and the number of queens per colony, a genetic component in worker size variability and specialization is suggested, based on comparisons of colonies with single and multiple matrilines.[84] But the interaction of genes and environment and the forces of ecological selection that have programmed plasticity in specific age-related labor changes remain mostly uninvestigated. Indeed, the analysis of plasticity of labor roles as a genetic adaptation remains one of the outstanding challenges of insect sociobiology.

It is reasonable, meanwhile, to suppose that colony-level natural selection has struck a balance between plasticity of labor roles and efficiency of the labor performed. An age cohort perfectly specialized by physiology and exocrine gland production to perform a particular task is correspondingly less able to switch to another task. At the opposite extreme, a jack-of-all-roles cohort is unlikely to be as capable as a cohort of specialists to fill any particular role. The compromise between specialization and plasticity may explain the existence of several well-documented phenomena in age-based division of labor. They include hysteresis (the lingering effect on behavior of individual experience); the age-related physiological changes that alter response thresholds; and, in at least honeybees and a few ant species, the existence of genetic variation due to multiple patrilines (and matrilines in ants) in the worker force, which predisposes workers at the outset of adult life to specialize to some degree on particular tasks.

---

81 | G. E. Robinson, "Genomics and integrative analyses of division of labor in honeybee colonies," *American Naturalist* 160(Supplement): S160–S172 (2002); and G. J. Hunt, E. Guzmán-Novoa, J. L. Uribe-Rubio, and D. Prieto-Merlos, "Genotype-environment interaction in honeybee guarding behaviour," *Animal Behaviour* 66(3): 459–467 (2003).

82 | R. J. Stuart and R. E. Page Jr., "Genetic component to division of labor among workers of a leptothoracine ant," *Naturwissenschaften* 78(8): 375–377 (1991).

83 | R. Blatrix, J.-L. Durand, and P. Jaisson, "Task allocation depends on matriline in the ponerine ant *Gnamptogenys striatula* Mayr," *Journal of Insect Behavior* 13(4): 553–562 (2000).

84 | T. Schwander, H. Rosset, and M. Chapuisat, "Division of labour and worker size polymorphism in ant colonies: the impact of social and genetic factors," *Behavioral Ecology and Sociobiology* 59(2): 215–221 (2005).

How might patterns of plasticity be conceptualized to advance the analysis of division of labor? The logical point of entry is the sociogram, comprising as complete a list as possible of social behaviors displayed by a given species or caste within a species.[85] The list is constructed to allow objective comparisons with the sociograms of other species. At first, such correspondence across species may seem impracticable, since almost any behavioral "act" can be divided into the multiple elements that compose it. For example, "brood care" can be egg care or larval care or pupal care, and each of these acts can be broken further into grooming, transport, feeding, and so forth. But the problem is soluble once it is realized that two or more lists (for example, two or more sociograms of the same caste of the same species by different investigators or lists for different castes or different species) can be mapped onto one another, with the elements matched one on one, many on one, or one on many. For example:

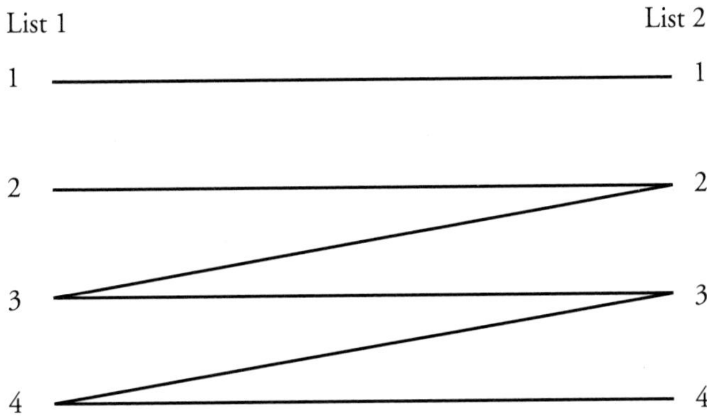

As individual elements become more finely divided, they also come to be unavoidably bundled, such as picking up a larva and carrying the larva. So beyond a certain level of splitting in the sociogram, relatively little information can be added.

In a similar manner, with more objectivity if less tightness, a larger number of elements can be bundled into roles, including those often recognized in the literature: nursing, nest construction, midden work (including corpse removal), defense, and foraging. Plasticity of a given caste would be measured and represented as

85 | See R. M. Fagen and R. N. Goldman, "Behavioural catalogue analysis methods," *Animal Behaviour* 25(2): 261–274 (1977).

the probability of a switch from one role to another by workers of that caste who encounter an unfilled task within the second role. One example would be a laden forager coming upon a misplaced larva; another would be a nurse caring for brood recruited by a forager. Let r_1 represent the current role and an arrow the change to the new, competing role, with the length of the arrow corresponding to the probability of a shift occurring under specified circumstances. Suppose, for example, there are five roles. If the ends of the arrows are connected to create a polygon, the plasticity of a primitive species might resemble the following:

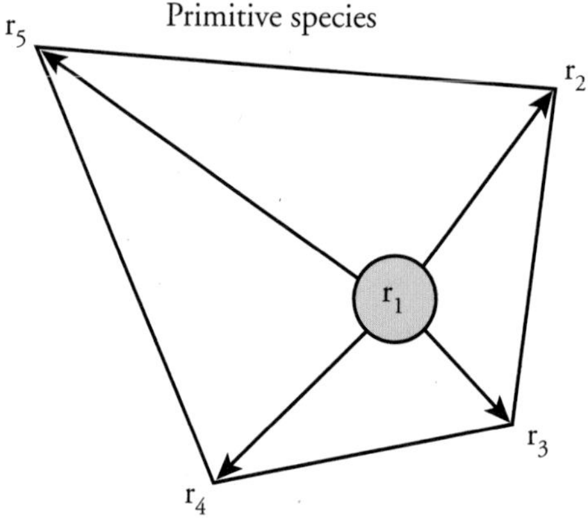

Thus, in the primitive species, division of labor is weak. A distracted worker is likely to switch to another role of any kind. In contrast, the labor polygon of an advanced species, with more elaborate social organization and more specialized castes, can be expected to be more constricted and less symmetrical, as in the following imaginary example:

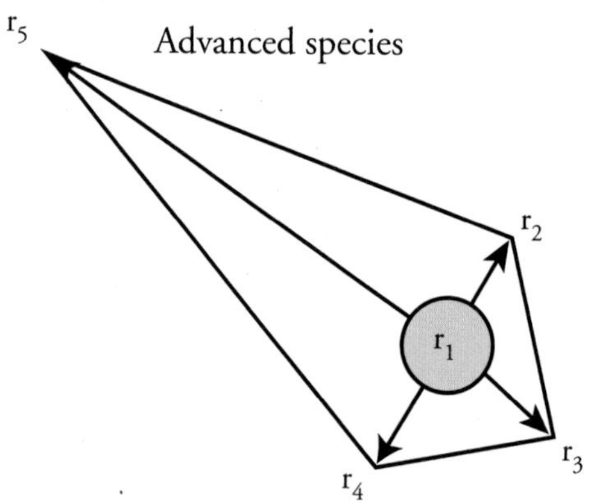

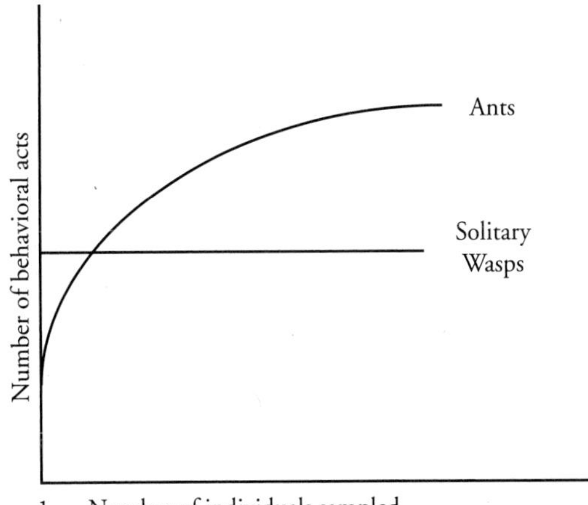

FIGURE 5-8. Because of the division of labor, the number of behavioral acts increases with an increase in the number of ants sampled, but this is not the case for solitary wasps.

Whatever the precise shape of the labor polygons, the evolution of labor division among species of social insects has been one of shrinking plasticity of individual behavior. With an increase in complexity comes more specialized castes. The tightening of the temporal distribution of labor is likely mediated in each colony member by cycles of information flow.

With an increase in the division of labor during the course of evolution, as evidenced by the comparison of species, the average repertoire size of each worker has decreased. At the same time, the repertoire of the colony as a whole has typically increased only slightly beyond that of the queen who founded the colony alone. The repertoire of the founding queen resembles that of solitary species with otherwise similar life cycles.[86] The principle can be represented abstractly for the case of an ant colony versus solitary wasps of a species that builds nests, then raises the young birdlike with repeated feedings (Figure 5-8).

Social life has not been built on increases in the complexity of individual behavior. It has been built instead on specialization among individuals to achieve with greater force and effectiveness a similar degree of complexity as that already programmed in the solitary ancestors. In evidence is the fact that most of the ecological breakthroughs achieved by colonies of eusocial insects have also been attained by solitary insects, as illustrated in Table 5-2.

CHILD LABOR

A common trend among the social insects is toward a tightening of temporal specialization. The ultimate form such evolution takes is not among the adults but

86 | G. F. Oster and E. O. Wilson, *Caste and Ecology in the Social Insects* (Princeton, NJ: Princeton University Press, 1978).

TABLE 5-2. *Similar adaptations in superorganisms and solitary organisms*

ADAPTATION	SUPERORGANISMS	SOLITARY ORGANISMS
Growing fungi for food within nests	Gardening ants (Attini) and termites (Macrotermitinae)	Scolytid beetles (3 genera)
Keeping homopteran "cattle" for their honeydew excretions	Many genera of ants	Silvanid beetles (*Coccidotrophus, Eunousibus*)
Mass predation	Army ants	Many nonsocial predators
Novel, elaborate defense techniques	Many kinds of eusocial insects	Many kinds of nonsocial insects

between life stages. Such caste systems are the rule in termites, where individual development is gradual, such that immature colony members are basically similar in body form to adults, possessing functional antennae and legs upon hatching from the egg. Termite child labor is made more complex in many of the species of the "higher" termites (members of the phylogenetically advanced family Termitidae) by a further division of labor based on sex. In one of the most elaborate systems, found in a species of the tropical African genus *Trinervitermes*, an immature male in the first instar can molt to become a small, ordinary worker and then proceed on through two molts to become a large soldier. The first-instar *Trinervitermes* male can also molt into a soldier larva, proceed through two more molts to become a small presoldier, and then another that turns it into a fully developed small soldier. All females of the *Trinervitermes* society become large workers.[87]

In contrast to the versatile termites, the social hymenopterans start life as nearly helpless larvae lacking antennae and legs. As a result, colonies have much less opportunity to recruit their children into labor. Further, metamorphosis to the adult form is accomplished by passage through a quiescent pupal stage. Yet even with these limitations, larvae of many hymenopterous species participate in cooperation and division of labor, and in some cases they deserve to be considered distinct castes. Wasp larvae

---

87 | C. Noirot, "Formation of castes in the higher termites," in K. Krishna and F. M. Weesner, eds., *Biology of Termites*, Vol. 1 (New York: Academic Press, 1969), pp. 311–350; and T. Abe, D. E. Bignell, and M. Higashi, eds., *Termites: Evolution, Sociality, Symbioses, Ecology* (Boston: Kluwer Academic, 2000).

of the common temperate-zone genera *Vespa* and *Vespula* supply the adult workers with salivary secretions rich in trehalose, glucose, and other carbohydrates, which they convert from the tissues of insect prey fed them by the workers.[88] In addition, salivary glands of many social insect larvae contain free amino acids.[89] This metabolic division of labor is especially advanced in the Oriental hornet *Vespa orientalis*, whose adults are incapable of such enzymatic conversion of tissue and hence depend on the larvae for the sugars vital for their daily energy needs.[90] Workers of the pharaoh ant, a tropical species adapted to dry habitats (and a common household pest), harvest salivary secretions from the larvae. During times of extreme water shortage, they survive longer if given access to even a small number of these immature forms.[91]

The offering of regurgitated liquid food, called stomodeal trophallaxis, by ant larvae has been observed in nearly all of the major subfamilies. (The single exception is Dolichoderinae.[92]) The larval secretions of at least one species, *Temnothorax* (formerly *Leptothorax*) *curvispinosus*, an American woodland ant, are avidly consumed by queens who "graze" from one larva to another, and it appears probable that the queens receive much of their sustenance by this means.[93] The last-instar larvae of the Australian harvester ant *Monomorium* (formerly *Chelaner*) *rothensteini* convert seeds given them to secretions that are fed back to the workers.[94] The origin of these secretions is most likely the paired salivary glands, the only well-developed exocrine glands that open into the mouth of larvae. Deby Cassill and her colleagues recently discovered a novel form of digestive cooperation between larvae and their adult nestmates. In the ant *Pheidole spadonia*, native to the southwestern United States,

---

88 | U. Maschwitz, "Das Speichelsekret der Wespenlarven und seine biologische Bedeutung," *Zeitschrift für vergleichende Physiologie* 53(3): 228–252 (1966).

89 | J. H. Hunt, "Adult nourishment during larval provisioning in a primitively eusocial wasp, *Polistes metricus* Say," *Insectes Sociaux* 31(4): 452–460 (1984); and J. H. Hunt, "Nourishment and social evolution in wasps *sensu lato*," in J. H. Hunt and C. A. Nalepa, eds., *Nourishment and Evolution in Insect Societies* (Boulder, CO: Westview Press, 1994), pp. 211–244.

90 | J. Ishay and R. Ikan, "Gluconeogenesis in the Oriental hornet *Vespa orientalis* F.," *Ecology* 49(1): 169–171 (1968).

91 | M. Wüst, "Stomodeal und proctodeal Sekrete von Ameisenlarven und ihre biologische Bedeutung," *Proceedings of the Seventh Congress of the International Union for the Study of Social Insects, London* (1973), pp. 412–417.

92 | B. Hölldobler and E. O. Wilson, *The Ants* (Cambridge, MA: The Belknap Press of Harvard University Press, 1990).

93 | E. O. Wilson, "Aversive behavior and competition within colonies of the ant *Leptothorax curvispinosus*," *Annals of the Entomological Society of America* 67(5): 777–780 (1974).

94 | E. A. Davison, "Seed utilization by harvester ants," in R. C. Buckley, ed., *Ant-Plant Interactions in Australia* (The Hague: Dr. W. Junk, 1982), pp. 1–6.

workers cut prey into small fragments and place them into little hairy depressions near the mouth of larvae in a late stage of development. By secreting enzymes, presumably from the labial gland, onto the pieces and employing further mastication, the larvae dissolve the food into a liquid, or slurry, for distribution among other colony members. The larvae apparently do not imbibe the nutritious liquid themselves. Instead, workers collect the prey tissue dissolved by the larvae in their crops and then regurgitate it to other larvae and to workers. Cassill and her colleagues measured time budgets invested for preparing the nutritious liquid and found that on average 5 workers and 22 larvae invested 12.8 hours into cutting up and predigesting one fly carcass. The larvae contributed a mean total of 9.5 hours to dissolve the solid fragments.[95]

A very different mode of trophallaxis between larvae and adult ants was discovered by Keichi Masuko in the tiny migratory ant *Leptanilla japonica*.[96] The larvae of this species have a specialized duct organ on each side of the third abdominal segment. Adult ants imbibe larval hemolymph directly from the organ. The queen in particular seems to feed exclusively on this material. Larval hemolymph also provides a source of nutrients for the queen of the amblyoponine species *Amblyopone silvestrii*. In this case, no special organ is involved. The queen punctures the larval skin and imbibes hemolymph from the bleeding wound. This kind of exploitation of larval blood has earned the species that practice it the name Dracula ants.[97]

A wholly different form of interaction between adults and larvae has evolved in the tropical weaver ants. In the genera *Dendromyrmex* and *Oecophylla* and species of *Camponotus* and *Polyrhachis*, workers construct arboreal nests in vegetation with silk drawn from late-instar larvae. The workers hold these larger individuals by the middle of their body and swing them back and forth like shuttles. Thus handled, the larvae release sticky threads of silk.[98]

---

95 | D. L. Cassill, J. Butler, S. B. Vinson, and D. E. Wheeler, "Cooperation during prey digestion between workers and larvae in the ant, *Pheidole spadonia*," *Insectes Sociaux* 52(4): 339–343 (2005).

96 | K. Masuko, "*Leptanilla japonica*: the first bionomic information on the enigmatic ant subfamily Leptanillinae," in J. Eder and H. Rembold, eds., *Chemistry and Biology of Social Insects* (Proceedings of the Tenth Congress of the International Union for the Study of Social Insects, Munich, 18–22 August 1986) (Munich: Verlag J. Peperny, 1987), pp. 597–598.

97 | K. Masuko, "Larval hemolymph feeding: a nondestructive parental cannibalism in the primitive ant *Amblyopone silvestrii* Wheeler (Hymenoptera: Formicidae)," *Behavioral Ecology and Sociobiology* 19(4): 249–255 (1986).

98 | The large literature on the weaver ants is reviewed in B. Hölldobler and E. O. Wilson, *The Ants* (Cambridge, MA: The Belknap Press of Harvard University Press, 1990).

Other than a few such strange and scattered adaptations, however, the physical castes of the hymenopterans are (as far as we know) limited to the social imago. The most fundamental and nearly universal of the caste systems is the division between reproductive females, called queens or gynes, and nonreproductive females, or workers.[99]

GENETIC CASTE DETERMINATION

What destines females in hymenopterous colonies to become queens or workers? The striking differences between these two castes in so many species, lacking any trace of intermediate forms, at first glance suggests genetic control. But genetic determination has proved to be very sparsely distributed and, when present, subject to considerable plasticity in response to environmental conditions. The most notable—and still controversial—example occurs in the tropical stingless bees of the genus *Melipona*, whose queens appear to be full heterozygotes of an unlinked, independently segregating two-locus system. In accord with elementary Mendelian principles, queens emerge as a random one-fourth of the female population of each colony. Yet significant deviations occur when the larvae are underfed, and the plasticity appears adaptive. The turning of genetically determined queens into workers whenever food is scarce is clearly an emergency tactic of obvious benefit to the colony.[100]

It is to be expected that natural selection will ordinarily oppose a genetic control of the queen-versus-worker divergence, even if it remains somewhat plastic, because

---

99 | In many species of social hymenopterans, workers are capable of reproduction, but if so are ordinarily less productive, active only in the absence of the queen, and then often able to lay only unfertilized eggs, which produce males. Those able to replace the queen entirely and then to produce both male and female nestmates are called gamergates; they are relatively common in the ponerine ants (see Chapter 8). A small minority of species possess forms intermediate between queens and workers, called ergatogynes or intercastes. Males are typically transient in the nest and do not contribute labor; hence, they are drones, not true castes. The known exceptions are the assistance of long-lived males of some carpenter ants (*Camponotus*) in the intestinal storage of liquid food, which is passed out to other colony members, and the contribution by male larvae of silk used in nest construction by *Oecophylla* weaver ants.

100 | W. E. Kerr, "Genetic determination of castes in the genus *Melipona*," *Genetics* 35(2): 143–152 (1950); W. E. Kerr and R. A. Nielsen, "Evidences that genetically determined *Melipona* queens can become workers," *Genetics* 54(3): 859–866 (1966); and R. Darchen and B. Délage-Darchen, "Nouvelles expériences concernant le déterminisme des castes chez les Mélipones (Hyménoptères)," *Compte Rendu de l'Academie des Sciences, Paris* 278: 907–910 (1974). The system as described, however, remains controversial: F. L. W. Ratnieks, "Heirs and spares: caste conflict and excess queen production in *Melipona* bees," *Behavioral Ecology and Sociobiology* 50(5): 467–473 (2001).

the sterile worker caste cannot easily transmit its genes to the next generation. Two rare exceptions are the harvester ants *Pogonomyrmex barbatus* and *Pogonomyrmex rugosus* of the American southwestern desert (Plate 11). This case is so unusual and potentially instructive as to deserve a detailed account.

In certain populations of the two harvester ant species, caste is rigidly determined by genotype, and the genetic components responsible for worker-gyne caste differentiation are maintained in two distinct and reproductively isolated lineages.[101] Young queens must mate with at least one male from each lineage to produce the worker caste and the reproductive queen caste. The worker caste results from matings between two different genetic lineages, and it is composed entirely of hybrids (F_1 hybrids) of the two lineages. In contrast, the queen caste results from matings within the same lineage. If a young queen happens to mate only with males of her own lineage, she will never be able to raise a colony, because all her female offspring will be genetically determined to develop into queens, unable to yield the workers needed to build a functional mature colony. On the other hand, if a queen happens to mate only with males from the other lineage, she will be able to produce workers, and the colony can reach maturity. But the queen will only be able to produce male reproductives (which develop from unfertilized eggs), and she will lack the capacity to generate female reproductives (because the queen will not carry sperm of males of her own lineage). Thus, both lineages must be sustained in the population to generate functional colonies.[102] Such lineages are dependent lineages.

Three competing hypotheses have been proposed for the origin of such a dependent lineage system. The first involves interspecific hybridization.[103] It states that dependent lineages are the result of incompatibilities between two interacting nuclear loci that influence caste determination. This explanation suggests that

---

101 | G. E. Julian, J. H. Fewell, J. Gadau, R. A. Johnson, and D. Larrabee, "Genetic determination of the queen caste in an ant hybrid zone," *Proceedings of the National Academy of Sciences USA* 99(12): 8157–8160 (2002); V. P. Volny, and D. M. Gordon, "Genetic basis for queen-worker dimorphism in a social insect," *Proceedings of the National Academy of Sciences USA* 99(9): 6108–6111 (2002); and S. H. Cahan, J. D. Parker, S. W. Rissing, R. A. Johnson, T. S. Polony, M. D. Weiser, and D. R. Smith, "Extreme genetic differences between queens and workers in hybridizing *Pogonomyrmex* harvester ants," *Proceedings of the Royal Society of London B* 269: 1871–1877 (2002).

102 | K. E. Anderson, B. Hölldobler, J. H. Fewell, B. M. Mott, and J. Gadau, "Population-wide lineage frequencies predict genetic load in the seed-harvester ant *Pogonomyrmex*," *Proceedings of the National Academy of Sciences USA* 103(36): 13433–13438 (2006).

103 | S. H. Cahan and L. Keller, "Complex hybrid origin of genetic caste determination in harvester ants," *Nature* 424: 306–309 (2003).

PLATE 11. Mating in the harvester ant *Pogonomyrmex rugosus.* Like most harvester ants studied, queens mate with multiple males. *Above*: A queen is in copula with one male; another male holding onto the female with his mandibles awaits his turn. *Below*: This queen is covered by at least five males struggling to copulate with her.

past hybridization between the environmental caste-determining species *Pogonomyrmex rugosus* and *Pogonomyrmex barbatus* produced two distinct F_3 generations that then stabilized as dependent lineages. Under this model, the two stable hybrid lineages are each fixed for a combination of alleles from each parental species, and worker development occurs when each allele has a partner. However, it is unclear how two distinct F_3 generations would stabilize in a sympatric setting because this requires a transient stage in which F_1 double heterozygotes (interlineage genomes) develop as fully fertile gynes.[104]

A second model states that dependent lineage systems originated via interactions between the cytoplasm and the nuclear genes, such that some cytonuclear combinations develop into gynes while others develop into workers.[105] But in many dependent lineage populations, nearly identical mitochondrial DNA sequences indicate that the cytoplasm may function similarly among lineages. The evidence suggests that both queen and worker castes can develop in the same cytoplasm.[106]

A third model suggests that the origins of the genetic caste determination and dependent lineage systems are due to mutations of a genetic segment with a major influence on caste determination.[107] Because this mutation precludes worker development and results in a queen phenotype, it becomes disproportionately represented in reproductive individuals. Thus, it behaves as a "selfish" genetic element, promoting its own survival at the expense of other parts of the genome. Within a population, this mutation would result in strong selection to retain the worker caste, and one resolution may be the evolution of dependent lineages.

What makes this remarkable genetic caste determination even more intriguing is the fact that there also exist populations of the harvester ant species that have an environmental caste determination. In these populations, colonies depend on the quality and quantity of food the larvae receive from the workers at specific larval

---

104 | T. A. Linksvayer, M. J. Wade, and D. M. Gordon, "Genetic caste determination in harvester ants: possible origin and maintenance by cyto-nuclear epistasis," *Ecology* 87(9): 2185–2193 (2006).

105 | T. A. Linksvayer, M. J. Wade, and D. M. Gordon, "Genetic caste determination in harvester ants: possible origin and maintenance by cyto-nuclear epistasis," *Ecology* 87(9): 2185–2193 (2006).

106 | K. E. Anderson, J. Gadau, B. M. Mott, R. A. Johnson, A. Altamirano, C. Strehl, and J. H. Fewell, "Distribution and evolution of genetic caste determination in *Pogonomyrmex* seed-harvester ants," *Ecology* 87(9): 2171–2184 (2006).

107 | K. E. Anderson, J. Gadau, B. M. Mott, R. A. Johnson, A. Altamirano, C. Strehl, and J. H. Fewell, "Distribution and evolution of genetic caste determination in *Pogonomyrmex* seed-harvester ants," *Ecology* 87(9): 2171–2184 (2006).

stages, which induce them to develop into either worker or queen castes.[108] This interpopulation heterogeneity raises the question of how the dependent lineage system is maintained, considering the genetic load that arises from such a system, which includes the risk of queens producing only reproductive females, or only workers, or of colonies producing reproductive offspring at times of the year when no mating flights occur.

In all *Pogonomyrmex* species studied, whether they have environmental or genetic determination, virgin queens mate with multiple males in large mating aggregations. During the annual cycle of both *Pogonomyrmex barbatus* and *Pogonomyrmex rugosus*, colonies produce a discrete pulse of reproductive winged females and males that depart the nests following summer rains. Males and females from many nests in the population fly to specific mating areas, where females mate with several males.[109] Once mated, the female departs from the mating aggregations, breaks off her wings, and then founds a new colony independently. She does not forage, but instead raises the first workers with reserves fed her by workers in the colony of her birth. Now she metabolizes these reserves, comprising materials stored in the fat body and the wing muscles.[110]

It is obvious that the perpetuation of a dependent lineage system relies on insemination by multiple males, because queens must mate with an intralineage male to generate gynes (females able to become queens) and an interlineage male to generate workers. Although both queens and males might increase their fitness by recognizing and mating with their own lineages, there is no evidence that such mate choice occurs. On the contrary, all existing data suggest that queens both mate and use sperm randomly. Thus, the genetic caste determination is governed by population dynamics in which colony success is a function in large part of the relative frequency of each lineage and levels of polyandry.

As already indicated, dependent lineage populations are subject to severe

---

108 | D. E. Wheeler, "Developmental and physiological determinants of caste in social Hymenoptera: evolutionary implications," *American Naturalist* 12(1)8: 13–34 (1986).

109 | B. Hölldobler, "The behavioral ecology of mating in harvester ants (Hymenoptera: Formicidae: *Pogonomyrmex*)," *Behavioral Ecology and Sociobiology* 1(4): 405–423 (1976); and A. J. Abell, B. J. Cole, R. Reyes, and D. C. Wiernasz, "Sexual selection on body size and shape in the western harvester ant, *Pogonomyrmex occidentalis* Cresson," *Evolution* 53(2): 535–545 (1999).

110 | D. A. Hahn, R. A. Johnson, N. A. Buck, and D. E. Wheeler, "Storage protein content as a functional marker for colony-founding strategies: a comparative study within the harvester ant genus *Pogonomyrmex*," *Physiological and Biochemical Zoology* 77(1): 100–108 (2004).

selection at two points in the life cycle. The first is founding, the success of which increases with the rarity of a lineage (negative frequency dependence).[111] The second is reproduction: some proportion of rare-lineage queens will mate only with alternate-lineage males, and all diploid (fertilized) eggs are genetically predestined to develop as workers. Because gyne-destined genotypes impose a huge cost during colony founding, the colonies produced have no genetic load early in the life cycle, but produce only haploid males at colony maturity. This suggests that lineage-specific sex ratios are an important factor in the recovery of the rare lineage and may establish a deterministic set point from which the rare lineage cannot recover.[112]

The colony-founding costs associated with the genetic caste determination in harvester ants are predicted by the relative frequency of each lineage, which is stabilized in part by negative dependent selection. The rare lineage enjoys greater colony-founding success because it randomly acquires more interlineage sperm and produces a strong initial workforce. However, following colony maturity, gyne production depends on the ratio of same-lineage sperm to alternate-lineage sperm acquired by the queen during the mating swarm. As lineage frequencies become increasingly skewed, the common lineage can be expected to acquire gyne-biased rather than worker-biased sperm stores and have low colony-founding success, but nearly all such colonies surviving to maturity will have mated with a same-lineage male and can produce gynes.

The cost of producing queen-destined genotypes during the ergonomic (exclusively worker-producing) phase of colony development may be offset if heterozygous workers develop more rapidly or are more resistant to disease. However, there is no current evidence for worker heterosis in *Pogonomyrmex*. Still, it remains possible that a "hybrid" workforce may have a competitive advantage in some environments. In fact, Glennis Julian and Sara Helms Cahan did find distinct behavioral differences between "normal" and genetically determined *Pogonomyrmex* colonies whenever there was no clear ecological advantage to having an interlineage workforce.[113]

---

111 | S. H. Cahan, G. E. Julian, T. Schwander, and L. Keller, "Reproductive isolation between *Pogonomyrmex rugosus* and two lineages with genetic caste determination," *Ecology* 87(9): 2160–2170 (2006).

112 | K. E. Anderson, B. Hölldobler, J. H. Fewell, B. M. Mott, and J. Gadau, "Population-wide lineage frequencies predict genetic load in the seed-harvester ant *Pogonomyrmex*," *Proceedings of the National Academy of Sciences USA* 103(36): 13433–13438 (2006).

113 | G. Julian and S. Helms Cahan, "Behavioral differences between *Pogonomyrmex rugosus* and dependent lineage (H1/H2) harvester ants," *Ecology* 87(9): 2207–2214 (2006).

Thus, at present (and to conclude the account of this bizarre and puzzling caste system), we do not have a good explanation for why there are harvester ant populations with an environmental caste determination system and others with a genetic caste determination system. The system may in fact be nonadaptive and a rapidly deteriorating state that has resulted from the overlapping of two distinct but only partly isolated species.

In contrast to the rarity and possibly aberrant causation of genetic queen-worker differences, the genetic determination of different forms of queens is quite compatible with natural selection theory. It is evidently common, at least among species of the myrmicine tribe Leptothoracini. In the European slave-making ant *Harpagoxenus sublaevis*, ergatogynes (worker-like reproductive females, also called intercastes or intermorphs) differ from regular winged queens by what appears to be a single recessive allele.[114] The difference between large and small forms of winged queens of the closely related ant genus *Temnothorax* have also been found to have a genetic basis. The results have been explained by the hypothesis that these species employ multiple mating and multiple colony-forming strategies fitted to different degrees of dispersal ability.[115] Genetically mediated queen polymorphism has also been demonstrated in the myrmicine ant species *Myrmecina graminicola*.[116]

Still another remarkable system of caste determination has been described for the formicine ant *Cataglyphis cursor*, where the queen caste is almost exclusively produced by thelytokous parthenogenesis (development from unfertilized diploid eggs), whereas workers develop from fertilized eggs. It can be argued that producing workers (the somatic units of the colony) by bisexual reproduction ensures genetic diversity of the somatic body of the superorganism, and the parthenogenetic production of the female reproductive caste ensures the full propagation of the germ line of the superorganism.[117]

---

114 | A. Buschinger, "Eine genetische Komponente im Polymorphismus der dulotische Ameise *Harpagoxenus sublaevis*," *Naturwissenschaften* 62(5): 239–240 (1975).

115 | O. Rüppell, J. Heinze, and B. Hölldobler, "Genetic and social structure of the queen size dimorphic ant *Leptothorax* cf. *andrei*," *Ecological Entomology* 26(1): 76–82 (2001); O. Rüppell, J. Heinze, and B. Hölldobler, "Complex determination of queen body size in the queen size dimorphic ant *Leptothorax rugatulus* (Formicidae: Hymenoptera)," *Heredity* 87(1): 33–40 (2001).

116 | A. Buschinger, "Experimental evidence for genetically mediated queen polymorphism in the ant species *Myrmecina graminicola* (Hymenoptera: Formicidae)," *Entomologia Generalis* 27(3–4): 185–200 (2005).

117 | M. Pearcy, S. Aron, C. Doums, and L. Keller, "Conditional use of sex and parthenogenesis for worker and queen production in ants." *Science* 306: 1780–1782 (2004).

NONGENETIC CASTE DETERMINATION

To summarize to this point, in the overwhelming majority of social insect species, only rarely are differences between reproductive queens and nonreproductive workers based on genetic differences among members of the same colony. In general, all colony members have the same genotype for caste formation, from which differences in the environment launch a newborn individual on a developmental pathway that leads to its final adult destination as either a queen or a worker. The genes, in other words, determine not castes but caste plasticity: in response to environmental conditions, they either turn on or turn off growth along the development of immature individuals, and by this commitment or default they guide progress, step by step, toward one caste or another. The most primary such dichotomy in social insects leads the individual to maturity as a queen or as a worker.[118]

Which environmental factors moderate gene expression in the formation of castes? Five decades of research have implicated six kinds of stimuli that operate singly or in various combinations according to species.[119]

1 | *Larval nutrition.* In honeybees, the future queens are housed in royal cells and shunted to this caste by frequent meals of royal jelly, a secretion

---

118 | J. D. Evans and D. E. Wheeler, "Differential gene expression between developing queens and workers in the honey bee, *Apis mellifera*," *Proceedings of the National Academy of Sciences USA* 96(10): 5575–5580 (1999). Some of the genetic control of the divergence of the queen and worker pathways, that producing wings (for queens) or winglessness (for workers), has been worked out for ants by E. Abouheif and G. A. Wray, "Evolution of the gene network underlying wing polyphenism in ants," *Science* 297: 249–252 (2002). Certain genes within the network needed for growth are shut down as part of worker development, and which genes are thus labile vary from one genus of ant to another. A similar pathway of gene expression in the production of soldiers in a Japanese termite has been identified by T. Miura, A. Kamikouchi, M. Sawata, H. Takeuchi, S. Natori, T. Kubo, and T. Masumoto, "Soldier caste-specific gene expression in the mandibular glands of *Hodotermopsis japonica* (Isoptera: Termopsidae)," *Proceedings of the National Academy of Sciences USA* 96(24): 13874–13879 (1999).

119 | Reviewed in more detail, for example, in E. O. Wilson, *The Insect Societies* (Cambridge, MA: The Belknap Press of Harvard University Press, 1971); D. E. Wheeler, "Developmental and physiological determinants of caste in social Hymenoptera: evolutionary implications," *American Naturalist* 128(1): 13–34 (1986); J. D. Evans and D. E. Wheeler, "Gene expression and the evolution of insect polyphenisms," *BioEssays* 23(1): 62–68 (2001); and B. Hölldobler and E. O. Wilson, *The Ants* (Cambridge, MA: The Belknap Press of Harvard University Press, 1990). An especially large debt is owed the British myrmecologist Michael V. Brian and his coworkers, whose pioneering and meticulously detailed work on *Myrmica* ants extended from 1951 through the mid-1980s, pointing the way for much that was to follow. For a thorough recent review of many of the questions concerning phenotypic plasticity, see M. J. West-Eberhard, "Wasp societies as microcosms for the study of development and evolution," in S. Turillazzi and M. J. West-Eberhard, eds., *Natural History and Evolution of Paper-Wasps* (New York: Oxford University Press, 1996), pp. 290–317.

produced primarily in the hypopharyngeal glands of the nurse workers. In some ant species, female larvae fed well enough to reach a threshold size by a certain age move on to the queen pathway. The others continue by default to maturity as workers.

2 | *Temperature.* The female larvae in the north temperate ant genera *Formica* and *Myrmica* tend to develop into queens more readily if reared at a temperature close to optimal for larval growth.

3 | *Winter chilling.* Eggs of *Formica* and larvae of *Myrmica* tend to mature into queens if chilled for a time, simulating a period of overwintering and making it more likely that their colonies will produce a crop of new queens in the spring.

4 | *Caste self-inhibition.* The presence of a mother queen inhibits the production of new queens in ants, honeybees, and termites. Similarly, in at least some species of ants and termites, the presence of a large number of soldiers inhibits the production of additional soldiers.[120] These negative-feedback loops, mediated by pheromones, work to stabilize caste ratios in colonies as a whole.

5 | *Egg size.* In at least several genera of ants, namely *Formica, Myrmica,* and *Pheidole,* the more yolk in the eggs at the outset, the higher proportion of eggs that develop into queens rather than workers.

6 | *Age of queen.* In at least one genus, *Myrmica,* young mother queens tend to produce fewer queens than do old mother queens.

From existing evidence, there is every reason to believe, and no reason to doubt, that sensitivity of the several proximate factors has evolved in response to natural selection at the colony level. They guide the manufacture of virgin queens to fit the life cycle of individual species. The result is the massing of the virgin queens, along with males, for release in the time of year favorable to the founding of new colonies (Figure 5-9). For many species of ants, and likely all species of honeybees and termites, the factors also provide the means to create new nest queens when the older ones die or else leave with part of the worker force to new sites.

---

120 | For example, see D. E. Wheeler and H. F. Nijhout, "Soldier determination in *Pheidole bicarinata*: inhibition by adult soldiers," *Journal of Insect Physiology* 30(2): 127–135 (1984).

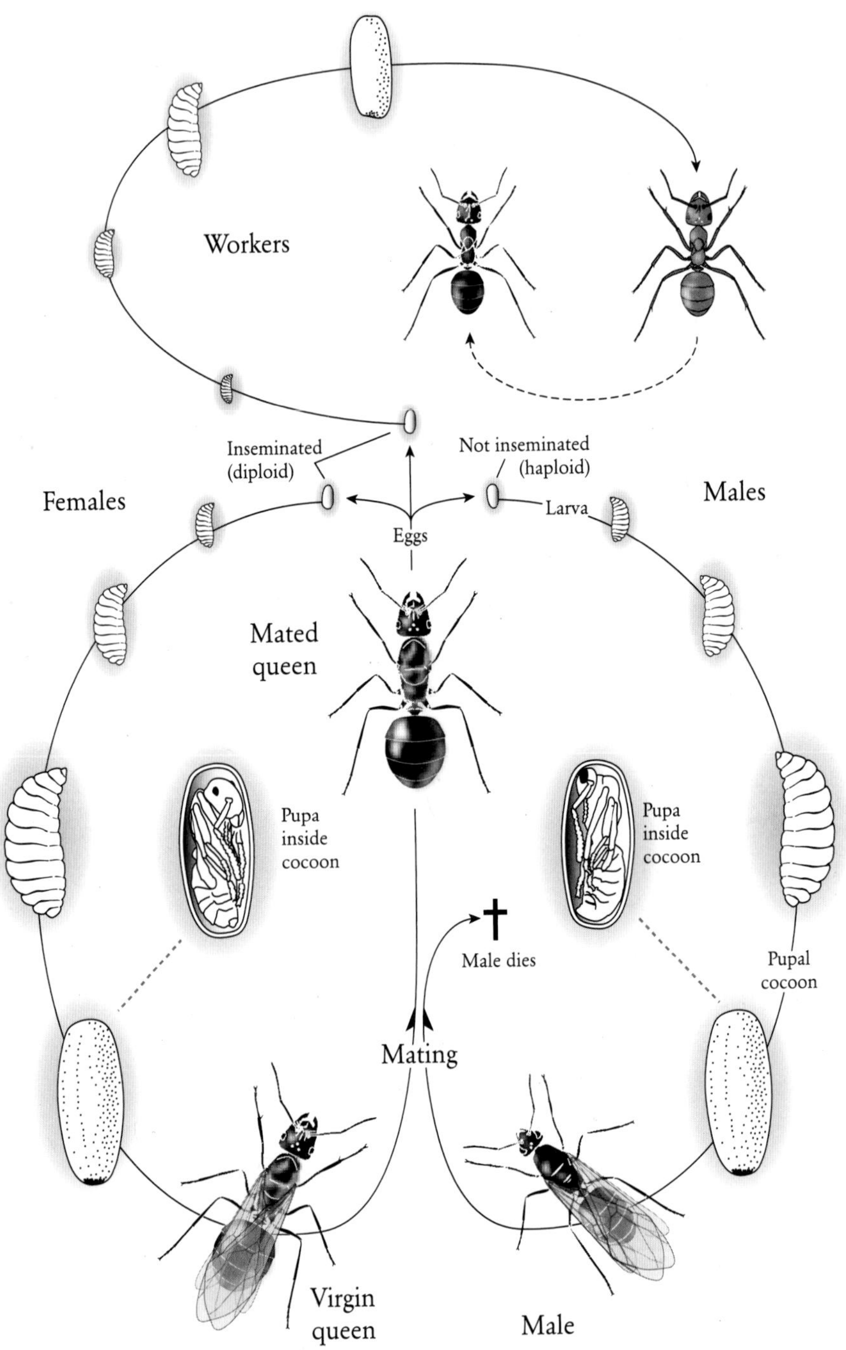

FIGURE 5-9. A basic ant colony life cycle, as illustrated by the European species *Formica polyctena*. Based on an original drawing by Turid Hölldobler-Forsyth from K. Gösswald, *Die Waldameise*, Vol. 1 (Wiesbaden: Aula Verlag, 1989).

WORKER SUBCASTES

The final step in the evolution of caste in the social insects—extending beyond the division of labor based on age—is the origin of physical worker subcastes. At this level, the division of labor is tightened further. In termites, physical subcastes comprise, according to species, various combinations of small workers, large workers, small soldiers, large soldiers, and worker-like but potentially reproductive intercastes (intermorphs). In many ant species, physical subcastes comprise combinations of minor workers, majors (also called soldiers), and supermajors (also called supersoldiers). As a rule, the soldiers are specialized for defense or other restricted functions; moreover, they do not pass through a succession of labor roles with age, as do ordinary workers (Plates 12 to 14).

Physical worker subcastes are not universal in the social insects. In fact, they are the exception rather than the rule and spottily distributed among different phylogenetic groups. Physical subcastes are virtually absent in the bees and wasps. They occur throughout the termites but have disappeared secondarily in the genus *Anoplotermes*. Of the 296 known living genera of ants recognized in 1995,[121] only 46, or 15 percent, have anatomical subcastes.[122] The most elaborate systems, containing supermajors in addition to majors, occur in only 5 genera: *Atta*, *Daceton*, *Eciton*, *Pheidologeton*, and *Pheidole* (and in the last, fewer than 2 percent of the hundreds of species).

The ecological forces favoring worker subcastes are only partly understood. For those species with strongly differentiated majors, whether accompanied by supermajors or not, the origin of the system is generally linked to extreme environmental adaptations by the species. In ants, the ecological pressures are the following, according to species:[123]

---

121 | The number of living ant genera is taken from Barry Bolton's monumental *A New General Catalogue of the Ants of the World* (Cambridge, MA: Harvard University Press, 1995). The number of taxonomic tribes these genera composed at publication time of the *Catalogue* was 16, and the number of species was 9,536. The roster of species has since risen to above 14,000 and may eventually reach twice this number.

122 | The genera known to have physical subcastes are listed by G. F. Oster and E. O. Wilson, *Caste and Ecology in the Social Insects* (Princeton, NJ: Princeton University Press, 1978).

123 | Reviewed more fully in B. Hölldobler and E. O. Wilson, *The Ants* (Cambridge, MA: The Belknap Press of Harvard University Press, 1990).

PLATE 12. The South American army ant *Eciton hamatum*. *Above*: Workers in an emigration column carrying brood. *Below*: A soldier with the characteristic sickle-shaped mandibles of her caste.

PLATE 13. Some dacetine species differentiate castes by an unusual form of allometry. *Above*: *Daceton armigerum*. *Below*: *Orectognathus versicolor*. The *Orectognathus* queen is at the far left; to her front are two soldiers (major workers), possessing heads larger than that of the queen and mandibles that are broadened and flattened; another soldier is at the extreme upper right.

PLATE 14. The greatest size variation within the worker caste of all known ants (and social insects generally) is that of the Asian marauder ant *Pheidologeton diversus*. *Above*: Minor, major, and supermajor. *Below*: A minor worker licks the head of a supermajor.

- *Unusually complex operations.* The processing of fresh vegetation as a substrate for fungus growing, a unique adaptation of gardening ant genera *Acromyrmex* and *Atta*, requires physical subcastes.

- *Pursuit of large quantities of diverse prey.* Long-mandibled soldiers employed to hunt large prey and defend nestmates during group foraging are important adaptations in the African driver ants (*Dorylus*) and New World tropical army ants (*Eciton*).

- *"Hoplite defense".* Soldiers are often equipped with strong anatomical devices for combat, such as the saber-shaped mandibles in *Cataglyphis bombycina*; the shield-shaped or stopper-shaped heads of majors of *Colobopsis* and some species of *Cephalotes* and *Pheidole*, used as living gates to protect nests from invasion; and the swollen muscle-filled heads, often equipped with horns and spikes, used in combat by species of *Pheidole*, *Camponotus*, and other genera.

- Milling seeds. Majors equipped with large heads, swollen with adductor muscles, are used to break open seeds in species of *Acanthomyrmex*, *Pheidole*, and at least one species of fire ant, *Solenopsis geminata*.

- *Liquid food storage.* Distensible crops (posterior division of the foregut) of majors, in which liquid food is stored, then regurgitated to nestmates in times of shortage, occur in the ant genera *Oligomyrmex* (formerly *Erebomyrma* in part) and *Myrmecocystus* (Plate 15) and in some species of *Camponotus* and *Pheidole*. In some species, for example *Acanthomyrmex ferox*[124] and *Crematogaster smithi*,[125] majors produce trophic eggs that are fed to larvae and the queen.

- *Weaving.* Major workers of *Oecophylla*, besides guarding the colony with suicidally aggressive behavior, build the nest with silk donated by mature larvae held in their jaws.

Only careful studies of the anatomy and entire life cycle and social behavior of a species may suffice to reveal the full significance of its caste system. As an example, consider the "big-headed ants" of the worldwide ant genus *Pheidole*, among the

---

124 | B. Gobin and F. Ito, "Sumo wrestling in ants: major workers fight over male production in *Acanthomyrmex ferox*," *Naturwissenschaften* 90(7): 318–321 (2003).

125 | J. Heinze, S. Foitzik, B. Oberstadt, O. Rüppell, and B. Hölldobler, "A female caste specialized for the production of unfertilized eggs in the ant *Crematogaster smithi*," *Naturwissenschaften* 86(2): 93–95 (1999).

PLATE 15. Honeypots of *Myrmecocystus mendax.* Liquid food is stored in the distensible crops of some of the majors. Their gasters can reach the size of a pea or cherry. (Photos: Turid Hölldobler-Forsyth.)

largest of all ant genera, with over 650 species known in the Western Hemisphere alone.[126] Its species are also collectively among the most abundant ants in many localities of the warm temperate and tropical zones of both hemispheres. The one trait that distinguishes all known species of *Pheidole* is their unusual large-headed soldiers. This diagnostic feature is associated with a second trait that appears coadapted with it: the reduction of the sting in both the soldier and minor worker subcastes to a vestigial condition. A similar reduction of the sting occurs in the large and prominent worldwide subfamilies Dolichoderinae and Formicinae.

This suite of changes has allowed *Pheidole* colonies to achieve an extreme division of labor in which an unusually heavy reliance for defense is placed on the major caste. The highly specialized soldiers are the surrogate for the missing sting and weaponry of the colony. They constitute a highly mobile strike force (Figure 5-10).

A third trait of *Pheidole* is the absence of ovaries in both the minors and the soldiers. Combined with the thinness of their exoskeleton, the sparseness of their chemical armament, and absence of a functional sting, the minors in particular have the appearance of a "throwaway" caste—small, light, cheaply manufactured, and short-lived.

A fourth trait of *Pheidole* is the ability of the soldiers to serve as an emergency standby substitute for the minors. They are able to take over the labor roles of the minors, albeit clumsily, when the ranks of the smaller caste have been severely depleted.

The typical *Pheidole* colony, in short, is reasonably viewed as a highly resilient superorganism able to expend and replace cheaply manufactured minor workers readily while utilizing the expensive soldier subcaste for defense and as an emergency replacement labor force. The soldiers are especially effective as a defense force. They are mobilized swiftly by minor worker scouts, thus reducing the need for stings or elaborate, energetically expensive chemical systems. These traits, when joined with the relatively small size of the ants and their short colony reproductive time, are thought responsible for the remarkable ecological success and hyperdiversity of *Pheidole*.[127]

---

126 | E. O. Wilson, Pheidole *in the New World: A Dominant, Hyperdiverse Ant Genus* (Cambridge, MA: Harvard University Press, 2003).

127 | E. O. Wilson, *Pheidole in the New World: A Dominant, Hyperdiverse Ant Genus* (Cambridge, MA: Harvard University Press, 2003).

FIGURE 5-10. The genus *Pheidole* is thought to be highly successful at least in part because of soldiers with large heads and sharp mandibles that can be organized into a fast-acting strike force by lightweight, expendable minor workers. This figure depicts soldiers, accompanied by their minor worker nestmates, defending their nest by chopping invader fire ants (*Solenopsis invicta*) to pieces. Based on an original drawing by Sarah Landry in E. O. Wilson, "The organization of colony defense in the ant *Pheidole dentata* Mayr (Hymenoptera: Formicidae)," *Behavioral Ecology and Sociobiology* 1(1): 63–81 (1976).

THE PHYSIOLOGY AND EVOLUTION OF PHYSICAL CASTES

The evolution of physical castes in the ants has been opportunistic in the appropriation of preadaptations and parsimonious in the pathways followed thereafter. This evolution is based on three properties of growth basic to ants and all other holometabolous insects, defined as those whose immature, larval stage of growth is radically different from the final, reproductive adult stage (think of a caterpillar and the butterfly it produces). The properties of caste-determining growth are as follows:

- The size attained by the larva prior to transformation into the adult form is critical.
- The larva-to-adult metamorphosis consists of the growth of masses of tissue that proliferate to form the legs, eyes, gonads, and other body parts of the adult. These imaginal disks, as the tissue masses are called, grow at different rates during the metamorphosis. As a result, the greater the size the larva reaches before metamorphosing, the longer the disks grow, and hence the greater the disproportion of the final organs and other body parts in the adult. Thus, a big larva may produce an adult with a larger head relative to the rest of its body, as compared with the adult produced by a small larva. This process of differential growth, or allometry as it is usually called, can be expressed as the equation $y = bx^a$, or $\log y = \log b + a\log x$, where y is one dimension (say the width of the head in the adult ant), x is a second dimension (such as the width of the adult thorax), and a and b are fitted constants. If the allometry parameters remain the same with increase in size, the relation of the two adult dimensions (for example, widths of head and thorax) forms a monotonically curving line on a regular plot but a straight line when drawn on a logarithmic plot. In more advanced stages of evolution, decision points are added to the growth algorithm, so that when a larva has reached a certain size or greater by a particular time in its overall growth, the allometry parameters increase or decrease (Figure 5-11). As a final result, the smallest minor workers usually end up even more different from the largest major workers than if the magnitude of allometry had stayed constant.

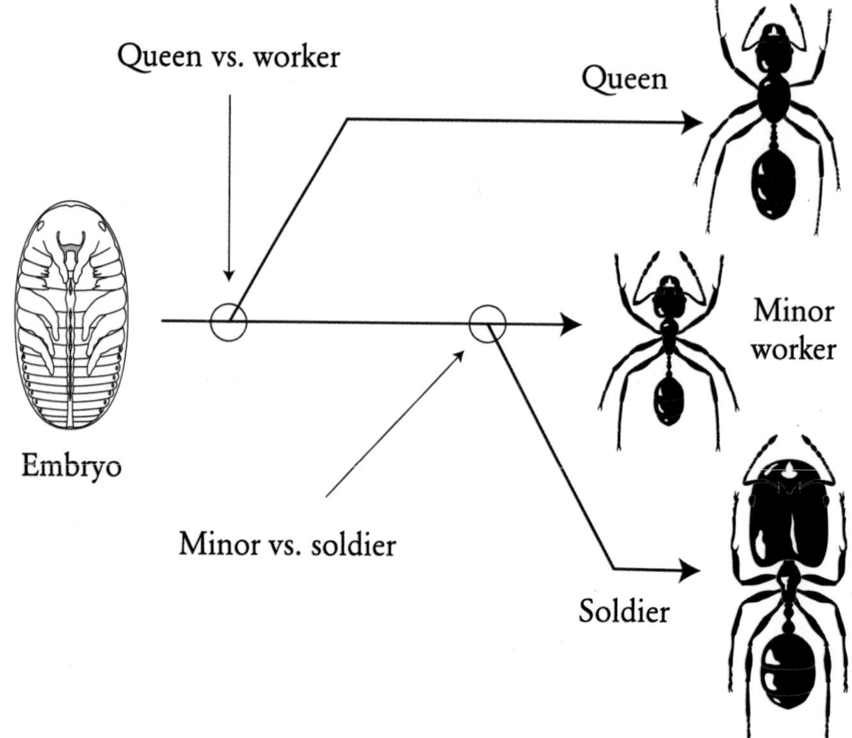

FIGURE 5-11. Worker subcastes are created in evolution by the addition of a second decision point in larval growth. Two periods of hormonal integration during development mediate the development of the queen caste and the two worker subcastes of *Pheidole dentata* ants from one totipotent genome. Based on G. Bloch, D. E. Wheeler, and G. E. Robinson, "Endocrine influences on the organization of insect societies," in D. W. Pfaff, A. P. Arnold, A. M. Etgen, S. E. Fahrbach, and R. T. Rubin, eds., *Hormones, Brain and Behavior*, Vol. 3 (New York: Academic Press, 2002), pp. 195–235.

- At any stage of growth, from the egg through the larval stages, a divergence in the overall growth rate can occur, resulting in a skewing of the final overall size-frequency distribution of the adult members of the colony. The greater this divergence in the adult size-frequency distribution and the greater the allometry that accompanies it, the more distinct are the adult subcastes. The upper two panels of Figure 5-12 represent early stages of such evolution. In them, the size classes—minors, medias, and majors—are overlapping and can be delimited only arbitrarily. In advanced stages of the evolution, which are illustrated in the two lower panels, two modes appear. At first, the two distributions are overlapping. And then (lowermost curve) the medias decline or disappear, leaving discrete minors and majors. In a very small number of evolutionary lines within the ants,

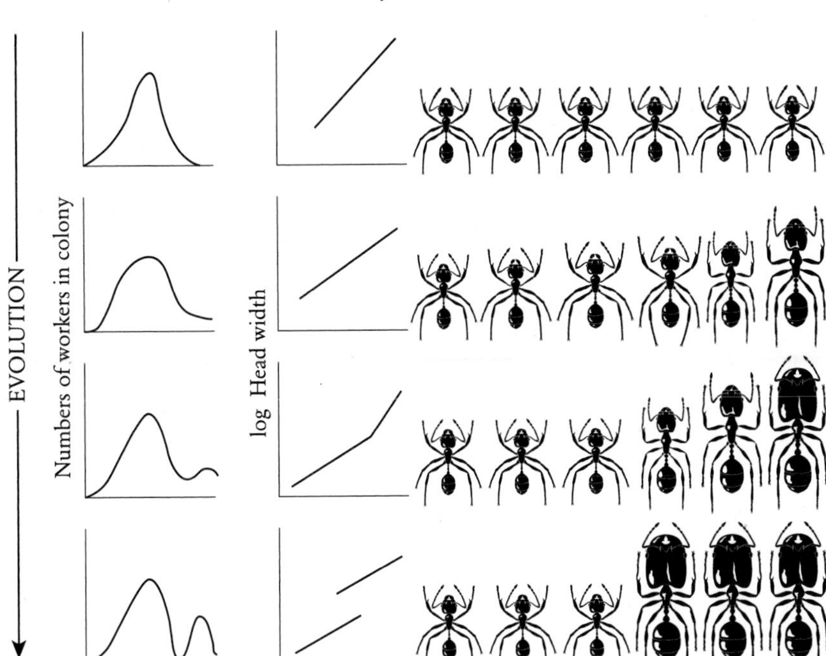

FIGURE 5-12. The evolution of worker subcastes (minors, medias, and majors) occurs by simple changes in the size-frequency distribution, accompanied by differential growth of imaginal (adult-destined) tissue disks during metamorphosis to the adult stage. Based on E. O. Wilson, "The origin and evolution of polymorphism in ants," *The Quarterly Review of Biology*, 28(2): 136–156 (1953).

an extension of the same process has created a supermajor caste to accompany the minor and major castes.[128]

Cesare Baroni Urbani and Luc Passera proposed (in 1996) an alternative explanation for the origin of discrete majors, which they prefer to call soldiers, arguing

---

128 | Charles Darwin, in *The Origin of Species* (1859), suggested that ant castes are based on plasticity of a single genotype, thus removing what he considered the single greatest obstacle to the theory of evolution by natural selection. J. H. Huxley, in *Problems of Relative Growth* (New York: Dial Press, 1932), suggested the role of allometric growth as a simple algorithm of caste differentiation. E. O. Wilson, in 1953, was the first to connect the processes of allometry and size-frequency distribution as the key to caste evolution and to work out the full range of the evolutionary stages reviewed here. These and many later contributions by other authors are reviewed in G. F. Oster and E. O. Wilson, *Caste and Ecology in the Social Insects* (Princeton, NJ: Princeton University Press, 1978), and in B. Hölldobler and E. O. Wilson, *The Ants* (Cambridge, MA: The Belknap Press of Harvard University Press, 1990).

that this caste was derived in evolution directly from the queen caste and not as the end product of secondary allometric evolution within the worker caste.[129] In quick response, however, Philip S. Ward pointed to abundant and far stronger morphometric, developmental, and phylogenetic evidence that supports the long-established reconstruction of a worker caste origin,[130] the earlier evidence for which we have also cited here.

We now come to the origin of the queen and worker castes, which is even more fundamental than the origin of worker subcastes. This event, which occurred in mid-Cretaceous times over 100 million years ago, evidently followed the same steps of coupled allometry (differential growing of different body parts according to size) and size-frequency molding that were later to generate the worker subcastes. When intercastes between workers and queens occur today as an alternative form of the reproductive caste, they display similar properties of allometry and size divergence during larval growth. The same evolutionary sequence appears to have been responsible for the origin of the queen-worker caste systems of the social bees and wasps.[131]

There is, of course, much more to the allometry and size-frequency divergence of castes than simple changes in body proportions. Allometry of imaginal disks and segments of disks occurs throughout the prepupal body to both large and small effect. It shapes multiple traits that distinguish one caste from another, such as the hair patterns, digestive system, sculpturing of the exoskeleton, and circuitry of the brain and sensory organs. It differentiates the defensive secretions, the pheromones used in communication, and the food given the queen and larvae.

A striking example of multiplicity born from the simple rules of growth and development is provided by the worker subcastes of the leafcutter ant *Atta sexdens* (Plate 16). In this most complex of insect social systems (see Chapter 9), the maximum size of the poison gland (containing substances used to recruit nestmates) in proportion to total body size occurs in precisely the size group (median head width 2.2 millimeters) containing the most active outdoor foragers and leafcutters and hence in most need of recruitment ability. Exactly the same peak occurs in

---

129 | C. Baroni Urbani and L. Passera, "Origin of ant soldiers," *Nature* 383: 223 (1996).

130 | P. S. Ward, "Ant soldiers are not modified queens," *Nature* 385: 494–495 (1997).

131 | Reviewed by E. O. Wilson, *The Insect Societies* (Cambridge, MA: The Belknap Press of Harvard University Press, 1971).

PLATE 16. Worker caste polymorphism in the leafcutter ant *Atta cephalotes*. Shown are medium and minor ants in the colony fungus garden.

the length of body spines, used in defense. Among other exocrine glands, the postpharyngeal gland, which produces secretions used in larval feeding, and the paired metapleural glands, which produce antibiotics, are proportionately largest—again as predicted—in the smallest workers, which specialize in brood care and fungus gardening.[132]

Where exactly are the decision points in the development of caste in the immature social hymenopterans? They can arise during evolution almost anywhere from the egg to the prepupal and early pupal stages. At each point, at each fork in the road, the immature stage moves on to one developmental path or the other. If sufficient juvenile hormone is present in the blood (JH+), it ordinarily continues or

132 | E. O. Wilson, "Caste and division of labor in leaf-cutter ants (Hymenoptera: Formicidae: *Atta*), I: The overall pattern in *A. sexdens*," *Behavioral Ecology and Sociobiology* 7(2): 143–156 (1980). No such clear pattern of proportionate size was found in the mandibular gland and Dufour's gland.

even accelerates growth toward the caste with the larger adult body size. If a lower amount of the hormone is present (JH-), maturity is reached sooner, and the destination becomes the smaller adult body size.[133, 134]

ADAPTIVE DEMOGRAPHY

The evidence is conclusive that the developmental programs leading to the physical caste systems of ants are genetically determined (albeit not the caste of any particular worker). Indications are also very strong that the size-frequency distributions undergirding the more complex caste systems are adaptive for the colony as a whole. In support of the latter interpretation, for example, is the correlation of the systems with ecological adaptation among the various species of leafcutter ants. The workers of mature or nearly mature colonies of *Atta* vary continuously over an extremely wide size range. Within this spread, a relatively broad array of medium-sized workers forage for fresh vegetation to be used as substrate for the fungus gardens, and they attack a correspondingly wide variety of leaves and flowers. The largest among them cut the thickest and toughest parts. Colonies of *Acromyrmex coronatus* and small, immature colonies of *Atta* dispatch into the field a narrow range of smaller workers, which concentrate on the soft leaves of herbaceous plants. Finally, colonies of two other species of the same genus, *Acromyrmex octospinosus* and *Acromyrmex volcanus*, possess a distinct two-mode size-frequency distribution. The small workers remain in the nest, while the uniform-sized large

---

133 | D. E. Wheeler and H. F. Nijhout, "Soldier determination in *Pheidole bicarinata*: effect of methoprene on caste and size within castes," *Journal of Insect Physiology* 29(11): 847–854 (1983); D. E. Wheeler and H. F. Nijhout, "Soldier determination in ants: new role for juvenile hormone," *Science* 213: 361–363 (1981); D. E. Wheeler, "The developmental basis of worker caste polymorphism in ants," *American Naturalist* 138(5): 1218–1238 (1991); and G. Bloch, D. E. Wheeler, and G. E. Robinson, "Endocrine influence on the organization of insect societies," in D. W. Pfaff, A. P. Arnold, A. M. Etgen, S. E. Fahrbach, and R. T. Rubin, eds., *Hormones, Brain and Behavior*, Vol. 3 (New York: Academic Press, 2002), pp. 195–235.

134 | The principle of larger body size in queens is not inviolable. Workers and queens of polistine wasps differ substantially in body proportions but very little in overall size; see R. L. Jeanne, C. A. Graf, and B. S. Yandell, "Non-size-based morphological castes in a social insect," *Naturwissenschaften* 82(6): 296–298 (1995). Queens of a few ants, such as those in the Australian predatory ant genus *Orectognathus* and the temporarily parasitic *Formica microgyna* group of North America, are actually smaller than the largest workers. An Australian weaver ant species close to *Polyrhachis doddi* has two types of queens, with one type distinctly smaller than the workers; see J. Heinze, and B. Hölldobler, "Queen polymorphism in an Australian weaver ant *Polyrhachis* cf. doddi," *Psyche* (Cambridge, MA) 100: 83–92 (1993).

workers forage outside to harvest vegetation from small herbaceous plants along with fallen leaves, fruits, and flowers.[135]

An even more persuasive example of adaptive demography of physical castes is provided by developmental changes in colonies of the *Atta* leafcutters during the colony's growth toward maturity, as depicted in Figure 5-13. Beginning colonies, started by a single queen from her own body reserves, have a nearly uniform size-frequency distribution across a relatively narrow head width range of 0.8 to 1.6 millimeters. The key to the arrangement is that workers in the span of 0.8 to 1.0 millimeter are required as gardeners of the symbiotic fungus on which the colony depends, whereas workers with head widths of 1.6 millimeters are the smallest that can cut vegetation of average toughness. This range also embraces the worker size groups most involved in brood care. Thus, remarkably, the queen produces about the minimum number of individuals who together can perform all of the essential colony tasks. As the colony continues growing, the worker size variation broadens

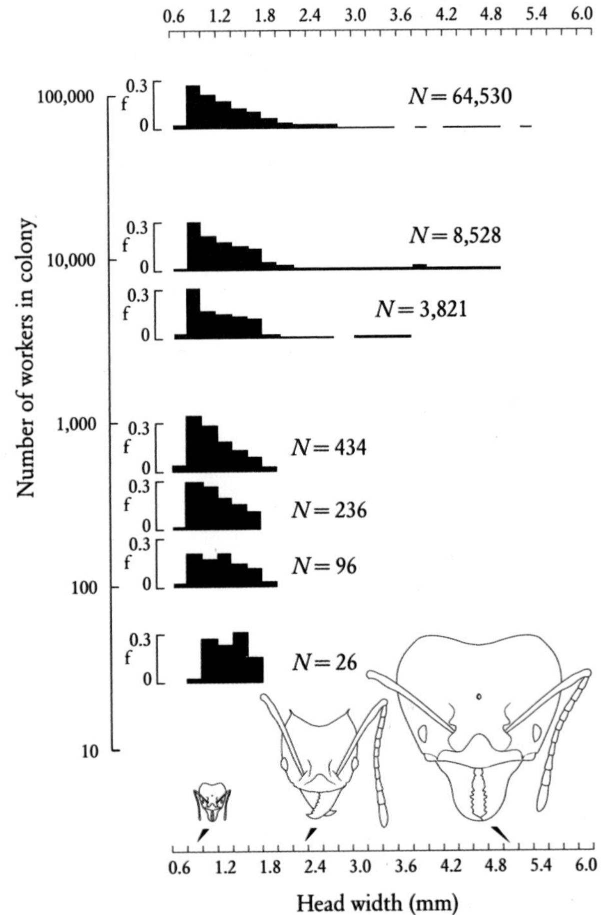

FIGURE 5-13. Sociogenesis of the leafcutter ant *Atta cephalotes*: the ontogeny of the caste system is illustrated here by seven representative colonies collected in the field or reared in the laboratory. The worker caste is differentiated into subcastes by continuous size variation associated with disproportionate growth in various body parts. The number of workers in each colony (*N*) is based on complete censuses; *f* is the frequency of individuals according to size class. The heads of three sizes of workers are shown to illustrate the disproportionate growth. Modified from E. O. Wilson, "The sociogenesis of insect colonies," *Science* 228: 1489–1495 (1985).

135 | J. K. Wetterer, "The ecology and evolution of worker size-distribution in leaf-cutting ants (Hymenoptera: Formicidae)," *Sociobiology* 34(1): 119–144 (1999).

in both directions, to head width 0.7 millimeter or slightly less at the lower end to more than 5.0 millimeters at the upper end, and the frequency distribution becomes more sharply peaked in the smallest size classes. A more physiological question immediately arises from the observed sociogenesis (colony ontogeny) of *Atta*: Which is the more important in determining the size-frequency distribution, the size of the colony or its age? To learn the answer, one of us (Wilson) selected four colonies 3 to 4 years old with about 10,000 workers and reduced the population of each to 236 workers, giving them a size-frequency distribution characteristic of natural young colonies of the same size collected in Costa Rica. It turned out that worker pupae produced at the end of the first brood cycle possessed a size-frequency distribution like that of small, young colonies rather than of larger, older ones. Thus, colony size—and, indirectly, the amount of food produced—are more important than colony age.[136]

Adding yet more strength to the colony-level adaptation hypothesis of physical castes is the existence of dwarf workers, or nanitics or minims as they are often called, in many or perhaps most ant species in which colonies are founded by solitary queens. These tiny workers predominate in the first brood raised by the queens, especially in species in which the queen is fully "claustral," meaning that she remains hidden in a closed chamber while raising the first generation of workers from her fat deposits and catabolized wing muscles. As tiny as the smallest minor workers of subsequent broods or even smaller, the nanitics are also exceptionally timid in comportment. They are the generation that must perform all the tasks, without the help of medias, majors, or even full-sized minors.[137]

Because of their small numbers, nanitics form the generation in which the colony is at highest risk. The loss of even a few could result in the death of the queen and extinction of the incipient colony, a circumstance that probably explains their timidity. Why do they exist at all, instead of a full array of castes? The answer is that the founding queen, with her limited reserves, cannot afford such a balanced portfolio of investments in her first brood. She must produce as many workers as possible, because the loss of even a single worker is very costly to the total productive

---

136 | E. O. Wilson, "Caste and division of labor in leaf-cutter ants (Hymenoptera: Formicidae: *Atta*), IV: Colony ontogeny of *A. cephalotes*," *Behavioral Ecology and Sociobiology* 14(1): 55–60 (1983).

137 | L. A. Wood and W. R. Tschinkel, "Quantification and modification of worker size variation in the fire ant *Solenopsis invicta*," *Insectes Sociaux* 28(2): 117–128 (1981); B. Hölldobler and E. O. Wilson, *The Ants* (Cambridge, MA: The Belknap Press of Harvard University Press, 1990).

capacity of the colony. Also, tiny workers, by collecting small food particles over a limited foraging range, are enough, even in the absence of larger nestmates, to continue growth of the colony in its earliest stages.

All of the evidence accumulated thus far indicates that the physical castes of individual ants are determined predominantly by environmental factors. In species with very complex systems, however, genetic differences among immature colony members may play a role in the division of labor. Queens of the leafcutter ant *Acromyrmex echinatior* and of the Florida harvester ant *Pogonomyrmex badius* mate many times, so that the workers of a given colony of both species are daughters not just of their mother but also of multiple males. It turns out that certain genes vary within the *Acromyrmex* and *Pogonomyrmex* colonies that predispose larvae to reach different sizes at maturity, so that some average notably larger and others smaller than their nestmates. This variation in turn influences the tasks undertaken by the workers.[138] The multiple genetic predispositions may allow colonies to respond to changing needs more quickly and effectively, as suggested earlier for colonies with multiple labor genotypes in the domestic honeybee *Apis mellifera*.[139] Multiple groups of brood that respond to caste-determining factors at different threshold levels could create a labor system more flexible than broods all of which respond uniformly.

Thus, a mixed portfolio of genes can prime colonies to respond quickly to unpredictable needs. Whether such innate readiness occurs widely among the social insects or is limited to a small set of species with highly evolved caste systems and multiple paternity of workers remains to be seen. Either way, there can be little doubt that the target of selection molding caste systems is the emergent properties of the colony. The colony as a whole—the superorganism—contains feedback loops of communication among nestmates that regulate both the proportions of castes and the tasks they undertake. In ants of the genus *Pheidole*, for example, a shortage of major workers activates a loop that stimulates a higher than usual rate of soldier production.[140] The same or a parallel loop is also triggered in colonies of *Pheidole*

---

138 | W. O. H. Hughes, S. Sumner, S. Van Borm, and J. J. Boomsma, "Worker caste polymorphism has a genetic basis in *Acromyrmex* leaf-cutting ants," *Proceedings of the National Academy of Sciences USA* 100(16): 9394–9397 (2003); and F. E. Rheindt, C. P. Strehl, and J. Gadau, "A genetic component in the determination of worker polymorphism in the Florida harvester ant *Pogonomyrmex badius*," *Insectes Sociaux* 52(2): 163–168 (2005).

139 | R. H. Crozier and R. E. Page Jr., "On being the right size: male contributions and multiple mating in social Hymenoptera," *Behavioral Ecology and Sociobiology* 18(2): 105–115 (1985).

140 | D. E. Wheeler and H. F. Nijhout, "Soldier determination in *Pheidole bicarinata*: inhibition by adult soldiers," *Journal of Insect Physiology* 30(2): 127–135 (1984).

pallidula when they are stressed by the presence of alien colonies of the same species, also resulting in an increased investment in soldiers.[141] Moreover, when minor workers of *Pheidole pubiventris* are removed until the minor-to-major ratio in colonies has fallen to about 1:1 or less, the majors switch from avoiding the brood to assuming the role of nurses.[142]

Despite convincing theoretical arguments and some experimental evidence with laboratory colonies that strongly suggest adaptive demography, there has been little empirical evidence for its occurrence in nature. Andrew Yang and his coworkers recently presented the first such evidence.[143] These scientists demonstrated that geographically distinct populations of *Pheidole morrisi* differ in worker subcaste ratio and worker body sizes in a manner consistent with microevolutionary divergence. When competitors and resources in the environment change, the adaptive demography of a colony is predicted to change in such a way as to optimize colony performance in the new environment. Thus, the investigators reasoned that colonies in different geographical populations will exhibit different phenotypes as a result of evolutionary divergence, as earlier suggested by S. N. Beshers and J. F. A. Traniello.[144]

Yang and his colleagues investigated three populations of *Pheidole morrisi* located in Florida, North Carolina, and New York. First they found that these several geographical populations exhibit characteristic differences in worker subcaste ratios. The ratios persisted even when the colonies were reared in the same environment. Colonies from Florida produce more majors than colonies from North Carolina and New York, and both worker subcastes are smaller in Florida than in the other populations, which suggests that increasing the number of majors has to

---

141 | L. Passera, E. Roncin, B. Kaufmann, and L. Keller, "Increased soldier production in ant colonies exposed to intraspecific competition," *Nature* 379: 630–631 (1996). In contrast, no increase in the production of majors was observed in colonies of *Pheidole dentata* chronically exposed to fire ants (*Solenopsis invicta*), a principal enemy to which they otherwise react quickly and violently; see A. B. Johnston and E. O. Wilson, "Correlates of variation in the major/minor ratio of the ant, *Pheidole dentata* (Hymenoptera: Formicidae)," *Annals of the Entomological Society of America* 78(1): 8–11 (1985).

142 | E. O. Wilson, "Between-caste aversion as a basis for division of labor in the ant *Pheidole pubiventris* (Hymenoptera: Formicidae)," *Behavioral Ecology and Sociobiology* 17(1): 35–37 (1985).

143 | A. S. Yang, C. H. Martin, and H. F. Nijhout, "Geographic variation of caste structure among ant populations," *Current Biology* 14: 514–519 (2004).

144 | S. N. Beshers and J. F. A. Traniello, "The adaptiveness of worker demography in the attine ant *Trachymyrmex septentrionalis*," *Ecology* 75(3): 763–775 (1994).

be energetically compensated by reducing somewhat the body size of the ants. An important question now arises: What adaptive significance can be found in those demographic differences? Obviously, the increasing proportions of soldiers in the Florida populations could be due to the need to enhance colony defense. In fact, interspecific competition, especially with colonies of *Solenopsis* species (including the fire ant *Solenopsis invicta*) is common in the Florida populations but absent or rare in the other populations. Indeed, in a series of experiments, Yang and his coworkers demonstrated that the larger number of soldiers (majors) in *Pheidole morrisi* colonies from Florida gave them a decisive advantage in confrontations with *Solenopsis* ants. Although the soldiers of the Florida colonies are somewhat smaller than those of the other populations, their larger number was more important than body size in fights with *Solenopsis*.

This work is an impressive demonstration of adaptive demography, and it also shows how developmental mechanisms for caste determination account for the covariance of worker subcaste ratio and worker size. It is also a convincing example of an ant colony operating as a developmental and functional superorganismic unit.

The adaptiveness at the colony level of caste systems is consistent with ergonomic theory, which predicts that the more efficient the caste, the fewer the members of the caste. This result, illustrated in Figure 5-14, is best explained by colony-level selection.[145]

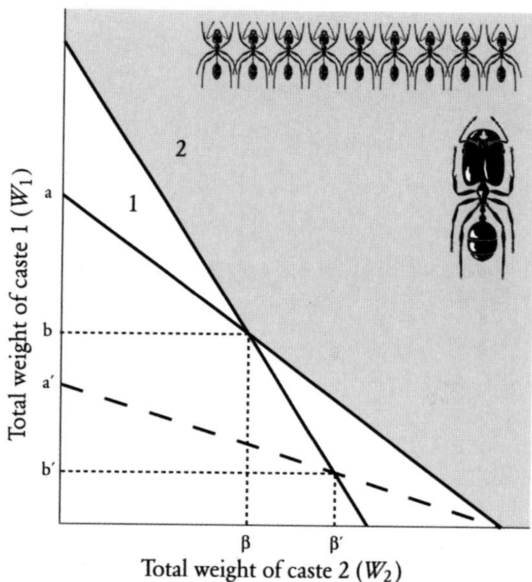

FIGURE 5-14. A counterintuitive principle of ergonomic theory is that when selection is at the colony level, a caste (1) that gains in efficiency (from ß to ß') drops in numbers (from a to a') rather than increases in numbers, the opposite expected if selection were at the level of the individual colony member. Modified from E. O. Wilson, *The Insect Societies* (Cambridge, MA: The Belknap Press of Harvard University Press, 1971).

145 | E. O. Wilson, "The ergonomics of caste in the social insects," *American Naturalist* 102: 41–66 (1968); G. F. Oster and E. O. Wilson, *Caste and Ecology in the Social Insects* (Princeton, NJ: Princeton Univerity Press, 1978).

In support, comparative studies of ten species of *Pheidole* belonging to seven phylogenetic lineages revealed that the fewer the number of behavioral acts in the repertoire of the major caste—hence the more specialized and presumably the more efficient it is in the labor roles it chooses—the smaller its representation in the colony.[146]

Colony-level selection, then, creates the physical caste system in response to selective forces in the environment, and it also adds plasticity in the size-frequency distribution of the caste system in response to variation in those forces. In Figure 5-15, an imaginary but typical size-frequency distribution is shown in conjunction with the survivorship curves of the various size classes. The survivorship curves are expected to reflect the optimized programs of temporal (age-based) division of labor. Do the survivorship curves in fact vary among physical castes and in a manner that reflects their labor roles? If that is the case, then survivorship schedules should vary in concert with roles under

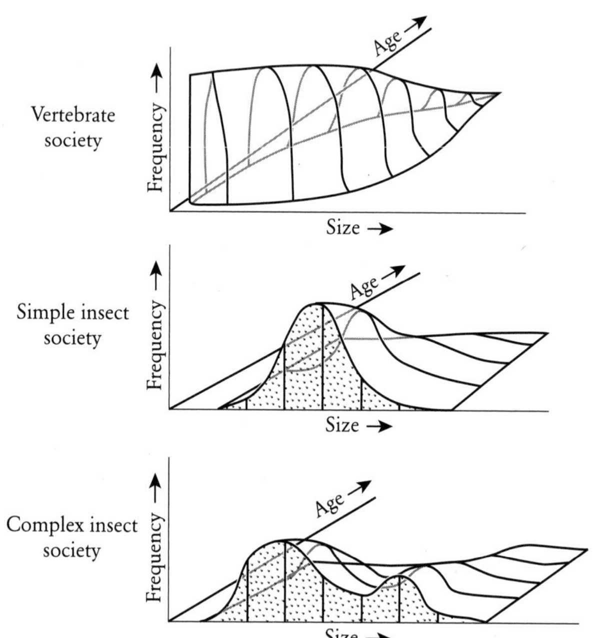

FIGURE 5-15. The concept of adaptive demography in ant colonies and other complex insect societies. In vertebrate societies, the overall size and frequency distributions are nonadaptive at the level of the population. In the simplest insect societies, such as those of the primitively eusocial bees, this remains the case. In complex insect societies, however, the proportion of individuals of various sizes and ages determines the efficiency of the division of labor and hence is adaptive at the level of the entire colony. In these imaginary but realistic examples, the ages shown for the two insect societies apply only to the final, adult stage, during which most or all of the labor is performed. Hence, no further increase in size occurs with aging. From E. O. Wilson, *Sociobiology: The New Synthesis* (Cambridge, MA: The Belknap Press of Harvard University Press, 1975).

146 | E. O. Wilson, "The relation between caste ratios and division of labor in the ant genus *Pheidole* (Hymenoptera: Formicidae)," *Behavioral Ecology and Sociobiology* 16(1): 89–98 (1984). The result has been supported by the addition of *Pheidole morrisi*, the majors of which are minor-worker-like in size and abundance and possessing correspondingly large repertoires; see A. D. Patel, "An unusualy broad behavioral repertory for a major worker in a dimorphic ant species: *Pheidole morrisi* (Hymenoptera: Formicidae)," *Psyche* (Cambridge, MA) 97(3–4): 181–191 (1990).

natural conditions, following basic theory of the genetic evolution of aging.[147] Furthermore, senescence of individuals kept under laboratory or other protected conditions should also vary in concert with labor roles. Such appears to be the case in weaver ants of the genus *Oecophylla*, one of the few species in which this matter has been explicitly addressed. Major workers do most of the foraging and are most active in defense, and many even disperse as they age to peripheral "barrack" nests of each colony, where they are exposed to maximum danger from invaders.[148] Minor workers remain in and close to the central nests, where they attend the brood, queen, and symbiotic scale insects. True to the genetic theory of senescence, and in further accordance with the observed division of labor, minor workers have been found to live longer than major workers when both castes are kept safe in laboratory nests and allowed to live to their maximum natural ages.[149]

TEAMWORK

The division of labor we have described so far is based on tasks performed by individual workers acting more or less on their own. They can be summoned to their roles by signals from nestmates, but essentially they perform the task from start to finish as individual actors. There are, in addition, several ways in which labor can be more complexly organized by teamwork. In such cases, a single task is divided into subtasks, each of which is performed by a different worker or set of workers. These elaborations are relatively simple as higher-order phenomena go, and the most clearly defined examples among the social insects have been reported so far in only a few species of ants.[150]

---

147 | G. C. Williams, "Pleiotropy, natural selection, and the evolution of senescence," *Evolution* 11(4): 398–441 (1957). Mainstream theory holds that if death by natural causes such as predation or accident normally occurs early, genes for early maturity, vigorous health, and reproduction will be selected; genes that include senescence in later periods of time will not be disfavored. Further, if they have the additional effect of promoting vigor in early maturity, they may be increased in the population by natural selection.

148 | B. Hölldobler, "Territorial behavior in the green tree ant (*Oecophylla smaragdina*)," *Biotropica* 15(4): 241–250 (1983).

149 | M. Chapuisat and L. Keller, "Division of labour influences the rate of ageing in weaver ant workers," *Proceedings of the Royal Society of London B* 269: 909–913 (2002).

150 | N. R. Franks, "The organization of worker teams in social insects," *Trends in Ecology and Evolution* 2(3): 72–75 (1987); B. Hölldobler and E. O. Wilson, *The Ants* (Cambridge, MA: The Belknap Press of Harvard

PLATE 17. This leaf tent nest is one of more than 100 such compartments spread over the entire canopy of several trees occupied by a single mature colony of the weaver ant *Oecophylla longinoda.*

Among teams that organize themselves to work together simultaneously, perhaps the most complex, and certainly the most striking in appearance, is nest building in the weaver ants of the genus *Oecophylla.* The aerial pavilions of these insects are constructed of leaves bound together by silk. One group of major workers cooperate to pull the edges of the leaves together. Another group brings out large, last-instar larvae and wags them bodily back and forth across the newly created seams. With each sweep, the larvae emit sticky threads of silk over the seams to hold the

University Press, 1990); C. Anderson and N. R. Franks, "Teams in animal societies," *Behavioral Ecology* 12(5): 534–540 (2001).

PLATE 18. Teamwork in the weaver ants of the genus *Oecophylla* during nest construction. *Above*: Workers form living chains to move a leaf into a suitable position. *Below*: Other crews of workers pull edges of the leaves together.

PLATE 19. Once the leaves are suitably arranged, another group of workers uses last-instar larvae as weaving shuttles to bind them together.

leaves in place. In other words, three task groups (including the larvae) must work simultaneously to build an *Oecophylla* nest (Plates 17 to 19).

A wholly different approach is used during the retrieval of large prey items by *Eciton burchelli*, the swarm-raider army ants of Central and South America.[151] Once a prey is subdued, some of the raiders cluster around and on top of it. The group often includes a large major that stands guard. As a worker starts to drag the prey along, a transport gang quickly forms to complete the transport back

151 | N. R. Franks, "Teams in social insects: group retrieval of prey by army ants (*Eciton burchelli,* Hymenoptera: Formicidae)," *Behavioral Ecology and Sociobiology* 18(6): 425–429 (1986). Similar team organization, evolutionarily convergent in origin, has recently been reported in African driver ants (*Dorylus wilverthi*) by N. R. Franks, A. B. Sendova-Franks, J. Simmons, and M. Mogie, "Convergent evolution, superefficient teams and tempo in Old and New World army ants," *Proceedings of the Royal Society of London B* 266: 1697–1701 (1999).

to the colony bivouac. Because exceptionally large objects can be moved only by strong workers, the first member of the gang is usually a submajor. After this heavy lifter has moved into position, smaller media workers join in until the prey is moving expeditiously homeward. These army ant teams are "superefficient": they carry items so large that if they were cut into pieces, the group members working separately would be unable to handle all the fragments. This effect is explained at least in part by the ability of the teams to overcome rotational forces. With sufficient coordination, the worker gang can support an object in such a way that these forces are balanced and disappear.

Teams can also be created by sequential cooperation, a phenomenon illustrated by the transport of food, building material, or refuse by relay from one worker to the next. The phenomenon, called task partitioning, evidently decreases cost per unit yield in time and energy.[152] The leafcutter ant *Atta vollenweideri*, for example, transports grass fragments along well-established trunk trails up to 150 meters in length. Larger workers scissor off the grass fragments, after which smaller ones carry them back to the nests. On long foraging trails, the transport chains consist of two to five carriers. J. Röschard and F. Roces, who studied this phenomenon, suggest that the primary function of the chains may be to speed communication about the material being harvested. Thus, the workers in the first segment can return to the caches deposited by the cutters, using freshly laid odor trails to guide them, while at the same time assessing the quality and amounts of material being deposited. As the attractiveness of the plant source rises or falls, the initial workers and thence the whole transport chain can respond more quickly and accurately to the change.[153] However, other causes of "bucket-brigade" organization may operate, either jointly with information enhancement or exclusively. They include greater ergonomic efficiency, based, for example, on the smallest workers near the food source yielding to the fastest near the nest or else the smallest workers, also being the weakest,

152 | R. L. Jeanne, "The organization of work in *Polybia occidentalis*: costs and benefits of specialization in a social wasp," *Behavioral Ecology and Sociobiology* 19(5): 333–341 (1986); J. L. Reyes and J. Fernández Haeger, "Sequential co-operative load transport in the seed-harvesting ant *Messor barbarus*," *Insectes Sociaux* 46(2): 119–125 (1999); C. Anderson and F. L. W. Ratnieks, "Task partitioning in insect societies: novel situations," *Insectes Sociaux* 47(2): 198–199 (2000); and C. Anderson, N. R. Franks, and D. W. McShea, "The complexity and hierarchical structure of tasks in insect societies," *Animal Behaviour* 62(4): 643–651 (2001).

153 | J. Röschard and F. Roces, "Cutters, carriers and transport chains: distance-dependent foraging strategies in the grass-cutting ant *Atta vollenweideri*," *Insectes Sociaux* 50(3): 237–244 (2003).

yielding to the largest, hence strongest, nearest the nest.[154] The study of such caste partitioning, through both ergonomic theory and empirical analysis, is in an early stage of development.

THE LARGER PICTURE

It seems appropriate to finish this treatment of caste and division of labor with a conclusion drawn by Thomas D. Seeley, whose work on foraging by the honeybee *Apis mellifera* is the most complete to date for any social insect. His formulation applies equally well across the more socially advanced bees, wasps, ants, and termites:

> Recent experimental analyses of honey bee colonies have revealed striking group-level adaptations that improve the foraging efficiency of colonies, including special systems of communication and feedback control. . . . These findings demonstrate that a colony of bees, like an individual bee or a cell within a bee, is an exceedingly intricate piece of biological machinery whose parts cooperate closely for the common good. They show, therefore, that biologists can analyze the functioning of some groups just as they long analyzed the functioning of cells and organisms. Multilevel selection theory shows that groups can evolve a high level of functional organization when between-group selection predominates over within-group selection.[155]

---

154 | C. Anderson and F. L. W. Ratnieks, "Task partitioning in insect societies, I: Effect of colony size on queueing delay and colony ergonomic efficiency," *American Naturalist* 154(5): 521–535 (1999); and C. Anderson, J. J. Boomsma, and J. J. Barthold, III, "Task partitioning in insect societies: bucket brigades," *Insectes Sociaux* 49(2): 171–180 (2002).

155 | T. D. Seeley, "Honey bee colonies are group-level adaptive units," *American Naturalist* 150(Supplement): S22–S41 (1997).

6

COMMUNICATION

The essence of social existence is reciprocal, cooperative communication. The study of communicative mechanisms is at the heart of research on social interactions, whether that communication occurs among the organelles of a cell, the cells and tissues of an organism, the organisms within a society, or the species within mutualistic symbioses. This fundamental principle of biology has been articulated by Thomas Seeley as follows: "The formation of a higher-level unit by integrating lower-level units will succeed only if the emerging organization acquires the appropriate 'technologies' for passing information among its members."[1]

This chapter will summarize our current knowledge of the technologies employed by social insect species, in particular ants and bees, to transmit information among the colony members. Information can be conveyed by cues as well as by signals, a distinction first proposed by James Lloyd, and characterized by Seeley thus: "Signals are information-bearing actions or structures that have been shaped by natural selection specifically to convey information. Cues are variables that likewise convey information, but have not been molded by natural selection to express this information."[2, 3]

As we will see, the flow of information from the group to the individual is achieved mainly by cues, whereas the reverse information flow, from the individual to the group, is more often by signals. Of course, the distinction between cues and signals is occasionally ambiguous. That circumstance does not affect the analysis of

1 | T. D. Seeley, *The Wisdom of the Hive: The Social Physiology of Honey Bee Colonies* (Cambridge, MA: Harvard University Press, 1995).

2 | J. E. Lloyd, "Bioluminescence and communication in insects," *Annual Review of Entomology* 28: 131–160 (1983).

3 | T. D. Seeley, *The Wisdom of the Hive: The Social Physiology of Honey Bee Colonies* (Cambridge, MA: Harvard University Press, 1995).

behavioral mechanisms of communication overall. But it does impose caution in how we interpret the evolutionary origins of particular cues and signals.

DANCE COMMUNICATION IN HONEYBEES

The celebrated "dance language" of honeybees is among the most intricate as well as most thoroughly studied communication behaviors in the animal kingdom, making it an ideal example for introducing the subject of communication in social insects and other superorganismic species. Ever since Karl von Frisch, the great Austrian zoologist, first decoded this language in 1947, the array of dances and the auxiliary signals that enrich them have been the subject of intensive study by a multitude of researchers.[4]

Upon discovering a new rich food source in the close vicinity of the hive, a foraging worker flies home, enters the hive, and regurgitates food to several nestmates. After a few trips back to the food source, she begins to conduct the so-called round dance, which alerts nestmates and stimulates them to fly out and search in all directions for the newly discovered food. The alerted bees perceive the odor of the source by sweeping their antennae over the body of the dancer, and they also receive samples of the food offered to them. These cues enable them to find the food sources in the close vicinity of the hive.

If the distance to the food site is greater than about 100 meters, the forager switches to the waggle dance (Figure 6-1). The "dance floor" on which she performs is located on the vertical surface of the combs inside the dark hive. Only bees returning from highly profitable food sources perform the dances. The essential element in the performance is the waggle run, or straight run; it is the middle piece of the figure-eight dance pattern, and it conveys the direction of the target during the outbound flight. Straight up on the vertical surface represents the direction of the sun the follower will see as she leaves the nest. If the target is on a line 40° to the right of the sun, say, the straight run is made 40° to the right of vertical on the comb (Figure

4 | K. von Frisch, *The Dance Language and Orientation of Bees* (Cambridge, MA: The Belknap Press of Harvard University Press, 1967); and T. D. Seeley, *The Wisdom of the Hive: The Social Physiology of Honey Bee Colonies* (Cambridge, MA: Harvard University Press, 1995).

FIGURE 6-1. The essential form of the honeybee waggle dance, usually performed on the vertical brood comb of the hive interior. The recruiting worker dances the figure eight indicated by the arrows. During the straight run, she waggles her body vigorously. The dancing bee is closely followed by stimulated nestmates. Based on K. von Frisch, *The Dance Language and Orientation of Bees* (Cambridge, MA: The Belknap Press of Harvard University Press, 1967).

6-2). The remainder of the figure eight consists of a circling back at the end of the run, during which the bee first goes left, then right to reach the departure point of the waggle run. During the straight run, the dancing bee waggles her body from side to side at a rate of about 15 times a second (15 hertz) while simultaneously vibrating her wings up and down to make a distinctive buzzing sound at 260 to 270 hertz.

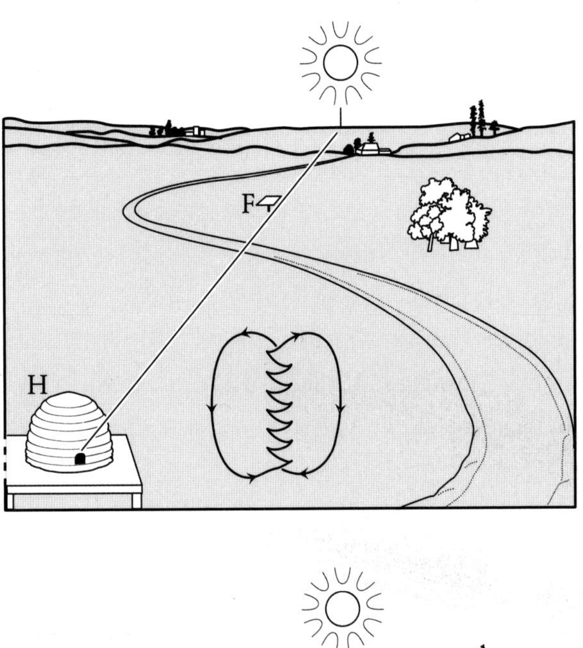

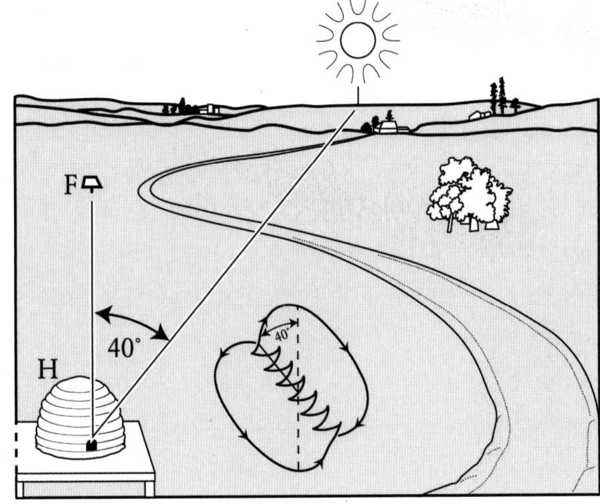

FIGURE 6-2. The code of the honeybee waggle dance: the angle of the zigzag vertical run to the line of gravity inside the nest represents the angle of the flight to the sun outside the hive (H) in the outbound flight to the food (F). The duration of the vertical run represents the distance to be flown. Based on an original illustration by Turid Hölldobler-Forsyth from B. Hölldobler, "Communication in social Hymenoptera," in T. A. Sebeok, ed., *How Animals Communicate* (Bloomington, IN: Indiana University Press, 1977), pp. 418–471.

The waggle dance conveys more information than just the direction of the outbound flight. The duration of the waggle run is correlated with the distance of the food site from the hive: the farther away the site, the longer each waggle run takes. Circumstantial evidence suggests that the key element in the signal is the duration of the buzzing sound.

Von Frisch and subsequent researchers have established that the forager bees, when employing the sun compass in their dances, also adjust the angle of the straight run to compensate for the movement of the sun through the sky. They continue to do so accurately even after spending hours in the dark interior of the hive.[5] Further, they can pinpoint the target when it is hidden by clouds, providing a patch of blue sky remains visible. The bees accomplish this remarkable feat by reading the pattern of the available polarized sky light.[6]

How, then, do bees measure the flight distance from the food source to the hive? For a long time it was assumed that they manage this feat by gauging the energy expended during the flight. In certain circumstances, this mechanism might still be a backup device for distance assessment, but the main odometer turns out to be very different. In a series of ingenious experiments, Harald Esch, Mandyam Srinivasan, and their colleagues demonstrated that honeybees measure distance by the quantity of optic flow, that is, the stream of visual cues encountered along the flight course.[7]

The experiments performed by these investigators to identify the mechanism are interesting in their own right, but they are also notable in illustrating the advantage social insects sometimes provide in studies of behavioral physiology. The researchers began by training bees to fly back and forth through a tunnel 8 meters in length. The wall had been painted with a variety of patterns. (Such training would be very difficult, if not impossible, with nonsocial insects.) At the end of the tunnel, they placed a feeder filled with concentrated sugar water. The bees, evidently perceiving the compact sequence of patterns as the equivalent of many signposts spread out in a normal environment, performed dances that signaled a distance significantly

---

5 | For example, in "marathon dances" reported by Martin Lindauer in "Dauertänze im Bienenstock und ihre Beziehung zur Sonnenbahn," *Naturwissenschaften* 41: 506–507 (1954).

6 | R. Wehner and S. Rossel, "The bee's celestial compass—a case study in behavioural neurobiology," in B. Hölldobler and M. Lindauer, eds., *Experimental Behavioral Ecology and Sociobiology* (In Memorium Karl von Frisch 1886–1982; International Symposium of the Akademie der Wissenschaften und der Literatur, 17–19 October 1983, Mainz) (Stuttgart: Gustav Fischer Verlag, 1985), pp. 11–53.

7 | H. E. Esch and J. E. Burns, "Honeybees use optic flow to measure the distance of a food source," *Naturwissenschaften* 82(1): 38–40 (1995); M. V. Srinivasan, S. Zhang, M. Altwein, and J. Tautz, "Honeybee navigation: nature and calibration of the 'odometer': *Science* 287: 851–853 (2000); H. E. Esch, S. Zhang, M. V. Srinivasan, and J. Tautz, "Honeybee dances communicate distances measured by optic flow," *Nature* 411: 581–583 (2001); and A. Si, M. V. Srinivasan, S. Zhang, "Honeybee navigation: properties of the visually driven 'odometer,'" *Journal of Experimental Biology* 206(8): 1265–1273 (2003).

greater than 8 meters. (In fact, such a short distance would normally be advertised with a round dance.) Moreover, when the tunnel was painted with horizontal stripes, giving no cue of vertical boundaries having been passed, the bees signaled much shorter distances—or else they performed round dances to denote a very short journey to the target.

Elizabeth Pennisi, in an article in *Science*, compared the phenomenon to a motorist who subconsciously estimates distance by the number of trees and other conspicuous objects passed during the drive. Like the subjective estimation of a human motorist, the honeybee's odometer is not absolute. Instead, it depends on the flight altitude and the density of visible settings through which the bee travels. Hence, according to Esch and his coworkers, "the accuracy of distance information gained from dances depends critically on the directional information received at the same time, for it is the direction of flight that determines the environment through which the recruits fly. As long as the recruit flies in the same direction as the dancer, she will translate the image motion signaled by the dancer into correct flight distance and find the goal."[8]

These elegant experiments yielded yet another important result about perception and communication. Forager bees returning from the feeder through the tunnel to the hive danced as though they had traveled 72 meters instead of the actual distance flown (the tunnel length plus the outside space at both ends).[9] After removing the tunnel, the researchers set up check posts at 35, 70, and 140 meters from the hive and counted the bees approaching them. About three-fourths of the 220 bees registered were tallied at the 70-meter post, indicating that they used the information transferred from the dances of the tunnel forager. This result is another striking piece of evidence that information contained in the waggle dance is understood by nestmates of the dancer.

How is information in the dance, which human observers can see, picked up by follower bees in the dark interior of the hive? There is some evidence that the followers depend on the buzzing sound produced during the waggle run, using their antennae as receptors. The resonant frequency of the antennal flagella is about 260

---

8 | H. E. Esch, S. Zhang, M. V. Srinivasan, and J. Tautz, "Honeybee dances communicate distances measured by optic flow," *Nature* 411: 581–583 (2001).

9 | The researchers had previously calibrated the flights in this particular direction and therefore knew how long the waggle dances should last for particular distances in this landscape.

to 280 hertz, corresponding to that emitted by the dancer bees. It also matches the maximal sensitivity of Johnston's organ, a vibration detector located at the base of each antennal flagellum.[10]

Substrate vibration on the comb may also transmit information to the followers. Jürgen Tautz found that structural properties of the dance floor have a considerable effect on the recruitment response. Dancers on combs with empty cells, for example, recruit three times as many nestmates to food sites as do dancers on capped brood cells.[11] Tautz and his collaborators later measured vibrations of 200 to 300 hertz on the comb generated by dancing bees. They also detected a peculiar alteration of vibration signals in certain comb cells that likely enhance the effect of the overall vibration communication. The bees perceive these seismic signals through highly sensitive vibration receptors in their legs, the so-called subgenual organs.[12]

In earlier reports, students of honeybee behavior postulated that the waggle movement during the straight run improves the travel of the 260- to 270-hertz vibration from the thorax of the dancer through the legs and onto the cell walls of the comb. True enough: it turns out that the sound burst is emitted by the dancing bee when she reaches the most lateral displacement of her movement. Harald Esch has noted that "the mass of the bee gains momentum through the lateral motion, and the bee must brace herself by gripping the walls of the cells; otherwise she would fall over."[13]

High-speed video recordings have further revealed that the straight run of the waggle dance consists of a single slow stride, during which the bee holds on to

---

10 | A. Michelsen, W. H. Kirchner, and M. Lindauer, "Sound and vibrational signals in the dance language of the honeybee *Apis mellifera*," *Behavioral Ecology and Sociobiology* 18(3): 207–212 (1986); A. Michelsen, W. F. Towne, W. H. Kirchner, and P. Kryger, "The acoustic near field of a dancing honeybee," *Journal of Comparative Physiology A* 161(5): 633–643 (1987); and C. Dreller and W. H. Kirchner, "Hearing in honeybees: localization of the auditory sense organ," *Journal of Comparative Physiology A* 173(3): 275–279 (1993).

11 | J. Tautz, "Honeybee waggle dance: recruitment success depends on the dance floor," *Journal of Experimental Biology* 199(6): 1375–1381 (1996).

12 | D. C. Sandeman, J. Tautz, and M. Lindauer, "Transmission of vibration across honeycombs and its detection by bee leg receptors," *Journal of Experimental Biology* 199(12): 2585–2594 (1996); J. C. Nieh and J. Tautz, "Behaviour-locked signal analysis reveals weak 200–300 Hz comb-vibrations during the honeybee waggle dance," *Journal of Experimental Biology* 203(10): 1573–1579 (2000); and J. Tautz, J. Casas, and D. Sandeman, "Phase reversal of vibratory signals in honeycomb may assist dancing honeybees to attract their audiences," *Journal of Experimental Biology* 204(21): 3737–3746 (2001).

13 | H. Esch, "Über die Schallerzeugung beim Werbetanz der Honigbiene," *Zeitschrift für vergleichendende Physiologie* 45(1): 1–11 (1961).

the comb most of the time, regardless of the length of the straight run.[14] As the waggling bee moves her body slowly forward, she stretches for shorter or longer distances in accordance with the length of the flight to be taken outside the hive. For example, when a food site is 200 meters away, the bee's stretch reaches 5 millimeters; when the site is 1,200 meters away, the bee's stretch reaches 8 millimeters.

Countless experiments, yielding overwhelming circumstantial evidence, have firmly established that the honeybee waggle dance all by itself conveys the direction and distance of targets to follower bees. It turns out that these nestmates are also able to recall and use this information over considerable periods of time.[15] Further, as Karl von Frisch reported as early as 1923, honeybees also rely on a variety of chemical cues and signals in recruitment communication. Floral scents are carried on the bodies of the returning bees. When combined with food regurgitated by the foragers, they inform nestmates of the nature and quality of newly discovered food sources. In addition to environmental odor cues, foragers release pheromones to help target the sources. Von Frisch observed and later researchers confirmed that secretions of Nasanov's gland, located between the sixth and seventh abdominal tergites, are sometimes released around targets (see Figure 5-6). The pheromones, a mix of geraniol, nerolic acid, and geranic acid, serve as homing devices for arriving workers recruited by the waggle dance. They are especially useful in locating targets that are odorless, such as water sources.

There is still more to the story of honeybee recruitment. As Thomas Seeley has demonstrated, foraging by the honeybee colony as a whole is based on a mass assessment of supply and demand, consisting of a "friendly competition" on the dance floor that allocates foragers among flower patches.

To see how the competition is staged, it is necessary to reintroduce from our earlier account two additional communicative displays performed by honeybees within the nest: the shaking dance and the tremble dance. These two displays are used to regulate, respectively, foraging and nectar storage. They further entail a basic division of labor in the colonies: foragers bring in food from the outside, while the storers receive it and either distribute it further among nestmates or place it within

---

14 | J. Tautz, K. Rohrseitz, and D. C. Sandeman, "One-strided waggle dance in bees," *Nature* 382: 32 (1996).

15 | T. D. Seeley, *The Wisdom of the Hive: The Social Physiology of Honey Bee Colonies* (Cambridge, MA: Harvard University Press, 1995).

FIGURE 6-3. The shaking display of the honeybee. One bee (black) signals a nestmate by shaking. The arrow indicates dorsoventral vibrations of the shaking bee's body. Based on T. D. Seeley, *The Wisdom of the Hive: The Social Physiology of Honey Bee Colonies* (Cambridge, MA: Harvard University Press, 1995).

the honeycomb cells. This specialization creates a problem for the honeybee colony: How can the rate of food collection, particularly of nectar, and the rate of food processing be kept in balance? If the rate of nectar processing exceeds the harvesting by foraging, the storage bees will be underemployed. Conversely, if there are not enough storers, the food delivered by the foragers will be backed up.

The problem is solved, and a balance of tasks achieved, by straightforward communication. When foragers have a prolonged period of successful trips or when they strike a rich source following a period of poor yield, they tend to perform the shaking dance. Moving about the hive, they vibrate their bodies up and down about 16 times a second, with each episode of vibration lasting 1 to 2 seconds (Figure 6-3). During this display, they often use their forelegs to grab hold of nearby nestmates. In a single minute, each forager approaches from 1 to 20 other workers. The shaking dance increases the number of foragers through either of two simple responses made by the nestmates: bees that encounter a shaker tend to move to the dance floor, where they are likely to meet workers performing the waggle dance, or else they fly without further instruction to food sites they themselves have visited earlier.[16] Overall, the shaking dance combines with the waggle dance to send more bees into the field as foragers.

---

16 | M. D. Allen, "The 'shaking' of worker honeybees by other workers," *Animal Behaviour* 7(3–4): 232–240 (1959); S. S. Schneider, J. A. Stamps, and N. E. Gary, "The vibration dance of the honey bee, II: The effects of foraging success on daily patterns of vibration activity," *Animal Behaviour* 34(2): 386–391 (1986); and T. D. Seeley, A. Weidenmüller, and S. Kühnholz, "The shaking signal of the honey bee informs workers to prepare for greater activity," *Ethology* 104(1): 10–26 (1998).

In the reverse imbalance, when the amount of food flowing into the hive exceeds the ability of the active food storers to handle it, the foragers perform what Karl von Frisch called the tremble dance. As described by Thomas Seeley, the bees run slowly over the comb in irregular patterns while "their bodies, as a result of quivering movements of the legs, constantly make trembling movements forward and backward, and right and left."[17] In addition, some of the tremble dancers emit a piping signal at irregular intervals while lunging forward to thrust their heads against other bees.[18]

The tremble dance was known but its function unclear until Seeley discovered a remarkable algorithm in response to the time required for a forager to locate a receptive storer. When the majority of foragers returning from a rich nectar source experience a search time of 20 seconds or less, they perform the waggle dance. That sends more bees out to the harvesting site. But foragers experiencing a search time of 50 seconds or more conduct the tremble dance.[19] Even when heard as a solitary signal, the piping sound induces workers performing the waggle dance to cease doing so, thus diminishing the outflow of foragers into the field.[20] When combined, the tremble dance and piping signals achieve a balance between nectar collecting and nectar processing by producing both a rise in the processing rate and a decrease in the collecting rate.

According to Seeley, foragers are the metaphorical "sensory units" of the honeybee colony.[21] The foragers respond to cues picked up inside the hive—for example, detecting the amount of protein in food regurgitated to them by nestmates or sensing phero-

---

17 | T. D. Seeley, *The Wisdom of the Hive: The Social Physiology of Honey Bee Colonies* (Cambridge, MA: Harvard University Press, 1995).

18 | J. C. Nieh, "The stop signal of honey bees: reconsidering its message," *Behavioral Ecology and Sociobiology* 33(1): 51–56 (1993).

19 | T. D. Seeley, *The Wisdom of the Hive: The Social Physiology of Honey Bee Colonies* (Cambridge, MA: Harvard University Press, 1995); T. D. Seeley, "The tremble dance of the honey bee: message and meanings," *Behavioral Ecology and Sociobiology* 31(6): 375–383 (1992); and W. H. Kirchner and M. Lindauer, "The causes of the tremble dance in the honeybee, *Apis mellifera*," *Behavioral Ecology and Sociobiology* 35(5): 303–308 (1994).

20 | W. H. Kirchner, "Vibrational signals in the tremble dance of the honeybee, *Apis mellifera*," *Behavioral Ecology and Sociobiology* 33(3): 169–172 (1993); J. C. Nieh, "The stop signal of honey bees: reconsidering its message," *Behavioral Ecology and Sociobiology* 33(1): 51–56 (1993); and C. Thom, D. C. Gilley, and J. Tautz, "Worker piping in honey bees (*Apis mellifera*): the behavior of piping nectar foragers," *Behavioral Ecology and Sociobiology* 53(4): 199–205 (2003).

21 | T. D. Seeley, "Honey bee foragers as sensory units for their colonies," *Behavioral Ecology and Sociobiology* 34(1): 51–62 (1994).

monal signals, including those emitted by hungry larvae.[22] The bees detect shortages of foragers and food storers by the increase or decrease of waggle and tremble dancers. Through the constant flow of such cues and signals, the colony—the superorganism—works remarkably well as a unified whole in securing its food supply.

The waggle dance, as von Frisch once put it, did not simply fall out of the sky. There must have been intermediate steps in evolution that led from the first rudimentary alerting signal of primitively social bees to the advanced dance repertoire of the honeybees. And researchers have indeed discovered, over more than half a century, such middle grades of recruitment and used them to derive a reconstruction of the most likely evolutionary pathway that led to the recruitment system of *Apis mellifera.*[23]

One important principle drawn from these comparative studies is that the straight run of the honeybee waggle dance is a highly ritualized flight from the hive to the target area. Further, the simpler motor displays observed in some species of tropical stingless bees and bumblebees, along with the vibrational and pheromonal signals used in a manner similar to that of honeybees, are now viewed as the likely system from which the waggle dance originated. Finally, the adaptation that led to the dance appears to be as follows: the *Apis* progenitors foraged well away from their nests, rendering recruitment and orientation by chemical signals less and less efficient. In response, the waggle dance evolved as a "technological breakthrough" that allowed rapid and accurate exploitation of food resources and new nest sites over what constitutes tremendous distances proportionate to the size of the bees.

COMMUNICATION IN ANT SOCIETIES

A radically different scenario from that of the honeybee unfolded during the evolution of the ant. Where social bees and wasps rule the air, ants are mistresses of the soil. Even outlier arboreal species living in the cavities of plants prefer a soil-like

---

22 | T. Pankiw and W. L. Rubink, "Pollen foraging response to brood pheromone by Africanized and European honey bees (*Apis mellifera* L.)," *Annals of the Entomological Society of America* 95(6): 761–767 (2002).

23 | M. Lindauer, *Communication Among Social Bees* (Cambridge, MA: Harvard University Press, 1961); M. Hrncir, F. G. Barth, and J. Tautz, "Vibratory and airborne-sound signals in bee communication (Hymenoptera)," in S. Drosopoulous and M. F. Claridge, eds., *Insect Sounds and Communication: Physiology, Behaviour, Ecology, and Evolution* (Boca Raton, FL: Taylor & Francis, 2006), pp. 421–436; and A. Dornhaus and L. Chittka, "Evolutionary origins of bee dances," *Nature* 40: 38 (1999).

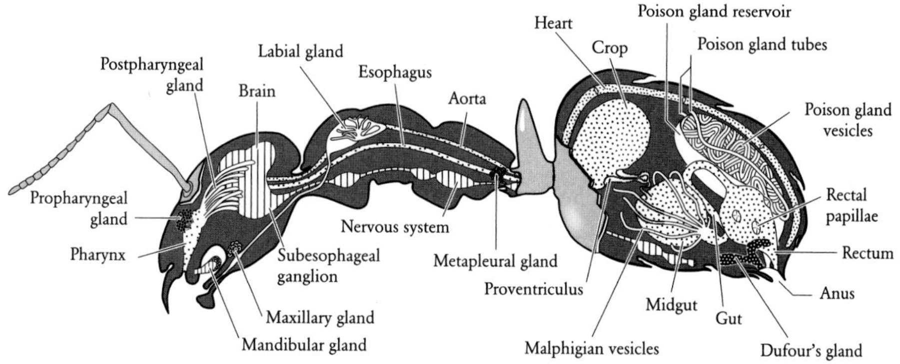

FIGURE 6-4. The major internal organs of a *Formica* worker ant. Adapted from B. Hölldobler and E. O. Wilson, *The Ants* (Cambridge, MA: The Belknap Press of Harvard University Press, 1990).

microenvironment. A crucial adaptation to accommodate this ground-dwelling lifestyle, the wingless worker caste, occurred over 100 million years ago. Colonies were thereby enabled to search efficiently for food over the ground, down into the leaf litter, and even deep beneath the soil surface. In so doing, they surrendered the kind of long-distance foraging practiced by the winged hymenopterans. The shift forced a heavy reliance on chemicals for all forms of communication and a corresponding restraint on the use of tactile signals and motor displays.[24]

Ants have thus become the insect geniuses of chemical communication. Their bodies are crowded with exocrine glands employed to manufacture pheromones, the chemical compounds used as signals (Figure 6-4). The ants have enhanced the chemical channel in several ways, variously by mixing pheromones from multiple glands, by giving separate meanings to different concentrations of the same pheromone, and by changing meanings according to context. They have simultaneously added auxiliary signals of touch and vibration.

Entomologists recognize at least 12 functional categories of communication in the repertoires of social insects, and almost all are primarily or entirely chemical in nature:[25]

---

24 | B. Hölldobler, "Multimodal signals in ant communication," *Journal of Comparative Physiology A* 184(2): 129–141 (1999).

25 | Detailed accounts of the kinds of communication are presented for ants in B. Hölldobler and E. O. Wilson, *The Ants* (Cambridge, MA: The Belknap Press of Harvard University Press, 1990); for honeybees, in T. D. Seeley, *The Wisdom of the Hive: The Social Physiology of Honey Bee Colonies* (Cambridge, MA: Harvard University Press,

1 | Alarm, as in response to an enemy invasion or breech in the nest wall
2 | Attraction, leading to assembly
3 | Recruitment, variously to food, new nest sites, and enemies
4 | Grooming, including assistance in molting, and brood care
5 | Trophallaxis, the exchange by feeding of oral, anal, and other body fluids, usually just for distribution of food but often in addition for sharing of pheromones
6 | Exchange of solid food particles
7 | Group effect: collectively either facilitating or inhibiting a particular activity
8 | Recognition of nestmates and different castes of nestmates, including fertility status and even individual recognition, as well as those injured or dead
9 | Caste determination, either by stimulation or by inhibition of the transformation of individuals into certain castes
10 | Control of competing reproductives
11 | Territorial and home range advertisement and orientation
12 | Sexual communication, including species recognition, gender recognition, synchronization of sexual activity, and responses to rivals and partners during sexual competition

More than 40 anatomically distinct exocrine glands have been found in all ant species taken together, and the total continues to climb. A clear majority of the glands are active in pheromone production.[26] The best-studied ant species are known to

---

1995), and in R. F. A. Moritz and E. E. Southwick, *Bees as Superorganisms: An Evolutionary Reality* (New York: Springer-Verlag, 1992); and more generally, in E. O. Wilson, *The Insect Societies* (Cambridge, MA: The Belknap Press of Harvard University Press, 1971), and in R. K. Vander Meer, M. D. Breed, K. E. Espelie, and M. L. Winston, eds., *Pheromone Communication in Social Insects: Ants, Wasps, Bees, and Termites* (Boulder, CO: Westview Press, 1998). For a general background on physiology, see M. F. Ryan, *Insect Chemoreception: Fundamental and Applied* (Boston: Kluwer Academic Publishers, 2002).

26 | J. Billen and E. D. Morgan, "Pheromone communication in social insects: sources and secretions," in R. K. Vander Meer, M. D. Breed, K. E. Espelie, and M. L. Winston, eds., *Pheromone Communication in Social Insects: Ants, Wasps, Bees, and Termites* (Boulder, CO: Westview Press, 1998), pp. 3–33. Other, nonsignal functions of exocrine glands in these insects include digestion, lubrication, and defense. The outstanding work of Johan Billen and his colleagues has built on a foundation of research that can be fairly said to have been put in place by the great French insect microscopist Charles Janet. A comprehensive atlas of ant exocrine glands is provided by B. Hölldobler and E. O. Wilson, *The Ants* (Cambridge, MA: The Belknap Press of Harvard University Press, 1990).

employ at least 10 to 20 kinds of signals, the great majority of which are pheromone mixtures presented with or without a smaller number of tactile or vibrational stimuli.[27] The number of pheromones is likely to grow as bioassay and chemical identifications become more sophisticated. The fire ant *Solenopsis invicta*, the species most intensively studied to date among ants,[28] is known to use about 20 signals, the exact figure depending on which of the most similar functions are distinguished. Only two of the signals are tactile, while the rest are chemical.[29] For comparison, the number of distinct signals in honeybees discovered to date, again mostly pheromonal, is 17, with at least twice as many additional modifying cues.[30]

What are the identities of the pheromone molecules? Among all organisms, ants have provided biologists with the most information concerning these chemical signals.[31] Alarm substances used by workers vary widely among species, with only a very loose correspondence to the evolutionary relationships inferred from anatomical traits. For example, hydrocarbons are, as far as we know, used only by members of the subfamily Formicinae.

In general, the exocrine sources of alarm substances are located, not surprisingly, at either the front ends of the ants (mandibular glands) or their rear ends (poison, Dufour's, and pygidial glands) (Figure 6-5). The molecules vary from species to species and spread during evolution, almost at random, across a large array of alcohols, aldehydes, aliphatic and cyclical ketones, esters, hydrocarbons, nitrogen heterocycles, sulfur compounds, terpenoids, and even formic acid, the latter the simplest of the carboxylic acids and also used as a poison.

The function of these compounds is to alert nestmates to either the nearby

---

27 | B. Hölldobler, "Multimodal signals in ant communication," *Journal of Comparative Physiology A* 184(2): 129–141 (1999).

28 | W. R. Tschinkel, *The Fire Ants* (Cambridge, MA: The Belknap Press of Harvard University Press, 2006).

29 | B. Hölldobler and E. O. Wilson, *The Ants* (Cambridge, MA: The Belknap Press of Harvard University Press, 1990); and R. K. Vander Meer, "The trail pheromone complex of *Solenopsis invicta* and *Solenopsis richteri*," in C. S. Lofgren and R. K. Vander Meer, eds., *Fire Ants and Leaf-Cutting Ants: Biology and Management* (Boulder, CO: Westview Press, 1986), pp. 201–210.

30 | T. D. Seeley, "Thoughts on information and integration in honey bee colonies," *Apidologie* 29(1–2): 67–80 (1998).

31 | The details of glandular sources and chemistry are given in B. Hölldobler and E. O. Wilson, *The Ants* (Cambridge, MA: The Belknap Press of Harvard University Press, 1990); and in R. K. Vander Meer, "The trail pheromone complex of *Solenopsis invicta* and *Solenopsis richteri*," in C. S. Lofgren and R. K. Vander Meer, eds., *Fire Ants and Leaf-Cutting Ants: Biology and Management* (Boulder, CO: Westview Press, 1986), pp. 201–210.

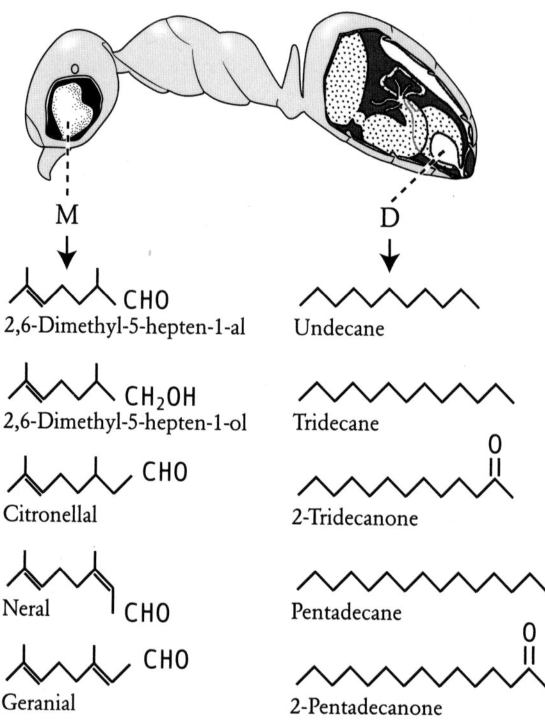

FIGURE 6-5. In the North American citronella ant *Lasius* (formerly *Acanthomyops*) *claviger*, different arrays of substances are released from the mandibular glands (M) and Dufour's gland (D) that variously and simultaneously alarm nestmates and repel enemies. Adapted from F. E. Regnier and E. O. Wilson, "The alarm-defence system of the ant *Acanthomyops claviger*," *Journal of Insect Physiology* 14(7): 955–970 (1968).

presence of enemies or the destruction of part of the nest. Other workers, upon detecting the pheromones with chemoreceptors located on their antennae, move toward the source. Then, upon reaching higher concentrations, they dash with open mandibles and fiercely attack any moving object that bears a foreign odor. If instead they discover a collapsed portion of the nest, they begin to excavate the material, thereby freeing nestmates trapped beneath.

Trail and recruitment substances are another major category of pheromones employed by ants. The roles they fill are multiple. According to species and circumstances, foragers use the substances to recruit nestmates variously to distant food sites, to new nest sites, and to enemy colonies. The glandular sources and chemical structures of these pheromones, like the sources and structures of alarm substances, are highly diverse. Differing from one species to the next, these molecules include alcohols, aldehydes, formic and nerolic acids, nitrogen heterocycles, terpenoids, lactones, and isocoumarins, among others.[32]

THE EVOLUTION OF ANT RECRUITMENT SIGNALS AND TRAIL GUIDES

Among ant species that are phylogenetically more basal ("primitive"), many species have no need of recruitment or trails because the workers are solitary huntresses and scavengers. A few, however, especially those that conduct hunting expeditions in groups and have adopted army ant lifestyles, have evolved such communication methods to a high degree of sophistication. These offshoot clades have emerged independently within the subfamilies Amblyoponinae, Leptanillinae, Ponerinae, and Cerapachyinae. Because of their diversity and uniqueness, we will treat examples from each separately. In this important domain of knowledge, we ask the reader's indulgence. Few generalizations can be made, and nothing can substitute for exact detail.

First, consider members of the remarkable amblyoponine genus *Onychomyrmex* (Plate 21). Restricted to the rain forests of northeastern Australia, they have independently evolved legionary (army ant) behavior, bivouacking temporarily in irregular leaf-litter nests and conducting nocturnal mass predation on centipedes and other arthropods. The raids are initiated by a small number of foraging workers, which lay recruitment trails to their often large, formidable prey. The trail pheromone is extruded from an unpaired sternal gland located at the median line between the fifth and sixth abdominal sternites (Figure 6-6). A successful *Onychomyrmex* scout engages in a distinctive locomotor behavior while laying a trail outward to the hunting field. She lowers her body and at irregular intervals stretches her hind legs backward, bringing the sternal gland on the ventral part of the abdomen into contact with the ground (Figure 6-7). Artificial trails made with extracts of the sternal gland and presented to colonies in the laboratory cause the characteristic mass outward move-

---

32 | B. Hölldobler and E. O. Wilson, *The Ants* (Cambridge, MA: The Belknap Press of Harvard University Press, 1990); M. F. Ryan, *Insect Chemoreception: Fundamental and Applied* (Boston: Kluwer Academic Publishers, 2002); and C. S. Lofgren and R. K. Vander Meer, eds., *Fire Ants and Leaf-Cutting Ants: Biology and Management* (Boulder, CO: Westview Press, 1986).

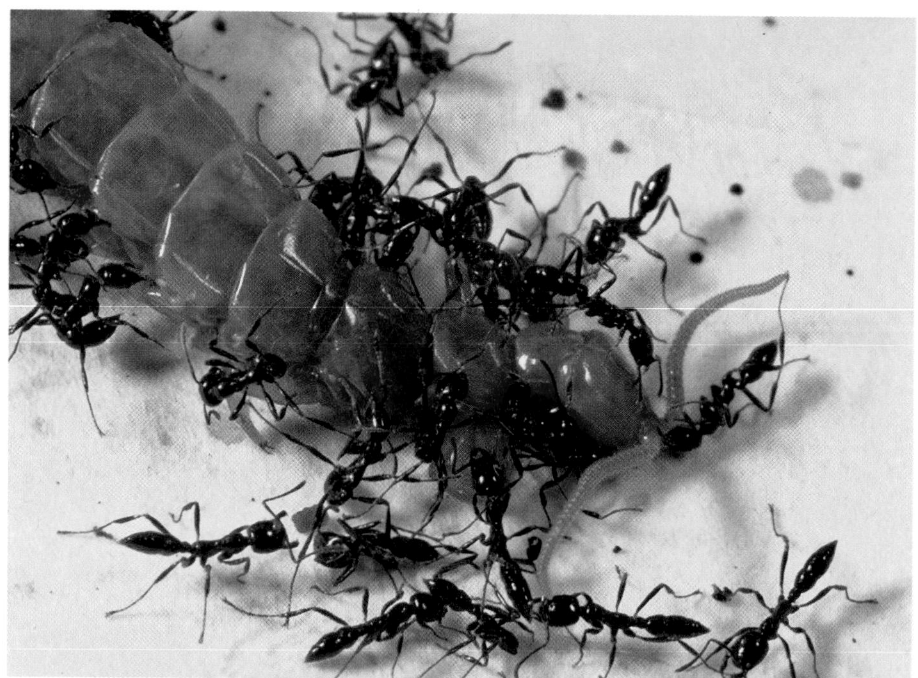

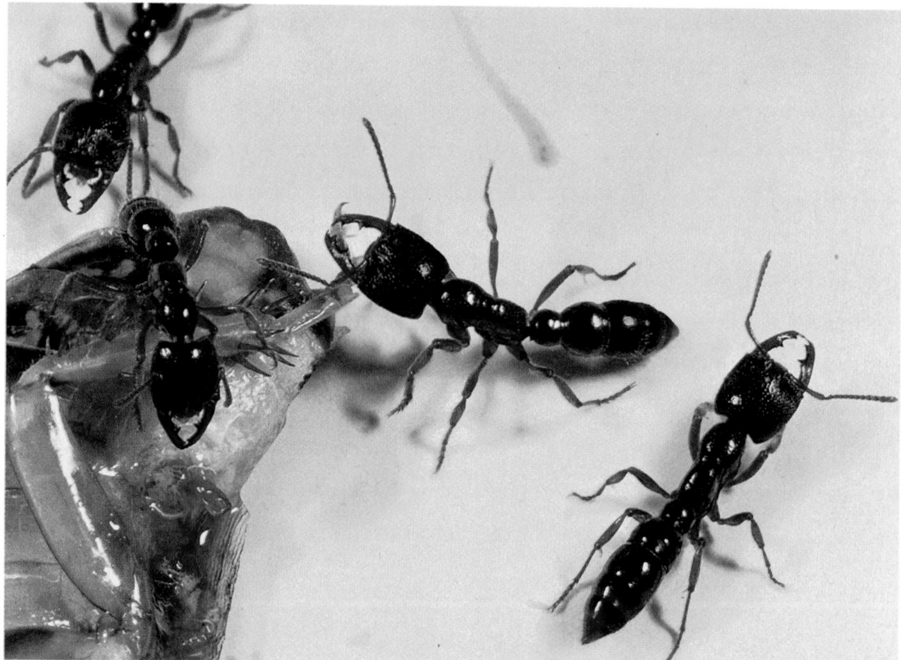

PLATE 21. Amblyoponine ants from Australia. *Above*: Workers of the genus *Onychomyrmex* attack a relatively gigantic centipede. Their lifestyle, typical of army ant species worldwide (including a highly efficient recruitment system), enables them to subdue large prey of the kind shown. *Below*: Although Australian ants of the species *Amblyopone australis* are usually solitary foragers, they do recruit a few nestmates to larger food items.

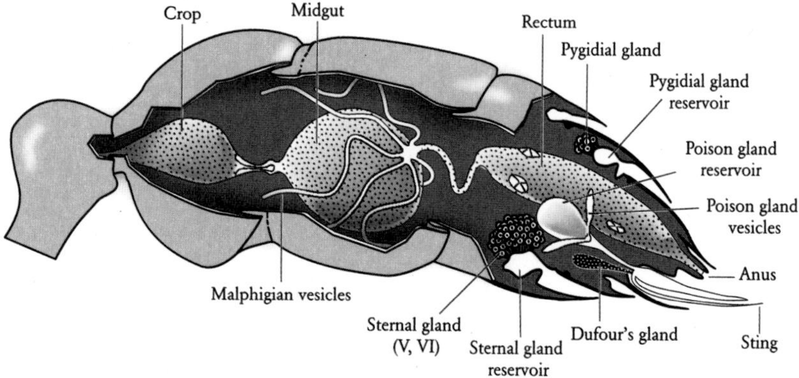

FIGURE 6-6. Gaster of a worker of the amblyoponine ant genus *Onychomyrmex*, showing the digestive organs and the major exocrine glands. (Not shown is the basitarsal gland; see Figure 6-7.) Adapted from B. Hölldobler and E. O. Wilson, *The Ants* (Cambridge, MA: The Belknap Press of Harvard University Press, 1990).

ment of workers ready for a raid.[33] The pheromone, whose chemical structure is still unknown, is highly volatile, so that after a few minutes the signal fades. Nevertheless, the *Onychomyrmex* ants return home, usually in the dark, by following in reverse more or less faithfully the outbound route. How is this accomplished? It turns out, as further experiments have revealed, that scouts lay not only the sternal gland recruitment trail but also a more persistent homing trail. The anatomical origin of the pheromone is in the hind basitarsi, the distalmost segments of the rear tarsi ("feet") of the ant. To release the pheromone, the scouts drag their hind legs so that the opening of the gland, a membranous villiform brush, touches the ground and paints the substance onto it.[34] Foragers that follow the scouts also drag their hind legs, but do not lay sternal gland trails. Thus, the trail pheromone serves as an Ariadne's thread that persists as a homing device when no other cues are available.[35]

The *Onychomyrmex* system illustrates a phenomenon common in ants: trails serve two functions, to attract nestmates to a target area and to direct them back home afterward. Pheromones performing these two functions may originate from

---

33 | B. Hölldobler, H. Engel, and R. W. Taylor, "A new sternal gland in ants and its function in chemical communication," *Naturwissenschaften* 69(2): 90–91 (1982).

34 | B. Hölldobler and J. M. Palmer, "A new tarsal gland in ants and the possible role in chemical communication," *Naturwissenschaften* 76(8): 385–386 (1989).

35 | In Greek mythology, Ariadne gave Theseus a thread to lay down and find his way out of the labyrinth after he had slain the Minotaur.

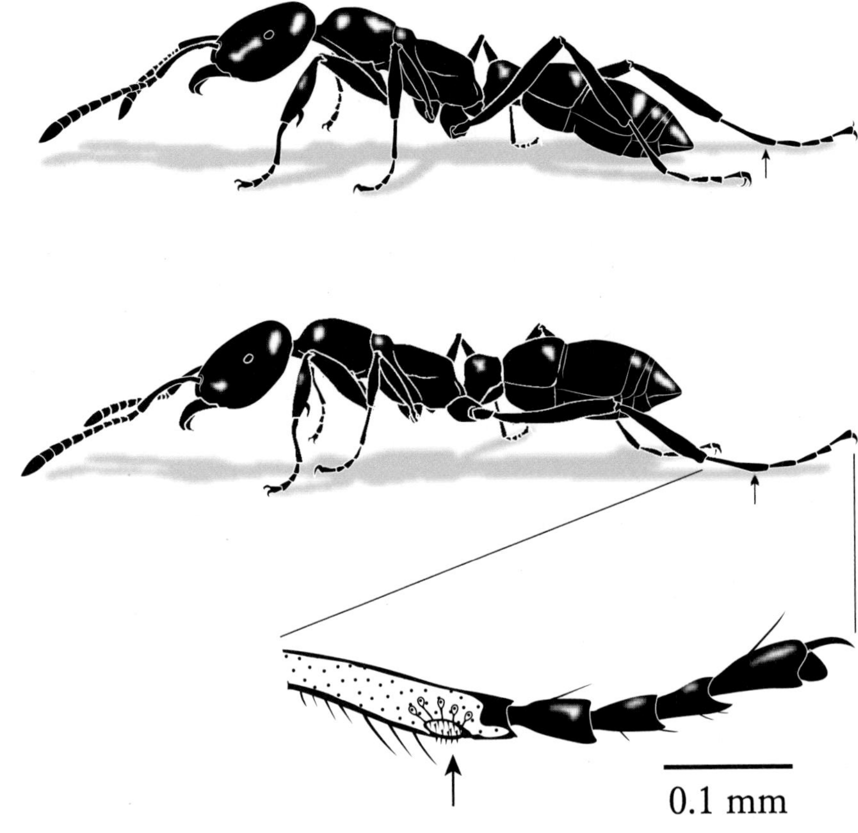

FIGURE 6-7. *Above*: A worker of the Australian amblyoponine ant *Onychomyrmex* laying a recruitment pheromone trail with secretions from the sternal gland (see Figure 6-6). Simultaneously, she drags one or the other hind leg, thereby touching the ground with the opening of a basitarsal gland (arrow). Recruited nestmates that follow the sternal gland trail also drag the basitarsal gland over the ground surface. *Below*: Location of the small, brush-like opening of the basitarsal gland, showing an enlargement of the gland.

the same exocrine gland, but occasionally, as found in *Onychomyrmex*, different glands are employed.

The Neotropical amblyoponine ant *Prionopelta amabilis* is anatomically similar to *Onychomyrmex*, but its natural history is very different. Colonies, which typically nest in pieces of rotting wood on forest floors, are not legionary. Instead, they prey on silverfish-like campodeid diplurans and a limited variety of other very small arthropods, which the huntresses carry back to the nests individually.[36] *Prionopelta amabilis* workers lack a sternal gland, but they have a basitarsal gland very similar to that of *Onychomyrmex*. In laboratory cultures, they exhibit similar foot dragging, whether exploring a new terrain, emigrating to a new nest location, or returning to the nest after encountering prey. Further, upon returning to the nest, they often shake their bodies in rapid, light vertical movements, each of which lasts from about 0.5 to 2 seconds.[37] Analysis of the shaking movement indicates that the behavior substantially enhances the effect of the basitarsal recruitment pheromone.

A third amblyoponine ant, the Australian *Amblyopone australis* (see Plate 21), possesses neither sternal gland nor basitarsal gland. Instead, it appears to lay "footprint trails" with secretions from the first segment of the hind tarsi (feet). Although solitary foraging appears to be the rule in this species, there is also evidence of recruitment by footprint trails, perhaps combined, at least under special circumstances, with body shaking.[38] The trails might also serve as orientation cues for this subterranean, notably cryptic ant species.

A final amblyoponine ant worth special attention is *Mystrium rogeri*, of Madagascar, which has also been found to employ pheromone trails for recruitment to food sources and new nest sites (Plate 22).[39] Solitary foragers lay chemical trails with secretions from an unpaired sternal gland located at the seventh abdominal sternite (Figure 6-8), a location very different from that of the sternal gland in *Onychomyrmex* and, as far as we know, limited to *Mystrium* within the subfamily Amblyoponinae. The recruiter deposits the pheromone by means of a specialized dragging of the gaster (part of the body behind the waist), enhanced in the presence of nestmates by rapid vertical body shaking (Figure 6-9).

In a class by itself is the subfamily Leptanillinae, possibly the most basal extant group of ants and yet paradoxically one of the most specialized and unusual in many anatomical and behavioral traits. Species of *Leptanilla*, the only leptanilline

---

36 | B. Hölldobler and E. O. Wilson, "Ecology and behavior of the primitive cryptobiotic ant *Prionopelta amabilis* (Hymenoptera: Formicidae)," *Insectes Sociaux* 33(1): 45–58 (1986).

37 | B. Hölldobler, M. Obermayer, and E. O. Wilson, "Communication in the primitive cryptobiotic ant *Prionopelta amabilis* (Hymenoptera: Formicidae)," *Journal of Comparative Physiology A* 170(1): 9–16 (1992).

38 | B. Hölldobler and J. M. Palmer, "Footprint glands in *Amblyopone australis* (Formicidae, Ponerinae)," *Psyche* (Cambridge, MA) 96: 111–121 (1989).

39 | B. Hölldobler, M. Obermayer, and G. D. Alpert, "Chemical trail communication in the amblyoponine species *Mystrium rogeri* Forel (Hymenoptera, Formicidae, Ponerinae)," *Chemoecology* 8(3): 119–123 (1998).

PLATE 22. *Above*: Workers of the amblyoponine species *Mystrium rogeri*, of Madagascar, are shown cooperatively transporting a pupa. *Below*: A worker of the African stink ant *Pachycondyla tarsata* has scooped termites together with her powerful jaws.

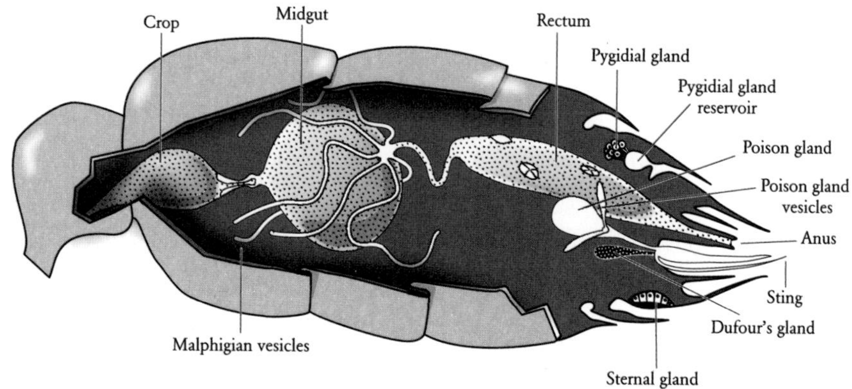

FIGURE 6-8. Gaster of a worker of the amblyoponine genus *Mystrium*, showing digestive organs and the major exocrine glands. Adapted from B. Hölldobler, M. Obermayer, and G. D. Alpert, "Chemical trail communication in the amblyoponine species *Mystrium rogeri* Forel (Hymenoptera, Formicidae, Ponerinae)," *Chemoecology* 8(3): 119–123 (1998).

genus studied in living colonies thus far, boast extremely small ants, less than 1 millimeter in length. (They are also rare and generally difficult to find; few entomologists have ever seen one outside of museums.) Their diminutive size allows them to move through tiny gaps between soil particles, where they hunt en masse and subdue centipedes.[40] Leptanilline raids are guided by pheromone trails, evidently from a large unpaired sternal gland in the seventh abdominal sternite. Whether this gland is homologous to that seen in *Mystrium* or its presence in the two genera represents independent products of convergent evolution is a question yet unresolved.[41]

In *Pachycondyla* (Plate 22), the globally dominant genus of the subfamily Ponerinae, we encounter extraordinary diversity among species in the glandular sources and chemistry of the trail pheromones. At least one species, the African stink ant *Pachycondyla tarsata* (formerly *Paltothyreus tarsatus*), produces trail substances from two pairs of sternal glands located respectively between the fifth and sixth and between the sixth and seventh abdominal sternites (lower abdominal segments).[42]

---

40 | K. Masuko, "*Leptanilla japonica*: the first bionomic information on the enigmatic ant subfamily Leptanillinae," in J. Eder and H. Rembold, eds., *Chemistry and Biology of Social Insects* (Proceedings of the Tenth Congress of the International Union for the Study of Social Insects, Munich, 18–22 August 1986) (Munich: Verlag J. Peperny, 1987), pp. 597–598.

41 | B. Hölldobler, J. M. Palmer, K. Masuko, and W. L. Brown, "New exocrine glands in the legionary ants of the genus *Leptanilla* (Hymenoptera, Formicidae, Leptanillinae)," *Zoomorphology* 108(5): 225–261 (1989).

42 | B. Hölldobler, "Communication during foraging and nest-relocation in the African stink ant, *Paltothyreus tarsatus* Fabr. (Hymenoptera, Formicidae, Ponerinae)," *Zeitschrift für Tierpsychologie* 65(1): 40–52 (1984).

FIGURE 6-9. *Above*: A worker of the Madagascar amblyoponine ant *Mystrium rogeri* lays pheromone trails from a unique sternal gland (arrow shows location of the gland). *Below*: In the nest, she enhances the signal with vertical shaking of the body. Based on an original drawing by Malu Obermayer from B. Hölldobler, M. Obermayer, and G. D. Alpert, "Chemical trail communication in the amblyoponine species *Mystrium rogeri* Forel (Hymenoptera, Formicidae, Ponerinae)," *Chemoecology* 8(3): 119–123 (1998).

The secretions are made up of ten compounds, of which one, 9-heptadecanone, induces trail following.[43] In several other *Pachycondyla* species, the trail pheromone originates in the pygidial gland, a paired organ located between the sixth and seventh abdominal tergites (dorsal abdominal segments), as illustrated in Figure 6-10. Such is the case notably in the Neotropical *Pachycondyla* (formerly *Termitopone*) *marginata*, whose workers conduct group raids on termite colonies. A scout that discovers a termite nest deposits trail substance by bending her gaster forward so

43 | E. Janssen, B. Hölldobler, and H. J. Bestmann, "A trail pheromone component of the African stink ant, *Pachycondyla (Paltothyreus) tarsata* Fabricius (Hymenoptera, Formicidae: Ponerinae)," *Chemoecology* 9(1): 9–11 (1999).

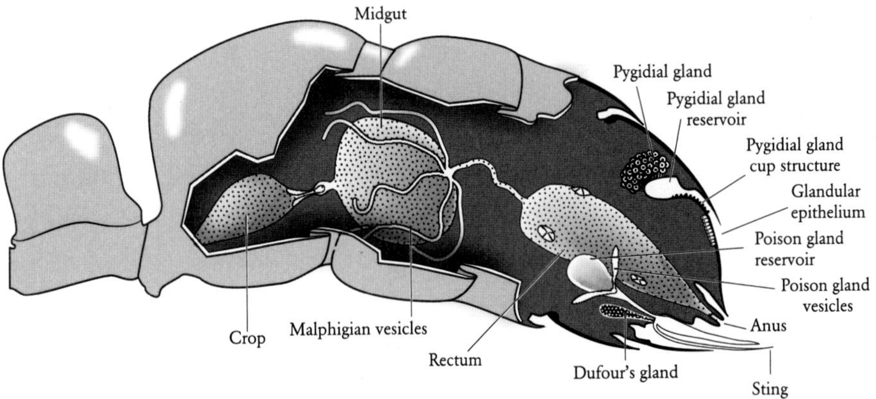

FIGURE 6-10. Gaster of a worker of the ponerine ant species *Pachycondyla marginata*, showing the digestive organs and major exocrine glands. Adapted from B. Hölldobler and E. O. Wilson, *The Ants* (Cambridge, MA: The Belknap Press of Harvard University Press, 1990).

that the surface of the tergal gland can be dragged over the ground (Figure 6-11). The gland secretion is a bouquet of compounds, among which citronellal (a familiar scent of citrus) is the most active component and isopulegol serves as a synergist.[44] In addition, the recruiting ant, upon arrival in the nest, performs a body-shaking display that facilitates the effect of the pheromones.

Another chapter in the *Pachycondyla* story is provided by the giant African hunting ant *Pachycondyla fochi* (formerly *Megaponera foetens*), whose workers use two glands to organize their raids. A short-term but powerful recruitment pheromone is released from the pygidial gland, while a more persistent orientation pheromone is deposited from the poison gland.[45] One of the pygidial gland substances that stimulates the ants is actinidine, but the identity of the trail-following stimulant from the same gland is not yet known. An orientation pheromone from the poison gland that has been identified is *N,N*-dimethyluracil.[46]

---

44 | B. Hölldobler, E. Janssen, H. J. Bestmann, I. R. Leal, P. S. Oliveira, F. Kern, and W. A. König, "Communication in the migratory termite-hunting ant *Pachycondyla (= Termitopone) marginata* (Formicidae, Ponerinae)," *Journal of Comparative Physiology A* 178(1): 47–53 (1996).

45 | B. Hölldobler, U. Braun, W. Gronenberg, W. H. Kirchner, and C. Peeters, "Trail communication in the ant *Megaponera foetans* (Fabr.) (Formicidae, Ponerinae)," *Journal of Insect Physiology* 40(7): 585–593 (1994).

46 | E. Janssen, H. J. Bestmann, B. Hölldobler, and F. Kern, "*N,N*-dimethyluracil and actinidine, two pheromones of the ponerine ant *Megaponera foetens* (Fab.) (Hymenoptera: Formicidae)," *Journal of Chemical Ecology* 21(12): 1947–1955 (1995).

FIGURE 6-11. *Above*: A worker of the New World tropical ponerine *Pachycondyla marginata* lays a trail by bending her gaster forward and depositing a secretion from the pygidial gland (see Figure 6-10). *Below*: Inside the nest, the recruiting ant enhances the effect by shaking her body vertically in the presence of nestmates. Based on B. Hölldobler, E. Janssen, H. J. Bestmann, I. R. Leal, P. S. Oliveira, F. Kern, and W. A. König, "Communication in the migratory termite-hunting ant *Pachycondyla* (= *Termitopone*) *marginata* (Formicidae, Ponerinae)," *Journal of Comparative Physiology A* 178: 47–53 (1996).

A few species of the large, cosmopolitan ponerine ant genus *Leptogenys* are group raiders, resembling army ants in their foraging behavior. At least one, *Leptogenys chinensis*, utilizes two exocrine glands to organize recruitment, in the same manner as *Pachycondyla fochi*.[47] From the poison gland comes a long-lasting orientation pheromone, and from the pygidial gland a powerful recruitment pheromone, which has been identified (at least in the closely related *Leptogenys diminuta*) as (3*R*,4*S*)-methyl-3-heptanol.[48]

The subfamily Cerapachyinae, close to and once considered only a tribe

47 | U. Maschwitz and P. Schönegge, "Forage communication, nest moving recruitment, and prey specialization in the oriental ponerine *Leptogenys chinensis*," *Oecologia* 57(1–2): 175–182 (1983).

48 | A. B. Attygalle, O. Vostrowsky, H. J. Bestmann, S. Steghaus-Kovaç, and U. Maschwitz, "(3*R*,4*S*)-Methyl-3–heptanol, the trail pheromone of the ant *Leptogenys diminuta*," *Naturwissenschaften* 75(6): 315–317 (1988).

(Cerapachyini) of Ponerinae, is notable for its dietary specialization, which for all member species thus far investigated consists exclusively of ants.[49] The heavily armored workers launch mass raids against their often formidable prey, killing or driving away the adults, then carrying the larvae and pupae back to their own nests for food. The attacks are initiated by single scouts who return to their home nests after detecting a nest of the prey species. Detailed studies of the Australian *Cerapachys turneri* have shown that the scouts lay trails with secretions from the poison gland, which then serve as both recruitment and orientation signals.[50] Another pheromone, apparently rousing in effect, is released from the pygidial gland. The dispersal occurs in the *Cerapachys* nests or on the trail when the running worker lifts its gaster in the slightly elevated "calling position." By this means, the pygidial gland secretion is dispensed directly into the air, as opposed to flat on the ground, the method employed by species of *Pachycondyla*.

The hyperdiversity of ponerine recruitment systems, which is matched by the extreme variety of their ecological adaptations (see Chapter 8), reflects the ancient provenance of the subfamily Ponerinae. To recap, sternal glands of various kinds that secrete trail pheromones have been found in Leptanillinae and in the poneromorph genera *Onychomyrmex*, *Mystrium*, and *Pachycondyla*. *Pachycondyla tarsata* is the only member of its genus known to use sternal glands in trail communication. Other *Pachycondyla* species employ the pygidial gland for this purpose, sometimes with amplification from poison gland secretions. The pygidial gland is an opening from the top near the rear end of the body. Similar patterns have been found in the ponerine genus *Leptogenys*.[51]

There is good reason to view the pygidial gland and its role in recruitment as ancestral traits in the evolution of ants as a whole: all basal ant subfamilies and also most derived subfamilies (a notable exception is Formicinae) possess a pygidial gland.[52] But only in the subfamily Ponerinae and the army ant subfamily Ecitoni-

---

49 | E. O. Wilson, "Observations on the behavior of the cerapachyine ants," *Insectes Sociaux* 5(1): 129–140 (1958).

50 | B. Hölldobler, "Communication, raiding behavior, and prey storage in *Cerapachys* (Hymenoptera: Formicidae)," *Psyche* (Cambridge, MA) 89: 3–21 (1982).

51 | U. Maschwitz and S. Steghaus-Kovac, "Individualismus versus Kooperation: Gegensätzliche Jagd- und Rekrutierungsstrategien bei tropischen Ponerinen (Hymenoptera: Formicidae)," *Naturwissenschaften* 78(3): 103–113 (1991).

52 | B. Hölldobler and H. Engel, "Tergal and sternal glands in ants," *Psyche* (Cambridge, MA) 85: 285–330 (1978). In Formicinae, only the genus *Polyergus* possesses a pygidial gland, which might be a convergently evolved

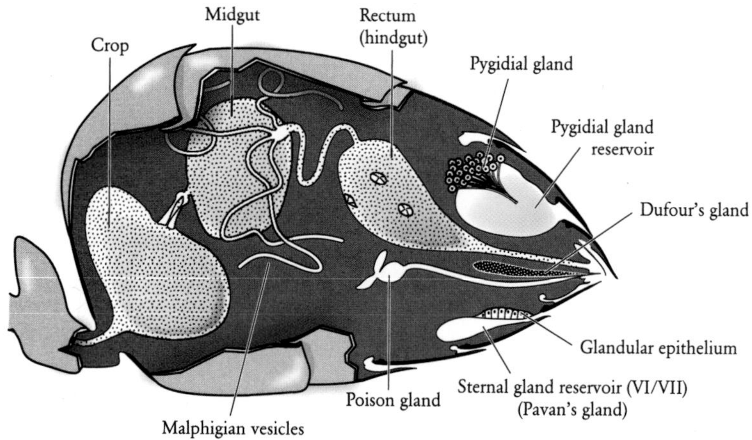

FIGURE 6-12. Gaster of a dolichoderine worker ant showing the major exocrine glands. Based on M. Pavan and G. Ronchetti, "Studi sulla morfologia esterna e anatomia interna dell'operaia di *Iridomyrmex humilis* Mayr e ricerche chimiche e biologiche sulla iridomirmecina," *Atti della Società Italiana di Scienze Naturali e del Museo civico di storia naturale in Milano* 94(3–4): 379–477 (1955).

nae have pygidial glands proved unequivocally to be a source of pheromonal recruitment signals. In other subfamilies, including Aneuretinae, Dolichoderinae, and Myrmicinae, the same glands manufacture repellent and alarm substances (Figures 6-12 and 6-13).

Of equal importance for evolutionary interpretation of the large picture, the pygidial gland serves in the very elementary recruitment technique called tandem running. This form of communication has been documented in the ponerine genera *Pachycondyla*, *Diacamma*, and *Hypoponera*, as well as in a medley of species from the "higher," nonponerine ant subfamilies Formicinae and Myrmicinae (Figure 6-14). Although workers of these species forage singly, they employ tandem running to recruit nestmates when the colony emigrates to new nest sites. The ponerine in which the full procedure has been analyzed is *Pachycondyla* (formerly *Bothroponera*) *tesserinoda*.[53] The recruiter invites a nestmate to follow by pulling it slightly with her mandibles, then turning and walking away in the direction of the target site. Thereafter, the follower ant stays close behind, frequently touch-

structure; see B. Hölldobler, "A new exocrine gland in the slave raiding ant genus *Polyergus*," *Psyche* (Cambridge, MA) 91: 225–235 (1984).

53 | U. Maschwitz, B. Hölldobler, and M. Möglich, "Tandemlaufen als Rekrutierungsverhalten bei *Bothroponera tesserinoda* Forel (Formicidae: Ponerinae)," *Zeitschrift für Tierpsychologie* 35(2): 113–123 (1974).

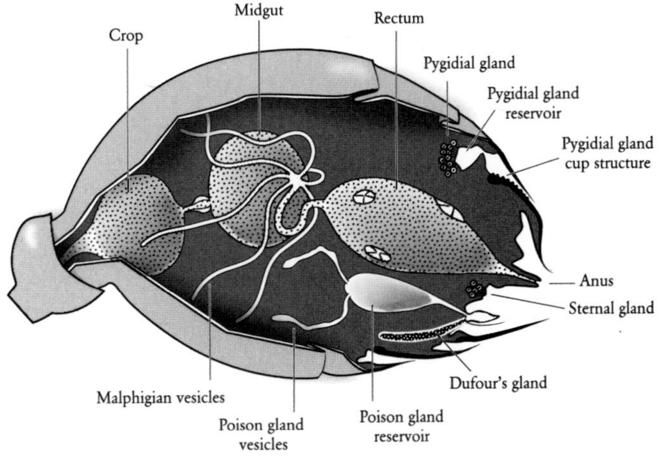

FIGURE 6-13. The location of the major abdominal exocrine glands of the myrmicine genus *Aphaenogaster*. Based on B. Hölldobler, and E. O. Wilson, *The Ants* (Cambridge, MA: The Belknap Press of Harvard University Press, 1990).

ing the leader with her antennae. As long as the leader feels the touch on her own gaster and hind legs, she keeps running. If these stimuli are interrupted, however, she stops and often then wipes her hind legs over her gaster. Meanwhile, the lost follower searches for her by running about in a looping pattern. When contact is reestablished, the tandem run proceeds on toward the goal. Experiments with dummies inserted between two ants have demonstrated that both tactile and chemical signals are necessary for tandem running. Other examples of tandem running, in the myrmicine ant genera *Leptothorax* and *Temnothorax*, and the formicine genus *Camponotus*, are shown in Plate 23 and Figure 10-2.

While conducting pheromonal research on *Pachycondyla tesserinoda*, Karla Jessen and Ulrich Maschwitz made a remarkable discovery about the discriminatory

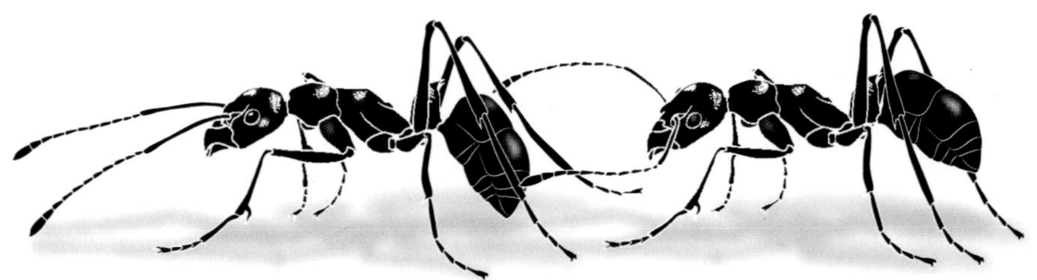

FIGURE 6-14. Recruitment by tandem running in the ponerine ant genus *Diacamma*.

PLATE 23. *Above*: Workers of the nocturnal carpenter ant *Camponotus ocreatus* recruit nestmates into new foraging areas by tandem running. *Below*: A forager of an African *Polyrhachis* species recruits a group of three nestmates to a newly discovered food source.

ability of the ants during communication.[54] Part of the repertoire of recruiting tandem leaders is to deposit orientation odor trails while exploring new terrain.[55] They lay the same kind of trails while searching for new nest sites. In coming and going, they display a preference for their own personal odor trails. Scout workers search singly for food and new nest sites. They report such important targets to the colony by means of tandem running.

---

54 | K. Jessen and U. Maschwitz, "Orientation and recruitment behavior in the ponerine ant *Pachycondyla tesserinoda* (Emery): laying of individual-specific traits during tandem running," *Behavioral Ecology and Sociobiology* 19(3): 151–155 (1986).

55 | Ants of the giant Neotropical *Paraponera clavata* also lay individual-specific orientation trails, independent of recruitment, as documented in M. D. Breed, J. H. Fewell, A. J. Moore, and K. R. Williams, "Graded recruitment in a ponerine ant," *Behavioral Ecology and Sociobiology* 20(6): 407–411 (1987).

So far, investigators have found the pheromones employed in trail laying and tandem running elusive. An attempt to locate the glandular source of tandem running in *Pachycondyla tridentata* has failed.[56] However, in *Pachycondyla obscuricornis*, such a mediating pheromone has been traced to its origin in the pygidial gland. The gland secretion appears to be transferred to the hind legs by a series of grooming movements prior to the initiation of the tandem run. During the run, the follower ant repeatedly touches the hind legs of the leader, allowing her to sense the pheromone almost continually.[57]

Research on other ponerine ants has revealed that recruitment behavior in their workers can be even more subtle and complex than that discovered by the early work on *Pachycondyla*. When Ulrich Maschwitz and colleagues first analyzed tandem running in the ponerine *Diacamma rugosum*, they found no indication that the pygidial gland is involved in this form of communication. The leader ant lays a trail with materials from its hindgut, but the pheromones serve for orientation only, not for recruitment.[58] In short, there existed clear evidence of a recruitment pheromone, yet its origin remained unknown.

The mystery may have been solved ten years later with the discovery of an entirely new kind of gland, now known to occur widely in ants, including the ponerine genera *Diacamma* and *Pachycondyla*.[59] This metatibial gland, as it came to be called, is located on the worker ant's hind "feet." In *Diacamma*, it is used by reproductive females as a sex attractant (see Chapter 8). Dominant individuals call to males by rubbing the metatibial gland secretion over their own gastral tergites. The glandular pores of the tibia are combined with a cuticular brush ideally suited to spread secretions onto the gaster. Tandem leaders in both *Diacamma* and *Pachycondyla* exhibit similar behavior whenever they lose the follower ant. It seems likely, at least from this amount of circumstantial evidence, that the metatibial secretion

---

56 | A. B. Attygalle, K. Jessen, H.-J. Bestmann, A. Buschinger, and U. Maschwitz, "Oily substances from gastral intersegmental glands of the ant *Pachycondyla tridentata* (Ponerinae): lack of pheromone function in tandem running and antibiotic effects but further evidence for lubricative function," *Chemoecology* 7(1): 8–12 (1996).

57 | J. F. A. Traniello and B. Hölldobler, "Chemical communication during tandem running in *Pachycondyla obscuricornis* (Hymenoptera: Formicidae)," *Journal of Chemical Ecology* 10(5): 783–793 (1984).

58 | U. Maschwitz, K. Jessen, and S. Knecht, "Tandem recruitment and trail laying in the ponerine ant *Diacamma rugosum*: signal analysis," *Ethology* 71(1): 30–41 (1986).

59 | B. Hölldobler, M. Obermayer, and C. Peeters, "Comparative study of the metatibial gland in ants (Hymenoptera, Formicidae)," *Zoomorphology* 116(4): 157–167 (1996).

binds the follower to the leader during tandem running. As further evidence, the metatibial gland is located precisely on the part of the hind legs touched most frequently by the antennae of the follower.

Working deeper into the recruitment behavior of *Diacamma*, the Maschwitz team stumbled on a bizarre auxiliary phenomenon. *Polyrhachis lama*, a formicine ant of southeastern Asia, was found to be a social parasite of a species of *Diacamma*. It was the first example of such a symbiosis across different ant subfamilies (in this case, a Formicinae parasitic on a member of Ponerinae). The relationship is all the more surprising because parasitic ant species usually evolve to exploit species closely related to themselves—the so-called Emery's rule (after Carlo Emery, the Italian entomologist who first proposed the idea).

The intrusion has been achieved by pheromonal deception and signal theft. *Polyrhachis lama* colonies inhabit their own nests independently where their own queens reside, but they also insinuate themselves into nearby *Diacamma* nests, where they bring their own brood from the mother nests to be reared by the unsuspecting *Diacamma* hosts.[60] The parasite species is highly integrated into the host colonies. Its workers never forage to obtain food for its own brood. When the *Diacamma* workers emigrate to a new nest site, they lead the *Polyrhachis* workers by tandem running in the same manner as their own nestmates. The follow-behind behavior of the *Polyrhachis* workers is not as smooth as that exhibited by those of *Diacamma*, but it is sufficient for the parasites to complete their journey. The *Polyrhachis* workers then return to the old *Diacamma* nest to retrieve any of their own brood left behind. Remarkably, during this last episode of the emigration, they follow private hindgut odor trails of their own.[61]

In any comprehensive review, a special look is needed at the Neotropical ant *Ectatomma ruidum*. The tribe to which this species belongs, Ectatommini, was long thought, following an early taxonomic revision by William L. Brown, to be part of the subfamily Ponerinae.[62] It has now been elevated to subfamily rank as

---

60 | U. Maschwitz, C. Go, E. Kaufmann, and A. Buschinger, "A unique strategy of host colony exploitation in a parasitic ant: workers of *Polyrhachis lama* rear their brood in neighbouring host nests," *Naturwissenschaften* 91(1): 40–43 (2004).

61 | U. Maschwitz, C. Liefke, and A. Buschinger, "How host and parasite communicate: signal analysis of tandem recruitment between ants of two subfamilies, *Diacamma* sp. (Ponerinae) and its inquiline *Polyrhachis lama* (Formicinae)," *Sociobiology* 37(1): 65–77 (2001).

62 | W. L. Brown, "Contributions toward a reclassification of the Formicidae, II: Tribe Ectatommini (Hymenoptera)," *Bulletin of the Museum of Comparative Zoology, Harvard* 118(5): 175–362 (1958).

Ectatomminae. DNA evidence places the ectatommines close to Myrmicinae—an opinion first expressed, in fact, by Brown himself.[63] *Ectatomma ruidum* ants lay chemical trails to food sources that are either especially productive or else difficult to retrieve. The trail pheromone originates in the Dufour's gland, an accessory gland of the sting apparatus.[64] Among its secreted products, the major pheromone is all-*trans*-geranylgeranyl acetate, which elicits trail following by the *Ectatomma ruidum* workers.[65] That result is consistent with the current phylogenetic placement close to the myrmicines: until recently, *Ectatomma* was the only known genus outside Myrmicinae to employ Dufour's gland secretions to lay recruitment trails; and all myrmicine species use the sting apparatus to lay the trails, with the pheromone originating from the Dufour's gland, the poison gland, or both[66] (see Figure 6-13).

As might be expected, other ectatommine species studied thus far—namely, members of the genus *Gnamptogenys*—lay recruitment trails.[67] *Gnamptogenys striatula* has been found more recently to use Dufour's gland secretion, which is laid out through the sting apparatus. The trail pheromone consists of 4-methylgeranyl esters.[68] There is, in addition, some evidence that members of the primarily Australian ectatommine genus *Rhytidoponera* use recruitment pheromones, in particular to collect larger food items.[69]

Enough data are now available to draw a rough sketch of phylogenetic trends in trail pheromone glands for the ants as a whole. In the "true" army ants, comprising Aenictinae, Dorylinae, and Ecitoninae, trail pheromones originate in the

---

63 | C. S. Moreau, C. D. Bell, R. Vila, S. B. Archibald, and N. E. Pierce, "Phylogeny of the ants: diversification in the age of angiosperms," *Science* 312: 101–104 (2006).

64 | S. C. Pratt, "Recruitment and other communication behavior in the ponerine ant *Ectatomma ruidum*," *Ethology* 81(4): 313–331 (1989).

65 | H. J. Bestmann, E. Janssen, F. Kern, B. Liepold, and B. Hölldobler, "All-*trans*-geranylgeranyl acetate and geranylgeraniol, recruitment pheromone components in the Dufour's gland of the ponerine ant *Ectatomma ruidum*," *Naturwissenschaften* 82(7): 334–336 (1995).

66 | B. Hölldobler and E. O. Wilson, *The Ants* (Cambridge, MA: The Belknap Press of Harvard University Press, 1990).

67 | C. A. Johnson, E. Lommelen, D. Allard, and B. Gobin, "The emergence of collective foraging in the arboreal *Gnamptogenys menadensis* (Hymenoptera: Formicidae)," *Naturwissenschaften* 90(7): 332–336 (2003).

68 | R. Blatrix, C. Schulz, P. Jaisson, W. Francke, and A. Hefetz, "Trail pheromone of ponerine ant *Gnamptogenys striatula*: 4-methylgeranyl esters from Dufour's gland," *Journal of Chemical Ecology* 28(12): 2557–2567 (2002).

69 | M. L. Thomas and V. W. Framenau, "Foraging decisions of individual workers vary with colony size in the greenhead ant *Rhytidoponera metallica* (Formicidae, Ectatomminae)," *Insectes Sociaux* 52(1): 26–30 (2005).

hindgut, the pygidial and postpygidial glands, and—at least in Ecitoninae—the sternal gland at the seventh abdominal segment.[70] In the subfamilies Aneuretinae and Dolichoderinae, which are phylogenetic sister groups, Pavan's gland, a sternal gland located between the sixth and seventh abdominal sternites (see Figure 6-12), serves as the source of the recruitment pheromone.[71] In most species of Formicinae, the pheromones originate in the rectal bladder. In some species of the carpenter ant genus *Camponotus*, orientation pheromones are emitted from the rectal sac and enhanced by a powerful recruitment effect of formic acid, which is released from the poison gland.[72] In yet another variation, scouts of the Australian *Camponotus ephippium* lead a group of 3 to 12 nestmates to the target with stimulating secretions from the cloacal gland, a paired cluster of glandular cells at the base of the seventh abdominal sternite,[73] and further guide them by chemical signposts deposited from the rectal bladder.

A notable deviation from the usual formicine pattern is that of the *Oecophylla* weaver ants. The trail orientation pheromone originates in the rectal gland, an invagination of the hindgut epithelium. This marker substance is discharged independently from that of the rectal bladder, which is applied in fecal droplets to mark the colony territory. The workers also employ a short-range alarm and recruitment trail substance that originates in yet another special sternal gland, this one located at the seventh abdominal sternite[74] (Figure 6-15).

Some variation in trail pheromone production also exists within the subfamily Myrmicinae, the chief rival of Formicinae in species diversity and global abundance. Whereas the great majority of myrmicine genera rely on either the Dufour's

---

70 | B. Hölldobler and H. Engel, "Tergal and sternal glands in ants," *Psyche* (Cambridge, MA) 85: 285–330 (1978); and J. Billen and E. D. Morgan, "Pheromone communication in social insects: sources and secretions," in R. K. Vander Meer, M. D. Breed, H. L. Winston, and K. E. Espelie, eds., *Pheromone Communication in Social Insects* (Boulder, CO: Westview Press, 1998), pp. 3–33.

71 | E. O. Wilson and M. Pavan, "Glandular sources and specificity of some chemical releasers of social behavior in dolichoderine ants," *Psyche* (Cambridge, MA) 66(4): 70–76 (1959).

72 | E. Kohl, B. Hölldobler, and H.-J. Bestmann, "Trail and recruitment pheromones in *Camponotus socius* (Hymenoptera: Formicidae)," *Chemoecology* 11(2): 67–73 (2001); and E. Kohl, B. Hölldobler, and H.-J. Bestmann, "Trail pheromones and Dufour gland contents in three *Camponotus* species (*C. castaneus*, *C. balzani*, *C. sericeiventris*: Formicidae, Hymenoptera)," *Chemoecology* 13(3): 113–122 (2003).

73 | B. Hölldobler, "The cloacal gland, a new pheromone gland in ants," *Naturwissenschaften* 69(4): 186–187 (1982).

74 | B. Hölldobler and E. O. Wilson, "The multiple recruitment systems of the African weaver ant *Oecophylla longinoda* (Latreille) (Hymenoptera: Formicidae)," *Behavioral Ecology and Sociobiology* 3(1): 19–60 (1978).

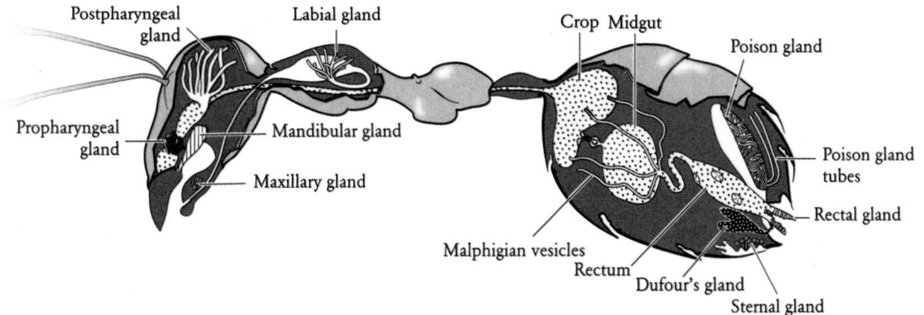

FIGURE 6-15. The location of some of the exocrine glands in the weaver ant genus *Oecophylla*, showing the location of digestive organs along with exocrine glands that secrete pheromones used in social organization. Adapted from B. Hölldobler and E. O. Wilson, *The Ants* (Cambridge, MA: The Belknap Press of Harvard University Press, 1990).

gland, the poison gland, or both, to generate recruitment and orientation pheromones, there is one outstanding exception: workers of *Crematogaster*, a mostly arboreal genus, secretes its trail pheromones from tibial glands in the hind legs.[75]

Throughout their more than 100-million-year evolutionary history, ants have "tinkered" with trail communication, producing in extant species an impressive array of biochemical and behavioral products across their extant phylogenetic clades. The primitive bulldog ants of the Australian subfamily Myrmeciinae evidently lack recruitment and trail signals. This mode of communication is also uncommon in Amblyoponinae and Ponerinae, a large majority of which are solitary predators and scavengers. A few species within these two subfamilies have acquired such pheromonal systems to facilitate group predation. In the highly successful "crown" subfamilies Dolichoderinae, Formicinae, and Myrmicinae, recruitment trail communication is the rule, and some clades within their ranks have reached the highest levels of complexity known in the ants. The techniques they employ make possible rapid food recovery, concerted attacks on enemies, and efficient emigration to new nest sites. In a few clades, the need for recruitment has been lost. For example, the far-ranging workers of the desert-dwelling formicine *Cataglyphis bicolor* have no need for pheromone orientation and have lost it except for chemical signposts at the

75 | R. H. Leuthold, "A tibial gland scent trail and trail-laying behavior in the ant *Crematogaster ashmeadi* Mayr," *Psyche* (Cambridge, MA) 75: 231–248 (1968); and D. J. C. Fletcher and J. M. Brand, "Source of the trail pheromone and method of trail laying in the ant *Crematogaster peringueyi*," *Journal of Insect Physiology* 14(6): 783–786 (1968).

nest entrance. Instead, they rely on an extraordinary capacity for visual orientation and memory.[76]

We do not understand why, through the twists and turns of evolution, particular glands have been chosen as the source of trail pheromones over other glands that are also seemingly well positioned. Nor is it clear why some ant clades modified preexisting glands, such as the poison and Dufour's glands, for this role in communication, while other ant clades evolved entirely new glands for the same purpose.

The specificity and effectiveness of the pheromone receptors that mediate trail communication are dependent on fine details of molecular structure in the pheromone molecules.[77] For example, the Neotropical ant species *Camponotus rufipes* uses 3,7-dimethylisocoumarin as a trail pheromone, while the related Neotropical ant species *Camponotus silvicola* uses a closely similar compound 3,5,7-trimethylisocoumarin. Each responds only to its own trail pheromone.[78] Thus, privacy in the communication system can be achieved by a relatively trivial change in the structure of one molecule.

The specificity of trail pheromones can be enhanced still further by isomerism, in which relatively minor differences in configuration of the same molecule generate new physical or chemical properties discernable by the ants. *Camponotus* workers significantly prefer the (*R*) enantiomer of its trail pheromone over the (*S*) enantiomer.[79] Similar results have been obtained with alarm pheromones: *Atta* leafcutter ants and *Pogonomyrmex* harvester ants prefer the natural molecules produced in their mandibular glands at hundreds of times lower concentration over the synthetic enantiomers of the same substances.

---

76 | R. Wehner, "The ant's celestial compass system: spectral and polarization channels," in M. Lehrer, ed., *Orientation and Communication in Arthropods* (Basel: Birkhäuser Verlag, 1997); R. Wehner, "Desert ant navigation: how miniature brains solve complex tasks," *Journal of Comparative Physiology A* 189(8): 579–588 (2003); and M. Knaden and R. Wehner, "Nest mark orientation in the desert ants *Cataglyphis*: what does it do to the path integrator?" *Animal Behaviour* 70(6): 1349–1354 (2005).

77 | B. Hölldobler and E. O. Wilson, *The Ants* (Cambridge, MA: The Belknap Press of Harvard University Press, 1990); and R. K. Vander Meer, M. D. Breed, K. E. Espelie, and M. L. Winston, eds., *Pheromone Communication in Social Insects: Ants, Wasps, Bees, and Termites* (Boulder, CO: Westview Press, 1998).

78 | E. Übler, F. Kern, H. J. Bestmann, B. Hölldobler, and A. B. Attygalle, "Trail pheromone of two formicine ants, *Camponotus silvicola* and *C. rufipes* (Hymenoptera: Formicidae)," *Naturwissenschaften* 82(11): 523–525 (1995).

79 | E. Kohl, B. Hölldobler, and H.-J. Bestmann, "Trail pheromones and Dufour gland contents in three *Camponotus* species (*C. castaneus, C. balzani, C. sericeiventris*: Formicidae, Hymenoptera)," *Chemoecology* 13(3): 113–122 (2003).

The differences in trail pheromones can be used to distinguish "sibling species," sets of closely related species so similar anatomically that taxonomists have difficulty distinguishing them. For example, *Camponotus floridanus* and *Camponotus atriceps* (once placed together by taxonomists under *Camponotus abdominalis*), turned out to be quite distinct when their behavior and trail pheromones were examined.[80] The pheromone for *Camponotus floridanus* is nerolic acid, while that for *Camponotus atriceps* is the lactone 3,5-dimethyl-6(1-methylpropyl)-tetrahydropyran-2-one.[81]

It is nevertheless true that ant species often follow the trails of other species. The relationship is usually asymmetrical. In experiments with *Camponotus rufipes*, it turned out that workers follow trails drawn with the hindgut contents of *Camponotus silvicola* workers, but the reverse does not occur. Why is that? It turns out that *Camponotus rufipes* manufactures both its own pheromone and that of *Camponotus silvicola,* but *Camponotus silvicola* manufactures only its own pheromone. Another example, from the subfamily Myrmicinae, is the ability of *Aphaenogaster albisetosus* to follow the pheromone of its species as well as that of *Aphaenogaster cockerelli*, but the reverse is not the case. *Aphaenogaster cockerelli* generates 1-phenylethanol (87.8 percent), its personal pheromone, combined with 4-methyl-3-heptanone (10.4 percent), the pheromone of *Aphaenogaster albisetosus. Aphaenogaster albisetosus*, however, produces only its own trail pheromone.[82]

In at least a few other cases, recruitment trail pheromones are not species specific. The key substance in many species of several genera of the subfamily Myrmicinae is 3-ethyl-2,5-dimethylpyrazine (EDMP), generated by the poison gland.[83] The same compound is the principal recruitment substance of at least four sympatric species of the myrmicine harvester genus *Pogonomyrmex*.[84] On the other hand,

---

80 | M. A. Deyrup, N. Carlin, J. Trager, and G. Umphrey, "A review of the ants of the Florida Keys," *Florida Entomologist* 71(2): 163–176 (1988).

81 | U. Haak, B. Hölldobler, H. J. Bestmann, and F. Kern, "Species-specificity in trail pheromones and Dufour's gland contents of *Camponotus atriceps* and *C. floridanus* (Hymenoptera: Formicidae)," *Chemoecology* 7(2): 85–93 (1996).

82 | B. Hölldobler, N. J. Oldham, E. D. Morgan, and W. A. König, "Recruitment pheromones in the ants *Aphaenogaster albisetosus* and *A. cockerelli* (Hymenoptera: Formicidae)," *Journal of Insect Physiology* 41(9): 739–744 (1995).

83 | E. D. Morgan, "Insect trail pheromones: a perspective of progress," in A. R. McCaffrey and I. D. Wilson, eds., *Chromatography and Isolation of Insect Hormones and Pheromones* (New York: Plenum Press, 1990), pp. 259–270.

84 | B. Hölldobler, E. D. Morgan, N. J. Oldham, and J. Liebig, "Recruitment pheromone in the harvester genus *Pogonomyrmex*," *Journal of Insect Physiology* 47(4–5): 369–376 (2001).

the longer-lasting trunk trails used by homing foragers contain species-specific cues.[85] The latter trails are apparently marked by species-specific Dufour's gland components. Even more important, the trunk trail blends are specific to individual colonies.[86] Hence, the trails are private, and they may even serve as long-lasting territorial signals.

Colony specificity in trail communication may prove to be a very common, if not general, phenomenon in ants. Judging from responses to material from the hindgut, this level of discrimination occurs in the formicine *Lasius neoniger*,[87] as well as in *Lasius japonicus*, where it apparently is also derived at least in part from tarsal glands.[88] Colony specificity in hindgut pheromones of trunk route and territorial markers has also been demonstrated in the African weaver ant *Oecophylla longinoda*.[89]

Even more surprising has been the discovery that workers of some ant species lay odor trails peculiar not just to colonies but to individual colony members. This level of discrimination has been documented in the ponerine *Pachycondyla tesserinoda*, the paraponerine *Paraponera clavata*, and the myrmicine *Temnothorax* (formerly *Leptothorax) affinis*.[90]

These various studies have led to the following principle of pheromonal communication: Although species specificity can sometimes be achieved by alteration of the structural design of the pheromone compound, such specificity and other kinds

---

85 | B. Hölldobler and E. O. Wilson, "Recruitment trails in the harvester ant *Pogonomyrmex badius," Psyche* (Cambridge, MA) 77: 385–399 (1970).

86 | B. Hölldobler, E. D. Morgan, N. J. Oldham, J. Liebig, and Y. Liu, "Dufour gland secretion in the harvester ant genus *Pogonomyrmex*," *Chemoecology* 14(2): 101–106 (2004).

87 | J. F. A. Traniello, "Colony specificity in the trail pheromone of an ant," *Naturwissenschaften* 67(7): 361–362 (1980).

88 | T. Akino, M. Morimoto, and R. Yamaoka, "The chemical basis for trail recognition in *Lasius nipponensis* (Hymenoptera: Formicidae)," *Chemoecology* 15(1): 13–20 (2005); and T. Akino and R. Yamaoka, "Trail discrimination signal of *Lasius japonicus* (Hymenoptera; Formicidae)," *Chemoecology* 15(1): 21–30 (2005).

89 | B. Hölldobler and E. O. Wilson, "Colony-specific territorial pheromone in the African weaver ant *Oecophylla longinoda* (Latreille)," *Proceedings of the National Academy of Sciences USA* 74(5): 2072–2075 (1977); and B. Hölldobler, "Territories of the African weaver ant (*Oecophylla longinoda* [Latreille]): a field study," *Zeitschrift für Tierpsychologie* 51(2): 201–213 (1979).

90 | K. Jessen and U. Maschwitz, "Individual specific trails in the ant *Pachycondyla tesserinoda* (Formicidae, Ponerinae)," *Naturwissenschaften* 72(10): 549–550 (1985); K. Jessen and U. Maschwitz, "Orientation and recruitment in the ponerine ant *Pachycondyla tesserinoda* (Emery): laying of individual-specific trails during tandem running," *Behavioral Ecology and Sociobiology* 19(3): 151–155 (1986); U. Maschwitz, S. Lenz, and A. Buschinger, "Individual specific trails in the ant *Leptothorax affinis* (Formicidae: Myrmicinae)," *Experientia* 42(10): 1173–1174 (1986); and M. D. Breed and J. M. Harrison, "Individually discriminable recruitment trails in a ponerine ant," *Insectes Sociaux* 34(3): 222–226 (1987).

of privacy are in most cases achieved by mixing compounds.[91] The phenomenon is clear in the hydrocarbon mixtures of Dufour's gland secretions and in the mixture of substances put into trail signals from multiple glands. In the leafcutter ant genus *Atta*, to take another notable example, the two key components of the trail pheromone produced by the poison gland are methyl 4-methylpyrrole-2-carboxylate (MMPC) and 3-ethyl-2,5-dimethylpyrazine (EDMP) (see Chapter 9). All *Atta* species tested thus far follow artificial trails drawn with MMPC but not those drawn with EDMP; the single exception is *Atta sexdens*, which follows EDMP but not MMPC. Experimental data suggest that the majority of species use MMPC as a trigger (stimulating signal), while EDMP plays a further role in creating species-level specificity. The ratio of the two compounds in the secretion defines the species and determines the optimum response from workers of each species.[92] Other, similar examples in ants have been reviewed by David Morgan.[93]

The most complex multicomponent chemical trail system known is that of the red imported fire ant *Solenopsis invicta*. When a colony is presented with a whole extract of Dufour's gland, the workers respond in three ways. They are attracted to the trail, they are excited by it, and they follow it outward from the nest.[94] Following this discovery, the first that identified the glandular source of an ant pheromone, the exact structure of the chemical substance proved extremely difficult to characterize, and the first attempts to do so failed. Then, in a classic study of pheromone chemistry, Robert K. Vander Meer and colleagues proved that the Dufour's gland structure is actually a medley of pheromones with interdependent effects.[95] The principal component for recruitment along the trail turned out to be *Z,E*-α-farnesene. But this substance in pure form is less active than whole extracts

---

91 | E. D. Morgan, "Chemical words and phrases in the language of pheromones for foraging and recruitment," in T. Lewis, ed., *Insect Communication* (New York: Academic Press, 1984), pp. 169–194; B. Hölldobler and N. F. Carlin, "Anonymity and specificity in the chemical communication signals of social insects," *Journal of Comparative Physiology A* 161(4): 567–581 (1987).

92 | J. Billen, W. Beeckman, and E. D. Morgan, "Active trail pheromone compounds and trail following in the ant *Atta sexdens sexdens* (Hymenoptera: Formicidae)," *Ethology Ecology & Evolution* 4(2): 197–202 (1992).

93 | E. D. Morgan, "Chemical words and phrases in the language of pheromones for foraging and recruitment," in T. Lewis, ed., *Insect Communication* (New York: Academic Press, 1984), pp. 169–194.

94 | E. O. Wilson, "Source and possible nature of the odor trail of fire ants," *Science* 129: 643–644 (1959).

95 | R. K. Vander Meer, "Semiochemicals and the red imported fire ant (*Solenopsis invicta* Buren) (Hymenoptera: Formicidae)," *Florida Entomologist* 66(1): 139–141 (1983); R. K. Vander Meer, "The trail pheromone complex of *Solenopsis invicta* and *Solenopsis richteri*," in C. S. Lofgren and R. K. Vander Meer, eds., *Fire Ants and Leaf-Cutting Ants: Biology and Management* (Boulder, CO: Westview Press, 1986), pp. 201–210.

of Dufour's glands unless combined with two homofarnesene synergists stored in the glands. Oddly, the synergists themselves remain inactive unless primed by yet another, still unidentified primer from the same gland.

The amount of trail pheromone carried by each worker varies greatly among species but is typically minute: in *Camponotus*, it is 2 to 10 nanograms; in leafcutters of the genera *Atta* and *Acromyrmex*, 0.3 to 3.3 nanograms; in *Myrmica rubra*, about 6 nanograms; and in *Pristomyrmex*, measurable only in picograms.[96]

Even such trace amounts, while wholly undetectable to human beings without the aid of microanalytical instruments, are sufficient to convey complete messages between ants. James H. Tumlinson and his coworkers, who in their pioneering work identified MMPC as the trail substance of *Atta texana*, estimated that 1 milligram of this substance, roughly the quantity in a single colony, could suffice, if laid out with maximum efficiency, to lead a worker of *Atta texana* 3 times around the world.[97] Even more astonishing is the subsequent finding that a milligram of the pheromone from the grass-cutting *Atta vollenweideri* suffices to lead a worker 60 times around the Earth, at least at the threshold level of response in half the worker ants tested in the laboratory.[98] In fact, the ants would have a good many molecules to guide them during their long hypothetical journey: 1 milligram spread out in world-girdling trails would still comprise 2 billion molecules in each meter of the trail.

DESIGN AND FUNCTIONAL EFFICIENCY OF PHEROMONES

We now turn to the nature of the chemical signals, which biologists have worked out by theory and experiment. Considerable progress has been made in understanding

---

96 | R. P. Evershed and E. D. Morgan, "The amounts of trail pheromone substances in the venom of workers of four species of attine ants," *Insect Biochemistry* 13(5): 469–474 (1983); R. P. Evershed, E. D. Morgan, and M.-C. Cammaerts, "3–Ethyl-2,5-dimethylpyrazine, the trail pheromone from the venom gland of eight species of *Myrmica* ants," *Insect Biochemistry* 12(4): 383–391 (1982); and E. Janssen, B. Hölldobler, F. Kern, H.-J. Bestmann, and K. Tsuji, "Trail pheromone of myrmicine ant *Pristomyrmex pungens*," *Journal of Chemical Ecology* 23(4): 1025–1034 (1997).

97 | J. H. Tumlinson, R. M. Silverstein, J. C. Moser, R. G. Brownlee, and J. M. Ruth, "Identification of the trail pheromone of a leaf-cutting ant, *Atta texana*," *Nature* 234: 348–349 (1971).

98 | C. J. Kleineidam, W. Rössler, B. Hölldobler, and F. Roces, "Perceptual differences in trail-following leaf-cutting ants relate to body size," *Journal of Insect Physiology* 53(12): 1233–1241 (2007).

the design features of the molecular clouds forming the signals and the contribution of the size and structure of the molecules that compose the clouds. The theory of design is based on the concept of the *active space*, which is the zone within which the concentration of a pheromone or any other biologically active chemical substance is at or above threshold concentration.[99] The active space is thus the signal itself. According to need, the active space can be made large or small, it can reach its maximum quickly or slowly, and it can last for brief or long periods of time. These properties depend primarily on three variables: the amount of substance released as a vapor into the air or evaporated from a liquid or solid chemical deposited on the ground; the sensitivity of the insect to the substance; and the rate at which the substance diffuses. Various adjustments in these variables have been made in the course of evolution, species by species, first by choice of the molecule and then by altering its Q/K ratio, that is, the ratio of the amount emitted (Q) to the threshold concentration at which the receiving animal responds (K). Q, the strength of the signal, is measured in number of molecules released, whereas K, the receiving capacity of the insect, is measured in the minimum concentration of molecules (per unit volume) needed to elicit a response.

The reach and rate of information transfer can be lowered either by reducing the emission rate Q or by raising the threshold concentration K or both. This adjustment achieves a shorter fade-out time of the signal and allows it to be more precisely located in time and space by the receiving insect. For example, an alarm substance, in order to have a short fade-out time and hence operate efficiently, must have an intermediate threshold concentration. That is, the threshold concentration should be neither very high nor very low in comparison with other pheromone systems. The principle is illustrated very well by the alarm pheromone system of the harvester ant *Pogonomyrmex badius*. When the effect of 4-methyl-3-heptanone, the main component, was presented to laboratory colonies, the Q/K ratio values obtained ranged from 939 to 1,800. This magnitude of 10^3 falls well below that calculated for moth sex attractants, at 10^{11}, and well above that for the trail pheromone of fire ants (*Solenopsis invicta*), at about 1 (10^0).

---

99 | W. H. Bossert and E. O. Wilson, "The analysis of olfactory communication among animals," *Journal of Theoretical Biology* 5(3): 443–469 (1963); and E. O. Wilson and W. H. Bossert, "Chemical communication among animals," *Recent Progress in Hormone Research* 19: 673–716 (1963). The subject has been reviewed and updated in B. Hölldobler and E. O. Wilson, *The Ants* (Cambridge, MA: The Belknap Press of Harvard University Press, 1990), from which the present account is mostly drawn.

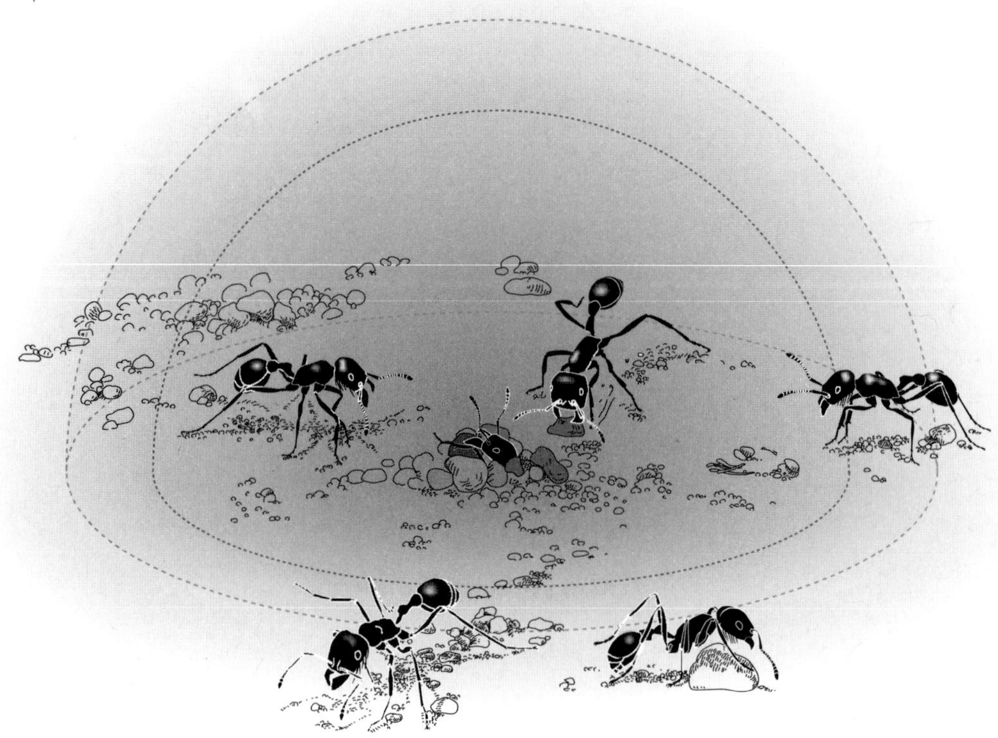

FIGURE 6-16. Zones of response to the alarm pheromone 4-methyl-3-heptanone in the harvester ant *Pogonomyrmex badius*. At low concentrations of the molecule (outer ring), workers are attracted to the source of the emission. At higher concentrations, the ants are both attracted and excited; at the source, they attract enemies encountered or, as shown here, help extricate trapped nestmates that emit the pheromone. Based on E. O. Wilson, "A chemical releaser of alarm and digging behavior in the ant *Pogonomyrmex badius* (Latreille)," *Psyche* (Cambridge, MA) 65: 41–51 (1958); and W. H. Bossert and E. O. Wilson, "The analysis of olfactory communication among animals," *Journal of Theoretical Biology* 5(3): 443–469 (1963).

As a consequence of the intermediate *Q/K* value, the entire contents of the paired mandibular glands of *Pogonomyrmex badius* provide a brief signal when discharged into the air. The active space generated, within which the concentration of the molecules is at or above response threshold concentration, remains small, attaining a maximum radius in still air of only about 6 centimeters. After only 30 seconds or so, diffusion of the molecules shrinks the active space to near zero. Hence, the signal vanishes. The shell of lower concentration toward the edge of the space first attracts workers inward as they walk into the space, and the inner hemisphere, which expands to a maximum radius of about 3 centimeters and lasts approximately 8 seconds, evokes either aggressive or rescue behavior, depending on circumstances (Figure 6-16).

Similar design features have been documented in the alarm systems of other ant species. The parameter values that have evolved seem well suited for pheromone-based alarm, provided the disturbance caused by enemies is local and short-lasting and only a small squad of workers is needed to handle it. If, on the other hand, the problem is greater in magnitude, more persistent, or both, a positive feedback sets in: some of the workers attracted to the scene release pheromones of their own, still more workers are attracted, and the responding force of ants accordingly builds up until the problem is reduced to a manageable level.

More precisely, the concentration-dependent system in *Pogonomyrmex badius* harvester ants is organized as follows. A worker, when excited—say, by a disturbance of the nest in her vicinity—discharges 1 to 3 micrograms of 4-methyl-3-heptanone from her paired mandibular glands. Nestmates in the vicinity are attracted when they come within a zone containing 10^{10} molecules per cubic centimeter. They then move up the concentration gradient toward the point of release. When the concentration reaches about an order of magnitude higher, the workers shift into an aggressive frenzy.[100] The active space can therefore be envisioned as two nested hemispheric shells. A very similar pattern of double zoning occurs in the alarm system of the leafcutter *Atta texana*, which employs the same pheromone, 4-methyl-3-heptanone.[101]

Across many phylogenetic groups of ants, alarm pheromone glands, like trail pheromone glands, produce mixtures of substances that often serve different roles. This enrichment of a crowded exocrine system is nicely exemplified by the "citronella ant" *Lasius* (formerly *Acanthomyops*) *claviger*, a wholly subterranean species of the eastern United States. The mandibular glands produce multiple terpenoid aldehydes and alcohols, which serve as both defensive secretions against enemies and alarm pheromones in the alerting of nestmates. The Dufour's gland produces undecane, which functions as an alarm pheromone, and other hydrocarbons and ketones serve mainly or wholly in defense.[102] A second example of the phenomenon is found

---

100 | E. O. Wilson, "A chemical releaser of alarm and digging behavior in the ant *Pogonomyrmex badius* (Latreille)," *Psyche* (Cambridge, MA) 65: 41–51 (1958); and K. W. Vick, W. A. Drew, E. J. Eisenbraun, and D. J. McGurk, "Comparative effectiveness of aliphatic ketones in eliciting alarm behavior in *Pogonomyrmex barbatus* and *P. comanche*," *Annals of the Entomological Society of America* 62(2): 380–381 (1969).

101 | J. C. Moser, R. C. Brownlee, and R. Silverstein, "Alarm pheromones of the ant *Atta texana*," *Journal of Insect Physiology* 14(4): 529–530 (1968).

102 | F. E. Regnier and E. O. Wilson, "The alarm-defence system of the ant *Acanthomyops claviger*," *Journal of Insect Physiology* 14(7): 955–970 (1968).

in the Australian bulldog ant *Myrmecia gulosa*. Workers defend their nests by pheromones from three sources: an alerting substance from the rectal sac, an activating pheromone from the Dufour's gland, and an attack pheromone from the mandibular glands.[103]

Still another and quite different system can be created through multiple alarm responses evoked by the contents of a single exocrine source. When agitated, workers of the African weaver ant (*Oecophylla longinoda*) expel the mixed contents of their mandibular glands, which evaporate and spread outward by diffusion. Then, due to differences in the diffusion rates and possibly also to differences in the ant's sensitivity to the varying components, the pheromones act at different distances from the point of their release and in a sequence of intensifying response toward the point of release. In the outermost shell of the concentric active space, hexanol alerts the workers; next, 1-hexanol attracts them; and finally, 3-undecanone and 2-butyl-2-octenal incite them to attack and bite any foreign object in the vicinity (Figure 6-17).[104] A comparable phenomenon has been documented in the myrmicine *Myrmicaria eumenoides* (also shown in Figure 6-17).

To return to the design of trail pheromones, it is logical that the pheromone deposit has to be either short-lived, as in recruitment trails, or long-lasting, as in orientation trails. In trail communication, the active space is most readily designed during evolution by adjustment of the chemoreceptive program of the ant and hence the response threshold itself. The ants do not follow the liquid trace laid onto the ground. Instead, they move through the above-threshold vapor "tunnel" created by diffusion of the odorant molecules off the trace. Moving their antennae and hence their odor chemoreceptor organs right and left, with the tips pointed close toward the ground, the ants detect the edge of the active space as they cross it and then move back within the space (Figure 6-18). If an active space were too wide, the ants would have difficulty following the trail. They must be able to stay on track by measuring the difference in the concentration of molecules detected from the antenna on one side to the

---

103 | P. L. Robertson, "Pheromones involved in aggressive behaviour in the ant, *Myrmecia gulosa*," *Journal of Insect Physiology* 17(4): 691–715 (1971).

104 | J. W. S. Bradshaw, R. Baker, and P. E. House, "Multicomponent alarm pheromones of the weaver ant," *Nature* 258: 230–231 (1975); and J. W. S. Bradshaw, R. Baker, and P. E. House, "Multicomponent alarm pheromones in the mandibular glands of major workers of the African weaver ant, *Oecophylla longinoda*," *Physiological Entomology* 4(1): 15–25 (1979).

Myrmicaria eumenoides

Limonene (circling)

β-Pinene (alerting and circling)

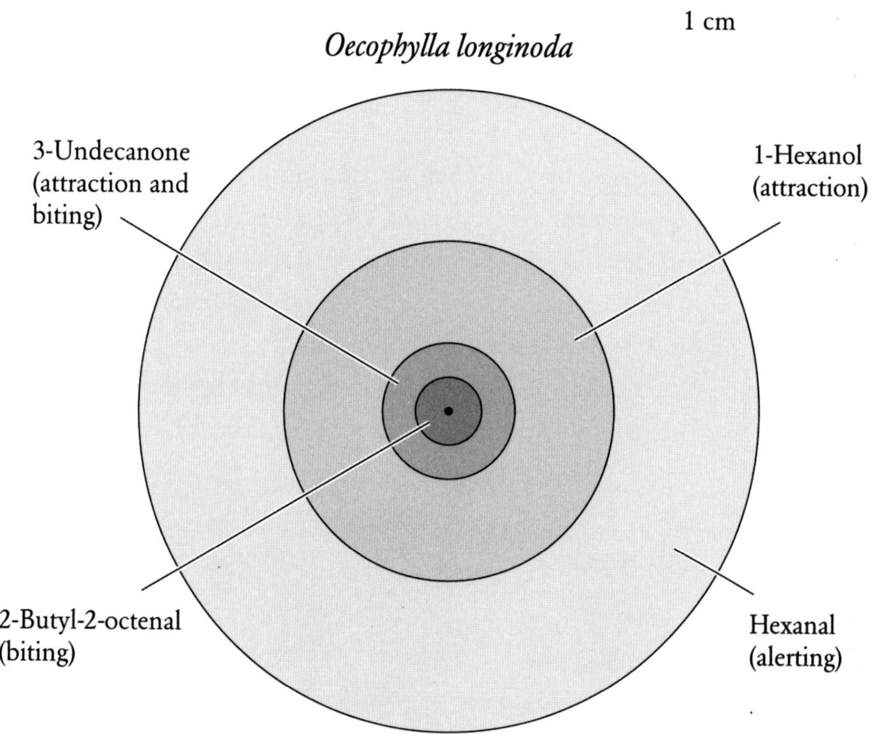

FIGURE 6-17. Concentric active spaces of multiple pheromone components released from the exocrine gland of two species of ants. The spaces are depicted as circles but are actually overlapping hemispheres that speed above the ground from the point of source. *Above*: Poison gland substances used in recruitment by *Myrmicaria eumenoides*, 40 seconds after deposition on a flat surface at the point of source. *Below*: Mandibular gland substances used in alarm recruitment by *Oecophylla longinoda*, 20 seconds after deposition on a flat surface at a point in the center. Modified from J. W. S. Bradshaw and P. E. Howse, "Sociochemicals of ants," in W. J. Bell and R. T. Cardé, eds., *Chemical Ecology of Insects* (London: Chapman and Hall, 1984), pp. 429–473.

antenna on the other side. This form of orientation (technically called osmotropotaxis) was experimentally demonstrated by Walter Hangartner in the European carton-building ant *Lasius fuliginosus*.[105]

Once launched on a trail, how do ants determine its polarity? In other words, which way is the food and which way is the nest? During the day at least, they can choose by sun compass orientation, using the angle to the sun to learn their direction during trips out from the nest and back again. More precisely, they use the pattern of polarized sunlight in the sky.[106] Even if an ant lingers for hours away from home, she can still adjust her reading by means of an internal clock that compensates for the arc of the sun's movement. That ants use the sun for orientation was discovered by the pioneering behavioral physiologist Felix Santschi in 1923. In his classic yet simple experiment, he let ants orient by the sun. Then, using blinds and mirrors, he reversed the apparent position of the sun in the sky. The worker ants that were deluded in this way switched their direction by 180°. Among the species of ant genera *Aphaenogaster*, *Cataglyphis*, *Messor*, and *Monomorium* tested by Santschi were some that use odor trails and those that orient entirely by vision.[107] We have duplicated this experiment in the laboratory with *Solenopsis* fire ants and *Pogonomyrmex* harvester ants. We placed table lamps on each side of columns of foragers in a darkened room, then allowed the ants to adapt to

FIGURE 6-18. Trail-following ants, such as workers of the formicine genus *Lasius* shown here, use comparative input from their two antennae to detect the edge of the active space and move back into a region of higher molecular concentration to stay on the trail. Based on an original illustration by Turid Hölldobler-Forsyth from B. Hölldobler, "Chemische Verständigung im Insektenstaat am Beispiel der Hautflügler (Hymenoptera)," *Umschau* 70(21): 663–669 (1970).

---

105 | W. Hangartner, "Spezifikät und Inaktivierung des Spurpheromons von *Lasius fuliginosus* Latr. und Orientierung der Arbeiterinnen im Duftfeld," *Zeitschrift für vergleichende Physiologie* 57(2): 103–136 (1967).

106 | R. Wehner, "Arthropods," in F. Papi, ed., *Animal Homing* (New York: Chapman & Hall, 1992), pp. 45–144; and R. Wehner, "The ant's celestial compass system: spectral and polarization channels," in M. Lehrer, *Orientation and Communication in Arthropods* (Basel: Birkhäuser Verlag, 1997), pp. 145–185.

107 | See the historical review, emphasizing the Santschi experiments, in R. Wehner, "On the brink of introducing sensory ecology: Felix Santschi (1872–1940)—Tabib-en-Neml," *Behavioral Ecology and Sociobiology* 27(4): 295–306 (1990).

one lighted lamp with the other turned off. When the lighted lamp was turned off and the inactive one switched on, most of the ants abruptly turned 180°.

In an equally classic series of studies by Rüdiger Wehner, the fine details of visual physiology and sun compass orientation were revealed using the desert ant *Cataglyphis bicolor*.[108] We know from this work and that of other investigators that the ant keeps in her brain a map of the pattern of polarized sky light, and the pattern changes with the sun's movement.

A major conclusion drawn by Wehner and other sensory physiologists is that sun compass orientation is only one part of an integrated navigational system, with different cues and even different sensory modalities called into play according to circumstance. Thus, in *Pogonomyrmex* harvester ants, trunk trails used in long-distance orientation originate in the Dufour's gland, and their pheromones are deposited with short-lived recruitment trail pheromones from the poison gland. Combined with these chemical signals are visual landmark cues and, typically at greater distances, sun compass orientation. The integration of all three cues yields a high level of accuracy in orientation.[109] A similar system has been documented in a harvester species belonging to a second genus, *Pheidole militicida*.[110]

The exact combination of cues used by foraging workers varies widely across species. Exclusively subterranean ants depend primarily on odor trails and likely require some directionality provided by odor gradients.[111] Mound-building species of the wood ant genus *Formica* depend on the visual memory of landmarks along the chemically marked trails.[112] Perhaps most surprising, workers of the large African stink ant *Pachycondyla tarsata* memorize patterns in the forest canopy after leaving the nest by gazing at the canopy pattern from the ground. They then reverse the image to find their way home.[113]

There have been a few observations and more than a little speculation on how

---

108 | R. Wehner, "The ant's celestial compass system: spectral and polarization channels," in M. Lehrer, ed., *Orientation and Communication in Arthropods* (Basel: Birkhäuser Verlag, 1997), pp. 145–185.

109 | B. Hölldobler, "Recruitment behavior, home range orientation and territoriality in harvester ants, *Pogonomyrmex*," *Behavioral Ecology and Sociobiology* 1(1): 3–44 (1976).

110 | B. Hölldobler and M. Möglich, "The foraging system of *Pheidole militicida* (Hymenoptera: Formicidae)," *Insectes Sociaux* 27(3): 237–264 (1980).

111 | E. O. Wilson, "Chemical communication among workers of the fire ant *Solenopsis saevissima* (Fr. Smith), 3: The experimental induction of social responses," *Animal Behaviour*, 10(1–2): 159–164 (1962).

112 | R. Rosengren, "Route fidelity, visual memory and recruitment behaviour in foraging wood ants of the genus *Formica* (Hymenoptera, Formicidae)," *Acta Zoologica Fennica* 133: 1–106 (1971).

113 | B. Hölldobler, "Canopy orientation: a new kind of orientation in ants," *Science* 210: 86–88 (1980).

ants might determine polarity with cues in the trails themselves. In the long, hugely populous columns of the Asiatic ant *Pheidologeton diversus*, workers are able to determine direction by contrasting the homeward movements of nestmates laden with food with the opposite movements of their outbound, unburdened nestmates.[114] This is, to our knowledge, the only such example documented to the present time, and it does not imply the use of any form of odor gradient along the trail.

Another possible directional cue is the presence of odor gradients emanating from territorial marking pheromones laid by ants in the vicinity of the nest. However, other than underground carbon dioxide gradients, no cases have been documented.[115] Yet another possibility, and one that has been experimentally documented, is the use of trail geometry to determine polarity. If, as shown in the household pest species *Monomorium pharaonis*, the trail bifurcation is Y-shaped outbound, with a narrow angle facing away from the nest, foragers can simply choose to move down the stem of the Y and safely homeward.[116]

BEHAVIORAL MODES OF RECRUITMENT COMMUNICATION

The behavioral programs of recruitment vary enormously among ant species, reflecting differences in their way of life, the size of their colonies, and their target of recruitment. Consider the phenomenon of mass communication—the transmission of information from one group of individuals to another group.[117] In the case of the red imported fire ant *Solenopsis invicta*, where the phenomenon was first documented, the number of workers leaving the nest is controlled by the amount of trail substance being emitted by foragers already in the field. Tests involving the use of

---

114 | M. W. Moffett, "Ants that go with the flow: a new method of orientation by mass communication," *Naturwissenschaften* 74(11): 551–553 (1987).

115 | E. O. Wilson, "Chemical communication among workers of the fire ant *Solenopsis saevissima* (Fr. Smith), 3: The experimental induction of social responses," *Animal Behaviour*, 10(1–2): 159–164 (1962); and W. Hangartner, "Carbon dioxide, a releaser for digging behavior in *Solenopsis geminata* (Hymenoptera: Formicidae)," *Psyche* (Cambridge, MA) 76: 58–67 (1969).

116 | D. E. Jackson, M. Holcombe, and F. L. W. Ratnieks, "Trail geometry gives polarity to ant foraging networks," *Nature* 432: 907–909 (2004).

117 | E. O. Wilson, "Chemical communication among workers of the fire ant *Solenopsis saevissima* (Fr. Smith), 2: An information analysis of the odour trail," *Animal Behaviour* 10(1–2): 148–158 (1962). (The name *Solenopsis saevissima* has been replaced by *Solenopsis invicta*.)

enriched trail pheromone have shown that the number of individuals drawn outside the nest is a linear function of the amount of the substance presented to the colony as a whole. Under natural conditions, this quantitative relation results in the adjustment of the outflow of workers to the level needed at the food source. A balance in number is then achieved in the following manner. The initial buildup of workers at a newly discovered food source is logistic. That is, after first accelerating at a nearly exponential rate, it decelerates toward a limit as workers become crowded on the food mass, because workers unable to reach the mass turn back without laying trails and because trail deposits made by single workers evaporate within a few minutes. As a result, the number of workers at food masses tends to stabilize at a level that is a linear function of the area of the food mass. Sometimes, as when the food find is of poor quality or far away or when the colony is already well fed, the workers do not cover the food mass entirely, but end up at a lower density. This mass communication of quality is achieved by means of an "electorate" response, in which individuals choose whether to lay trails after inspecting the food find. If they do lay trails, they adjust the quantity of pheromone according to circumstance. The more desirable the food find, the higher the percentage of positive responses, the greater the trail-laying effort by individuals, the more trail pheromones presented to the colony, and hence the more newcomer ants emerging from the nest.[118]

Subsequent research has revealed that even single ants can contribute to the flexibility of the mass communication system. Individual workers of *Solenopsis geminata* are able to adjust the amounts of their own pheromone emissions to the specific food needs of their colony and to the quality of the food source. Walter Hangartner, by inducing the homing foragers of *Solenopsis geminata* to lay their trail on a soot-coated glass plate, found that the continuity of the sting trail rises with increasing starvation time of the colony, an increased quality of the food source, and a decreased distance between the food and the nest.[119]

In addition, Debby Cassill discovered that individual response by *Solenopsis invicta* workers to recruitment signals is further fine-tuned by auxilia y tactile and motor display signals, whose intensity is correlated with the quality of the food as

118 | E. O. Wilson, "Chemical communication among workers of the fire ant *Solenopsis saevissima* (Fr. Smith), 2: An information analysis of the odour trail," *Animal Behaviour* 10(1–2): 148–158 (1962). (The name *Solenopsis saevissima* has been replaced by *Solenopsis invicta*.)

119 | W. Hangartner, "Structure and variability of the individual odor trail in *Solenopsis geminata* Fabr. (Hymenoptera, Formicidae)," *Zeitschrift für vergleichende Physiologie* 62(1): 111–120 (1969).

well as its distance from the nest and the degree of hunger in the colony. At least during the initiation phase of recruitment, the most effective trail is not the incoming trail deposited by the returning scout ant, but the outgoing trail. A trail-laying scout, upon returning from the newly discovered food source, stimulates her nestmates by touching them with her antennae, performing motor displays, and presenting taste samples. She then leads them along a freshly deposited outgoing recruitment trail.[120]

In a similar case, Madelein Beekman and her colleagues report that small colonies of pharaoh ants (*Monomorium pharaonis*), comprising about 600 ants or less, are unable to forage in an organized manner when the feeding dish is only 50 centimeters away. The researchers concluded that the trail is ineffective because the trail pheromone is so highly volatile and the number of trail layers so limited. As colony size increases, however, more foragers visit the food source and in turn more ants lay trails. Hence, the recruitment substances attain a critical concentration along the trail, and workers respond with full trail behavior. Subsequently, the trail becomes sufficiently reinforced to keep the foragers moving back and forth. It remains potent so long as the workers can fill their crops at the food source and the colony accepts the retrieved bounty. Beekman and her colleagues call this process a "phase transition" from disordered foraging (meaning the trail is not persistent enough because of high evaporation) to ordered foraging (trail-based foraging), and they suggest an analogy to first-order transition in physical systems, such as the discontinuous change from water to ice at a critical temperature.[121] Pointing out this analogy and adopting the terminology to the transition from disordered to mass communication trail foraging is original, but understanding the phenomenon is far from new. In fact, this phenomenon is intimately related to the active space concept, behavioral threshold concentration, and information transfer analyzed and modeled for *Solenopsis invicta* over 40 years before.[122]

---

120 | D. Cassill, "Rules of supply and demand regulate recruitment to food in an ant society," *Behavioral Ecology and Sociobiology* 54(5): 441–450 (2003).

121 | M. Beekman, D. J. T. Sumpter, and F. L. W. Ratnieks, "Phase transition between disordered and ordered foraging in Pharaoh's ants," *Proceedings of the National Academy of Sciences USA* 98: 9703–9706 (2001).

122 | E. O. Wilson, "Chemical communication among workers of the fire ant *Solenopsis saevissima* (Fr. Smith), 1: The organization of mass-foraging," *Animal Behaviour* 10(1–2): 134–147 (1962); E. O. Wilson, "Chemical communication among workers of the fire ant *Solenopsis saevissima* (Fr. Smith), 2: An information analysis of the odour trail," *Animal Behaviour* 10(1–2): 148–158 (1962); E. O. Wilson, "Chemical communication among workers of the fire ant *Solenopsis saevissima* (Fr. Smith), 3: The experimental induction of social responses," *Animal*

While the recruitment trail pheromones are generally highly volatile and hence short-lived, they are often combined with longer-lived orientation substances, which may be derived from either the same or another exocrine gland. For example, in *Monomorium pharaonis*, the short-lived recruitment substance comes from the Dufour's gland, but there is evidence that longer-lasting homing or orientation chemicals are added to the trails from the poison gland. The latter is also the source of repellent substances used by the ants for defense.[123]

Pharaoh ants do not differ greatly from *Solenopsis* or other mass-recruiting ant species in combining short-term and long-term signals in the same message delivery system.[124] An additional twist in odor trail communication is the recent report of "repellent pheromones" used by pharaoh ants to mark unrewarding branches on foraging trail bifurcations.[125] Despite experiments, however, no evidence has yet been found of any anatomical origin of such a pheromone aimed at repelling nestmates, including the poison and Dufour's glands tested for it.[126]

Overall, trail pheromones, through the mass effect, provide a control that is more complex than might have been assumed from knowledge of the relatively elementary individual responses that compose it. This complexity is increased still further by the pheromone acquiring different meanings in at least two different contexts. When colonies move from one nest site to another, a common event in the life of fire ants and many other kinds of ants, the new site is chosen by scout workers, who then lay

---

Behaviour 10(1–2): 159–164 (1962). (The name *Solenopsis saevissima* has been replaced by *Solenopsis invicta*.); W. H. Bossert and E. O. Wilson, "The analysis of olfactory communication among animals," *Journal of Theoretical Biology* 5: 443–469 (1963); E. O. Wilson and W. H. Bossert, "Chemical communication among animals," in G. Pincus, ed., *Recent Progress in Hormone Research*, Vol. 19 (New York: Academic Press, 1963), pp. 673–716; and E. O. Wilson, W. H. Bossert, and F. E. Regnier, "A general method for estimating threshold concentrations of odorant molecules," *Journal of Insect Physiology* 15(4): 597–610.

123 | B. Hölldobler, "Chemische Strategie beim Nahrungserwerb der Diebsameise (*Solenopsis fugax* Latr.) und der Pharaoameise (*Monomorium pharaonis* L.)," *Oecologia* 11(4): 371–380 (1973).

124 | D. E. Jackson, S. J. Martin, M. Holcombe, and F. L. W. Ratnieks, "Longevity and detection of persistent foraging trails in Pharaoh's ants, *Monomorium pharaonis* (L.)," *Animal Behaviour* 71(2): 351–359 (2006).

125 | E. J. H. Robinson, D. E. Jackson, M. Holcombe, and F. L. W. Ratnieks, " 'No entry' signal in ant foraging," *Nature* 438: 442 (2005). Similar results have been found in other myrmicine species—for example, in the Neotropical ant *Daceton armigerum*, where poison gland secretions have a long-lasting orientation effect and secretions from sternal glands elicit a short-lasting recruitmient effect; see B. Hölldobler, J. M. Palmer, and M. Moffett, "Chemical communication in the dacetine ant *Daceton armigerum* (Hymenoptera: Formicidae)," *Journal of Chemical Ecology* 16(4): 1207–1219 (1990).

126 | B. Hölldobler, "Chemische Strategie beim Nahrungserwerb der Diebsameise (*Solenopsis fugax* Latr.) und der Pharaoameise (*Monomorium pharaonis* L.)," *Oecologia* 11(4): 371–380 (1973); and B. Hölldobler, additional unpublished data.

odor trails back to the old nest. Other workers are drawn out by the pheromone. They investigate the new site and, if satisfied, add their own pheromone to the trail. In this fashion, the number of workers traveling back and forth builds up exponentially. In time, the brood is transferred, the queen walks over, and the emigration is complete. The pheromone also functions as an auxiliary signal in alarm communication. When a worker is seriously disturbed, she releases some of the trail substance simultaneously with alarm substance from her head, so that nearby workers are not only alarmed but also attracted to the threatened nestmate.

While impressive complexity can be built into the trail communication system of ants with pheromones alone, tactile signals such as that derived from "head waggling" can enhance nestmate response, especially during the first stages of recruitment.[127]

THE EXTREME MULTIPLE RECRUITMENT SYSTEM OF WEAVER ANTS

The African weaver ant (*Oecophylla longinoda*) possesses the most complex trail system thus far discovered in ants. Workers, which use one repertoire of signals to construct nests of larval silk in tree canopies, utilize no fewer than five recruitment mechanisms to draw nestmates from nests to the remainder of the nest trees and to the foraging areas beyond. These include:

1 | Recruitment to new food sources, under the stimulus of orientation odor trails produced by the scout from her rectal gland (Figure 6-19; see also Figure 6-15), together with a tactile and motor display. These signals are presented while the scout engages in slight head waving with mandibles opened and the lower lip (labium) extended, using the antennae to touch the receiving ant (Figure 6-20).

2 | Recruitment to new terrain or new construction sites of leaf nests, employing pheromones from the rectal gland, slight back-and-forth body jerking, and antennal play.

---

127 | D. Cassill, "Rules of supply and demand regulate recruitment to food in an ant society," *Behavioral Ecology and Sociobiology* 54(5): 441–450 (2003).

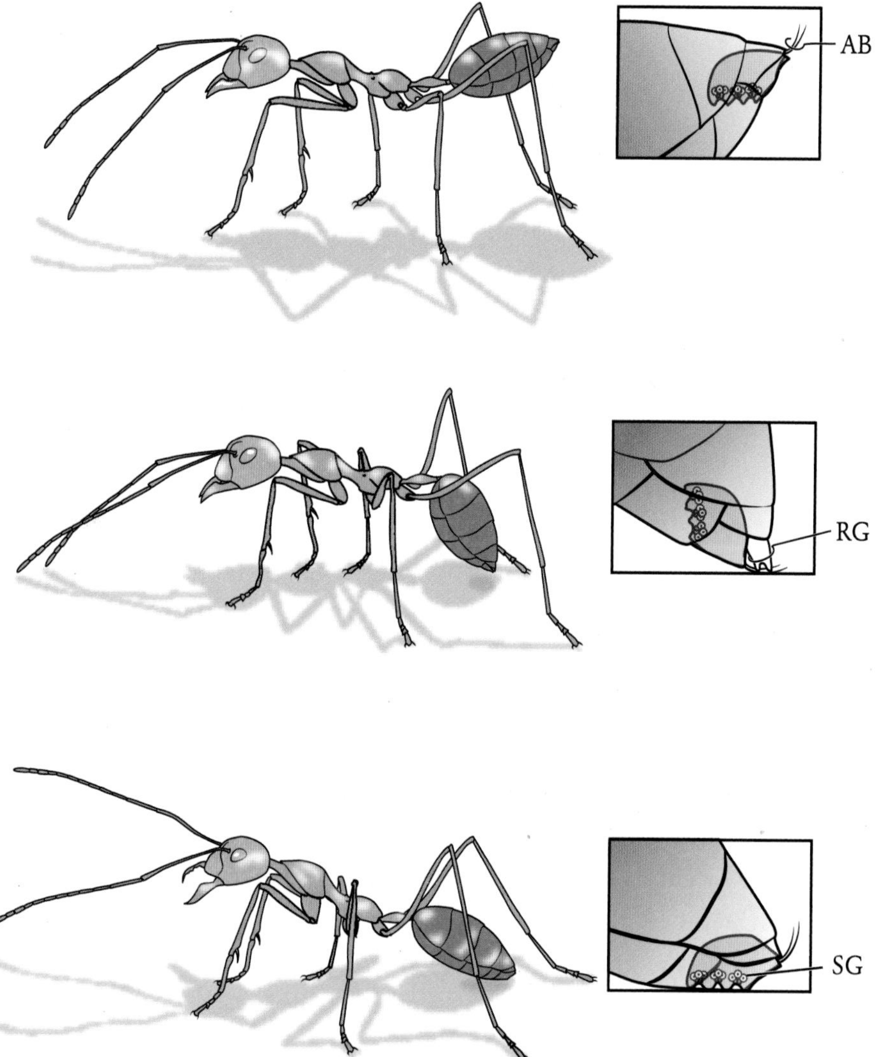

FIGURE 6-19. Workers of the African weaver ant *Oecophylla longinoda* use a trail substance from the rectal gland (RG, middle panel) applied with the aid of an anal brush (AB, top panel) when recruiting to new food sources, nest sites, and other targets (long-range recruitment). They rotate the anal pore and brush upward and apply a sternal gland secretion (SG, bottom panel) to alert nestmates to large prey or to territorial intruders (short-range recruitment). Based on an original illustration by Turid Hölldobler-Forsyth from B. Hölldobler and E. O. Wilson, "The multiple recruitment systems of the African weaver ant *Oecophylla longinoda* (Latreille) (Hymenoptera: Formicidae)," *Behavioral Ecology and Sociobiology* 3(1): 19–60 (1978).

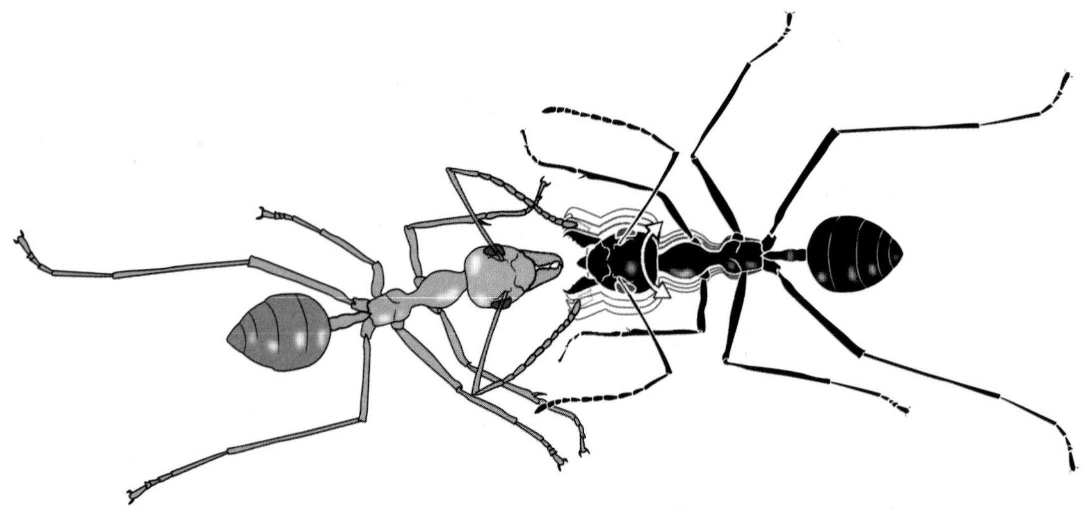

FIGURE 6-20. In the weaver ant *Oecophylla longinoda*, recruiting workers reinforce the rectal gland trail with the motor display shown here when recruiting nestmates to food sources.

3 | Emigration into newly built nests by mandible pulling and jerking, accompanied by the laying of rectal gland trails.

4 | Short-range recruitment to large prey or territorial intruders, during which the terminal abdominal sternite is maximally exposed and dragged for short distances over the ground, releasing an attractant from the sternal gland (see Figure 6-19 and also Figure 6-15). This action is most likely initiated by the discharge of alarm pheromones from the mandibular gland.

5 | Long-range recruitment to territorial intruders, mediated by antennation and intense body jerking with the gaster raised and mandibles aggressively opened (Figure 6-21) and trails laid with rectal gland material that lead to the location of the territorial violation (Plates 24 to 26).[128]

---

128 | B. Hölldobler and E. O. Wilson, "The multiple recruitment systems of the African weaver ant *Oecophylla longinoda* (Latreille) (Hymenoptera: Formicidae)," *Behavioral Ecology and Sociobiology* 3(1): 19–60 (1978).

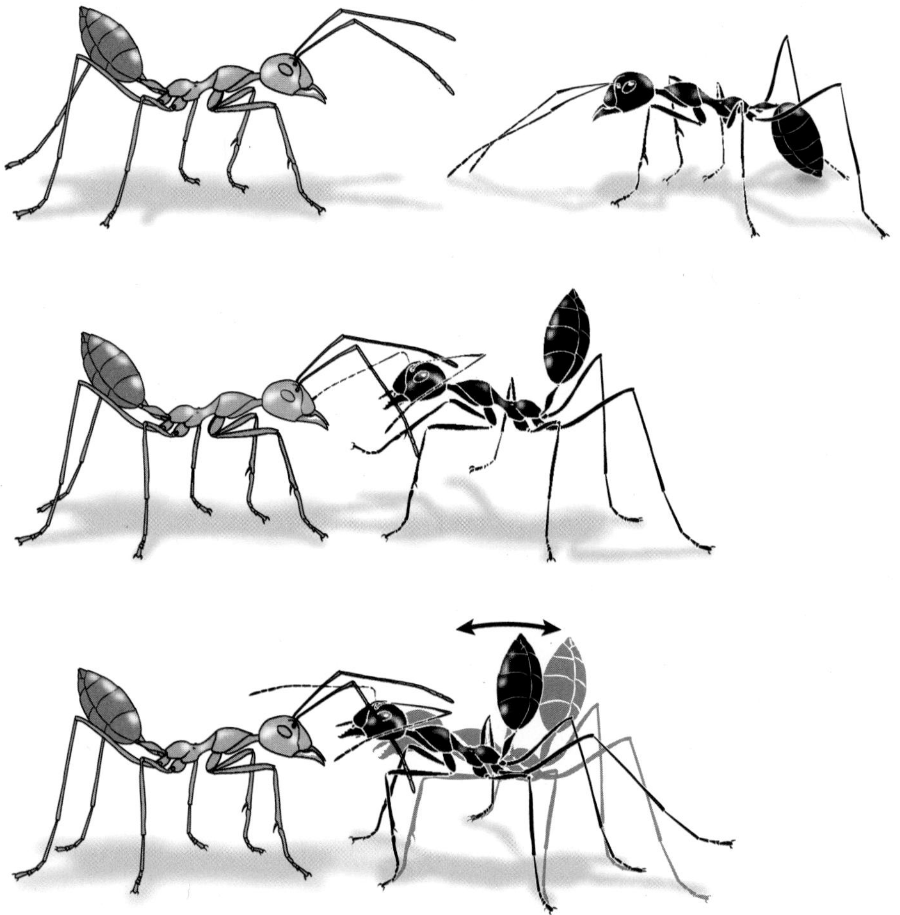

FIGURE 6-21. To alert nestmates to territorial intruders encountered at a longer distance from the home nests, African weaver ant workers lay trails from the intruders to the home nests, where they alert nestmates with antennal tapping and intense body jerking while assuming an aggressive posture.

MULTIMODAL SIGNALS, PARSIMONY, AND RITUALIZATION

In the vast majority of cases, the origin of communication signals is through "ritualization," the evolutionary process by which a phenotypic trait with a noncommunicative function is altered to serve as a signal. Ordinarily, the process begins when some movement, anatomical feature, or physiological process that is functional in another context requires a secondary role—and then perhaps a primary or even

PLATE 24. *Above*: After an encounter with enemies on the colony territory, a worker of the African weaver ant *Oecophylla longinoda* lays a trail with secretions from the rectal gland. *Below*: When the weaver ant worker encounters nestmates, she exhibits an aggressive display that helps initiate recruitment to defend against the intra- or interspecific invaders. The aroused nestmates respond with a similar display and subsequently follow the trail to the target area.

PLATE 25. *Above*: Workers of the African weaver ant attack a foreign conspecific ant from a neighboring colony. *Below*: Weaver ant workers attack a soldier of the African army ant of the genus *Dorylus*.

PLATE 26. Workers of weaver ants, in this case *Oecophylla smaragdina*, the so-called green tree ant of Australia, employ close-range recruitment with secretions from the sternal and mandibular glands, assembling cooperating groups that are able to capture and transport large prey objects.

exclusive role—as a signal.[129] Relatively few communication systems in ants have been analyzed explicitly with evolutionary origins in mind. Still, sufficient evidence has been adduced to pinpoint a few.

The five recruitment systems of the African weaver ant provide one example. They illustrate in striking manner one of the properties of ritualization: the economy of manufacture. Although the messages differ from one another, their signals have been built from pheromones that originated from only two organs, the rectal gland and the sternal gland. The information was then enhanced by the addition of an array of chemical and mechanical elements. The specificity of each weaver ant recruitment system comes principally from a combination of chemical and mechanical elements. A primitive form of syntax is achieved by this vocabulary. Recruitment of nestmates, variously to food, new terrain, new nest sites, and territorial intruders, are all guided by pheromones from the rectal gland. But the discovery of food is further specified by head waving, accompanied by opening of the mandibles with extension of the labium (see Figure 6-20). Long-range recruitment to territorial intruders, on the other hand, is specified by intensive body jerking, with the mandibles opened, the labium folded inward, and the gaster raised (see Figure 6-21)—the actions of a worker in a highly aggressive state. A close cinematographic analysis shows that these food recruitment movements precede actual regurgitation, not just in *Oecophylla* but in formicine ants generally. Similarly, the motor display signaling recruitment to territorial intruders is almost identical to the aggressive "intention movements" that often precede actual fights between members from different colonies (Figure 6-22). It thus appears that the intention movements have been ritualized. That is, they have been adapted to a second function within the recruitment communication systems. They are employed as symbolic icons to transmit specific messages to nestmates, informing them why they should proceed along the rectal gland orientation trail. In summary, food-offering behavior has been ritualized to become a foraging recruitment stimulus, and aggressive intention behavior has been ritualized to recruit nestmates for territorial defense.

Ritualization is one means employed by ants to achieve parsimony in evolutionary adaptation. There is evidently no need—at least under conditions thus far identified by researchers—to create a new biological system if an old system can be suitably retrofitted without more than equal loss of fitness.

---

129 | An extensive account of ritualization in the signal evolution of ants is given in B. Hölldobler and E. O. Wilson, *The Ants* (Cambridge, MA: The Belknap Press of Harvard University Press, 1990).

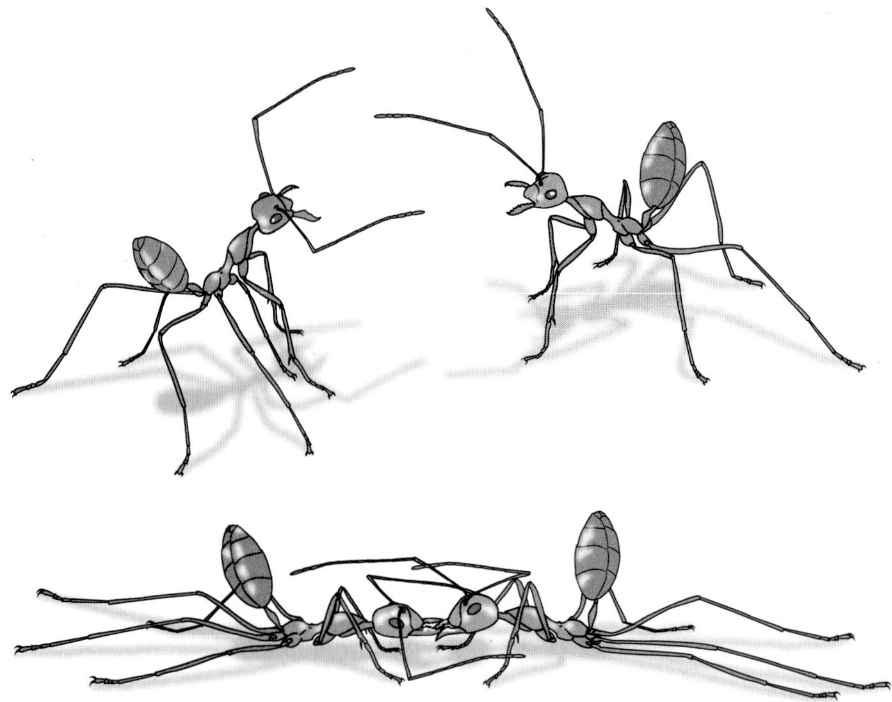

FIGURE 6-22. Workers from different colonies of African weaver ants threaten one another with characteristic aggressive intention movements (above) before locking mandibles in combat (below). Based on an original painting by Turid Hölldobler-Forsyth from B. Hölldobler and E. O. Wilson, "The multiple recruitment systems of the African weaver ant *Oecophylla longinoda* (Latreille) (Hymenoptera: Formicidae)," *Behavioral Ecology and Sociobiology* 3(1): 19–60 (1978).

Ritualization is not limited to tactile signaling of the kind that occurs in *Oecophylla* weaver ants. Chemical alarm communication in ants has evidently evolved from chemical defense behavior. Like many solitary insects, ants and other social insects use chemical secretions to repel predators and other enemies. In social insects, secretions are closely linked to alarm communication. Quite often, a single substance serves both functions. A well-documented example of a substance with the dual function is citronellal, the mandibular gland product of *Lasius* (formerly *Acanthomyops*) *claviger* (see Figure 6-5).[130]

Further, *Lasius* and other formicine ants commonly employ hindgut contents as trail pheromones, a function that seems likely to have originated by ritualization

---

130 | E. O. Wilson, W. H. Bossert, and F. E. Regnier, "A general method for estimating threshold concentrations of odorant molecules," *Journal of Insect Physiology* 15(4): 597–610 (1969).

of the defecation process. The final development of this propensity, exemplified by the extraordinary rectal gland of the *Oecophylla* weaver ants, is the origin of a wholly new structure to generate the trail substances. In fact, *Oecophylla* workers employ the hindgut in two ways that could have evolved as ritualized defecation. The first is recruitment by odor trails made with secretions from the rectal gland, which are excreted independently from the discharge of fecal material from the hindgut. The second hindgut employment is the use of fecal material in territorial marking. The ants deposit fecal droplets more or less uniformly along trunk routes and over the surface of the vegetation around their nests, rather than in refuse piles or other special areas in the manner of most other kinds of ants. The droplets contain substances specific to their colony, allowing the ants to determine whether they are in the vicinity of their own nests or on foreign terrain.[131] Further, there is reason to conclude that the sternal gland and its alarming close-range recruitment secretions evolved from a lubrication gland. If true, the original role would have been to smooth the movements of the sixth and seventh sternites, which occurs during defensive formic acid spraying, when the gaster is raised and the acidopore pointed forward (see Figure 6-22).

Closely related to the evolutionary process of ritualization of chemical signals is the assignment of multiple functions to individual signals. The first such versatility discovered was the close association of alarm communication with defensive behavior. Not only are the behavioral patterns frequently identical, but often the same substances function as both defensive secretions and alarm messengers (Figure 6-23).

The role of signal parsimony in the evolution of social insect communication, based on conservation and opportunism in evolution, is supported by abundant documentation.[132] A remarkable example is provided by *Formica subintegra*, a North American "slave-making" ant whose workers raid colonies of other *Formica* species and rob their pupae. When the captives eclose as adults, they work for the slave-raiding colony. Workers of *Formica subintegra* spray acetates from the Dufour's gland during the raids. These substances attract and assemble the invaders at the

131 | B. Hölldobler and E. O. Wilson, "Weaver ants: social establishment and maintenance of territory," *Science* 195: 900–902 (1977).

132 | B. Hölldobler and E. O. Wilson, *The Ants* (Cambridge, MA: The Belknap Press of Harvard University Press, 1990); and M. S. Blum, "Semiochemical parsimony in the Arthropoda," *Annual Review of Entomology* 41: 353–374 (1996).

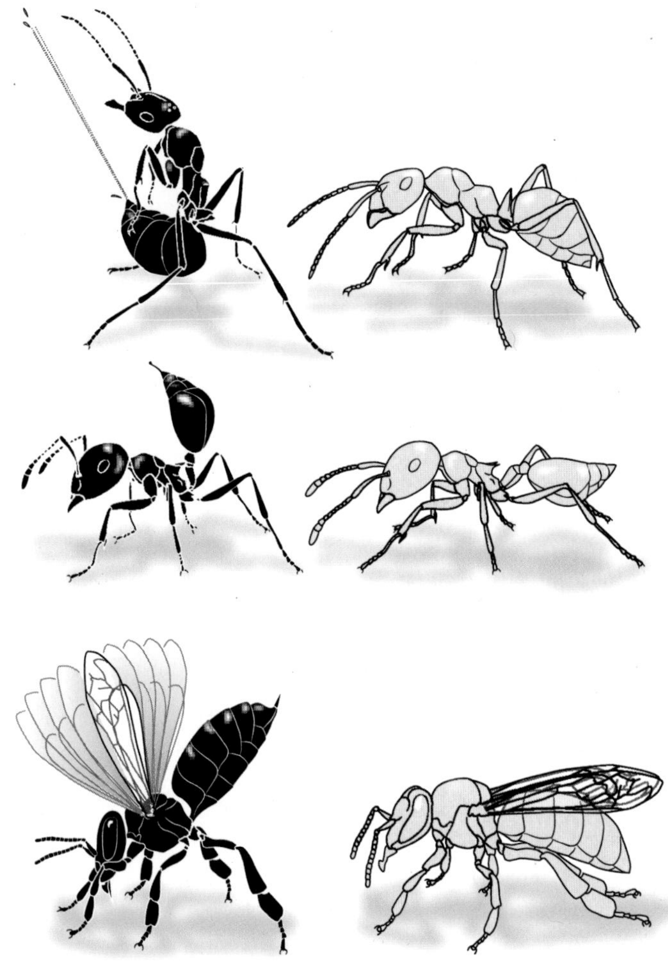

FIGURE 6-23. The alarm-defense behavior (black) is contrasted with the normal posture (white). *Above*: *Formica polyctena*. *Middle*: *Crematogaster ashmeadi*. *Below*: *Apis mellifera*. Based on an original illustration by Turid Hölldobler-Forsyth from B. Hölldobler, "Chemische Verständigung im Insektenstaat am Beispiel der Hautflügleer (Hymenopera)," *Umschau* 70(6): 663–669 (1970).

raiding site, but they also act as superefficient alarm or propaganda substances for the workers they assault, inducing their victims to panic and scatter.[133]

Workers of the fire ant *Solenopsis invicta*, to take a second example, dispense venom as an aerosol by forming a droplet on the tip of their extended stings and

---

133 | F. E. Regnier and E. O. Wilson, "Chemical communication and 'propaganda' in slave-maker ants," *Science* 172: 267–269 (1971).

vibrating ("flagging") the abdomen vertically. They perform this unusual act in two radically different circumstances and for two separate functions. When repelling other species of ants from their foraging grounds, each worker discharges a relatively massive load of up to 500 nanograms of the substance. Inside the nest, on the other hand, nurse workers dispense an aerosol of about 1 nanogram each over the surface of the brood pile, where it very likely serves as an antibiotic.[134] In addition, in queens of *Solenopsis invicta*, the poison gland secretions function first as pheromones that initiate and maintain worker retinue formation around the queen and second as primer pheromones that inhibit alate virgin queens from shedding their wings.[135]

Workers of the myrmicine *Xenomyrmex floridanus*, a common arboreal ant in the mangroves of the Florida Keys, lay recruitment trails with a secretion from their poison gland. In the radically different content of the nuptial flights, *Xenomyrmex* males are strongly attracted to poison gland secretions of females when these substances are dispersed into the air. In fact, the males gather on sticks contaminated with poison gland secretion during laboratory experiments; they even try to copulate with the stick. A similar dual function is found in several other myrmicine species in which the source of the recruitment trail pheromone is the same as that of the sex pheromone. We dare to predict that the chemical compounds are also the same.[136]

MESSAGE AND MEANING

The difference between message and meaning, an important aspect of semiotics, has become clear during research on ant communication. When an ant worker broadcasts alarm signals following the encounter with an enemy ant in her home territory,

---

134 | M. S. Obin and R. K. Vander Meer, "Gaster flagging by fire ants (*Solenopsis* spp.): functional significance of venom dispersal behavior," *Journal of Chemical Ecology* 11(12): 1757–1768 (1985); and R. K. Vander Meer and L. Morel, "Ant queens deposit pheromones and antimicrobial agents on eggs," *Naturwissenschaften* 82(2): 93–95 (1997).

135 | R. K. Vander Meer, B. M. Glancey, C. S. Lofgren, A. Glover, J. H. Tumlinson, and J. Rocca, "The poison sac of red imported fire ant queens: source of a pheromone attractant," *Annals of the Entomological Society of America* 73(5): 609–612 (1980); E. L. Vargo and M. Laurel, "Studies on the mode of action of a queen primer pheromone of the fire ant *Solenopsis invicta*," *Journal of Insect Physiology* 40(7): 601–610 (1994); and E. L. Vargo, "Reproductive development and ontogeny of queen pheromone production in the fire ant *Solenopsis invicta*," *Physiological Entomology* 24(4): 370–376 (1999).

136 | For a more detailed review, see B. Hölldobler and E. O. Wilson, *The Ants* (Cambridge, MA: The Belknap Press of Harvard University Press, 1990).

the message can induce different responses in nestmates as a result of changes in context. The phenomenon is well demonstrated in Ulrich Maschwitz's definitive comparison of alarm signals across the social hymenopterans. The response to alarm signals varies greatly in impact in time and space, as well as among different castes.[137] If, for example, the signal is received close to the nest, it triggers aggressive behavior; but if detected at a greater distance from the nest, it elicits escape behavior. Furthermore, young workers usually retreat into the nest when they receive the alarm signal, while older workers, especially those belonging to the soldier caste, move out aggressively.

Another such repertoire of responses dependent on the context of the signal has been documented in the alarm and agonistic behavior of the harvester ant genus *Pogonomyrmex*. The territory of each colony contains a core area, which is the site of the heaviest regular use within the home range. This space is absolutely forbidden to members of other conspecific colonies. Although the outer foraging areas of neighboring colonies often overlap, fighting between foreign workers is usually limited to two foragers that encounter each other. And there is a clear difference in fighting intensity at the core area versus the outer foraging ground. This discrimination holds even for the same individuals: if the confrontation is staged in the core area, the ant that fights on the home turf is highly aggressive; but when the same fighting pair is placed on the periphery of the home range, the aggression lasts only a few seconds, after which the combatants separate. Fighting in the core area is usually intense. The foreign ant often assumes a submissive posture and is subsequently dragged or carried by the local inhabitants away from the core area and then released.[138]

Similar locality-specific antagonistic reactions and outcomes have been described for several ant species, including members of the desert genus *Cataglyphis*,[139] some of which appear to mark the more frequently used hunting grounds with a colony-specific pheromone from the cloacal gland. Aggressive encounters with conspecific competitors are settled quickly, with the owner

---

137 | U. Maschwitz, "Gefahrenalarmstoffe und Gefahrenalarmierung bei sozialen Hymenopteren," *Zeitschrift für vergleichende Physiologie* 47(6): 596–655 (1964).

138 | B. Hölldobler, "Recruitment behavior, home range orientation and territoriality in harvester ants, *Pogonomyrmex*," *Behavioral Ecology and Sociobiology* 1(1): 3–44 (1976).

139 | M. Knaden and R. Wehner, "Nest defense and conspecific enemy recognition in the desert ant *Cataglyphis fortis*," *Journal of Insect Behavior* 16(5): 717–730 (2003).

of the territory usually prevailing.[140] Identical findings have been made in both laboratory and free-living colonies of the African weaver ant *Oecophylla longinoda*, whose workers chemically mark and defend huge three-dimensional territories. The defended core areas of these multiple leaf nests are dispersed throughout the canopy of several trees within the territory.[141]

In summary, it is critically important in the study of animal communication to distinguish message from meaning, particularly with reference to context, the behavioral states of sender and receiver, and the locality where the communicative behavior takes place.

MODULATORY COMMUNICATION

A striking enhancement of information in ants is achieved by modulatory communication, based on auxiliary signals that influence the behavior of the receiving insects, not by directing them into narrowly defined behavioral channels but by shifting the probabilities of responses induced by other signals.[142]

Modulatory communication is consistent with the broader principle just presented that information transfer in complex societies is seldom a direct, all-or-nothing response. Rather, it is usually contextual, that is, adjusted in a manner appropriate to the surrounding environment. Workers of the desert ants *Aphaenogaster albisetosus* and *Aphaenogaster cockerelli*, for example, use modulatory vibrations to enhance recruitment by pheromones. These slender, graceful insects are adept at retrieving large food objects, such as dead insects, within short periods of time following

---

140 | T. Wenseleers, J. Billen, and A. Hefetz, "Territorial marking in the desert ant *Cataglyphis niger*: does it pay to play bourgeois?" *Journal of Insect Behavior* 15(1): 85–93 (2002).

141 | B. Hölldobler and E. O. Wilson, "Colony-specific territorial pheromone in the African weaver ant *Oecophylla longinoda* (Latreille)," *Proceedings of the National Academy of Sciences USA* 74(5): 2072–2075 (1977); B. Hölldobler and E. O. Wilson, "The multiple recruitment systems of the African weaver ant *Oecophylla longinoda* (Latreille) (Hymenoptera: Formicidae)," *Behavioral Ecology and Sociobiology* 3(1): 19–60 (1978); and B. Hölldobler, "Territories of the African weaver ant (*Oecophylla longinoda* [Latreille]): a field study," *Zeitschrift für Tierpsychologie* 51: 201–213 (1979).

142 | H. Markl and B. Hölldobler, "Recruitment as food-retrieving behavior in *Novomessor* (Formicidae, Hymenoptera), II: Vibration signals," *Behavioral Ecology and Sociobiology* 4(2): 183–216 (1978); and H. Markl, "Manipulation, modulation, information, cognition: some of the riddles of communication," in B. Hölldobler and M. Lindauer, eds., *Experimental Behavioral Ecology and Sociobiology: In Memorium Karl von Frisch 1886–1982* (Fortschritte der Zoologie, no. 31) (Stuttgart: Gustav Fischer Verlag, 1985), pp. 163–194.

discovery. When a worker encounters an object too large to be carried or dragged by a single ant, she releases poison gland secretion into the air. The signal is extraordinarily effective: nestmates as far away as 2 meters are attracted and move toward the source. Time is of the essence for the *Aphaenogaster* foragers. The bonanza must be moved before less agile but more formidable ants arrive on the scene in large numbers. So in reinforcement of their strategy, the *Aphaenogaster* scouts not only disperse the poison gland attraction pheromone but also often stridulate (make a chirping noise) by rubbing a "scraper," the sharp posterior edge of the postpetiolar segment (second segment of the waist) against a file of horizontally arrayed ridges on the first gastral tergite (top of the segment just to the rear of the waist). The chirping sounds generated this way last up to a fifth of a second each and are repeated at frequencies of 0.1 to 10 kilohertz. Upon perceiving the vibration, nestmates of the successful forager remain in the vicinity up to twice as long as when the forager falls silent. Most importantly, ants sensing the signals also release their own attractive poison gland secretion sooner. Overall, both the recruitment of nestmates and the retrieval of the food object are speeded up by 1 to 2 minutes when stridulation accompanies the pheromone release—a time difference that can make the difference between success or failure in the race among ant colonies to secure the prize.

Put another way, vibrational signals serve as amplifiers in recruitment by the *Aphaenogaster* worker ants. A similar case of modulatory stridulation to enhance recruitment by odor trails has been reported in the myrmicine genus *Messor* and ponerine genus *Leptogenys*.[143, 144]

Modulatory stridulation, acting in concert with chemical recruitment signals, can be viewed as another form of multimodal communication. In the examples just described, the substrate-borne vibrations are part of a multimodal signal lowering the response threshold of the receiver for the releasing component (the chemical stimulus) of the signal. A recruitment response appears to be elicited by only one component of the multimodal signal.

---

143 | M. Hahn and U. Maschwitz, "Foraging strategies and recruitment behaviour in the European harvester ant *Messor rufitarsis* (F.)," *Oecologia* 68(1): 45–51 (1985); E. Schilliger and C. Baroni Urbani, "Morphologie de l'organe de stridulation et sonogrammes comparés chez les ouvrières de deux espèces de fourmis moissonneuses du genre *Messor* (Hymenoptera, Formicidae)," *Bulletin de la Société Vaudoise des Sciences Naturelles* 77(4): 377–384 (1985); and C. Baroni-Urbanai, M. W. Buser, and E. Schillige, "Substrate vibration during recruitment in ant social organization," *Insectes Sociaux* 35(3): 241–250 (1988).

144 | U. Maschwitz and P. Schönegge, "Recruitment gland of *Leptogenys chinensis*: a new type of pheromone gland in ants," *Naturwissenschaften* 64(11): 589–590 (1977).

The situation is different in the multimodal communication system of leaf-cutting ants of the genus *Atta* (see also Chapter 10). In his pioneering research on vibrational communication, the German physiologist Hubert Markl discovered that the substrate-borne stridulatory signals of *Atta* workers can release specific behavioral responses in the recipient nestmates,[145] but the semantic content—that is, the message signaled by the sender and the meaning of the signal for the receiver—varies according to its situational context. For example, *Atta* workers dispense an alarm pheromone (4-methyl-3-heptanone) from the mandibular glands. The effect of this chemical signal is enhanced when combined with stridulatory vibrations. Thus, ants held fast by an enemy ant or a pair of forceps release the alarm pheromone and simultaneously stridulate. *Atta* workers also stridulate when trapped under soil, for example, following a partial cave-in of the nest. In this circumstance, the ants are attracted to the source of substrate-borne vibrations alone and dig until the stridulating nestmate has been freed. Not all of the nestmates respond equally to the emergency. The soldier caste exhibits very little response, whereas workers currently occupied with digging and soil transport inside the nest respond best. Although alarm and rescue signals are multimodal in leaf-cutting ants, the workers react to either component presented separately.

The significance of context for modifying the message-to-meaning relationship in *Atta* communication has become particularly clear with the discovery that stridulatory vibrations can also function as close-range recruitment signals during foraging.[146] It is well known that foragers of leafcutter ants cut vegetation into small fragments that they transport to the nest, where the material is processed by the ant colony's fungal garden. In addition to a well-developed chemical communication system, leaf-cutting ants also use mechanical signals during recruitment communication. When

145 | H. Markl, "Die Verständigung durch Stridulationssignale bei Blattschneiderameisen, I: Die biologische Bedeutung der Stridulation," *Zeitschrift für vergleichende Physiologie* 57(3): 299–330 (1967); H. Markl, "Die Verständigung durch Stridulationssignale bei Blattschneiderameisen, II: Erzeugung und Eigenschaften der Signale," *Zeitschrift für vergleichende Physiologie* 60(2): 103–150 (1968); H. Markl, "Die Verständigung durch Stridulationssignale bei Blattschneiderameisen, III: Die Empfindlichkeit für Substratvibrationen," *Zeitschrift für vergleichende Physiologie* 69(1): 6–37 (1970); and W. M. Masters, J. Tautz, N. H. Fletcher, and H. Markl, "Body vibration and sound production in an insect (*Atta sexdens*) without specialized radiating structures," *Journal of Comparative Physiology* 150(2): 239–249 (1983).

146 | F. Roces, J. Tautz, and B. Hölldobler, "Stridulation in leaf-cutting ants: short-range recruitment through plant-borne vibrations," *Naturwissenschaften* 80(11): 521–524 (1993); and F. Roces and B. Hölldobler, "Vibrational communication between hitchhikers and foragers in leaf-cutting ants (*Atta cephalotes*)," *Behavioral Ecology and Sociobiology* 37(5): 297–302 (1995).

workers stridulate during the cutting of leaf fragments, the vibrations produced by the stridulatory organ travel forward along the body onto the head and down into the leaf being cut. Laser-Doppler vibrometry has made it possible to record the stridulation signals, which are inaudible to humans (unless the listener is a young person and holds the ant close to the ear). The substrate-borne vibrations consist of long series of repetitive pulse trains (chirps), with each pulse arising from the impact of the ant's scraper on a ridge of the file of the stridulation organ. The signal repetition rate varies between 2 and 20 chirps per second. The temporal pattern of the recorded chirps does not differ from that previously described during alarm vibrations in the same *Atta* species. Flavio Roces and his collaborators demonstrated that nearby workers respond to the vibrations by turning toward the source of the vibrations. Because they orient correctly in the absence of chemical signals, it is evident that stridulation acts as a short-range recruitment signal (see Chapter 9).

How does the stridulatory signal component interact with the chemical recruitment component? In one set of laboratory experiments, the ants responded to both components when presented with them separately. When the stridulatory and the chemical components were tested against each other in a choice experiment, the chemical component was always greatly preferred. However, when the ants were allowed to choose between the multimodal signal (chemical plus vibrational components) and a chemical component alone, the ants consistently chose the dual signal. In natural situations at foraging sites, the vibrational component appears to be superimposed on the wide-range chemical recruitment signal, thus serving to fine-tune the ants' close-range recruitment response.

On the other hand, it is evident that in social insects, acoustical communication (more accurately put, vibrational communication) is only weakly developed when compared with communication by pheromones. Even when present, it is usually combined in some manner or other with chemical signals. Most vibrational signals are transmitted primarily through the soil, nest walls, leaves and twigs, and other solid surfaces, rather than through the air. Several forms of vibration production have been discovered. They include rapping or drumming the body against the substrate, scraping the mandibles over the surface of nest walls or other solid materials, and stridulation with the use of the specially evolved scrapers and files.[147] Also

147 | H. Markl, "The evolution of stridulatory communication in ants," *Proceedings of the Seventh Congress of the International Union for the Study of Social Insects* (10–15 September 1973, London), pp. 258–265.

employed are body shaking, waggling, and jerking, as well as sounds and substrate vibrations caused by high-frequency wing muscle vibration, the latter as described for honeybees and stingless bees.[148]

MOTOR DISPLAYS IN RECRUITMENT COMMUNICATION

The cumulative studies during the past three decades have established that motor displays and tactile signaling play important roles during recruitment communication in many ant species, by which scouts lead nestmates to food, new nest sites, or enemies. When employed, they usually interact with chemical signals. The motor displays in *Oecophylla* are an example, serving as "icons" to convey messages about the specific target area. Motor displays contribute to the mass recruitment system of *Solenopsis invicta.* They also occur in the simple recruitment behaviors of amblyoponine and ponerine ants, in particular *Mystrium rogeri* and *Pachycondyla marginata.* Other examples reported widely in the literature suggest the prevalence of such communication systems among ants as a whole.[149]

Motor displays can function as graded signal components in multimodal communication during group recruitment, a common behavior in the formicine genera *Camponotus* and *Polyrhachis.*[150] The recruiting ant summons 2 to 30 nestmates at a time, which then follow closely behind the leader ant to the target area (Figure 6-24). The species best documented for the behavior is *Camponotus socius,* from the southeastern United States (Plate 27).[151] The scout ants first deposit signposts

---

148 | M. Hrncir, F. G. Barth, and J. Tautz, "Vibratory and airborne-sound signals in bee communication (Hymenoptera)," in S. Drosopoulous and M. F. Claridge, eds., *Insect Sounds and Communication: Physiology, Behaviour, Ecology, and Evolution* (Boca Raton, FL: Taylor & Francis, 2006), pp. 421–436.

149 | J. H. Sudd, "Communication and recruitment in Pharaoh's ant, *Monomorium pharaonis* (L.)," *Animal Behaviour* 5(3): 104–109 (1957); R. Szlep and T. Jacobi, "The mechanism of recruitment to mass foraging in colonies of *Monomorium venustum* Smith, *M. subopacum* spp. *phoenicium* Em., *Tapinoma israelis* For., and *T. simrothi* v. *phoenicium*," *Insectes Sociaux* 14(1): 25–40 (1967); R. H. Leuthold, "Recruitment to food in the ant *Crematogaster ashmeadi*," *Psyche* (Cambridge, MA) 75: 334–350 (1968); and R. Szlep-Fessel, "The regulatory mechanism in mass foraging and the recruitment of soldiers in *Pheidole*," *Insectes Sociaux* 17(4): 233–244 (1970).

150 | B. Hölldobler and E. O. Wilson, *The Ants* (Cambridge, MA: The Belknap Press of Harvard University Press, 1990); B. Hölldobler, "Multimodal signals in ant communication," *Journal of Comparative Physiology A* 184(2): 129–141 (1999); and C. Liefke, B. Hölldobler, and U. Maschwitz, "Recruitment behavior in the ant genus *Polyrhachis* (Hymenoptera, Formicidae)," *Journal of Insect Behavior* 14(5): 637–657 (2001).

151 | B. Hölldobler, "Recruitment behavior in *Camponotus socius* (Hym. Formicidae)," *Zeitschrift für vergleichende Physiologie* 75(2): 123–142 (1971).

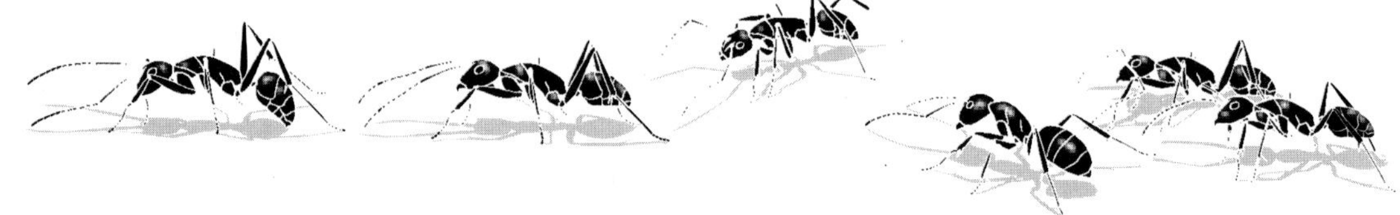

FIGURE 6-24. Group recruitment in *Camponotus socius*. The leader ant lays a longer-lasting orientation trail with contents from the hindgut and a short-lasting recruitment signal from the poison gland. Based on B. Hölldobler, "Recruitment behavior in *Camponotus socius* (Hym. Formicidae)," *Zeitschrift für vergleichende Physiologie* 75(6): 123–142 (1971).

around newly discovered food sources, then lay a trail with hindgut contents from the food source to the nest. The trail pheromone, a mixture of (*ZS*,4*R*,5*S*)-2,4-dimethyl-5-hexanolide and 2,3-dihydro-3,5-dihydroxy-6-methylpyran-4-one, does not by itself induce recruitment to any significant degree.[152] Instead, the recruiting ant inside the nest supplements it by a waggle display performed while facing nestmates head-to-head (Figure 6-25). Each waggle episode lasts 0.5 to 1.5 seconds and is repeated toward the same or other individuals. Nestmates, alerted by this behavior, then follow the recruiting ant to the food source.[153] The significance of the motor display inside the nest was demonstrated by experimentally closing the rectal opening of the recruiting ants with wax plugs, effectively separating the waggle display from the chemical signals. Significantly more workers followed the pheromone trail when first exposed to the waggle display. The intensity of the waggle display performed by individual scouts and the response of the nestmates are both positively correlated with the colony's need for food. Successful scouts from a starved colony were observed to exhibit more vigorous waggle displays. They contacted more nestmates and attracted more nestmates. When only the scouts were starved and the colony well fed, the scouts still performed a vigorous waggle display, although significantly fewer ants responded. On the other hand, when scouts that were well fed were allowed to return after 5 days from a food source to a starved colony, they exhibited at most only very weak motor displays but extensive regurgitation. Some scouts moved back to the food source, but then were not

152 | E. Kohl, B. Hölldobler, and H. J. Bestmann, "Trail and recruitment pheromones in *Camponotus socius* (Hymenoptera: Formicidae)," *Chemoecology* 11(2): 67–73 (2001).

153 | The major recruitment signal during group recruitment appears to be the highly volatile formic acid that is discharged from the poison gland and that "binds" the group of followers behind the recruiting leader ant.

PLATE 27. *Above*: Workers of the carpenter ant *Camponotus socius* inside the nest. *Below*: A trail-laying forager touching her abdominal tip to the ground while laying trail pheromone outside the nest.

followed by nestmates. When they returned again to the nest they typically performed a vigorous waggle display and were subsequently followed by a substantial number of nestmates.

These observations show that the waggle display of *Camponotus socius* is employed as a graded signal, with its intensity dependent on the motivational state of the recruiter as well as the responsiveness of the colony.

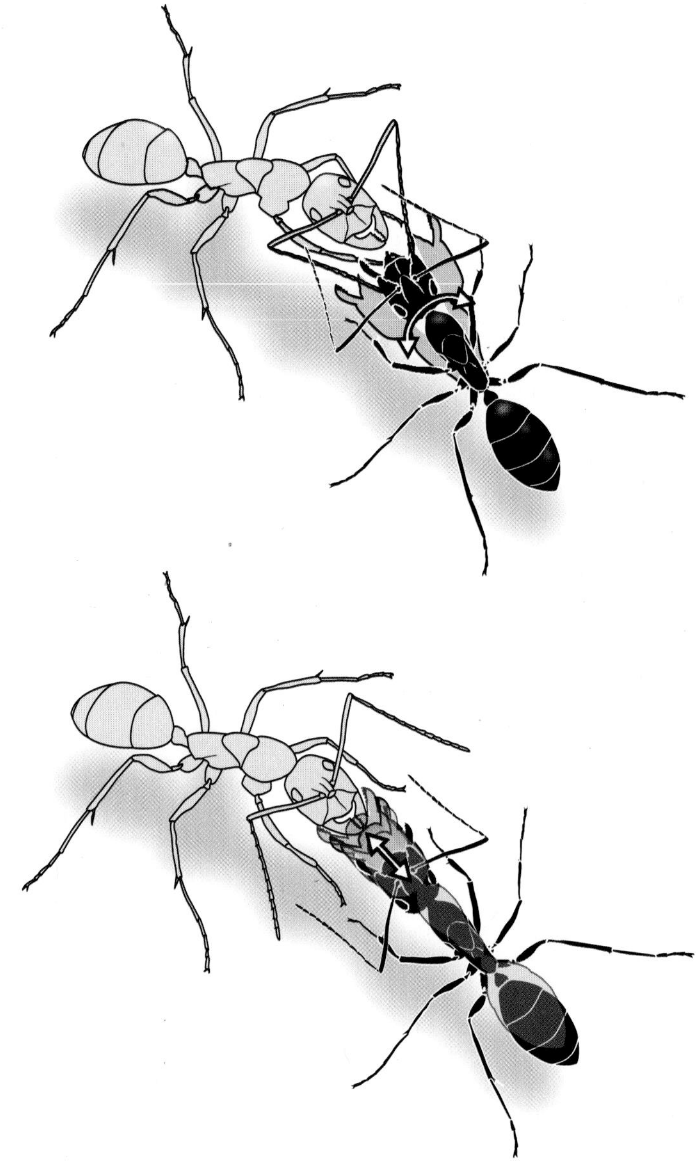

FIGURE 6-25. In *Camponotus socius*, the invitation signal employed during recruitment to food (above) is a lateral wagging of the body that appears to be a ritualization of food offering. It differs from the invitation signal employed during emigration (below), which is a back-and-forth jerking of the body, apparently a ritualization of the more "primitive" mandible pulling displayed by *C. sericeus* (see Figure 6-26). Based on an original illustration by Turid Hölldobler-Forsyth from B. Hölldobler, "Recruitment behavior in *Camponotus socius* (Hym. Formicidae)," *Zeitschrift für vergleichende Physiologie* 75(6): 123–142 (1971).

During the waggle display, the recruiting *Camponotus socius* ant usually holds her mandibles open and her labium extended. As in the *Oecophylla* weaver ants, this behavior resembles that of an ant offering food to a nestmate. In fact, the scout often appears to present food samples to the surrounding ants. Thus, the waggle display can reasonably be interpreted to constitute an intention movement preceding the social food exchange, which has been ritualized to serve as a communicative signal denoting the discovery of food. As in the case of ritualization in animals generally, the waggle display is much more repetitive and stereotyped than its precursor, the intention movement.

A different motor display is employed by *Camponotus socius* workers when recruiting nestmates to new nest sites. It comprises jerking movements applied to all sides of the receiver ant, most frequently when the recruiter faces a nestmate. Usually, the recruiter takes a position above the nestmate that has approached, seizes its head, and jerks it forward (see Figure 6-25). This particular motor display specifies recruitment to nest emigration. It evidently has been derived as an intention movement initiating adult transport, a social behavior frequently employed during colony emigrations from one nest site to another.

Overall, *Camponotus socius* workers that recruit nestmates to food sources or to new nest sites employ an orientation trail pheromone originating from the hindgut and mixed with formic acid from the poison gland as a stimulant. But the recruiters specify the recruitment context with different motor displays employed inside the nest. Further, whereas only foragers respond to the waggle display, a much larger worker cohort, comprising nurse ants, virgin queens, and even males, respond to the jerking display. Thus, the multimodal signals of *Camponotus socius* convey information about the external environment.

Ethologists call such signals "functionally referential." Marc Hauser, writing about birds and monkeys, considers an animal signal to be referential "if it is reliably associated with objects and events in the world. As a result of this association, the listener can accurately assess the range of potential contexts for signal emission. The breadth of the ranges depends, in part, on the specificity of the signal with regard to target objects and events."[154] This statement exclusively concerns vocalization in vertebrates, but it can be equally well applied to the waggle dance communication of honeybees and the recruitment communication of *Oecophylla longinoda*, *Camponotus socius*, and other ant species.

---

154 | M. D. Hauser, *The Evolution of Communication* (Cambridge, MA: MIT Press, 1996).

How does ritualization work in creating new signals during evolution? Relatively few communicative systems in the social insects have been analyzed explicitly with this evolutionary process in mind. An exception is the celebrated waggle dance in honeybees, where comparative studies of recruitment communication mechanisms in several honeybee species and in stingless bees have revealed a sequence of evolutionary grades from simple to complex. Examination of the sequence has persuasively confirmed the waggle dance behavior to be a ritualization of a guiding flight from the hive to the food source.[155]

Another relatively clear case of ritualization is provided by the invitation movements of workers of the ant genus *Camponotus* during recruitment of others to new food and nest sites. At the apparently most primitive levels is tandem running, exemplified in the recruitment system of *Camponotus sericeus*. As a successful *Camponotus sericeus* forager heads home, she deposits chemical signposts with material from the hindgut. Inside the nest, she performs short-lasting "fast" runs, interrupted by brief food exchanges and grooming. During individual recruitment episodes, these rituals are repeated 3 to 16 times. When the scout finally leaves the nest, she returns to the food source along the orientation trail she had laid. Several of the nestmates encountered inside the nest follow her. However, only the ant maintaining the closest antennal contact succeeds, by means of tandem running, in accompanying her all the way to the food source. Most followers that reach the food source in this manner soon return to the nest and begin to recruit nestmates on their own. Experiments have shown that the hindgut trail laid by the lead ant has no recruitment effect by itself. Only experienced ants follow the trail, and they appear to use it exclusively for orientation. Similarly, during tandem running, the trail pheromone appears to serve no significant communicative function. The leader and follower are bound together by a continuous exchange of tactile signals and the perception of persistent nestmate recognition substances on the surface of the body. Nothing indicates that the leader discharges content from the poison gland sac in the manner employed by the group-recruiting *Camponotus socius*.

For tandem running to succeed, it is important for the leader ant to be always informed of the presence of the ant behind her. The follower regularly accomplishes this task using her antennae and head to touch the leader's hind legs and gaster. If

---

155 | M. Lindauer, *Communication among Social Bees* (Cambridge, MA: Harvard University Press, 1961); see also E. O. Wilson, *The Insect Societies* (Cambridge, MA: The Belknap Press of Harvard University Press, 1971).

these contacts are interrupted, either because the follower goes astray or the pair is experimentally separated, the leader ant immediately halts and remains in place, like a mountaineer guide on a foggy slope, until the follower has contacted her again. A leader ant whose follower has been removed can be enticed to press on by regularly touching her with a human hair on the hind legs at a rate of about once per second. In this way, she leads the experimenter to the target area. Similarly, a follower that has lost her leader follows a dummy consisting of the severed gaster of a freshly killed ant mounted on a pin in any direction the experimenter chooses. This latter response is further evidence that neither the orientation trail laid by forager scouts nor the poison gland secretion plays a role in the recruitment process.[156]

Tandem running is also used in *Camponotus sericeus* in the recruitment of nestmates to new nest sites. A scout that discovers a superior nest site returns home, where she performs a recruitment behavior markedly different from that used in recruitment to new food sources. Facing a nestmate head-on, she grasps her on the mandibles and pulls her vigorously forward. For a short while then, she loosens her grip, turns around 180°, and presents her gaster to the nestmate. If the nestmate responds by touching her gaster at the hind legs, the recruiter begins tandem running. This behavioral sequence is very stereotyped and is regularly employed when nestmates are invited to follow the leader to a new nest (Figure 6-26).

After a few such tandem running episodes, many of the recruited ants become recruiters themselves. Nestmates, especially younger ones, are carried bodily to the nest by the "mover ants." Sometimes a mover ant transports one nestmate while simultaneously leading another in tandem (Figure 6-27). Alate queens and males are also led by tandem running. In addition, males are frequently carried in the transport mode typical for *Camponotus,* with the two sexes assuming different positions (see Figure 6-27). A comparison of the behaviors that initiate tandem running and transport during nest moving beautifully illustrates the ritualization process. The behavioral sequence that initiates carrying behavior is almost identical to that of the invitation behavior leading to tandem recruitment. The recruiting ant, however, keeps a firm grip when turning around. The nestmate is thereby slightly lifted, causing her to fold her legs tightly to the body and to roll her gaster inward. In this posture she is carried to the target area (Figure 6-28).

---

156 | B. Hölldobler, M. Möglich, and U. Maschwitz, "Communication by tandem running in the ant *Camponotus sericeus*," *Journal of Comparative Physiology A* 90(2): 105–127 (1974).

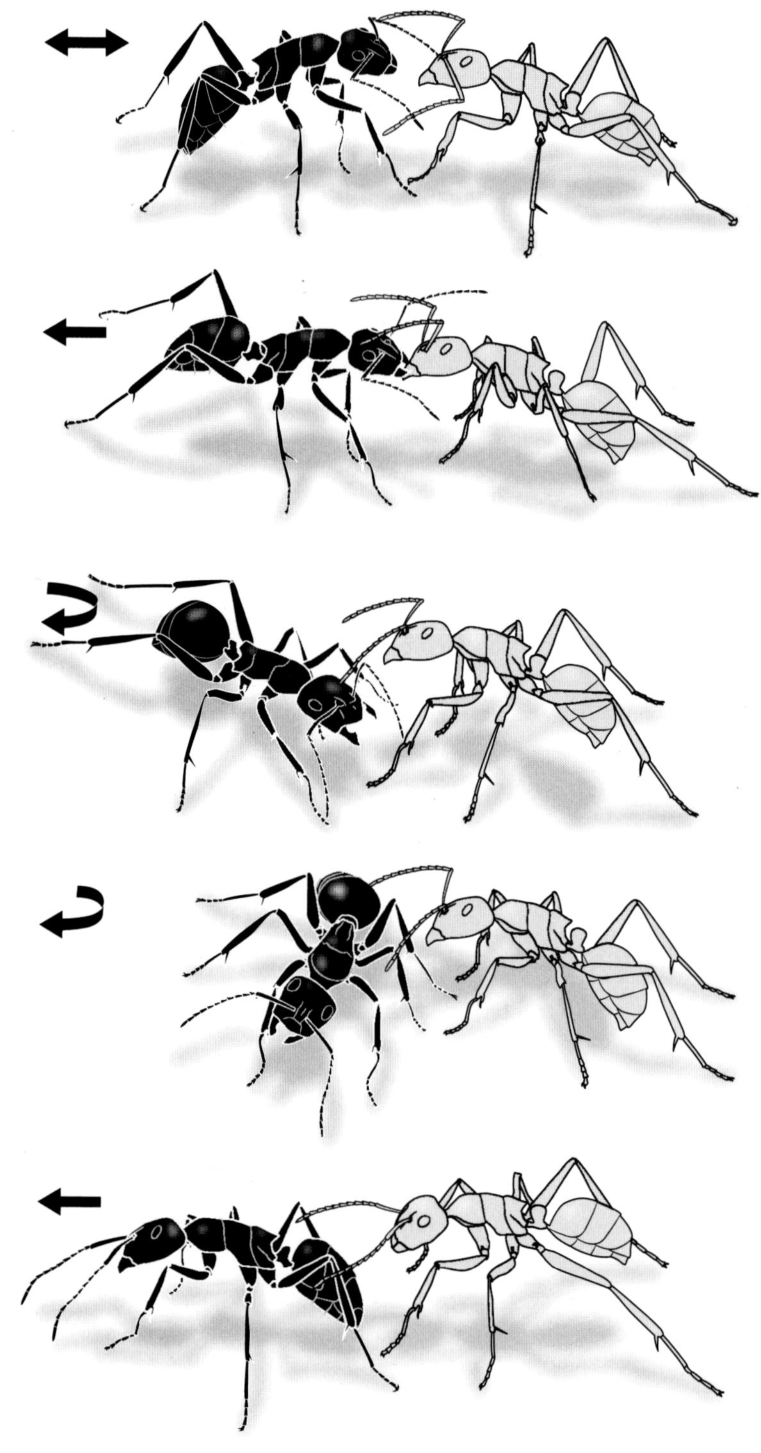

It seems reasonable to suppose that the initiation behavior that mediates nestmate transport is the precursor of the invitation behavior used in recruitment by tandem running during nest emigration. The jerking and slight pulling employed by a *Camponotus socius* ant for recruitment to a new nest site is likely derived from a behavior exhibited by *Camponotus sericeus.* Workers of this species recruit nestmates to a new food source by running in fast, short spurts broken by pauses to regurgitate food samples. No waggle display is used. Again, the evidence suggests that the "symbolic" food offering display in *Camponotus socius,* enhanced by the waggle behavior, is derived from the more elementary behavior of the kind seen in *Camponotus sericeus.*

In both of the aforementioned *Camponotus* species, hindgut trails serve as orientation trails, even though they remain uninvolved in recruitment communication. In *Camponotus socius,* however, the leader ant does not require direct physical contact with the group of followers; if separated, she continues to move on. She seems to discharge a highly volatile recruitment pheromone, likely formic acid from the poison gland, which keeps the group of nestmates moving in the right direction.

The next higher organizational level in the *Camponotus* panoply occurs in the carpenter ants *Camponotus ligniperdus* and *Camponotus herculeanus* of Europe, as well as in the similar *Camponotus pennsylvanicus* of North America. Group recruitment occasionally occurs in *Camponotus pennsylvanicus.* James Traniello, who has studied the recruitment behavior of this large black ant in great detail, found that scouts returning from newly discovered food sources lay odor trails.[157] These individuals

157 | J. F. A. Traniello, "Recruitment behavior, orientation, and the organization of foraging in the carpenter ant *Camponotus pennsylvanicus* DeGeer (Hymenoptera: Formicidae)," *Behavioral Ecology and Sociobiology* 2(1): 61–79 (1977).

FIGURE 6-26. Recruitment by tandem running during colony emigration in the Asiatic species *Camponotus sericeus* begins with the scout (black) employing a ritualized invitation behavior, causing the nestmate (gray) to follow her to the new nest site. The behavioral sequence is depicted from upper to lower: The recruiter approaches a nestmate and jerks her body back and forth for 2 to 3 seconds. She then grasps the nestmate by the mandibles and pulls her through a distance of about 2 to 20 centimeters. Next the recruiter loosens her grip and turns around. Finally, releasing a pheromone from the hindgut, she leads the nestmate toward the new nest site, while the nestmate maintains direct tactile contact by playing her antennae over the leader's abdomen and hind legs. Based on an original illustration by Turid Hölldobler-Forsyth from B. Hölldobler, M. Möglich, and U. Maschwitz, "Communication by tandem running in the ant *Camponotus sericeus,*" *Journal of Comparative Physiology* 90(2): 105–127 (1974).

FIGURE 6-27. Adult transport to a new nest site is accomplished by *Camponotus sericeus* colonies both by adult transport and tandem running. The characteristic posture of the ant being transported differs in the worker (above) and the male (below).

further stimulate nestmates with a waggle display. When nestmates are alerted by this signal, they follow the previously laid trail. The principal component of the trail substance is most likely 2,4-dimethyl-5-hexanolide, which has been identified in the closely related *Camponotus herculeanus* and shown to elicit trail following in *Camponotus pennsylvanicus*.[158] It is not known whether *Camponotus pennsylvanicus* also releases a short-lasting recruitment pheromone from the poison gland.

---

158 | H. J. Bestmann, U. Haak, F. Kern, and B. Hölldobler, "2,4-Dimethyl-5-hexanolide, a trail pheromone component of the carpenter ant *Camponotus herculeanus*," *Naturwissenschaften* 82(3): 142–144 (1995).

We now come to what may be viewed as the apex of the *Camponotus* evolutionary series, namely recruitment systems in species that appear to be analogous to the mass recruitment behavior exhibited by fire ants (*Solenopsis invicta*). An example at this level is *Camponotus sericeiventris,* a silvery pilose carpenter ant species found in the New World tropics. As in other species of the same genus, returning scout ants of *Camponotus sericeiventris* perform stimulating motor displays and antennation bouts inside the nest. But a full recruitment response can be elicited by nothing more than the trail and recruitment pheromones. The principal component of the hindgut trail pheromone is an isocoumarin (3,4-dihydro-8-hydroxy-3,5,7-trimethylisocoumarin). This compound does not by itself elicit a strong recruitment effect, but ants already persuaded to leave the nest readily follow trails drawn with it. The higher the concentration of the pheromone, the more workers that follow. However, once a certain level of concentration has been reached, the response does not improve further.

Although trails drawn with the isocoumarin alone elicit some trail following in ants roaming around in the foraging arena, it is quite obvious that the main function of these trails is not recruitment. Even after 24 hours, trails drawn with 15 ant equivalents of newly extracted trail pheromone evoked some trail following in *Camponotus sericeiventris* workers (Plates 28 and 29). These observations, and the fact that the hindgut trails alone do not elicit a strong recruitment response, suggest that a second, perhaps primary recruitment pheromone serves as a major "turn-on" signal that induces the ants to follow the more enduring hindgut trails. And indeed, a short-lasting powerful recruitment signal has been found that originates from the poison gland. When isocoumarin trails (as hindgut trails) were overlaid by trails drawn with minute amounts of formic acid, a product of the poison gland, *Camponotus sericeiventris* workers responded by rushing out of the nest. Many followed the trails beyond the formic acid trail and all the way to the end of the isocoumarin trail. Several additional experiments have confirmed that recruitment trails in *Camponotus sericeiventris* do consist of at least two functional components: the longer-lasting orientation component from the hindgut and the short-lived but highly stimulating recruitment component from the poison gland. Higher concentrations of formic acid release a higher mass recruitment exodus at the nest exit, and the emerging ants then follow the hindgut or isocoumarin trails with precision. In the initial phase of the natural recruitment process, ants leave the nest and move to the food source without a recruiting leader ant, but when trail-laying ants move out of the nest and return to the feeder, they are often

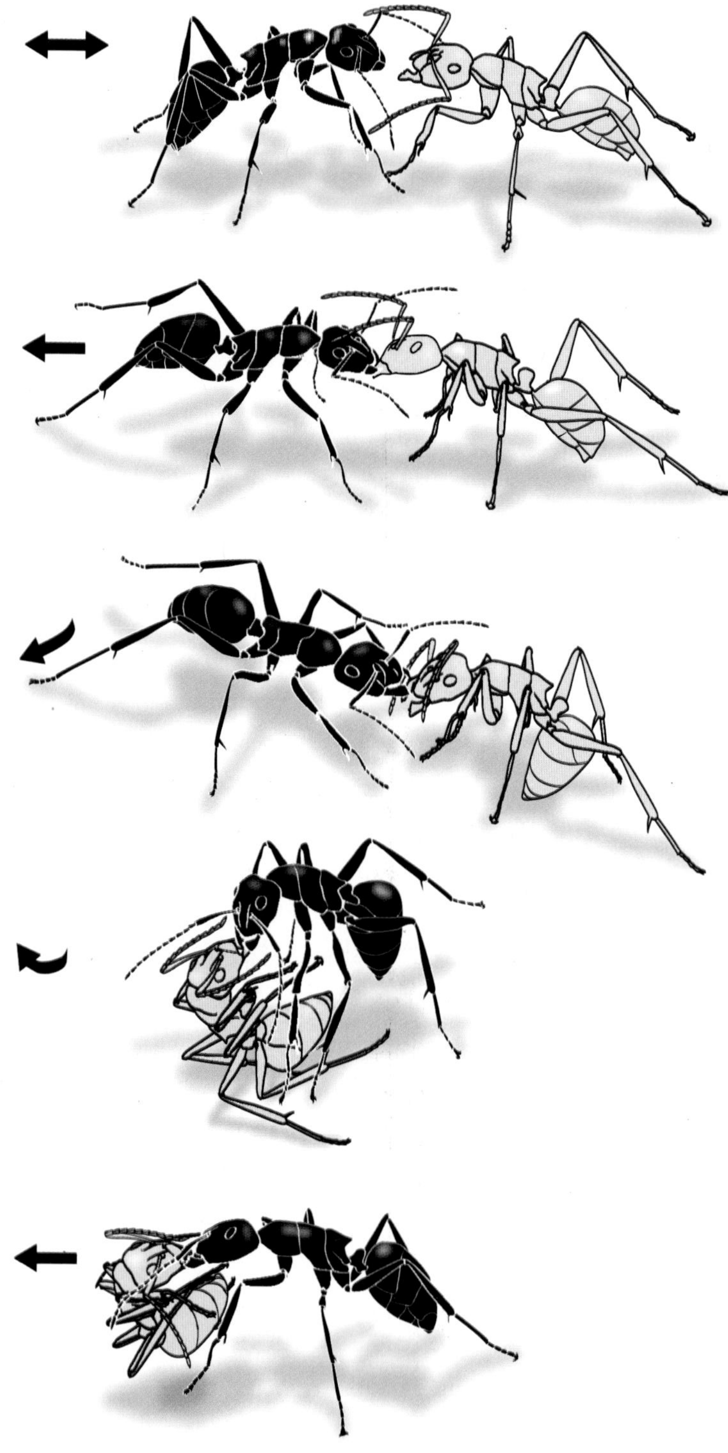

followed by a group of nestmates, which have obviously been stimulated by the release of formic acid.

Still, once such a recruitment process in *Camponotus sericeiventris* is in full swing, with ever more ants leaving the nest to travel to the food source, the loose "personal" group recruitment patterns disappear, and the ants start to rush along the trail in independent forays. As the food source dwindles, the number of trail-laying ants also declines, and the poison gland recruitment signal soon vanishes. That, in turn, leads to a drastic reduction of ants leaving the nest in the direction of the food source.[159] The dynamics of this recruitment process thus proves basically the same as that described for the fire ant *Solenopsis invicta*.

ENVIRONMENTAL CORRELATES OF RECRUITMENT SYSTEMS

Overall, extensive studies have thus shown that motor displays and tactile signaling play an important role during recruitment communication in many ant species. It appears, however, that during the course of evolution, these signals became less important with an increasing growth in sophistication of the chemical recruitment system. The biological correlates of this advance remain to be worked out, but it is already evident that they include the population size of colonies. The larger the mature colony among ant species, the more reliance is placed on chemical signals as opposed to motor displays in the initiation of recruitment. *Camponotus sericeus* colonies are relatively small, limited to hundreds of ants, whereas colonies of

---

159 | E. Kohl, B. Hölldobler, and H.-J. Bestmann, "Trail pheromones and Dufour gland contents in three *Camponotus* species (*C. castaneus, C. balzani, C. sericeiventris*: Formicidae, Hymenoptera)," *Chemoecology* 13(3): 113–122 (2003); and B. Hölldobler, unpublished results.

FIGURE 6-28. The initiation of adult transport in *Camponotus sericeus*. The recruiting worker (black) approaches a nestmate (gray) and jerks her body back and forth 2 to 3 seconds. She then grasps the nestmate on the mandibles and pulls her a distance of 2 to 20 centimeters. Holding onto the mandibles of her nestmate and lifting her slightly, the recruiter begins to turn. The nestmate folds her antennae and legs tightly against her body and rolls her gaster inward. The recruiter now carries the nestmate to the new nest site. Based on an original illustration by Turid Hölldobler-Forsyth from B. Hölldobler, M. Möglich, and U. Maschwitz, "Communication by tandem running in the ant *Camponotus sericeus*," *Journal of Comparative Physiology* 90(2): 105–127 (1974).

PLATE 28. Workers of the South American species *Camponotus sericeiventris* on the foraging trail.

PLATE 29. Two workers of *Camponotus sericeiventris* exchanging food.

Camponotus socius are quite large, with thousands of individuals. But *Camponotus socius* is polydomous, so that each subnest houses no more than several hundred ants. *Camponotus sericeiventris* colonies, on the other hand, comprise thousands of individuals.

Other correlates of the recruitment system design are the nature of the food sources to which the species is adapted, including the size of the food units and their degree of temporality, and the seasonal nutritional needs of the colony. These connections are based on casual observations. Only a few comparative field studies have been made that relate recruitment techniques to the nature of resources and colony needs.[160] Fortunately, *Camponotus* is the second largest ant genus in the world, smaller only than *Pheidole*, and its hundreds of species offer a large base for future comparison.

To this point, we have deliberately focused our narrative on the behavioral mechanisms of recruitment communication, because they are the behavioral traits shaped by natural selection at the colony level.

In recent years, a wealth of theoretical studies have been published on the self-organization of certain emergent patterns of foraging in ant colonies.[161] Their emphasis is on the following question: Do ant colonies rely on positive feedback when engaged in collective decision making? And when recruiters return from different foraging sites, how can colony-level foraging be directed at the richest source available—even though

---

160 | See, for example, C. Liefke, W. H. O. Dorow, B. Hölldobler, and H. Maschwitz, "Nesting and food resources of syntopic species of the ant genus *Polyrhachis* (Hymenoptera, Formicidae) in West-Malaysia," *Insectes Sociaux* 45(4): 411–425 (1998); also see the review in J. F. A. Traniello and S. K. Robson, "Trail and territorial communication in social insects," in R. T. Cardé and W. J. Bell, eds., *Chemical Ecology of Insects 2* (New York: Chapman & Hall, 1995), pp. 241–286; for some empirical and theoretical considerations, see R. Beckers, S. Goss, J. L. Deneubourg, and J. M. Pasteels, "Colony size, communication and ant foraging strategy," *Psyche* (Cambridge, MA) 96: 239–256 (1989); L. Edelstein-Keshet, J. Watmoügh, and G. B. Ermentrout, "Trail following in ants: individual properties determine population behaviour," *Behavioral Ecology and Sociobiology* 36(2): 119–133 (1995); E. Bonabeau, G. Theraulaz, and J.-L. Deneubourg, "Group and mass recruitment in ant colonies: the influence of contact rates," *Journal of Theoretical Biology* 195(2): 157–166 (1998); M. Beekman, D. J. T. Sumpter, and F. L. W. Ratnieks, "Phase transition between disordered and ordered foraging in Pharaoh's ants," *Proceedings of the National Academy of Sciences USA* 98(17): 9703–9706 (2001); and S. Portha, J.-L. Deneubourg, and C. Detrain, "Self-organized asymmetries in ant foraging: a functional response to food type and colony needs," *Behavioral Ecology* 13(6): 776–781 (2002).

161 | T. R. Stickland, N. F. Britton, and N. R. Franks, "Complex trails and simple algorithms in ant foraging," *Proceedings of the Royal Society of London B* 260: 53–57 (1995); see reviews in S. Camazine, J.-L. Deneubourg, N. R. Franks, J. Sneyd, G. Theraulaz, and E. Bonabeau, *Self-Organization in Biological Systems* (Princeton, NJ: Princeton University Press, 2001); and C. Detrain, J.-L. Deneubourg, and J. M. Pasteels, eds., *Information Processing in Social Insects* (Basel: Birkhäuser Verlag, 1999).

no individual makes comparative evaluations of food source quality? Several empirical studies show that it is highly probable that the colony eventually favors the most profitable resource. In mass recruitment systems, the amount of recruitment trail pheromone deposited on the trail is directly dependent on the number of trail-laying ants. That number depends in turn on three factors: first, how many ants can fill their crops at the food source; second, how quickly and successfully foragers can unload the food back at the nest; and finally, how quickly foragers can return to the food source while reinforcing the trail. The quality and accessibility (distance from the nest and space for foragers around the source) are key parameters that affect the trail-laying activity. Also crucial are the potency of the recruitment pheromone and hence the number of nestmates responding to the recruitment signals in the first place. In situations where trails lead to several food sources, the majority of recruitees choose the trail from which the strongest trail signal emanates.

When individual ants have a high recruitment influence on their nestmates, as in the group recruitment technique employed by *Camponotus socius*, the decision-making process is fast. But speed comes at a cost: the colony fixates on a single trail, and potential foraging sites can remain undiscovered.[162] In these colonies, however, several scouts might return from different food sources. The intensity of their motor displays can be correlated with their perception of resource quality. If so, groups of recruitees led by single recruiters to the respective food sources vary considerably in size. If we assume that recruitees subsequently also become recruiters, the most profitable food source will eventually be selected by foragers as a whole, because the concentration on the orientation trail increases proportionately.

Although group decisions are self-organizing processes, the underlying mechanisms are individual behavioral actions elicited by external stimuli. The cues used by each ant are the overall quality of resources and the readiness of nestmates to accept the resources. These in turn determine the intensity of the foragers' recruitment behavior.

Such collective decision processes arising from a set of simple behavioral rules have been modeled mathematically by a number of theoreticians. Such abstractions are helpful tools for empirical and experimental analysis. However, the real world

162 | T. R. Stickland, N. F. Britton, and N. R. Franks, "Complex trails and simple algorithms in ant foraging," *Proceedings of the Royal Society of London B* 260: 53–58 (1995).

in nature is usually much more complex than predictions made from the outcome of simple laboratory experiments and models based on the data obtained that way. Still, we appreciate this approach because it often reveals that seemingly complex features in collective actions can be reduced to simple feedback loops and behavioral rules of thumb.

THE MEASUREMENT OF INFORMATION

How much information is transmitted by the communication systems of the social insects? A fairly precise measure can be made in the case of scout workers recruiting nestmates to new food sites. As first pointed out by the great English evolutionist J. B. S. Haldane and his wife and coworker Helen Spurway, honeybees following the directions given them in waggle dances performed by foragers do not all arrive at the exact flower bed or other honey source toward which they were guided back at the nest.[163] Rather, most arrive in a predictable scatter on both sides of the target. The degree of precision can be translated with amounts of information measured in bits.[164] The amount of information in the honeybee waggle dance concerning direction of the target is about 4 (the equivalent of 16 equiprobable choices). The same amount is transmitted for distance. Distance information in the fire ant (*Solenopsis invicta*) odor trail laid down by foragers after the discovery of food is roughly the same as in honeybees, while that for direction is also about the same but increases with the length of the trail.[165]

The information transfer in both the honeybee and fire ant systems can be understood intuitively by imagining that our own audiovisual brain and communication system allows us to transmit any one of 16 directions with perfect accuracy. Then the message "go north by northwest," one of 16 equally likely messages,

---

163 | J. B. S. Haldane and H. Spurway, "A statistical analysis of communication in 'Apis mellifera' and a comparison with communication in other animals," *Insectes Sociaux* 1(3): 247–283 (1954).

164 | A bit is the amount of information needed to make a choice between two equiprobable alternatives, as in the flip of a coin. If n alternatives are available, a choice provides $H = \log_2 n$ amounts of information; thus, 4 choices provide 2 bits, 8 choices 3 bits, and so on.

165 | E. O. Wilson, "Chemical communication among workers of the fire ant *Solenopsis saevissima* (Fr. Smith), 2: An information analysis of the odour trail," *Animal Behaviour* 10(1–2): 148–158 (1962). (The name *Solenopsis saevissima* has been replaced by *Solenopsis invicta*.)

conveys log 16 = 4 bits of information for direction. This amount of precision is in fact about the amount that humans do provide when they think about spatial locations and express them verbally. Thus, we can just about place "north by northwest" mentally but cannot be more precise with any confidence. That just happens to be the capacity of the honeybee and fire ant as well.

Why don't the social insects do better? Paradoxically, this level of accuracy may constitute not the best the insects have managed in the course of evolution, but instead the optimal amount of error. As Haldane and Spurway expressed, "Natural selection is always acting so as to reduce the error of the mean direction, while acting less intensely, if at all, to reduce individual errors which lead to a scatter around this direction." They cite the analogous strategy in naval gunnery, "where a superior force pursuing ships with less fire power should fire salvoes with a considerable scatter, in the hope that at least one shell will hit a hostile ship and slow it down." The analogy is even more apt in the case of fire ants. One of the chief problems faced by this species in the course of foraging is the recruitment of sufficient numbers of workers in time to immobilize small animals detected as they pass through the colony territory. In at least the case of laboratory colonies, fire ants often succeed in capturing insects only because they deviate from the odor trails that have been rendered inaccurate by the continued movement of the prey.[166]

TACTILE COMMUNICATION AND TROPHALLAXIS

The exact significance of most tactile signals, such as antennal touch, is not known. They may serve in modulatory communication, as demonstrated separately for stridulation in some forms of pheromone-based recruitment. On the other hand, it is now generally understood that most antennal exchanges serve to receive information rather than send it. Ants antennate the bodies of other nestmates to smell

---

166 | E. O. Wilson, "Chemical communication among workers of the fire ant *Solenopsis saevissima* (Fr. Smith), 2: An information analysis of the odour trail," *Animal Behaviour* 10(1–2): 148–158 (1962). (The name *Solenopsis saevissima* has been replaced by *Solenopsis invicta*.); J.-L. Deneubourg, J. M. Pasteels, and J. C. Verhaeghe later drew the same conclusion for other species of ants in "Probabilistic behaviour in ants: a strategy of errors?" *Journal of Theoretical Biology* 105(2): 259–271 (1983).

their identifying odors and other pheromones, not to inform them by means of a tactile code. This now universally accepted conclusion is a radical departure from the famous assessment made in 1899 by the German entomologist Erich Wasmann, who interpreted the play of the antennae on the bodies of sister ants as a complex "antennal language."[167]

Nevertheless, a role for touch by the antennae and forelegs has been documented in the invitation behaviors that entrain most forms of recruitment. Typically, one ant runs up to a nestmate and taps her body very lightly and rapidly with her antennae, sometimes raising one or both forelegs to touch the nestmate with these appendages as well. The ant seems almost to be saying, "Pay attention to the message about to follow." The recruiter then turns and follows a recently laid odor trail or lays a new one. In the case of tandem running, during which leader and follower remain in tight contact on the outbound run, the leader requires the frequent touch of the follower in order to start and finish the journey.[168]

The best-documented mode of tactile communication is that mediating oral trophallaxis, the exchange of liquid food from the crop of one ant (the anterior storage segment of the foregut) to the alimentary tract of another ant (Figure 6-29; Plate 30). The ability of some myrmecophile beetles and other social parasites to induce ant workers to regurgitate to them, despite their outlandishly different anatomies, suggests that there must be some simple trick involved in the soliciting procedure. Realizing this, Bert Hölldobler was able, with nothing but the tip of a human hair, to induce regurgitation by workers of *Formica* and *Myrmica*.[169] The full exchange between the ants can be summarized as follows. The most susceptible worker ant is one that has just returned to the nest with a full crop after foraging and is searching for nestmates to unload part of it by regurgitation. To gain the searching donor's attention, the recipient nestmate or social parasite has only to tap the donor's body lightly with her antennae or forelegs. This causes the donor to turn and face the individual that gave the signal. Sometimes this signal suffices to elicit regurgitation of a droplet from the donor's

---

167 | E. Wasmann, "Die psychischen Fähigkeiten der Ameisen," *Zoologica* (Stuttgart) 11(26): 1–133 (1899).

168 | These and other aspects of tactile communication are reviewed in detail in B. Hölldobler and E. O. Wilson, *The Ants* (Cambridge, MA: The Belknap Press of Harvard University Press, 1990), from which this account has been condensed and modified in part.

169 | B. Hölldobler, "Communication between ants and their guests," *Scientific American* 224(3): 86–93 (1971).

FIGURE 6-29. Two workers of a *Formica* species exchange liquid food. *Above*: The worker on the left induces regurgitation by striking the donor's head with her forelegs and antennae. *Below*: The donor on the right passes liquid from her crop (C), the storage organ that serves as a "social stomach," through her esophagus into the mouth and crop of the recipient. Small amounts of the food are also passed from the crop into the midgut (M) to serve as nourishment for the donor. Waste material is passed out through the rectal bladder (R). Based on an original drawing by Turid Hölldobler-Forsyth from B. Hölldobler, 16mm Film E2013, Encyclopaedia Cinematographica, Göttingen, 3-11 (1973).

crop. If not, the soliciting ant employs a stronger stimulus by tapping the donor's labium with antennae and forelegs, an action that triggers the regurgitation reflex. Regular play of the recipient's palpi on the donor's labium keeps the liquid flowing until the food exchange is interrupted by either one or both trophallaxis partners (see Figure 6-29).

PLATE 30. Oral trophallaxis in two very different ant species. *Above*: *Camponotus castaneus*, from North America; the donor ant is on the right. *Below*: *Daceton armigerum*, from South America; the minor on the right presents food to the major on the left.

Because of the superficial complexity of tactile behavior, the observer is always at risk of overinterpreting its informational content, as happened in the early analysis by Erich Wasmann.

Measurements of the transition probabilities in the shift from one antennal movement to another have revealed very little transmission of information, at least in the species most carefully examined, *Camponotus vagus* and *Myrmica rubra*.

There is negligible autocorrelation between the movements. That is, any one antennal position does not consistently lead to another within or across the repertoires of the signaler and recipient.[170] A. Bonavita-Cougourdan, in one important experiment, used radioactive gold to monitor the flow of liquid food regurgitated between paired *Camponotus* workers while antennal positions were recorded moment by moment. Varying positions were not associated with changes in the flux.[171] Overall, it seems unlikely that any of the postures convey particular messages to the receiver or donor ants.

On the other hand, Zhanna Reznikova and Boris Ryabko have claimed that workers of *Formica polyctena* may be capable, by varying duration and frequency of antennation, of transmitting information to nestmates about the position of a particular trail branch, and in the absence of directional pheromone signals. They state that the ants achieve this by "presenting the number of a branch in a way similar to that for Roman figures." They conclude that "the ants seem to be able, first, to put the duration of a message into agreement with its frequency and, second, to add and subtract small numbers, as people do when they use the Roman figures."[172] This astounding claim of transmission of abstract information by antennation behavior obviously has to be confirmed by additional studies before being accepted.

The sharing of food in some form or other is one of the elementary bonds of insect colonies. Prey objects and seeds brought into the nest by foragers are usually consumed directly or after some processing by many individuals, including other workers as well as the queen and larvae. On the other hand, liquid food, after being first stored in the forager's crop, is partially regurgitated to nestmates, who pass fractions of what they receive to still others, and so on until

---

170 | A. Lenoir, "An informational analysis of antennal communication during trophallaxis in the ant *Myrmica rubra*," *Behavioural Processes* 7(1): 27–35 (1982).

171 | A. Bonavita-Cougourdan, "Activité antennaire et flux trophallactique chez la fourmi *Camponotus vagus* Scop. (Hymenoptera, Formicidae)," *Insectes Sociaux* 30(4): 423–442 (1983).

172 | Zh. I. Reznikova and B. Y. Ryabko, "Experimental study of ant capability for addition and subtraction of small numbers," *Zhurnal Vysske • Nervno • Deyatelnosti Imeni i p Pavlova* 49(1): 12–21 (1999) [English abstract]; Zh. I. Reznikova and B. Y. Ryabko, "A study of ants' numerical competence," *Electronic Transactions on Artificial Intelligence B* 5: 111–126 (2001) [this paper can be downloaded from http://reznikova.net/Publications.html]; and Zh. I. Reznikova, *Animal Intelligence* (New York: Cambridge University Press, 2007).

large portions of the colony possess fragments of the original bounty.[173] Oral trophallaxis is by far the more common form of liquid food exchange (see Figure 6-29). The crops of most ant species that feed on nectar and on the honeydew excreta of aphids and other homopterous insects are capable of considerable distension. Individual foragers are consequently able to carry home large loads of carbohydrates laced with amino acids. Some groups of workers serve as living reservoirs for the colony during lean periods. The storage of liquid food has been carried to extreme by the "replete" caste of certain species in the myrmicine genus *Oligomyrmex*, the dolichoderine genus *Leptomyrmex*, and the formicine genera *Camponotus*, *Melophorus*, *Myrmecocystus*, *Plagiolepis*, *Prenolepis*, and *Proformica*. Workers in this specialized role have abdomens so distended with liquid that they have difficulty moving and are forced to remain permanently in the nest as living storage castes (Plate 31).[174]

The liquid is passed freely from one ant to another. The crops of all the workers taken together thus serve as a social stomach, from which all draw common nourishment. Oral trophallaxis serves two other roles, both with a more purely communicative function. The first is to inform individual colony members of the nutritional state of the colony as a whole. Carbohydrate crop loads tend to become more uniformly distributed in ant colonies. In contrast, proteinaceous food is shunted mainly to the nurse ants, larvae, and queen. The difference in food flow reflects the social demography of the ant society. As a result, it provides cues to foragers about what kind of food is needed at a particular time. A second communicative function of food exchange is the transmission of pheromones added to the liquid as it is ingested or regurgitated.[175] Each worker picks up pheromones in her

---

173 | The present account of food exchange is drawn largely from the more detailed review of the subject in B. Hölldobler and E. O. Wilson, *The Ants* (Cambridge, MA: The Belknap Press of Harvard University Press, 1990). For a thorough history of the origins of the concept of trophallaxis and its history to 1930, see C. Sleigh, "Brave new worlds: trophallaxis and the origin of society in the early twentieth century," *Journal of the History of the Behavioral Sciences* 38(2): 133–156 (2002).

174 | A more comprehensive review of the replete caste is provided in E. O. Wilson, *The Insect Societies* (Cambridge, MA: The Belknap Press of Harvard University Press, 1971).

175 | E. O. Wilson and T. Eisner, "Quantitative studies of liquid food transmission in ants," *Insectes Sociaux* 4(2): 157–166 (1957); K. Gösswald and W. Kloft, "Tracer experimentation food exchange in ants and termites," in *Proceedings of a Symposium on Radiation and Radioisotopes Applied to Insects of Agricultural Importance* (Vienna: International Energy Agency, 1963), pp. 25–42; R. Lange, "Die Nahrungsverteilung unter den Arbeiterinnen das Waldameisenstaates," *Zeitschrift für Tierpsychologie* 24: 513–545 (1967); A. A. Sorenson, T. M. Busch, and S. B.

PLATE 31. The honeypots, the living storage containers of *Myrmecocystus mimicus*. *Above*: The honeypot chamber of an excavated nest in the New Mexico desert. *Below*: The honeypot chamber in a laboratory nest.

mouth during oral grooming of herself and her nestmates and passes these substances on with the food during regurgitation.

Liquid food exchange by regurgitation, a highly evolved form of social behavior, is common among species belonging to the phylogenetically advanced subfamilies Myrmicinae, Dolichoderinae, and Formicinae. It is also a widespread practice among the eusocial bees, wasps, and termites.[176] A likely evolutionary precursor to the behavior in the ants is still practiced by some members of the subfamilies Ponerinae and Ectatomminae. Most ponerine ants are primarily predators and scavengers, foraging for particulate food, but some species collect liquid material as well. Workers of a West African species of *Odontomachus*, for example, after gathering sugary excreta ("honeydew") from aphids and scale insects, carry the liquid home as droplets between their mandibles. Other large poneromorph ants transport liquid in a similar manner, including the giant paraponerine "bullet ant" *Paraponera clavata* and the ectatommine *Ectatomma ruidum* of American tropical forests, whose workers search shrubs and trees for liquid extrafloral nectaries and decaying fruit.

THE SOCIAL BUCKET

Not just transport but also transmission of the droplets from one worker to another has been observed in *Pachycondyla obscuricornis* and *Pachycondyla villosa*, which are large ponerine ants inhabiting tropical forests of the New World. When a forager enters her nest laden with liquid food, she stands still for a period of time, swinging her head from side to side while waiting for a nestmate to approach; or else she moves directly toward nestmates and presents them with the food droplet held between her widely opened mandibles. If the colony is already well fed, a forager may have to wait as long as 30 minutes before a nestmate responds. Sometimes she is wholly ignored and is not able to share her booty. In this case, she imbibes a

---

Vinson, "Control of food influx by temporal subcastes in the fire ant, *Solenopsis invicta*," *Behavioral Ecology and Sociobiology* 17(3): 191–198 (1985); and D. L. Cassill and W. R. Tschinkel, "Information flow during social feeding in ant societies," in C. Detrain, J. L. Deneubourg, and J. M. Pasteels, eds., *Information Processing in Social Insects* (Basel: Birkhäuser Verlag, 1999), pp. 69–82.

176 | The evolution of trophallaxis in the termites and social bees and wasps is reviewed in E. O. Wilson, *The Insect Societies* (Cambridge, MA: The Belknap Press of Harvard University Press, 1971).

portion of the droplet herself and wipes off the residue against the floor and walls of the nest. Most of the time, however, nestmates readily accept the liquid food and even actively solicit it from the forager. While jerking her head rapidly up and down, each solicitor approaches the food carrier head-on and intensively antennates the front of the donor's head and mandibles. She makes a "spooning" or licking motion with her labium and slowly transfers part of the standing drop to the space between her own mandibles. All the while she continues to antennate the head and mandibles of the donor. When about half the liquid has been transferred, the ants pull apart. After the separation, the solicitor appears to imbibe a small fraction of the liquid. The remainder she shares with other nestmates, until as many as ten or more have received a portion.[177]

In short, the *Pachycondyla* workers do not share food by regurgitation in the characteristic manner of most other ants. Rather, they employ a "social bucket" system in which they first collect liquid food, then pass portions into the gaping mandibles of nestmates. The bucket itself is formed by the mandibles on each side and by inwardly curving setae and the extruded labium underneath. The liquid is held in place by surface tension (Plate 32).

The whole social bucket procedure, while crude, nevertheless bears a striking similarity to the kind of liquid food exchange by regurgitation employed by species of Formicinae and other phylogenetically advanced ants. In the latter case, the food is collected in the crop. In response to very similar antennal signals and the mechanical stimulation of the food carrier's labium, the donor regurgitates a droplet of liquid from her crop. At the same time, she opens her mandibles widely, pulling her antennae backward and out of the way and extruding her labium. Occasionally, when a large droplet is regurgitated all at once, it is held between the mandibles in the ponerine manner. In contrast to the typical ponerine exchange, however, the soliciting ant imbibes all the food she receives and stores it in her crop. A small amount of this food passes through the proventriculus into the midgut, where it is digested. The major portion, however, is distributed by regurgitation to nestmates.

With all this evidence at hand, we may reasonably suppose that the social bucket

---

177 | B. Hölldobler, "Liquid food transmission and antennation signals in ponerine ants," *Israel Journal of Entomology* 19: 89–99 (1985).

PLATE 32. *Above*: A *Pachycondyla villosa* ant carrying liquid food between her gaping mandibles. *Below*: A nestmate (left) imbibes food from the food carrier while exhibiting the tactile signaling that occurs during oral food exchange.

method of liquid food exchange is a precursor to stomodeal (mouth-to-mouth) regurgitation. It is currently the only evolutionary entry conceivable and thus seems the most plausible. The hypothesis gains further support from the fact that *Ectatomma*, which employs the social bucket technique, is generally considered close to the stock that gave rise to the subfamily Myrmicinae, among the master users of regurgitation.

Although trophallaxis by regurgitation is almost unknown in ponerine ant species, there is unequivocal evidence of its occurrence in two species: a *Hypoponera* species from Japan and the European *Ponera coarctata*.[178] It is difficult to observe regurgitation of a liquid droplet as long as the mouthparts of both ants are touching. However, after the two workers part, a droplet emerging at the top of the labium of the donor ant indicates that regurgitation has taken place. In fact, it might very well be that what has been called "pseudoregurgitation" in *Ponera pennsylvanica* is actually true trophallaxis.[179]

Nevertheless, the droplet is tiny, which casts some doubt on the importance of trophallaxis for food exchange in the two ponerine species. Perhaps the function of trophallaxis is not regurgitation of food but the exchange of hydrocarbons shared in the postpharyngeal gland. Abraham Hefetz and his coworkers have found that in at least several ant species, workers take up cuticular hydrocarbons from nestmates by mutual grooming. These hydrocarbon mixtures are then loaded into the postpharyngeal gland. By spreading the substances among colony members, the workers create a "homogeneous" colony recognition label.[180] With this hypothesis in mind, occasional past reports about behavior resembling regurgitation in other ponerine species and in myrmeciine (bulldog ant) species should be reconsidered and the behavior investigated in greater depth.

Why is food sharing by regurgitation and the storage of liquid food in the crop so poorly developed in ponerine ants? Christian Peeters has suggested that the fusion of the second top and bottom plates of the gaster (hind part of the body), a characteristic of poneromorph ants, prevents the workers from expanding their gasters. This restriction very much limits the capacity of the crop, which is located in the gaster. The same anatomical constraint exists in the myrmeciine bulldog ants. In all these species, the ability to share liquid is greatly reduced compared with ant species that do not possess

---

178 | Y. Hashimoto, K. Yamauchi, and E. Hasegawa, "Unique habits of stomodeal trophallaxis in the ponerine ant *Hypoponera* sp.," *Insectes Sociaux* 42(2): 137–144 (1995); and J. Liebig, J. Heinze, and B. Hölldobler, "Trophallaxis and aggression in the ponerine ant, *Ponera coarctata*: implications for the evolution of liquid food exchange in the Hymenoptera," *Ethology* 103(9): 707–722 (1997).

179 | S. C. Pratt, N. F. Carlin, and P. Calabi, "Division of labor in *Ponera pennsylvanica* (Formicidae: Ponerinae)," *Insectes Sociaux* 41(1): 43–61 (1994).

180 | S. Lahav, V. Soroker, A. Hefetz, and R. K. Vander Meer, "Direct behavioral evidence for hydrocarbons as ant recognition discriminators," *Naturwissenschaften* 86(5): 246–249 (1999); and R. Boulay, T. Katzav-Gozansky, A. Hefetz, and A. Lenoir, "Odour convergence and tolerance between nestmates through trophallaxis and grooming in the ant *Camponotus fellah* (Dalla Torre)," *Insectes Sociaux* 51(1): 55–61 (2004).

the same abdominal "tubulation."[181] Unlike other ants, all ponerine and myrmeciine species that supplement their diet with extrafloral nectar or honeydew must carry liquid in the external "social bucket"—a droplet held between opened mandibles.

If this interpretation is correct, it must be considered a major evolutionary breakthrough by "higher" ant clades to have overcome the anatomical constraints. The reason is that part of the foregut then became an internal transport container for liquid food and regurgitation, allowing liquid food to be rapidly distributed through the entire colony. In one striking example, after honey water was labeled with a radioactive tracer and fed to a small group of foragers at a large natural colony of the European carpenter ant *Camponotus herculeanus*, the food was distributed among all the thousands of workers of this colony housed in 20 tree trunks within 24 hours.[182] Similar but variable patterns of distribution have been documented in other species of formicine and myrmicine ants.[183]

This evolutionary advance, in our interpretation, freed mandibles of liquid food carriers for other tasks during forage expeditions. That opened up a tremendous wealth of nutritional niches and gave rise to a stunning radiation in Dolichoderinae, Myrmicinae, and Formicinae. Without liquid exchange by regurgitation, species of these subfamilies would probably never have evolved symbiotic relationships with aphids, scale insects, mealybugs, treehoppers, and caterpillars of lycaenid and riodinid butterflies (the "blues" and "metalmarks"), which give sugary secretions to the ants for food (Plate 33). In return, these insects are protected from enemies by the ants. Such high-level "trophobioses" depend on food exchange by regurgitation. Curiously, the symbiosis has evidently regressed during further evolution in species that have adapted to yet other major food niches, such as seeds and, in the attine fungus-growing ants, fresh vegetation.[184]

---

181 | C. Peeters, "Morphologically 'primitive' ants: comparative review of social characters, and the importance of queen-worker dimorphism," in J. C. Choe and B. J. Crespi, eds., *The Evolution of Social Behavior in Insects and Arachnids* (New York: Cambridge University Press, 1997), pp. 372–391; and R. W. Taylor, *"Nothomyrmecia macrops*: a living-fossil ant rediscovered," *Science* 201: 979–985 (1978).

182 | W. Kloft and B. Hölldobler, "Untersuchungen zur forstlichen Bedeutung der holzzerstörenden Rossameisen unter Verwendung der Tracer-Methode," *Anzeiger für Schädlingskunde* 38: 163–169 (1964).

183 | E. O. Wilson and T. Eisner, "Quantitative studies of liquid food transmission in ants," *Insectes Sociaux* 4(2): 157–166 (1957).

184 | Other aspects of the ecological and evolutionary implications of liquid food intake and regurgitation capabilities of workers of 77 ant species are presented by Diane Davidson and coworkers in D. W. Davidson, S. C. Cook, and R. R. Snelling, "Liquid-feeding performances of ants (Formicidae): ecological and evolutionary implications," *Oecologia* 139(2): 255–266 (2004).

PLATE 33. Trophobiosis between the Australian meat ant *Iridomyrmex purpureus* and a eurymeline leafhopper nymph (above) and adult (below). The leafhoppers offer droplets of sugary excretion from their anuses in exchange for protection by the ants.

There remains one last aspect of tactile signaling and regurgitative exchange to be considered. Aggressive behavior associated with liquid food exchange has been observed in numerous hymenopteran species.[185] Regurgitation, for example, is occasionally involved during ritualized aggressive territorial interactions between colonies of the Australian meat ant *Iridomyrmex purpureus* (Plate 34).[186] Food exchange during aggressive confrontations between conspecific alien workers has been observed in several other ant species and reported to occur even between workers belonging to different genera and subfamilies.[187] This behavior has been generally interpreted to be a form of appeasement behavior, because the apparent subordinate individual usually donates a droplet of crop content to the dominant. In the myrmicine ant *Chalepoxenus muellerianus*, aggression within the colony often elicits trophallactic food offering in nestmates, varying in degree from exchange of visible food droplets to ritualized trophallaxis with no apparent food flow.[188]

Food exchange by regurgitation is called stomodeal trophallaxis because the material originates from the mouth. In contrast, procteal trophallaxis involves material originating from the anus. The latter phenomenon has been reported within colonies of only two genera of myrmicine ants in the aberrant tribe Cephalotini, *Cephalotes* and *Procryptocerus*. The function of the behavior remains unknown, but one promising candidate is the transmission of symbiotic bacteria or other microorganisms. Outside Cephalotini, anal trophallaxis has been observed between the myrmicine slave-maker *Protomognathus americanus* and its host species in the related genus *Temnothorax* (formerly classified as part of the genus *Leptothorax*). Workers of *Protomognathus* occasionally assume a stereotyped posture, standing quietly

---

185 | A review and list of species is provided in J. Liebig, J. Heinze, and B. Hölldobler, "Trophallaxis and aggression in the ponerine ant, *Ponera coarctata*: implications for the evolution of liquid food exchange in the Hymenoptera," *Ethology* 103(9): 707–722 (1997).

186 | G. Ettershank and J. A. Ettershank, "Ritualized fighting in the meat ant *Iridomyrmex purpureus* (Smith) (Hymenoptera: Formicidae)," *Journal of the Australian Entomological Society* 21(2): 97–102 (1982).

187 | R. Lange, "Über die Futterweitergabe zwischen Angehörigen verschiedener Waldameisen," *Zeitschrift für Tierpsychologie* 17(4): 389–401 (1960); C. De Vroey and J. M. Pasteels, "Agonistic behaviour in *Myrmica rubra*," *Insectes Sociaux* 25(3): 247–265 (1978); N. F. Carlin, and B. Hölldobler, "The kin recognition system of carpenter ants (*Camponotus* spp.), I: Hierarchical cues in small colonies," *Behavioral Ecology and Sociobiology* 19(2): 123–134 (1986); and A. P. Bhatkar and W. J. Kloft, "Evidence, using radioactive phosphorus, of interspecific food exchange in ants," *Nature* 265: 140–142 (1977).

188 | J. Heinze, "Reproductive hierarchies among workers of the slave-making ant, *Chalepoxenus muellerianus*," *Ethology* 102(2): 117–127 (1996).

PLATE 34. *Above*: Australian meat ants (*Iridomyrmex purpureus*) engaged in ritualized territorial interactions. *Below*: During such confrontations, the opposing ants sometimes touch each other with their antennae or forelegs, a signal that elicits brief regurgitations.

with their abdomens raised, and extrude a droplet of liquid, which is eaten by the slaves. The phenomenon is of additional importance because it is a rare instance of a social parasite donating something to her host. Its function remains unknown, however, and may prove to be a signal mediating dominance or some other form of exploitation.

VISUAL COMMUNICATION

Evidence for visual communication in ants and other social insects is very weak, in sharp contrast to the rich documentation of chemical and tactile communication. Not a single case has been solidly documented, despite the existence of many species of large-eyed ants and also despite evidence that at least a few of the species use vision to detect moving prey. For example, the large-eyed formicines *Cataglyphis* and *Gigantiops* appear to spot prey with their eyes. *Gigantiops*, like the ponerine *Harpegnathos saltator* and certain species of *Myrmecia* appear to stalk and jump on their moving prey.[189] But there is only limited evidence that other ants are enticed by the vision of the hunting actions to join in the hunt. In particular, workers of two species, *Formica nigricans*, a European mound builder and general insect predator, and *Daceton armigerum*, a large-eyed hunter of South American forests, appear to be aroused by the sight of nestmates attacking prey, after which they rush to join the action (Plates 35 and 36).[190] Yet such behavior does not prove visual communication. The same response might equally well be evoked by alarm pheromones, which are known to be used by some other large-eyed ant species, notably the weaver ants *Oecophylla longinoda* and *Oecophylla smaragdina*. Still, it is entirely possible that communication displays performed by *Oecophylla* workers during recruitment to territorial defense, which often take place outside the leaf nests, are visually perceived by nestmates.

More clear-cut in function, but less easy to analyze, is the class of communication referred to by social psychologists as facilitation: "an increase of response merely from the sight or sound of others making the same movement."[191] The same expression can legitimately be applied to behavior in which the stimulus is an odor. A relatively clear example of facilitation in the ants has been reported in the European formicine ant *Lasius emarginatus*, whose workers excavate the soil and attend larvae at a higher rate when in large groups. When workers of this soil-dwelling

---

189 | T. M. Musthak Ali, C. Baroni Urbani, and J. Billen, "Multiple jumping behaviours in the ant *Harpegnathos saltator*," *Naturwissenschaften* 79(8): 374–376 (1992); and J. Tautz, B. Hölldobler, and T. Danker, "The ants that jump: different techniques to take off," *Zoology* 98(1): 1–6 (1994).

190 | S. A. Sturdza, "Beobachtungen über die stimulierende Wirkung lebhaft beweglicher Ameisen auf träge Ameisen," *Bulletin de la Section Scientifique de l'Académie Roumaine* 24: 543–546 (1942); E. O. Wilson, "Behavior of *Daceton armigerum* (Latreille), with a classification of self-grooming movements in ants," *Bulletin of the Museum of Comparative Zoology, Harvard* 127(7): 401–421 (1962).

191 | F. H. Allport, *Social Psychology* (Boston: Houghton Mifflin, 1924).

PLATE 35. *Above*: The Neotropical *Daceton armigerum* has large eyes and most likely detects moving prey visually. It has been suggested that huntresses might also respond by sight to the hunting movements of their nestmates. *Below*: Workers have captured a termite soldier.

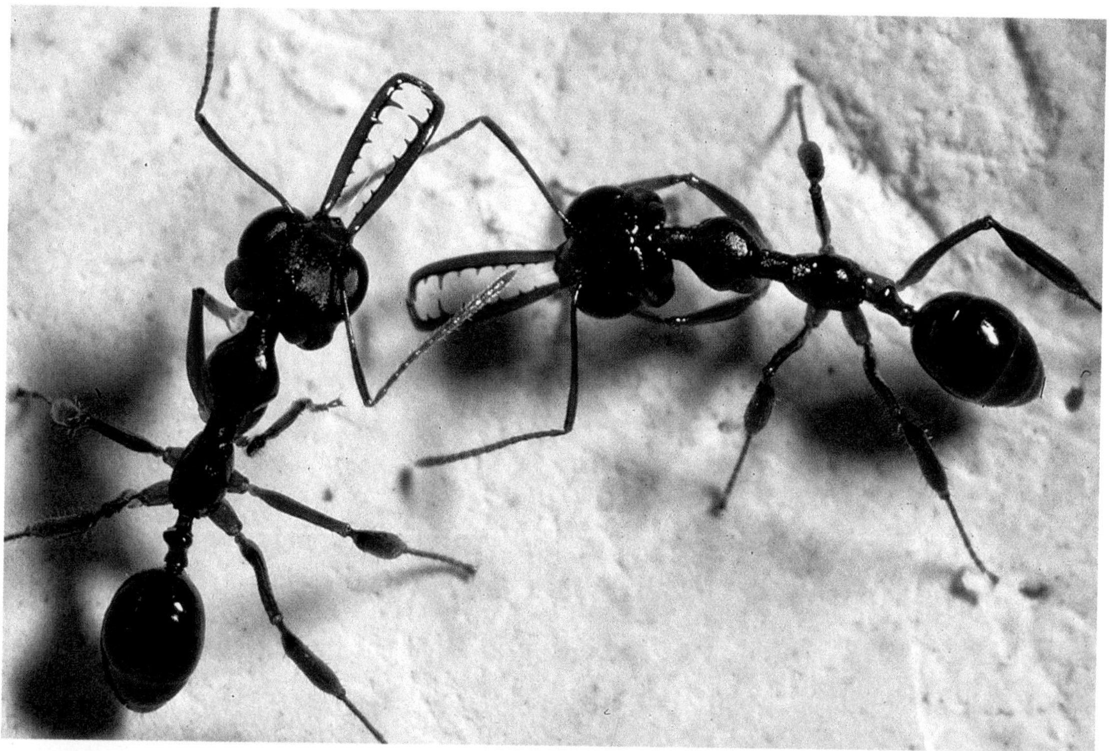

PLATE 36. *Above*: The Asian formicine ant *Myrmoteras toro* has specialized trap mandibles and large eyes. Presumably, the huntresses locate prey by sight. *Below*: This has been documented for the Australian bulldog ant of the genus *Myrmecia*. (Photo: Vincent Dietemann.)

species were broken into groups of four to six each and the groups then separated only by a gauze barrier, activity rates remained high. But when a glass barrier was substituted, the activity rates dropped. The result can quite reasonably be interpreted as being due to a pheromone, but proof of such a substance has not been adduced.[192]

ANONYMITY AND SPECIFICITY OF CHEMICAL SIGNALS

Identification and discrimination are major features of all kinds of biological systems, from embryogenesis and immune responses to social interactions among kin groups, societies, and ecological communities. Every form of recognition requires distinguishable signals that vary in ways linked to the evolutionary advantages they provide.

The key features of any signal are anonymity and specificity.[193] These properties of complex chemical signals may be clarified with an analogy from the field of artificial intelligence, which is concerned with (among other things) programming computers to distinguish among different types of objects. Such machine-based discriminations are comparable to those made by insects. In the technique known as object-oriented programming, objects are characterized by both "class variables" and "instance variables." Class variables are common to all members of the same class, while instance variables are specific to each object. In addition, classes may themselves be instances of higher-level classes; for example, my car is an instance of "cars," and those in turn are an instance of "motor vehicles." A higher class is characterized by all the class and instance variables contained in its component classes. However, just as each instance differs from the others in the same class, the class variables of each component class differ from those of other members of the same higher class.[194]

We define the anonymous properties of a chemical communication signal as those that identify the signaler as a member of a class or organizational level, but do

---

192 | R. Francfort, "Quelques phénomênes illustrant l'influence de la fourmilière sur les fourmis isolées," *Bulletin de la Société Entomologique de France* 50(7): 95–96 (1945).

193 | B. Hölldobler and N. F. Carlin, "Anonymity and specificity in the chemical communication signals of social insects," *Journal of Comparative Physiology A* 161(4): 567–581 (1987).

194 | P. H. Winston, *Artificial Intelligence*, 2nd ed. (Reading, MA: Addison-Wesley, 1984).

not distinguish it from other instances within the same class or level. Anonymous cues are uniform or invariant among all instances of a class. Diagnostic properties are those that vary so as to identify the signaler as an instance of its class, in other words, belonging to one class among others that together make up a higher class.

These hierarchical terms are relative, and their application depends on the level being examined. As a concrete example, consider an ant following a chemical recruitment trail. At the species level, it orients with respect to the species-specific trail substance and does not usually respond to trails of any other species. At the colony level, this response may be anonymous—that is, a trail laid by any conspecific individual will be followed; or it may be colony specific. At the individual level, no distinction may be made among the anonymous trails of different nestmates, or each individual may specifically recognize its own trail as different.

For example, a single-molecule pheromone, such as the recruitment signal from the poison gland of the genus *Pogonomyrmex* (a pyrazine compound), is obviously an anonymous signal and in fact is uniform throughout several species of the harvester ant genus. However, the pheromone used to mark trunk routes, which originates from the Dufour's gland, is a blend of hydrocarbons in which a variety of chemical properties differ between different instances of a class of signals. It contains species-specific properties, hence an anonymous signal uniform throughout the species (or at least a local population), and it contains properties specific for a particular colony but anonymous for the members of this colony. Whether it also contains individual specific properties is unknown.[195]

Exocrine products of social insects exhibit a high level of intrinsic complexity, providing considerable opportunity for such variation. The Dufour's gland secretions of the European carpenter ant *Camponotus ligniperdus* include at least 41 compounds,[196] while the mandibular glands of the weaver ant *Oecophylla longinoda* contain over 30 compounds. In both species, the substances occur in various proportions specific to colonies.[197]

---

195 | B. Hölldobler, E. D. Morgan, N. J. Oldham, and J. Liebig, "Recruitment pheromone in the harvester ant genus *Pogonomyrmex*," *Journal of Insect Physiology* 47(4–5): 369–374 (2001); and B. Hölldobler, E. D. Morgan, N. J. Oldham, J. Liebig, and Y. Liu, "Dufour gland secretion in the harvester ant genus *Pogonomyrmex*," *Chemoecology* 14(2): 101–106 (2004).

196 | G. Bergström and J. Löfqvist, "Similarities between the Dufour gland secretions of the ants *Camponotus ligniperdus* (Latr.) and *Camponotus herculeanus* (L.) (Hym.)," *Entomologica Scandinavica* 3(3): 225–238 (1972).

197 | J. W. S. Bradshaw, R. Baker, and P. E. Howse, "Multicomponent alarm pheromones in the mandibular glands of major workers in the African weaver ant, *Oecophylla longinoda*," *Physiological Entomology* 4(1): 15–25 (1979).

What properties convey specificity? Although some or all constituents may be shared between a given pair of chemical signals, their relative quantities can vary, producing idiosyncratic patterns. These patterns may be specific at the species level. However, when a set of components of a signal is species specific, still finer levels of specificity can be attained by further varying component ratios. Three nested levels of compositional variation are strikingly illustrated in the Dufour's gland secretions of the halictine bee *Evylaeus malachurum*, characterized by a species-wide profile of lactones, isopentyl esters, and hydrocarbons. The relative amounts of these components are more similar among nestmates than among non-nestmates, and some details of the pattern are unique to individuals.[198]

Following is a list of examples where recognition and discrimination are essential for the normal organization of an insect society:

- Recognition of nestmates and discrimination of foreigners
- Recognition of kin groups
- Recognition of age cohorts, task groups, and castes
- Recognition of reproductive states
- Recognition of social rank and of individuals
- Recognition of developmental stages of brood
- Recognition of dead nestmates

All recognition and discrimination behaviors in social insects are based on distinguishable sets of anonymous and specific stimuli. It is, however, important to note that behaviors are not solely dependent on signals. In addition, specific cues emanating from individuals but not evolved explicitly for communication can also be employed by ants for recognition, identification, and discrimination. This is strikingly illustrated by the recognition and removal of dead nestmates.

---

198 | A. Hefetz, G. Bergström, and J. Tengö, "Species, individual and kin specific blends in Dufour's gland secretions of halictine bees," *Journal of Chemical Ecology* 12(1): 197–208 (1986).

NECROPHORIC BEHAVIOR

It is one of the fascinating and enduring stories that ants carry their dead to "cemeteries." Ants of most species recognize dead nestmates and quickly remove them from the colony to dispose of them at litter dumps or other remote areas away from the colony. Obviously, this is an adaptive behavior. It prevents the spread of germs and parasites that grow on corpses.

Identification of the dead is not communication in the strict sense, but it has some features in common with communication, particularly in its dependence on stereotyped responses triggered by narrowly specific chemical cues. The removal of dead nestmates and other decomposing material from the nest serves the hygiene of the colony as a whole. The interiors of the nests, and particularly the brood chambers, are kept meticulously clean. Workers drag alien objects, including particles of waste material and defeated enemies, out of the nest and dump them onto the ground nearby. They carry waste liquid and meconia (the pellets of accumulated solid waste voided by larvae at pupation) to the nest perimeter or beyond. They respond to disagreeable but immovable objects by covering them with pieces of soil and nest material. When a researcher allows the corpse of a *Pogonomyrmex badius* worker to decompose in the open air for a day or more and then places it in the nest or outside near the nest entrance, the first sister worker to encounter it ordinarily investigates it briefly by repeated antennal contact, then picks it up and carries it directly away from the main nest chambers toward the refuse piles. In laboratory tests of *Pogonomyrmex*, in which the phenomenon was discovered,[199] the most distant walls of the foraging arenas were less than a meter from the nest entrances, and the ants had built the refuse piles against them. The distance was evidently inadequate to allow the consummation of the corpse removal response because workers bearing corpses frequently wandered for many minutes back and forth along the distant wall before dropping their burdens on the refuse piles. Others were seen to approach the distant wall unburdened, pick up the corpses already on the piles, and transport them in similarly restless fashion before redepositing them. This behavior constituted a distinctive and easily repeated bioassay. It was soon established that

---

199 | E. O. Wilson, N. I. Durlach, and L. M. Roth, "Chemical releasers of necrophoric behavior in ants," *Psyche* (Cambridge, MA) 65: 108–114 (1958); D. M. Gordon, "Dependence of necrophoric response to oleic acid on social context in the ant, *Pogonomyrmex badius*," *Journal of Chemical Ecology* 9(1): 105–111 (1983).

bits of paper treated with acetone extracts of *Pogonomyrmex* corpses were treated just like intact corpses. Separation and behavioral assays of principal components of the extract implicated long-chain fatty acids and their esters. Furthermore, it turned out that oleic acid, a common decomposition product in insect corpses, is fully effective. (The same substance has been implicated in the fire ant *Solenopsis invicta.*[200]) On the other hand, many other principal products of insect decomposition, including short-chain fatty acids, amines, indoles, and mercaptans, were ineffective. When *Pogonomyrmex* corpses were thoroughly leached out in solvents, dried, and presented to colonies, they were seldom transported as corpses, but were more commonly eaten instead.

Thus, the worker ants appear to recognize corpses on the basis of a limited array of chemical breakdown products. They are, moreover, very "narrow-minded" on the subject. Almost any object possessing an otherwise inoffensive odor is treated as a corpse when daubed with oleic acid. This classification even extends to living nestmates. When a small amount of the substance is daubed on live workers, they are picked up and carried, unprotesting, to the refuse pile. After being deposited, they clean themselves and return to the nest. If the cleaning has not been thorough enough, they are sometimes mistaken a second or third time for corpses and taken back to the refuse pile yet again.

Necrophoric behavior based on specific chemical cues appears to be almost universal in ants. Many anecdotal accounts report such behavior. This work has been reviewed by Dennis Howard and Walter Tschinkel in context with their thorough behavioral analysis of necrophoric behavior in the fire ant, *Solenopsis invicta*. They unequivocally demonstrate and confirm that necrophoric behavior depends on specific chemical cues.[201]

Perhaps even more remarkable than the simplicity of this control of necrophoric behavior is the tendency of the workers of some ant species to remove themselves from the nest when they are about to die. We have repeatedly observed that injured and dying ants loiter more in the vicinity of the nest entrance or outside the nest than do normal workers. Injured *Solenopsis invicta* fire ants and *Pogonomyrmex* harvester ants, particularly those that have lost their abdomens or one or more

---

200 | M. S. Blum, "The chemical basis of insect sociality," in M. Beroza, ed., *Chemicals Controlling Insect Behavior* (New York: Academic Press, 1970), pp. 61–94.

201 | D. F. Howard and W. R. Tschinkel, "Aspects of necrophoric behavior in the red imported fire ant, *Solenopsis invicta*," *Behaviour* 56(1–2): 157–180 (1976).

appendages, tend to leave the nest more readily when the nest is disturbed. Workers of *Formica rufa* fatally infected with the fungus *Altermaria tenuis* leave the nest and cling fast to blades of grass with their mandibles and legs just before dying.[202]

NESTMATE RECOGNITION

A profoundly important form of communication in all social insects is simple recognition—of alien species, of members of colonies of the same species, and of nestmates belonging to various castes and immature stages.

Consider nestmate recognition. Just as a human being identifies another person by scanning face and body form, so does an ant classify another ant by the bouquet of odors on and around its body.[203] The transaction occurs in a split second. When two ants meet in the nest or on an odor trail, each sweeps its antennae over part of the body of the other. They are not signaling each other; rather, they are testing body odors. If they belong to the same colony and hence possess familiar odors, they pass on by with no further response. Within the nest they may cluster, groom,

---

202 | P. I. Marikovsky, "On some features of behavior of the ants *Formica rufa* L. infected with fungous disease," *Insectes Sociaux* 9(2): 173–179 (1962).

203 | Over the past hundred years, the study of recognition odors has attracted the attention of many gifted experimentalists, with progress accelerating toward the end of the twentieth century. Key reviews at different intervals during the history of the subject include W. M. Wheeler, *Ants: Their Structure, Development and Behavior* (New York: Columbia University Press, 1910), including coverage of the classic colony-mixing experiments of Adele M. Fielde; E. O. Wilson, *The Insect Societies* (Cambridge, MA: The Belknap Press of Harvard University Press, 1971); B. Hölldobler and C. D. Michener, "Mechanisms of identification and discrimination in social Hymenoptera," in H. Markl, ed., *Evolution of Social Behavior: Hypotheses and Empirical Tests* (Weinheim: Verlag Chemie, 1980), pp. 35–50; N. F. Carlin, "Species, kin, and other forms of recognition in the brood discrimination behavior of ants," in J. C. Trager, ed., *Advances in Myrmecology* (Leiden: E. J. Brill, 1988), pp. 267–295; D. J. C. Fletcher and C. D. Michener, eds., *Kin Recognition in Animals* (New York: John Wiley, 1987), with separate reviews concerning social insects by R. H. Crozier, C. D. Michener and B. H. Smith, and M. D. Breed and B. Bennett. Also B. Hölldobler and E. O. Wilson, *The Ants* (Cambridge, MA: The Belknap Press of Harvard University Press, 1990); R. K. Vander Meer, M. D. Breed, M. L. Winston, and K. E. Espelie, eds., *Pheromone Communication in Social Insects: Ants, Bees, Wasps, and Termites* (Boulder, CO: Westview Press, 1998), with reviews by M. D. Breed, R. K. Vander Meer and L. Morel, T. L. Singer, K. E. Espelie, G. L. Gamboa, J.-L. Clément, and A.-G. Bagnères; P. Jaisson, "Kinship and fellowship in ants and social wasps," in D. G. Hepper, eds., *Kin Recognition* (New York: Cambridge University Press, 1991), pp. 60–93; A. Lenoir, D. Fresneau, C. Errard, and A. Hefetz, "Individuality and colonial identity in ants: the emergence of the social representation concept," in C. Detrain, J. L. Deneubourg, and J. M. Pasteels, eds., *Information Processing in Social Insects* (Basel: Birkhäuser Verlag, 1999), pp. 219–237; and Z. B. Liu, S. Yamane, K. Tsuji, and Z. M. Zheng, "Nestmate recognition and kin recognition in ants," *Entomologia Sinica* 7(1): 71–96 (2000).

and exchange food by regurgitation. On the other hand, if one of the pair is from another colony of the same species, and the species is one of the large majority in which colony boundaries are recognized, the intruder is treated very differently. The response now falls somewhere along a broad gradient of rejection. At one extreme, intruders are accepted but offered less food than nestmates until they have time to acquire the colony odor. At the other extreme, they are attacked and quickly killed. At intermediate levels, they are variously avoided, threatened with open mandibles, nipped, and dragged out of the nest and dumped.

What are the colony labels, or "discriminators"? In general, they are hydrocarbons carried in the waxy coating that covers the body cuticle.[204] Their role is indicated by several lines of evidence, mostly the correlations between differences in hydrocarbon profiles (of colonies and individuals) and aggressive interactions. It has proved almost universally true that the greater the difference of the profiles, the stronger the aggression. But there is also direct experimental evidence. When the profiles were altered by applying appropriate isolated hydrocarbons to the cuticle of the Mediterranean desert ant *Cataglyphis niger*, aggression was increased toward nestmate workers and lessened toward alien workers.[205]

In retrospect, hydrocarbons are ideal discriminators, not just for ants but for insects generally. They are cheap to manufacture and small enough to disperse readily. They are also easily adsorbed into the epicuticle, the outermost lipophilic layer of the cuticle—in fact, they compose much of it. Not least, hydrocarbons can be diversified both in chain length and in patterns of methane branching and double bonding. In *Myrmica incompleta*, 111 hydrocarbon compounds have been found, while 242 have been identified in seven species of *Cataglyphis*.[206] The particularity of

204 | The hydrocarbon source of ant colony odors was first reported in 1987 by two independent teams: in *Camponotus vagus* by A. Bonavita-Cougourdan, J. L. Clément, and C. Lange, "Nestmate recognition: the role of cuticular hydrocarbons in the ant *Camponotus vagus* Scop.," *Journal of Entomological Science* 22(1): 1–10 (1987); and in *Camponotus floridanus* by L. Morel and R. K. Vander Meer, "Nestmate recognition in *Camponotus floridanus*: behavioral and chemical evidence for the role of age and social experience," in J. Eder and H. Rembold, eds., *Chemistry and Biology of Social Insects* (Proceedings of the Tenth Congress of the International Union for the Study of Social Insects, 18–22 August 1986 Munich) (Munich: Verlag J. Peperny, 1987), pp. 471–472.

205 | S. Lahav, V. Soroker, A. Hefetz, and R. K. Vander Meer, "Direct behavioral evidence for hydrocarbons as ant recognition discriminators," *Naturwissenschaften* 86(5): 246–249 (1999).

206 | A. Lenoir, C. Malosse, and R. Yamaoka, "Chemical mimicry between parasitic ants of the genus *Formicoxenus* and their host *Myrmica* (Hymenoptera, Formicidae)," *Biochemical Systematics and Ecology* 25(5): 379–389 (1997); A. Dahbi, A. Lenoir, A. Tinaut, T. Taghizadeh, W. Francke, and A. Hefetz, "Chemistry of the postpharyngeal gland secretion and its implication for the phylogeny of Iberian *Cataglyphis* species (Hymenoptera: Formicidae)," *Chemoecology* 7(4): 163–171 (1996).

the mix (and hence the information it conveys) is enormously enhanced by variations in the proportions of constituent hydrocarbons. Ants create bouquets that are unique in the same sense that wines and perfumes are unique. The potential is such that workers from different colonies and from within colonies can possess their own signatures, and even workers monitored at different times can bear distinct odor signatures. This principle applies even when the kinds of hydrocarbons in the mix are relatively few. Such is the case for the European mound-building ant *Formica truncorum*, in which pentacosane, heptacosane, nonacosane, and hentriacosane dominate the mixture, accompanied by traces of their corresponding alkenes.[207]

Cuticular compounds, and in particular hydrocarbons, have also been implicated in honeybees, *Polistes* paper wasps, and termites. Although not as well documented as in ants, there is every reason to expect that they will prove to be the source of colony odors throughout these other social insects.[208]

In the ants at least, the colony odor has proved invariably to be a blend of hydrocarbons produced and somehow distributed by the colony in a more or less homogeneous state. It composes an odoriferous gestalt to which the colony members predictably respond.[209] They learn the colony-wide blend; as far as we know, they do not learn the idiosyncratic hydrocarbon signature of their nestmates or memorize any membership register, although individual recognition has recently been demonstrated in a ponerine ant species to be discussed later. The hydrocarbons are manufactured within the bodies of the ants, transported into the hemolymph (blood), and then passed either directly through the epidermis onto the cuticle or to the postpharyngeal gland for storage and later distribution (Figure 6-30). The postpharyngeal gland, a large organ located in the head and comprising two bilateral glove-shaped halves, is unique to ants. So far as we know, it occurs in all species of

---

207 | J. Nielsen, J. J. Boomsma, N. J. Oldham, H. C. Petersen, and E. D. Morgan, "Colony-level and season-specific variation in cuticular hydrocarbon profiles of individual workers in the ant *Formica truncorum*," *Insectes Sociaux* 46(1): 58–65 (1999).

208 | This generalization is supported in reviews of nestmate recognition authored variously by A.-G. Bagnères, M. D. Breed, J.-L. Clément, K. E. Espelie, G. J. Gamboa, and T. L. Singer in R. K. Vander Meer, M. D. Breed, K. E. Espelie, and M. L. Winston, eds., *Pheromone Communication in Social Insects: Ants, Wasps, Bees, and Termites* (Boulder, CO: Westview Press, 1998), pp. 57–155.

209 | The gestalt (or colony blend) model was first predicted by R. H. Crozier and M. W. Dix on theoretical grounds in "Analysis of two genetic models for the innate components of colony odor in social Hymenoptera," *Behavioral Ecology and Sociobiology* 4(3): 217–224 (1979), and later abundantly demonstrated by experimental evidence as reviewed in the present account.

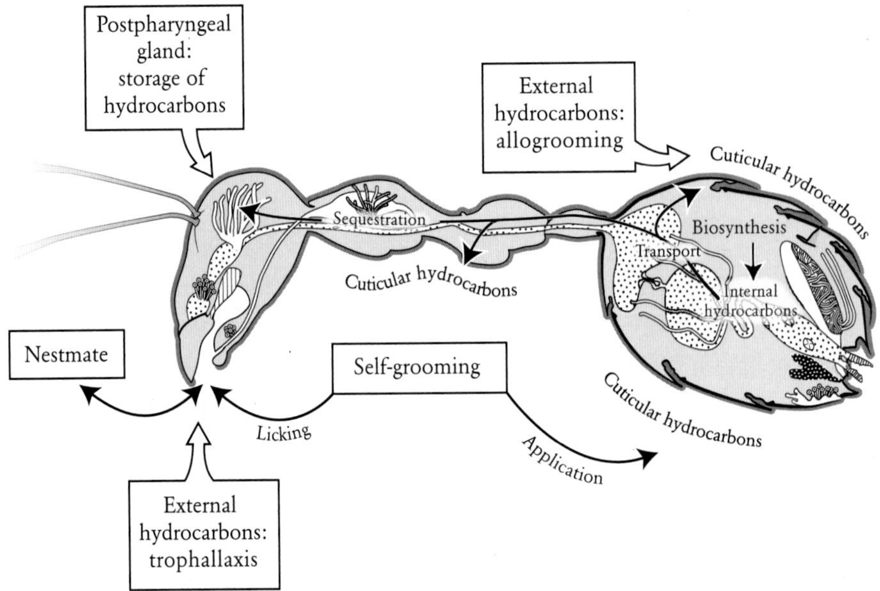

FIGURE 6-30. Depicted here are the pathways of generation and distribution of the cuticular hydrocarbons used by ants to identify colony membership. Modified from A. Lenoir, D. Fresneau, C. Errard, and A. Hefetz, "Individuality and colonial identity in ants: the emergence of the social representation concept," in C. Detrain, J. L. Deneubourg, and J. M. Pasteels, eds., *Information Processing in Social Insects* (Basel: Birkhäuser Verlag, 1999), pp. 219–237.

ants and all adult castes, as well as males. The organ is an important source not only of cuticular hydrocarbons but also of liquid food fed to the larvae.[210] Because many ponerine and other phylogenetic primitive species do not engage in regurgitation

---

210 | The evidence for the nutritive role of postpharyngeal secretions is reviewed in B. Hölldobler and E. O. Wilson, *The Ants* (Cambridge, MA: The Belknap Press of Harvard University Press, 1990), pp. 165–166; A. G. Bagnères and E. D. Morgan, "The postpharyngeal glands and the cuticle of Formicidae contain the same characteristic hydrocarbons," *Experientia* 47(1): 106–111 (1991); V. Soroker, C. Vienne, A. Hefetz, and E. Nowbahari, "The postpharyngeal gland as a "Gestalt" organ for nestmate recognition in the ant *Cataglyphis niger*," *Naturwissenschaften* 81(11): 510–513 (1994); V. Soroker, C. Vienne, and A. Hefetz, "Hydrocarbon dynamics within and between nestmates in *Cataglyphis niger* (Hymenoptera: Formicidae)," *Journal of Chemical Ecology* 21(3): 365–378 (1995); and V. Soroker and A. Hefetz, "Hydrocarbon site of synthesis and circulation in the desert ant *Cataglyphis niger*," *Journal of Insect Physiology* 46(7): 1097–1102 (2000). But note somewhat diverging results obtained with the African species *Myrmicaria eumenoides*, where only young workers show a congruence between postpharyngeal gland blends and cuticular blends of hydrocarbons; E. Schoeters, M. Kaib, and J. Billen, "Is the postpharyngeal gland in *Myrmicaria* ants the source of colony specific labels?" in M. P. Schwarz and K. Hogendoorn, eds., *Social Insects at the Turn of the Millennium* (Proceedings of the Thirteenth Congress of the International Union for the Study of Social Insects, 29 December 1998 to 3 January 1999, Adelaide) (Adelaide, Australia: Flinders University Press, 1998), p. 426.

when feeding larvae, it seems to follow that the postpharyngeal gland originated as a storage organ for colony odor hydrocarbons. It is, in other words, a genuine colony-level social organ.[211]

The hydrocarbons are spread over the body of each ant as she grooms herself and her nestmates by licking motions of her tongue. The famous cleanliness of ants, achieved by frequent bouts of grooming, has generally been assumed to be hygienic in function, removing bacteria and fungus spores that could cause disease. But the almost frenetic level it often reaches might better be explained as a means of homogenizing and distributing the colony odor. When worker ants are separated from their nestmates for a few days and returned, they are often closely inspected and then subjected to heightened levels of grooming. It has been suggested that ants (in particular, *Camponotus japonicus*) are equipped with a special chemosensory sensillum that responds exclusively to non-nestmate cuticular hydrocarbon profiles.[212]

The source of the odor is primarily endogenous; that is, the hydrocarbons are produced by metabolism and circulated within the body. The biochemical pathways may be genetically prescribed, producing for each ant a fixed set of hydrocarbons, with these products then being mixed with those of nestmates to create the colony-specific mix. Or else they may in some manner be determined by the food and other particular qualities of the environment in which the colony lives.

Strong evidence of a genetic component in species odors is provided by the fire ant *Solenopsis invicta* and its sibling species *Solenopsis richteri* in the southern United States: each of the two species possesses a distinctive hydrocarbon profile, while

---

211 | According to J. V. Hernández, H. López, and K. Jaffe, the main source of the colony odor in the leafcutter ant *Atta laevigata* is not the postpharyngeal gland but the mandibular gland; see their paper "Nestmate recognition signals of the leaf-cutting ant *Atta laevigata*," *Journal of Insect Physiology* 48(3): 287–295 (2002). This is an extraordinary claim worthy of further research, since the *Atta* mandibular gland is also the source of methylheptanone alarm pheromones. Klaus Jaffé and collaborators implicated the mandibular gland as the source of the chemical colony label also for *Atta cephalotes*, *Odontomachus bauri*, *Solenopsis geminata*, and *Camponotus rufipes*; literature cited in Z. B. Liu, S. Yamane, K. Tsuji, and Z. M. Zheng, "Nestmate recognition and kin recognition in ants," *Entomologia Sinica* 7(1): 71–96 (2000).

212 | M. Ozaki, A. Wada-Katsumata, K. Fujikawa, M. Iwasaki, F. Yokohari, Y. Satoji, T. Nisimura, and R. Yamaoka, "Ant nestmate and non-nestmate discrimination by a chemosensory sensillum," *Science* 309: 311–314 (2005).

their hybrids have an intermediate profile.[213] Similarly, honeybee workers in the same colony but with different fathers have different odor profiles.[214]

Thus, in general, it is to be expected that the closer the kinship of social insect workers, the more similar their colony odors and the more likely they are to live together without aggression. Such is the case in encounters between ants from different nests of the same colony of the European mound-building ant *Formica pratensis* and the Australian meat ant *Iridomyrmex purpureus*,[215, 216] as well as toleration of intruders by nest guards in the primitively eusocial sweat bee *Lasioglossum zephyrum*.[217]

Among the environmental cues documented in colony odor are nest materials in bees and wasps.[218] The same is true of the small European ant *Temnothorax* (formerly *Leptothorax*) *nylanderi* of Europe, which nests in decaying twigs on the ground. Differences in colony odor are based, at least in part, on whether the colonies live in pine or oak twigs.[219] It seems likely that in each of these cases, the odorant molecules are absorbed directly into the outer hydrocarbon layer of the cuticle, and they may prove to comprise compounds other than hydrocarbons.

A similar finding has been reported for the Argentine ant *Linepithema humile*, where hydrocarbon blends make up the cues used in nestmate recognition, but where these substances, or at least a major fraction of them, are acquired from insect prey. A difference in diet alters the hydrocarbon recognition cue on the cuticle and elicits

---

213 | R. K. Vander Meer, C. L. Lofgren, and F. M. Alvarez, "Biochemical evidence for hybridization in fire ants," *Florida Entomologist* 68(3): 501–506 (1985); K. G. Ross, R. K. Vander Meer, D. J. C. Fletcher, and E. L. Vargo, "Biochemical phenotypic and genetic studies of two introduced fire ants and their hybrid (Hymenoptera: Formicidae)," *Evolution* 41(2): 280–293 (1987).

214 | G. Arnold, B. Quenet, J.-M. Cornuet, C. Masson, B. De Schepper, A. Estoup, and P. Gasqui, "Kin recognition in honeybees," *Nature* 379: 498 (1996).

215 | C. W. W. Pirk, P. Neumann, R. F. A. Moritz, and P. Pamilo, "Intranest relatedness and nestmate recognition in the meadow ant *Formica pratensis* (R.)," *Behavioral Ecology and Sociobiology* 49: 366–374 (2001); and M. Beye, P. Neumann, M. Chapuisat, P. Pamilo, and R. F. A. Moritz, "Nest-mate recognition and the genetic relatedness of nests in the ant *Formica pratensis*," *Behavioral Ecology and Sociobiology* 43(1): 67–72 (1998).

216 | M. L. Thomas, L. J. Parry, R. A. Allan, and M. A. Elgar, "Geographic affinity, cuticular hydrocarbons and colony recognition in the Australian meat ant *Iridomyrmex purpureus*," *Naturwissenschaften* 86(2): 87–92 (1999).

217 | L. Greenberg, "Genetic component of bee odor in kin recognition," *Science* 206: 1095–1097 (1979). The pioneering nature and early significance of this study is especially notable.

218 | M. D. Breed, M. F. Garry, A. N. Pearce, B. E. Hibbard, L. B. Bjostad, and R. E. Page Jr., "The role of wax comb in honey bee nestmate recognition," *Animal Behaviour* 50(2): 489–496 (1995); and G. J. Gamboa, "Kin recognition in social wasps," in S. Turillazzi and M. J. West-Eberhard, eds., *Natural History and Evolution of Paper-Wasps* (New York: Oxford University Press, 1996), pp. 161–177.

219 | J. Heinze, S. Foitzik, A. Hippert, and B. Hölldobler, "Apparent dear-enemy phenomenon and environment-basal recognition cues in the ant *Leptothorax nylanderi*," *Ethology* 102(6): 510–522 (1996).

aggressive discrimination toward nestmates fed with another diet.[220] As noted repeatedly by many authors, intercolony aggression in Argentine ants is not very pronounced in areas where they have invaded and formed gigantic unicolonial nest populations.[221] But in their native land, different colonies are much more distinct and have been observed to exhibit aggressive discrimination against members from alien colonies.[222]

An important circumstance affecting the origin of colony odors is that many ant species, principally those of the phylogenetically advanced subfamilies Dolichoderinae and Formicinae, exchange food and alimentary pheromones by regurgitation. As a consequence, the colony possesses a communal stomach and by this means alone a more nearly uniform shared hydrocarbon mix.[223, 224] Species that do not regurgitate food among adults obtain the same result by grooming; the hydrocarbons pass from the postpharyngeal gland and alimentary tract to the body of the ant during self-grooming and then (possibly with other odorants, such as fatty acids and contaminants from nest materials) are spread among her nestmates during mutual grooming. Workers of many ant species have a dense brush of cuticular hairs on the first segment of the front tarsus, and they use this brush as an aid in grooming. There is circumstantial evidence that workers of the ponerine *Pachycondyla apicalis* use the brush to apply hydrocarbons during the grooming process; the exocrine source may be the front basitarsal glands.[225] On the other hand, a few ponerine species may exchange postpharyngeal gland secretions by oral trophallaxis.

The recognition of nestmates and discrimination of foreigners are related to the social organizations of the colonies. A difference in degree of discrimination behavior

220 | D. Liang and J. Silverman, "'You are what you eat': diet modifies cuticular hydrocarbons and nestmate recognition in the Argentine ant, *Linepithema humile,*" *Naturwissenschaften* 87(9): 412–416 (2000).

221 | D. A. Holway, "Competitive mechanisms underlying the displacement of native ants by the invasive Argentine ant," *Ecology* 80(1): 238–251 (1999); and A. V. Suarez, D. A. Holway, D. Liang, N. D. Tsutsui, and T. J. Case, "Spatiotemporal patterns of intraspecific aggression in the invasive Argentine ant," *Animal Behaviour* 64(5): 697–708 (2002).

222 | D. A. Holway, A. V. Suarez, and T. J. Case, "Loss of intraspecific aggression in the success of a widespread invasive social insect," *Science* 282: 949–952 (1998); and A. V. Suarez, N. D. Tsutsui, D. A. Holway, and T. J. Case, "Behavioral and genetic differentiation between native and introduced populations of the Argentine ant," *Biological Invasions* 1(1): 43–53 (1999).

223 | E. O. Wilson and T. Eisner, "Quantitative studies of liquid food transmission in ants," *Insectes Sociaux* 4(2): 157–166 (1957).

224 | A. Dahbi, A. Hefetz, X. Cerdá, and A. Lenoir, "Trophallaxis mediates uniformity of colony odor in *Cataglyphis iberica* ants (Hymenoptera, Formicidae)," *Journal of Insect Behavior* 12(4): 559–567 (1999).

225 | A. Hefetz, V. Soroker, A. Dahbi, M. C. Malherbe, and D. Fresneau, "The front basitarsal brush in *Pachycondyla apicalis* and its role in hydrocarbon circulation," *Chemoecology* 11(1): 17–24 (2001).

has been found in monogynous colonies (with only one egg-laying queen) as opposed to polygynous colonies (with many egg-laying queens). The young alate females and males of polygynous colonies show a marked tendency to mate inside the mother nest or at least close to it, with the subsequent adoption of the freshly mated queens by their worker nestmates. In contrast, the young alate females and males of monogynous colonies leave the mother nest and embark on mating flights. Alates from many other colonies in the area join them to form swarms, where a high degree of outbreeding takes place. Usually, the freshly mated queens do not return to their nest of origin, but instead start their own colony in unoccupied ground elsewhere.

Queens of truly polygynous species are usually not aggressive toward one another. In fact, they often cluster together in large numbers and, so far as we know, do not form dominance orders. Further, all lay eggs. Queens of monogynous colonies, in contrast, do not tolerate other queens in their vicinity, and this exclusion is maintained within mature colonies. Workers of polygynous colonies usually display less aggression toward conspecific neighboring colonies. Sometimes the whole population of nests can be considered one large colony, because workers, brood, and even queens are freely exchanged between different nests. In contrast, workers of most monogynous species discriminate aggressively against members of conspecific neighbor colonies, and they are typically highly territorial, at least in the immediate vicinity of their nests.[226]

The observed biological differences between the two types of colonies have led to the hypothesis that queens of monogynous colonies might have a significant influence on colony odor. The reasoning is as follows. If genetic differences play a role in the determination of colony odor in ants, the simplest mechanism would appear to be for the queen herself to provide the essential ingredients. In the simplest conceivable case, where two alleles influence odor at each locus, a mere ten such loci could generate $3^{10} = 59{,}049$ diploid combinations; three alleles at ten loci could yield 9^{10} such combinations. Monogyny would make such a system easily operable, while polygyny would tend to break it down.[227]

---

226 | B. Hölldobler and E. O. Wilson, *The Ants* (Cambridge, MA: The Belknap Press of Harvard University Press, 1990).

227 | B. Hölldobler and E. O. Wilson, "The number of queens: an important trait in ant evolution," *Naturwissenschaften* 64(1): 8–15 (1977); B. Hölldobler and C. D. Michener, "Mechanisms of identification and discrimination in social Hymenoptera," in H. Markl, ed., *Evolution of Social Behavior: Hypotheses and Empirical Tests* (Dahlem Konferenzen, 18–22 February 1980, Berlin) (Weinheim: Verlag Chemie, 1980), pp. 35–57.

Colony odor determination by queens would also be functionally simple, given the central importance of the queen to colony social organization. A queen of a monogynous colony is highly attractive to workers. She is continuously licked and groomed, thus facilitating the chemical advertisement of her presence.

Norman Carlin and Bert Hölldobler conducted a series of experiments with several monogynous carpenter ant species to test the hypothesis that the queen is the colony odor generator. They first manipulated small colonies so that each comprised a queen and two worker groups, the latter comprising five of the queen's daughters and five workers from a different *Camponotus* species added to the experimental colony in the pupal stage.[228]

The results demonstrated clearly that at least in colonies of this size, the queen odors acquired by the workers serve as the recognition cues for all of the colony members. This proved to be the case not only for colonies composed of the same species but even for colonies composed of different species. The effect was strong enough to cause rejection of genetic sisters reared by heterospecific queens. Workers of *Camponotus ferugineus* and *Camponotus pennsylvanicus* groomed and fed each other when they had been raised by the same queen, whereas genetic sisters raised separately by *Camponotus americanus* and *Camponotus pennsylvanicus* queens fiercely attacked each other. Further experiments with small colonies showed that the strong aggression observed between *Camponotus* workers adopted by foreign queens and their unfamiliar sisters from pure colonies is independent of the proportion of each represented in the mixed colony. If discriminators are transferred homogeneously among all nestmates to form a collective label (an olfactory "gestalt"), colonies containing similar proportions of kin groups should have similar composite labels and be more mutually recognizable than those of differing proportions. Yet as Ross Crozier has pointed out, queens are expected to contribute more to a common "gestalt odor" when colonies are small.[229] Thus, the influ-

---

228 | N. F. Carlin and B. Hölldobler, "Nestmate and kin recognition in interspecific mixed colonies of ants," *Science* 222: 1027–1029 (1983); N. F. Carlin and B. Hölldobler, "The kin recognition system of carpenter ants (*Camponotus* spp.), I: Hierarchical cues in small colonies," *Behavioral Ecology and Sociobiology* 19(2): 123–134 (1986); N. F. Carlin and B. Hölldobler, "The kin recognition system of carpenter ants (*Camponotus* spp.), II: Larger colonies," *Behavioral Ecology and Sociobiology* 20(3): 209–217 (1987); and N. F. Carlin, B. Hölldobler, and D. S. Gladstein, "The kin recognition system of carpenter ants (*Camponotus* spp.), III: Within-colony discrimination," *Behavioral Ecology and Sociobiology* 20(3): 219–227 (1987).

229 | R. H. Crozier, "Genetic aspects of kin recognition: concepts, models, and synthesis," in D. J. C. Fletcher and C. D. Michener, eds., *Kin Recognition in Animals* (New York: John Wiley, 1987), pp. 55–73.

ence of the queen on worker recognition cues might be greater in newly founded colonies and decline as the colony grows. In fact, queen-derived cues could be important in the elimination of incipient colonies by nearby mature nests or in brood raiding among incipient colonies, but less significant in worker encounters at established territorial borders.

In a second series of experiments conducted by Carlin and Hölldobler, this time with colonies of *Camponotus floridanus*, each containing about 200 workers adopted by a foreign conspecific queen, a considerable influence of the queen presence was still detected on the worker recognition label, but only if the queen was highly fertile. Even in the latter case, however, tests revealed that the queen is by no means the only source of shared recognition cues. For example, workers removed as pupae from a single colony and reared separately in the absence of queens were relatively tolerant of one another, but exhibited stronger aggressive behavior toward nonrelatives. Furthermore, differences in diet slightly enhanced aggression among separately reared kin. When a queen was present, however, workers attacked both unfamiliar kin and nonkin with equal vigor. The response was unaffected by food odors.

Hence, cues derived from queens with active ovaries apparently are sufficient to label all workers in experimental *Camponotus floridanus* colonies containing approximately 200 workers each. The workers' personal odors become more important when their queen is infertile or when the colony reaches a sufficient size to render queen-derived cues insufficient to label all workers. In fact, the size of the experimental colonies used by Carlin and Hölldobler was still relatively small compared to mature *Camponotus floridanus* colonies found in the wild, whose worker populations can reach into the thousands (Plates 37 and 38).

Further experiments have shown that in mature *Camponotus floridanus* colonies, like those of other *Camponotus* species and the formicine genus *Cataglyphis* and other ant species, the main component of the colony recognition label is a blend of multiple colony-specific cuticular hydrocarbons.[230] Of course, it is still

---

230 | L. Morel, R. K. Vander Meer, and B. K. Lavine, "Ontogeny of nestmate recognition cues in the red carpenter ant (*Camponotus floridanus*): behavioral and chemical evidence for the role of age and social experience," *Behavioral Ecology and Sociobiology* 22(3): 175–183 (1988); A. Bonavita-Cougourdan, J. L. Clément, and C. Lange, "Nestmate recognition: the role of cuticular hydrocarbons in the ant *Camponotus vagus*, Scop.," *Journal of Entomological Science* 22(1): 1–10 (1987); A. Dahbi and A. Lenoir, "Queen and colony odour in the multiple nest ant species, *Cataglyphis iberica* (Hymenoptera, Formicidae)," *Insectes Sociaux* 45(3): 301–313 (1998); and

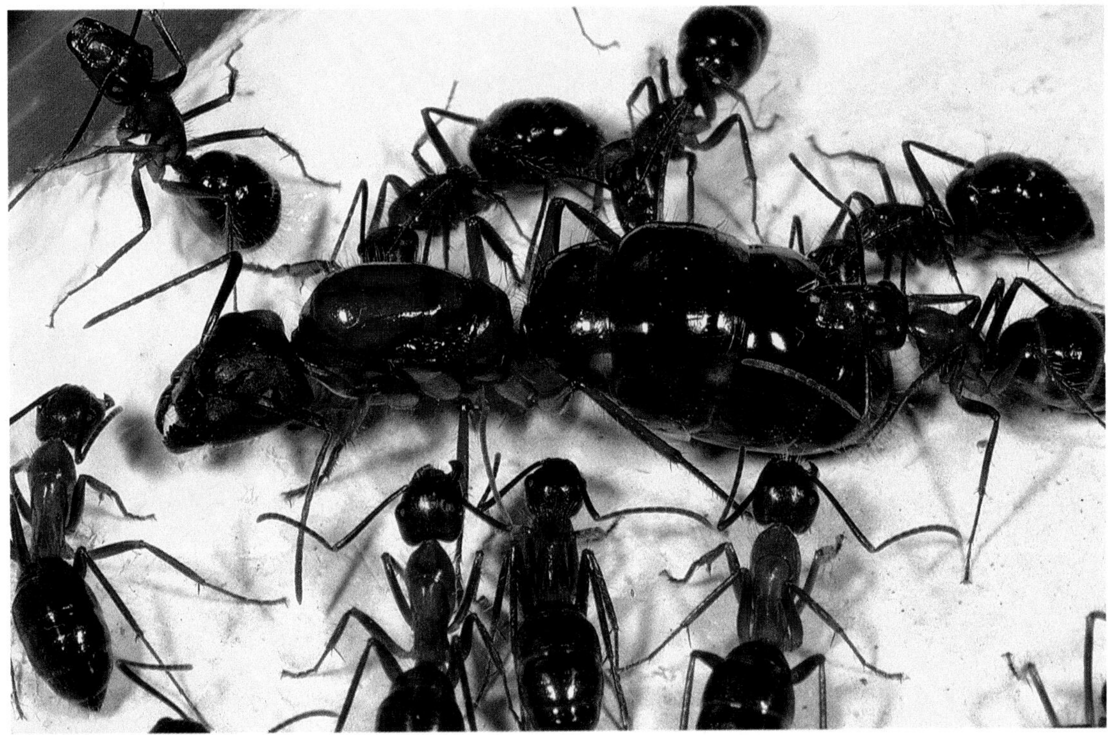

PLATE 37. *Above*: Incipient colony of *Camponotus floridanus* with queen and retinue of workers. *Below*: Two sister workers raised by different queens confront each other with threat displays.

PLATE 38. Escalated aggression between sister workers of *Camponotus floridanus* raised by different queens.

possible that in addition to this mix, workers closest to the queen, and in particular nurses and queen attendants, carry a stronger proportion of the queen-derived hydrocarbons. Some evidence from cuticular hydrocarbon studies support just this supposition.[231]

Why, then, if the personal odors of queens are important, do polygynous colonies display less pronounced nestmate recognition? Most likely, the members of such colonies are genetically close at the start, and so the collective labels diverge to a limited degree. Unicolonial populations from more distant localities, when tested in arena experiments, exhibit stronger discrimination behavior against locals. Nevertheless, monogynous colonies typically exhibit a much stronger colony-level discrimination than do most polygynous colonies, possibly due to closer genetic relatedness among workers within each monogynous colony.

In all social hymenopterans whose colony odor has been examined in sufficient detail, the workers learn the odor by an imprinting process during the first days of their eclosion as adults,[232] when they simultaneously acquire the colony label by mutual grooming and liquid food exchange with nestmates. An experimental analysis of *Cataglyphis iberica* demonstrated that the newly emerged workers have few hydrocarbons to start but soon pick up the full colony mix from the older workers.[233]

It is further possible that during territorial engagements, ants become familiar with the odors of neighboring colonies and in the process become less aggressive toward those colonies. This dear-enemy effect, as it is generally called in studies of vertebrate animal behavior, can be interpreted as an accommodation with the

---

A. Lenoir, A. Hefetz, T. Simon, and V. Soroker, "Comparative dynamics of gestalt odour formation in two ant species *Camponotus fellah* and *Aphaenogaster senilis* (Hymenoptera: Formicidae)," *Physiological Entomology* 26(3): 275–283 (2001); also see E. Provost, G. Riviere, M. Roux, A.-G. Bagneres, and J. L. Clément, "Cuticular hydrocarbons whereby *Messor barbarus* ant workers putatively discriminate between monogynous and polygynous colonies: are workers labeled by queens?" *Journal of Chemical Ecology* 20: 2985–3003 (1994).

231 | Preliminary evidence from interspecific mixed colonies suggests that workers that interact very aggressively with unfamiliar kin acquired some of the hydrocarbons characteristic of their adoptive queen (R. Vander Meer and N. F. Carlin, unpublished results). Analysis of cuticular hydrocarbon profiles in small colonies of *Camponotus floridanus* indicates that workers' cuticular hydrocarbon blends contain a significant portion of this queen's label (A. Endler, personal communication).

232 | B. Hölldobler and E. O. Wilson, *The Ants* (Cambridge, MA: The Belknap Press of Harvard University Press, 1990).

233 | A Dahbi, X. Cerdá, and A. Lenoir, "Ontogeny of colonial hydrocarbon label in callow workers of the ant *Cataglyphis iberica*," *Comptes Rendus de l'Academie des Sciences, Paris* 321(5): 395–402 (1998).

enemy you know in preference to the potentially more dangerous enemy you do not know.[234] In sharp contrast to this effect, which is seen in the ant genera *Temnothorax, Iridomyrmex, and Pheidole,* is the reverse effect discovered in *Pristomyrmex pungens.* This case has been interpreted as reflecting the greater danger from attacks by near neighbors.[235, 236]

As an additional response, colonies of some ant species, detecting the presence of enemies, can use the information to adjust their overall organization. When *Camponotus floridanus* colonies sense the presence of alien colonies of the same species, evidently by smell, they cut back on brood production, presumably as part of the adaptation to reduce adult mortality by curtailment of foraging.[237] A similar effect has been discovered in the formicine *Lasius pallitarsus*; unless food quality in the foraging area is high, the colonies decrease foraging when the presence of a dangerous competitor is sensed.[238] Yet another example of the phenomenon has been documented in the honey ant *Myrmecocystus mimicus.*[239]

WITHIN-COLONY RECOGNITION

If the queen is not the dominating source of the nestmate recognition label in mature insect colonies, she is nevertheless otherwise the central player. Her presence

---

234 | J. Heinze, S. Foitzik, A. Hippert, and B. Hölldobler, "Apparent dear-enemy phenomenon and environment-based recognition cues in the ant *Leptothorax nylanderi,*" *Ethology* 102(6): 510–522 (1996); M. L. Thomas, L. J. Parry, R. A. Allan, and M. A. Elgar, "Geographic affinity, cuticular hydrocarbons and colony recognition in the Australian meat ant *Iridomyrmex purpureus,*" *Naturwissenschaften* 86(2): 87–92 (1999); and T. A. Langen, F. Tripet, and P. Nonacs, "The red and the black: habituation and the dear-enemy phenomenon in two desert *Pheidole* ants," *Behavioral Ecology and Sociobiology* 48(4): 285–292 (2000). That foragers of the harvester ant *Pogonomyrmex barbatus* can distinguish members of more distant foreign colonies from those of neighboring colonies was discovered by D. M. Gordon, "Ants distinguish neighbors from strangers," *Oecologia* 81(2): 198–200 (1989).

235 | S. Sanada-Morimura, M. Minai, M. Yokoyama, T. Hirota, T. Satoh, and Y. Obara, "Encounter-induced hostility to neighbors in the ant *Pristomyrmex pungens,*" *Behavioral Ecology* 14(5): 713–718 (2003).

236 | For a second review of most issues in nestmate recognition, see Z. B. Liu, S. Yamane, K. Tsuji, and Z. M. Zheng, "Nestmate recognition and kin recognition in ants," *Entomologia Sinica* 7(1): 71–96 (2000).

237 | P. Nonacs and P. Calabi, "Competition and predation risk: their perception alone affects ant colony growth," *Proceedings of the Royal Society of London B* 249: 95–99 (1992).

238 | P. Nonacs, "Death in the distance: mortality risk as information for foraging ants," *Behaviour* 112(1–2): 24–35 (1990).

239 | B. Hölldobler, "Foraging and spatiotemporal territories in the honey ant *Myrmecocystus mimmicus* Wheeler (Hymenoptera: Formicidae)," *Behavioral Ecology and Sociobiology* 9(4): 301–314 (1981).

affects the workers' behavior in diverse and fundamental ways, and most intensely in monogynous colonies. In species of *Camponotus* and the fire ant *Solenopsis invicta*, the removal of the queen decreases worker aggressiveness toward workers of other colonies. After a period of days without a queen, formerly hostile worker groups can be merged with a subsequent high rate of mutual grooming and liquid food exchange. Even foreign queens may then be adopted.[240]

In addition, the presence of a fertile queen affects the workers' own fertility. It is well known that in many species of ants and other social insects, young workers activate their ovaries and commence laying eggs once their colony has lost its queen.[241] Hence, there must be a signal from the queen that announces her presence to the entire colony. In Chapter 8, we will describe several studies of ponerine ants involving recognition and fertility signals that emanate from queens and other physiologically distinct reproductive individuals, such as gamergates. These pheromones regulate the reproductive division of labor. They also play a major role in reproductive conflict within the ponerine societies.[242]

It has been generally believed that conflict over reproduction is less intense in evolutionarily advanced social insect species, where a greater caste dimorphism between workers and queen exists and where mature colonies are typically much larger.[243] Annett Endler, Jürgen Liebig, and their coworkers recently set out to analyze the behavioral mechanisms that regulate reproduction in such a species,

---

240 | B. Hölldobler, "Zur Frage der Oligogynie bei *Camponotus ligniperda* Latr. und *Camponotus herculeanus* L. (Hym. Formicidae)," *Zeitschrift für Angewandte Entomologie* 49(4): 337–352 (1962); E. O. Wilson, "Behaviour of social insects," in P. T. Haskell, ed., *Insect Behaviour* (Symposium of the Royal Entomological Society, no. 3) (London: Royal Entomological Society, 1966), pp. 81–96; A. Benois, "Étude experimentale de la fusion entre groupes chez la fourmi *Camponotus vagus* Scop., mettant en evidence la fermeture de la société," *Comptes Rendus de l'Academie des Sciences, Paris* 274: 3564–3567 (1972); R. K. Vander Meer and L. E. Alonso, "Queen primer pheromone affects conspecific fire ant (*Solenopsis invicta*) aggression," *Behavioral Ecology and Sociobiology* 51(2): 122–130 (2002); and R. Boulay, T. Katzav-Gozansky, R. K. Vander Meer, and A. Hefetz, "Colony insularity through queen control on worker social motivation in ants," *Proceedings of the Royal Society of London B* 270: 971–977 (2003).

241 | B. Hölldobler and E. O. Wilson, *The Ants* (Cambridge, MA: The Belknap Press of Harvard University Press, 1990). Worker-laid eggs are haploid because in most species of "phylogenetically advanced" subfamilies, workers are unable to mate and store sperm. Such haploid eggs will develop into males.

242 | J. Heinze, "Reproductive conflict in insect societies," *Advances in the Study of Behavior* 34: 1–57 (2004).

243 | J. Heinze, B. Hölldobler, and C. Peeters, "Conflict and cooperation in ant societies," *Naturwissenschaften* 81(11): 489–497 (1994); A. F. G. Bourke, "Colony size, social complexity and reproductive conflict in social insects," *Journal of Evolutionary Biology* 12(2): 245–257 (1999); and K. R. Foster, "Diminishing returns in social evolution: the not-so-tragic commons," *Journal of Evolutionary Biology* 17(5): 1058–1072 (2004).

PLATE 39. A *Camponotus floridanus* worker destroying eggs laid by nestmate workers. As long as the queen is fertile, workers do not tolerate the laying of viable eggs by nestmate workers. (Photo: Jürgen Liebig.)

Camponotus floridanus, whose colonies are monogynous and the worker population of mature colonies reach, as noted, into the several thousands. It turns out that queens of large colonies produce a highly distinct hydrocarbon signal that regulates reproduction in the colony. This pheromone is present on both the cuticle of the queen and the queen-laid eggs. It notifies the workers of the queen's presence and fertility, and so long as they encounter it, they refrain from producing their own eggs. When one or the other tries to sneak in its own production, the eggs are destroyed and eaten by nestmates (Plate 39). The reason is simply that worker-laid eggs do not carry the queen signal.[244] Obviously, such policing is a trait selected at

---

244 | A. Endler, J. Liebig, T. Schmitt, J. E. Parker, G. R. Jones, P. Schreier, and B. Hölldobler, "Surface hydrocarbons of queen eggs regulate worker reproduction in a social insect," *Proceedings of the National Academy of Sciences*

the colony level. If collateral kin selection is occurring, worker egg laying should be favored in these colonies. At least, that is expected if queens are singly inseminated and the colony is monogynous. Workers would then be more closely related to their sons (0.5) and to their nephews (0.375) than to their brothers (0.25); thus, workers should, in theory, be selected to raise sons and nephews instead of brothers. However, substantial worker reproduction would negatively affect colony efficiency and would therefore be detrimental to colony-level production of reproductives. Hence, policing in this case, as probably in most cases, appears to be based on natural selection at the colony level.

A problem remains with *Camponotus floridanus.* With its colonies so large and so often spread across far-flung nests, how is the queen signal broadcast? It turns out that the workers distribute the pheromone, at least in part, by carrying large numbers of the queen's eggs out from her abode to the satellite nests. Worker groups deprived of access to the queen but regularly exposed to queen-laid eggs do not become fertile.

Cuticular hydrocarbon structures specific to queens are probably quite common. They have been identified in two additional formicines, *Cataglyphis iberica* and *Formica fusca,*[245, 246] as well as in two species of the myrmicine genus *Leptothorax* and several ponerines (see Chapter 8) and myrmeciines.[247, 248] Further, as shown by Patrizia D'Ettorre and Jürgen Heinze,[249] founding queens of the ponerine *Pachycondyla villosa* use chemical cues to recognize each other individually. Aggression between two queens that had previously interacted was significantly lower than between queens with comparable social background but no previous experience with each other. The individual

---

USA 101(9): 2945–2950 (2004); and A. Endler, J. Liebig, and B. Hölldobler, "Queen fertility, egg marking and colony size in the ant *Camponotus floridanus,*" *Behavioral Ecology and Sociobiology* 59(4): 490–499 (2006). Egg policing has also been reported in ponerine ants; for example, see T. Monnin and C. Peeters, "Cannibalism of subordinates' eggs in the monogynous queenless ant *Dinoponera quadriceps,*" *Naturwissenschaften* 84(11): 499–502 (1997).

245 | A. Dahbi and A. Lenoir, "Queen and colony odour in the multiple nest ant species, *Cataglyphis iberica* (Hymenoptera, Formicidae)," *Insectes Sociaux* 45(3): 301–313 (1998).

246 | M. Hannonen, M. F. Sledge, S. Turillazzi, and L. Sundström, "Queen reproduction, chemical signalling and worker behaviour in polygyne colonies of the ant *Formica fusca,*" *Animal Behaviour* 64(3): 477–485 (2002).

247 | J. Tentschert, H.-J. Bestmann, and J. Heinze, "Cuticular compounds of workers and queens in two *Leptothorax* ant species—a comparison of results obtained by solvent extraction, solid sampling, and SPME," *Chemoecology* 12: 15–21 (2002).

248 | V. Dietemann, C. Peeters, and B. Hölldobler, "Role of the queen in regulating reproduction in the bulldog ant *Myrmecia gulosa*: control or signalling?" *Animal Behaviour* 69(4): 777–784 (2005).

249 | P. D'Ettorre and J. Heinze, "Individual recognition in ant queens," *Current Biology* 15(23): 2170–2174 (2005).

cuticular hydrocarbon badge stands alone: it is associated with neither dominance nor fertility and hence is not just a marker of reproductive status.

Cuticular hydrocarbon blends that signal reproductive status and are involved in the establishment of reproductive dominance have also been found in the social wasp *Polistes dominulus*. So far as we know, these fertility signals do not carry badges specific to individuals.[250]

Other pheromones, originating in various glandular structures and hence not necessarily part of the cuticular hydrocarbon bouquet, also signal the presence of a fertile queen. In so doing, they affect worker physiology and behavior. Examples have been documented in many social hymenopterans and most thoroughly in honeybees, in the fire ant *Solenopsis invicta*, and in the pharaoh ant *Monomorium pharaonis*.[251, 252, 253] Their behavioral role in the regulation of worker reproduction has been analyzed in other species as well, especially the formicines *Plagiolepis pygmaea* and *Oecophylla longinoda* (weaver ant).[254]

Not just reproductive individuals are chemically labeled. Various worker task groups also appear to carry particular cuticular hydrocarbon signatures. In the

---

250 | M. F. Sledge, F. Boscaro, and S. Turillazzi, "Cuticular hydrocarbons and reproductive status in the social wasp *Polistes dominulus*," *Behavioral Ecology and Sociobiology* 49(5): 401–409 (2001); M. F. Sledge, I. Trinca, A. Massolo, F. Boscaro, and S. Turillazzi, "Variation in cuticular hydrocarbon signatures, hormonal correlates and establishment of reproductive dominance in a polistine wasp," *Journal of Insect Physiology* 50(1): 73–83 (2004); J. Liebig, T. Monnin, and S. Turillazzi, "Direct assessment of queen quality and lack of worker suppression in a paper wasp," *Proceedings of the Royal Society of London B* 272: 1339–1344 (2005).

251 | S. E. R. Hoover, C. I. Keeling, M. L. Winston, and K. N. Slessor, "The effect of queen pheromones on worker honey bee ovary development," *Naturwissenschaften* 90(10): 477–480 (2003); T. Katzav-Gozansky, V. Soroker, F. Ibarra, W. Francke, and A. Hefetz, "Dufour's gland secretion of the queen honeybee (*Apis mellifera*): an egg discriminator pheromone or a queen signal?" *Behavioral Ecology and Sociobiology* 51(1): 76–86 (2001); S. J. Martin, N. Châline, B. P. Oldroyd, G. R. Jones, and F. L. W. Ratnieks, "Egg marking pheromones of anarchistic worker honeybees (*Apis mellifera*)," *Behavioral Ecology* 15(5): 839–844 (2004); R. Dor, T. Katzav-Gozansky, and A. Hefetz, "Dufour's gland pheromone as a reliable fertility signal among honeybee (*Apis mellifera*) workers," *Behavioral Ecology and Sociobiology* 58(3): 270–276 (2005).

252 | E. L. Vargo and C. D. Hulsey, "Multiple glandular origins of queen pheromones in the fire ant *Solenopsis invicta*," *Journal of Insect Physiology* 46(8): 1151–1159 (2000); R. K. Vander Meer and L. Morel, "Ant queens deposit pheromones and antimicrobial agents on eggs," *Naturwissenschaften* 82(2): 93–95 (1995).

253 | J. P. Edwards and J. Chambers, "Identification and source of a queen-specific chemical in the Pharaoh's ant, *Monomorium pharaonis* (L.)," *Journal of Chemical Ecology* 10(12): 1731–1747 (1984).

254 | L. Passera, "La fonction inhibitrice des reines de la fourmi *Plagiolepis pygmaea* Latr.: role de pheromones," *Insectes Sociaux* 27(3): 212–225 (1980); and B. Hölldobler and E. O. Wilson, "Queen control in colonies of weaver ants (Hymenoptera: Formicidae)," *Annals of the Entomological Society of America* 76(2): 235–238 (1983); see also the revision by D. J. C. Fletcher and K. G. Ross, "Regulation of reproduction in eusocial Hymenoptera," *Annual Review of Entomology* 30: 319–343 (1985).

harvester ant *Pogonomyrmex barbatus,* scouts on patrol differ significantly from nest maintenance workers in the relative proportions of particular classes of hydrocarbons as well as in individual compounds.[255] Although colonies also vary significantly in overall hydrocarbon patterns,[256] the various task groups differ in consistent ways across each colony investigated. Because division of labor follows a more or less age-based trajectory, the task-related hydrocarbon differences are most likely connected with the aging process. There is some experimental evidence that workers sense their own age cues and respond by adjusting their work activities according to that information.[257]

In fact, cuticular hydrocarbon changes are probably widespread in ant societies. In *Camponotus vagus,* brood tenders and foragers were found to be clearly distinguishable on the basis of their cuticular hydrocarbon profiles.[258] In the ponerine *Harpegnathos saltator,* where division of labor among nonreproductive workers is not very pronounced, there is nevertheless a clear difference in cuticular hydrocarbon profiles, not only between reproductive and nonreproductive individuals but also between the inside-nest and outside-nest workers.[259] Other behavioral evidence suggests that colonies of ant species with majors ("soldiers") and minors can distinguish these subcastes by differences in their surface chemistry. As a result, changes in the minor-to-major ratio leads to differential rearing of the subcaste that has been reduced, with a return to the original balance.[260]

Can workers belonging to the same colony recognize their own close kin? This is an important issue, because often workers and reproductive offspring inside the

---

255 | D. Wagner, M. J. F. Brown, P. Broun, W. Cuevas, L. E. Moses, D. L. Chao, and D. M. Gordon, "Task-related differences in the cuticular hydrocarbon composition of harvester ants, *Pogonomyrmex barbatus,*" *Journal of Chemical Ecology* 24(12): 2021–2037 (1998).

256 | D. Wagner, M. Tissot, W. Cuevas, and D. M. Gordon, "Harvester ants utilize cuticular hydrocarbons in nestmate recognition," *Journal of Chemical Ecology* 26(10): 2245–2257 (2000).

257 | M. J. Greene and D. M. Gordon, "Cuticular hydrocarbons inform task decisions," *Nature* 423: 32 (2003).

258 | A. Bonavita-Cougourdan, J.-L. Clément, and C. Lange, "Functional subcaste discrimination (foragers and brood-tenders) in the ant *Camponotus vagus* Scop.: polymorphism of cuticular hydrocarbon patterns," *Journal of Chemical Ecology* 19(7): 1461–1477 (1993).

259 | J. Liebig, C. Peeters, N. J. Oldham, C. Markstädter, and B. Hölldobler, "Are variations in cuticular hydrocarbons of queens and workers a reliable signal of fertility in the ant *Harpegnathos saltator?" Proceedings of the National Academy of Sciences USA* 97: 4124–4131 (2000).

260 | E. O. Wilson, "Between-caste aversion as a basis for division of labor in the ant *Pheidole pubiventris* (Hymenoptera: Formicidae)," *Behavioral Ecology and Sociobiology* 17(1): 35–37 (1985); A. B. Johnston and E. O. Wilson, "Correlates of variation in the major/minor ratio of the ant, *Pheidole dentata* (Hymenoptera: Formicdae)," *Annals of the Entomological Society of America* 78(1): 8–11 (1985).

colony are not closely related to each other, thus differing from the classic arrangement of an insect colony headed by a lone, singly mated queen. In this latter case, the workers and young reproductive females would all be related to each other by a coefficient of relatedness of 0.75. However, if the queen has mated with several males or if the colony is polygynous, especially with some queens having originated from other colonies, then the average coefficient of relatedness would be much lower. There would exist several patrilines (or matrilines) among the females born in the same colonies. This circumstance has serious consequences for traditional kin selection theory; in particular, we should expect mechanisms of kin recognition to exist in such colonies.

To distinguish among nestmates that share exogenously derived odors for between-colony discrimination, each individual must retain some degree of endogenous cue variation. That is, kinship-correlated odor specificity must be nested within colony-level anonymity. Discrimination among equally familiar nestmates is also assumed to require endogenous or "self-based" recognition templates either genetically encoded or learned from one's own endogenous label.

A great deal of research has therefore been devoted to the search for kin recognition in colonies of ants, bees, and wasps. At first the evidence proved consistently either ambiguous or negative; or, if positive, it was judged to be due to experimental artifacts. Some positive evidence was reported from the honeybees, but in most cases, the findings were challenged and alternative interpretations offered.[261] Robert Page and his colleagues then conducted tests of unprecedented rigor in search of the phenomenon. In laboratory-reared colonies that contained only two worker patrilines, workers had cuticular hydrocarbon profiles that were more similar between full sisters than between half sisters.[262] However, it has been pointed out that natural honeybee colonies have from 7 to more than 20 patrilines. That plus the continuous exchange of hydrocarbons and other cuticle components

---

261 | For a summary and review, see T. D. Seeley, *Wisdom of the Hive: The Social Physiology of Honey Bee Colonies* (Cambridge, MA: Harvard University Press, 1995); M. D. Breed, "Chemical cues in kin recognition: criteria for identification, experimental approaches, and the honey bee as an example," in R. K. Vander Meer, M. D. Breed, K. E. Espelie, and M. L. Winston, eds., *Pheromone Communication in Social Insects: Ants, Wasps, Bees, and Termites* (Boulder, CO: Westview Press, 1998), pp. 57–78; N. F. Carlin and P. C. Frumhoff, "Nepotism in the honeybee," *Nature* 346: 706–707 (1990); and N. F. Carlin, "Discrimination between and within colonies of social insects: two null hypotheses," *Netherlands Journal of Zoology* 39(1–2): 86–100 (1989).

262 | R. E. Page Jr., R. A. Metcalf, R. L. Metcalf, E. H. Erikson, and R. L. Lampman, "Extractable hydrocarbons and kin recognition in honeybees (*Apis mellifera* L.)," *Journal of Chemical Ecology* 17(4): 745–756 (1991).

through direct contact among workers and out from the comb wax diminish patriline order differences and may in fact render them biologically insignificant. Nevertheless, a subsequent study demonstrated the existence of patriline cuticle profiles within several natural honeybee colonies, and these proved more clearly conserved than previously assumed. Hence, cuticle hydrocarbons might, after all, possess the necessary prerequisites of sufficient variability and genetic determinism for use as labels for subfamily recognition.[263] What the sociobiological significance of such within-colony recognition or discrimination is, however, remains unclear. Further, observations of most nepotistic behaviors, such as preferential rearing of full-sister queens, are still controversial. As Michael Breed and his coauthors have concluded, "few areas in sociobiology have received as much experimental attention, yet yielded so little in the way of supportable conclusions, as the question of subfamily nepotism in honey bees."[264]

Rather ambiguous or mostly negative results have also resulted in the search for kin recognition in colonies of social wasps, mostly of the genera *Polistes* and *Ropalidia*.[265, 266]

In ants, experiments with *Camponotus floridanus* at first appeared promising in the search. Colonies contain a single queen that is singly mated. By inducing adoption of worker pupae from multiple alien colonies, cohorts of workers were created that originated from very different sources. Members of mixed groups, drawn from two colonies, evoked less aggression from their unfamiliar genetic sisters than from the sisters of their unrelated nestmates, thus suggesting that colony members retain

---

263 | G. Arnold, B. Quenet, J.-M. Cornuet, C. Masson, B. De Schepper, A. Estoup, and P. Gasqui, "Kin recognition in honeybees," *Nature* 379: 498 (1996).

264 | M. D. Breed, C. K. Welch, and R. Cruz, "Kin discrimination within honey bee (*Apis mellifera*) colonies: an analysis of the evidence," *Behavioural Processes* 33(1–2): 25–40 (1994).

265 | D. C. Queller, C. R. Hughes, and J. E. Strassmann, "Wasps fail to make distinctions," *Nature* 344: 388 (1990); D. C. Queller, J. E. Strassmann, and C. R. Hughes, "Microsatellites and kinship," *Trends in Ecology and Evolution* 8(8): 285–288 (1993); and J. E. Strassmann, P. Seppä, and D. C. Queller, "Absence of within-colony kin discrimination: foundresses of the social wasp, *Polistes carolina*, do not prefer their own larvae," *Naturwissenschaften* 87(6): 266–269 (2000). For a review, see T. L. Singer, K. E. Espelie, and G. J. Gamboa, "Nest and nestmate discrimination in independent founding paper wasps," in R. K. Vander Meer, M. D. Breed, M. L. Winston, and K. E. Espelie, eds., *Pheromone Communication in Social Insects: Ants, Wasps, Bees, and Termites* (Boulder, CO: Westview Press, 1998), pp. 104–125.

266 | R. Gadagkar, *The Social Biology of* Ropalidia marginata (Cambridge, MA: Harvard University Press, 2001). See also G. Gamboa, "Sister, aunt-niece, and cousin recognition by social wasps," *Behavior Genetics* 18(4): 409–423 (1988); and G. J. Gamboa, H. K. Reeve, and D. W. Pfennig, "The evolution and ontogeny of nestmate recognition in social wasps," *Annual Review of Entomology* 31: 431–454 (1986).

differential kinship labels.[267] Another study revealed that within mixed colonies, again from two other colonies, workers and young unmated adult queens (gynes) received more antennal touches and, in the absence of an egg-laying queen, more "subtle" aggression from unrelated nestmates.[268] However, this result has little or no significant social consequence, because no differences at all were found in food exchange and grooming. Those colonies with mixed cohorts were artificially created. To eliminate any potential artifact caused by the manipulators, research was continued with a polygynous *Camponotus* species, *Camponotus planatus* (Plate 40). No evidence could be found that workers preferentially direct social behavior toward their mother queen, sister workers, or sister virgin queens.[269] Very similar results were obtained from studies of polygynous colonies of *Formica argentea*. Some bias in interaction among workers might be correlated with particular tasks performed by different subgroups, but no indication of nepotism could be found. On the other hand, Minttumaaria Hannonen and Liselotte Sundström conclude from their quantitative analysis of interactions between different matrilines in *Formica fusca* colonies "that ant workers can apparently discriminate kin accurately and that they capitalize on this ability, thereby enhancing their genetic contribution to future generations, even in the presence of several queens."[270]

The larger issue of social evolution addressed by these studies is as follows. In colonies with a sole singly mated queen, workers share 75 percent of their genes by immediate descent with their sisters but only 25 percent with their brothers. According to traditional kin selection theory, workers will therefore be selected to invest three-quarters of the colony's reproductive efforts in the rearing of sister sexuals and only one-quarter in raising brothers. The situation is different, however, in colonies headed by multiply mated queens. The average relatedness of sisters is

---

267 | L. Morel and M. S. Blum, "Nestmate recognition in *Camponotus floridanus* callow worker ants: are sisters or nestmates recognized?" *Animal Behaviour* 36(3): 718–725 (1988).

268 | N. F. Carlin, B. Hölldobler, and D. S. Gladstein, "The kin recognition system of carpenter ants (*Camponotus* spp.), III: Within-colony discrimination," *Behavioral Ecology and Sociobiology* 20(3): 219–227 (1987); and N. F. Carlin, "Discrimination between and within colonies of social insects: two null hypotheses," *Netherlands Journal of Zoology* 39(1–2): 86–100 (1989).

269 | N. F. Carlin, H. K. Reeve, and S. P. Cover, "Kin discrimination and division of labour among matrilines in the polygynous carpenter ant, *Camponotus planatus*," in L. Keller, ed., *Queen Number and Sociality in Insects* (New York: Oxford University Press, 1993), pp. 362–401; similar results were obtained with a polygynous *Myrmica* species in L. E. Snyder, "Non-random behavioural interactions among genetic subgroups in a polygynous ant," *Animal Behaviour* 46(3): 431–439 (1993).

270 | M. Hannonen and L. Sundström, "Worker nepotism among polygynous ants," *Nature* 421: 910 (2003).

PLATE 40. *Above*: A colony fragment of the polygynous ant *Camponotus planatus*. One of the multiple queens is shown at the center of the photograph. *Below*: The queens and their worker offspring are marked with individual color codes in this photograph.

much lower and can reach a coefficient of relatedness close to 0.25; thus, the asymmetry of relatedness of workers to their sisters and brothers is almost erased. With a population-wide reproductive strategy, workers of all colonies would maximize their inclusive fitness if colonies with a singly mated queen specialized in producing reproductive females, while colonies with a multiply mated queen (or multiple queens) specialized in producing males. So far so good, but the workers have a problem: their queen is in any case equally related to her sons and daughters. That means that she should be selected to enforce equal investment in both sexes of her reproductive offspring, whether she is singly or multiply mated.

The data appear to indicate that the workers win this sex allocation conflict, and the queen's optimal reproductive interests are not realized. How do the workers control the sex allocation? Sundström and her collaborators discovered that in colonies headed by a singly mated queen, the proportion of males decreased significantly between the egg and pupal stages. Thus, workers must commit selected brood cannibalism, eliminating much of the male brood. Such is not the case headed by a multiply mated queen.[271]

The question then arises of how workers recognize male brood. The answer is not yet known. An even more interesting question is how workers raising the reproductive offspring know whether their colony has one or several fathers. The analysis of the quantitative variation in cuticular hydrocarbon blends of workers in colonies with multiply mated queens suggests that patriline differences (distinctions due to different fathers) in hydrocarbon profiles may be the essential cue for workers to decide to raise males instead of virgin queens.[272] The chemical profiles show that these differences are very subtle, yet perhaps sufficient to inform the workers whether their nestmates are a mix of full and half sisters, as opposed to exclusively full sisters. On the other hand, differences in individual cuticular hydrocarbon profiles are not distinct enough to foster nepotistic behavior, which would reduce colony productivity. Avoidance of such behavior is likely to be favored by colony-level selection.

These several correlations, while interesting and suggestive, need to be tested experimentally. That, admittedly, will not be an easy task.

---

271 | L. Sundström, M. Chapuisat, and L. Keller, "Conditional manipulation of sex ratios by ant workers: a test of kin selection theory," *Science* 274: 993–995 (1996).

272 | J. J. Boomsma, J. Nielsen, L. Sundström, N. J. Oldham, J. Tentschert, H. C. Petersen, and E. D. Morgan, "Informational constraints in optimal sex allocation in ants," *Proceedings of the National Academy of Sciences USA* 100(15): 8799–8804 (2003).

Meanwhile, the results of another study suggest that some kind of within-colony kin recognition may exist in ants. Workers of the European formicine *Formica truncorum* appear to specialize in rearing reproductive females or males, depending on whether the mother queen has mated with only one male or with several males.[273] This phenomenon is called a "split sex ratio," and the population study with *Formica truncorum* is to date the best documentation of such a split between colonies raising males (queens multiply mated) and colonies raising females (queens singly mated), just as traditional kin selection theory predicts, and at the same time impressively demonstrates colony-level selection.[274]

RECOGNITION OF BROOD

Egg marking, now known to be an important regulatory behavior in honeybees and some species of ants, has also been reported in *Polistes* wasps and is probably quite common in social insects generally. On reflection, this discovery should not be very surprising. The ability of social insects to identify the brood as to sex, caste, and developmental stage is otherwise essential for workers that raise the colony's offspring.

Accordingly, the mechanisms and adaptive advantages of brood recognition have attracted the interest of a number of investigators. Evidence for kin-based brood recognition has been generally weak, or absent, as concluded by Norman Carlin in his comprehensive review.[275] It has nonetheless been implicated in several ant genera, including *Atta*, *Acromyrmex*, *Tapinoma*, *Lasius*, *Cataglyphis*, and *Camponotus*. In particular, behavioral studies of *Camponotus floridanus* and species of *Cataglyphis* have demonstrated that learning is involved in colony-specific brood recognition.[276] Fur-

---

273 | L. Sundström, "Sex ratio bias, relatedness asymmetry and queen mating frequency in ants," *Nature* 367: 266–268 (1994); and L. Sundström, "Sex allocation and colony maintenance in monogyne and polygyne colonies of *Formica truncorum* (Hymenoptera: Formicidae): the impact of kinship and mating structure," *American Naturalist* 146(2): 182–201 (1995).

274 | J. J. Boomsma and A. Grafen, "Colony level sex ratio selection in the eusocial Hymenoptera," *Journal of Evolutionary Biology* 3(4): 383–407 (1991).

275 | N. F. Carlin, "Species, kin, and other forms of recognition in the brood discrimination behavior of ants," in J. C. Trager, ed., *Advances in Myrmecology* (Leiden: E. J. Brill, 1988), pp. 267–295.

276 | N. F. Carlin and P. H. Schwartz, "Pre-imaginal experience and nestmate brood recognition in the carpenter ant, *Camponotus floridanus*," *Animal Behaviour* 38(1): 89–95 (1989); M. Isingrini, A. Lenoir, and P. Jaisson, "Prei-

thermore, analysis of surface labels of larvae of *Camponotus vagus*, which turn out to be distinctive to this life stage, suggest that recognition of larval hydrocarbons could be involved.[277] Because ants lick and move their brood around constantly, numerous observers have proposed that this intimate relationship between nurse workers and brood must entail chemical communication.[278] The larval pheromones releasing brood-tending behavior in nurse workers appear to countermand colony recognition cues; ant larvae, pupae, and callow workers (newly eclosed from pupae) can easily be transferred from one colony to another—and sometimes even from one species to another. However, after a certain period following eclosion, usually about a week, young adult workers are no longer accepted by foreign colonies.

There remains a paradox in the phenomenon as envisioned. If the colony odor is generated by adsorption of specific mixtures of hydrocarbons and environmental odorants into the cuticle, it is not clear why larvae and pupae should lack the odor of the rest of the colony. How do they escape aggression from members of foreign colonies? This puzzle might be resolved if colony odor is masked in the brood stages by brood-tending pheromones that are not specific to colonies. It is also conceivable that these pheromones have high position in a hierarchical order of sensitivity to pheromone systems and thus dominate any other colony-specific odorous cues.[279]

The absence of brood discrimination at the colony level has actually been exploited by slave-making ant species, which attack colonies of closely related neighboring species and kidnap their brood. When the captured pupae eclose in a slave raider's nest, the young workers are cognitively imprinted to the odor of their captors' colony and in the future behave in a hostile manner toward their real sisters remaining behind in the natal nest.

To posit a high position of the brood pheromone in the overall pheromone hierarchy also implies that the *Q*/*K* ratio (the ratio of odor molecules released to the response threshold concentration) should be very low. A high *Q*/*K* would saturate

---

maginal learning as a basis of colony-brood recognition in the ant *Cataglyphis cursor*," *Proceedings of the National Academy of Sciences USA* 82(24): 8545–8547 (1985).

277 | A. Bonavita-Cougourdan, J.-L. Clément, and C. Lange, "The role of cuticular hydrocarbons in recognition of larvae by workers of the ant *Camponotus vagus*: changes in the chemical signature in response to social environment (Hymenoptera: Formicidae)," *Sociobiology* 16(2): 49–74 (1989).

278 | B. Hölldobler and E. O. Wilson, *The Ants* (Cambridge, MA: The Belknap Press of Harvard University Press, 1990).

279 | B. Hölldobler, "Communication in social Hymenoptera," in T. A. Sebeok, ed., *How Animals Communicate* (Bloomington, IN: Indiana University Press, 1977), pp. 418–471.

the nest with the dominant signal, and colony odors and other chemical signals could become almost ineffective. The observations that the brood pheromones are nonvolatile (or have very low volatility) and are effective only in very close range support this conjecture.

Communication between adult and brood is somewhat different in honeybees, because brood care is localized to the brood combs. Here the larvae are fed, the cells are capped, and the brood area in the hive is thermoregulated. These behavioral actions apparently are also evoked in part by chemical signals. In particular, a blend of ten fatty acid methyl and ethyl esters produced in the larval salivary glands have been identified as larval pheromones that attract and induce clustering in worker bees.[280] These esters also stimulate hypopharyngeal glands in nurse bees and inhibit ovary development in workers. Larval pheromones also seem to stimulate honeybee foragers to forage for pollen.[281] Further, honeybee workers distinguish not only between female larvae and male larvae,[282] but also between worker larvae and queen larvae.[283]

COMMUNICATING RESOURCE-HOLDING POTENTIAL AMONG COLONIES

So far, we have mainly discussed so-called mutualistic communication, during which information is shared within the colony to the benefit of all members of the society. The cooperative functioning and collective fitness of the colony of course depend on just such mutualistic communication. The social interactions mediated by such

---

280 | Y. Le Conte, G. Arnold, J. Trouiller, C. Masson, and B. Chappe, "Identification of a brood pheromone in honeybees," *Naturwissenchsaften* 77(7): 334–336 (1990).

281 | J. B. Free, "Factors determining the collection of pollen by honeybee foragers," *Animal Behaviour* 15(1): 134–144 (1967); and T. Pankiw and W. L. Rubink, "Pollen foraging response to brood pheromone by Africanized and European honey bees (*Apis mellifera* L.)," *Annals of the Entomological Society of America* 95(6): 761–767 (2002); see also Y. Le Conte, A. Mohammedi, and G. E. Robinson, "Primer effects of a brood pheromone on honeybee behavioural development," *Proceedings of the Royal Society of London B* 268: 163–168 (2001); and Y. Le Conte, J.-M. Bécard, G. Costagliiola, G. Vaublanc, M. El Maâtaoui, D. Crauser, E. Plettner, and K. N. Slessor, "Larval salivary glands are a source of primer and releaser pheromone in honey bee (*Apis mellifera* L.)," *Naturwissenschaften* 93(5): 237–244 (2006).

282 | M. H. Haydak, "Do the nurse honey bees recognize the sex of larvae?" *Science* 127: 1113 (1958).

283 | J. Woyke, "Correlations between the age at which honeybee brood was grafted, characteristics of the resultant queens, and results of insemination," *Journal of Apicultural Research* 10(1): 45–55 (1971).

exchange are an important part of the "extended phenotype" of the colony.[284] As a consequence, populations of colonies can be expected to show genetically based variations in their patterns for the following elementary reason: colonies compete with one another for resources. Those that variously establish and maintain territories in the most economical manner, employ the most effective recruitment systems to retrieve food, and deploy the most powerful colony defense against enemies will raise the largest number of reproductive females and males each generation. Their colony-level genotype will prevail.[285] Although reproductive conflicts exist among individuals or groups of individuals within each society, the fitness of individuals depends on that effectiveness of the colony as a whole. In general, the competition between colonies with advanced social organization outweighs decisively the competition of nestmates within a colony (see also Chapter 2).

Animals engaged in aggressive competition also commonly communicate information to their opponents about their fighting ability, called resource-holding potential (RHP) by behavioral ecologists. Such information includes body size, strength of teeth, and presence of horns or antlers. As behavioral ecologists have frequently noted, if the RHP of opponents is very unequal, the contest is quickly decided; the weaker individual yields. However, if the opponents are similar in their projected RHP, the contestants engage in elaborate and often sustained signaling, which nonetheless lacks reliable information about the intent of each signaler to escalate the aggression or to flee. In such situations, as expressed by Mark Hauser, "the outcome of a competitive interaction must be decided by a volley of signals, with each individual attempting to extract the most useful information with regards to the relative probability of winning or losing a fight."[286]

Territories of ant societies are part of their extended phenotype. They are defended cooperatively by the workers of the owner colony. Because of the division of labor between reproductive individuals and the usually sterile workers, fatalities caused by territorial defense have a different qualitative significance for social insects

---

284 | R. Dawkins, *The Extended Phenotype: The Genes as the Unit of Selection* (San Francisco: W. H. Freeman, 1982).

285 | B. Hölldobler, "Vom Verhalten zum Gen: Die Soziobiologie eines Superorganismus," *Nova Acta Leopoldina* NF76: 205–223 (1997); B. Hölldobler, "Multimodal signals in ant communication," *Journal of Comparative Physiology A* 184: 129–141 (1999); and H. K. Reeve and B. Hölldobler, "The emergence of a superorganism through inter-group competition," *Proceedings of the National Academy of Sciences USA* 104(23): 9736–9740 (2007).

286 | M. D. Hauser, *The Evolution of Communication* (Cambridge, MA: MIT Press, 1996).

as compared to solitary animals. The death of a sterile worker represents an energy or labor debit, rather than the destruction of a reproductive unit. A worker death might more than offset its costs by protecting resources and the colony itself.[287]

Nevertheless, ritualized combat is also known to exist in a few ant species.[288] The ecological significance has been analyzed in considerable depth in the honey ant *Myrmecocystus mimicus*.[289] These ants conduct display tournaments in which tens or hundreds of ants participate and during which almost no actual physical fighting occurs. Instead, individual ants exchange highly stereotyped aggressive displays (Plates 41 and 42). The tournaments are used to defend territories, and through them the opposing colonies evidently assess each other's strength. Depending on the outcome of this mutual evaluation, the opponents may continue the ritualized combat while the tournament site gradually shifts toward the nest of the weaker colony, a change that increasingly interferes with its foraging. If one colony is very much the stronger, the contest quickly escalates into raiding and possible enslavement of the weaker colony, during which its queen is killed. We postulate that numerous threat displays between individual workers at the tournament site are integrated into the massive group display between opposing colonies.

In parallel to the procedure followed by solitary animals, the group's "strategic decision" must aim for one of three alternatives: to retreat, to recruit reinforcements in order to continue to fight by display, or to launch an escalated attack. The colony decides by using information about the strength of the opposing colony obtained during the ritualized combat at the tournament site. The behavioral patterns involved suggest that this information is based on complex multimodal communication.

During a tournament, the ants walk on stilt legs while raising the head and

---

287 | B. Hölldobler and C. J. Lumsden, "Territorial strategies in ants," *Science* 210: 732–739 (1980); E. S. Adams, "Territory size and shape in fire ants: a model based on neighborhood interactions," *Ecology* 79(4): 1125–1134 (1998); and E. S. Adams, "Experimental analysis of territory size in a population of the fire ant *Solenopsis invicta*," *Behavioral Ecology* 14(1): 48–53 (2003).

288 | B. Hölldobler and E. O. Wilson, *The Ants* (Cambridge, MA: The Belknap Press of Harvard University Press, 1990); J. F. A. Traniello and S. K. Robson, "Trail and territorial communication in social insects," in R. T. Cardé and W. J. Bell, *Chemical Ecology of Insects 2* (New York: Chapman & Hall, 1995), pp. 241–286.

289 | B. Hölldobler, "Tournaments and slavery in a desert ant," *Science* 192: 912–914 (1976); and B. Hölldobler, "Foraging and spatiotemporal territories in the honey ant *Myrmecocystus mimicus* Wheeler (Hymenoptera: Formicidae)," *Behavioral Ecology and Sociobiology* 9(4): 301–314 (1981). The account in this section is based in part on B. Hölldobler, "Multimodal signals in ant communciation," *Journal of Comparative Physiology A* 184(2): 129–141 (1999).

PLATE 41. Ritualized territorial interactions in *Myrmecocystus mimicus*. A head-on encounter is shown above and a lateral display below.

PLATE 42. Ritualized territorial interactions in *Myrmecocystus mimicus*. *Above*: A worker engages in a lateral display while mounted on a stone, thereby seeming larger in size than its opponent. *Below*: This photograph illustrates the difference in size between displaying ants from a small and a large colony.

gaster, thus maximizing their height. When two hostile workers meet, they initially turn to confront each other head-on. Subsequently, they engage in more prolonged lateral displays, during which they raise the gaster even higher and point it at their opponent. Simultaneously, they drum their antennae intensely on and around each other's abdomen and frequently kick their legs against their opponent. In addition, each ant pushes sideways as if trying to dislodge the other. After several seconds, one of the ants usually yields and the encounter ends. The ants then continue to move along on stilt legs. They soon meet other opponents, and the ritual combat movements are repeated. When two nestmates meet, the encounter lasts only 1 or 2 seconds and is terminated by brief jerking movements of the body. The ants instantly discriminate foreigners from nestmates by antennal sweeps over one another's body, most likely sensing the cuticular hydrocarbon signatures. One feature that appears important during these exchanges is the size of the individual ants. If a large and a small ant are matched in the display encounter, usually the smaller yields.

Displaying ants not only walk "on stilts" while raising the gaster and head, but also inflate the gaster, so that the tergites seem larger. There is, moreover, a tendency during the tournament for the ants to mount pebbles in order to display down to their opponents. Magnified motion picture analyses of the various displays suggest that the contestants gauge each other's size, and while doing so, they bluff by pretending to be larger than they really are.

From these observations, Charles Lumsden and Bert Hölldobler developed two models of how *Myrmecocystus mimicus* ants may assess one another's true strength during the tournaments.[290] Individual workers may use the rate of encounters with nestmates and opponents ("head-counting" model) to gain a rough measure of the enemy's strength. Alternatively, individuals may determine whether a low or high percentage of the opponents are major workers and use this information to estimate the opposing colony's strength, since a high percentage of this caste is a reliable index of large colony size. In fact, majors are more frequently represented among tournamenting ants than among groups of foragers. Among colonies reared in the laboratory from queens, those younger than 4 years have a disproportionately smaller group of majors in their worker population.

Field experiments on *Myrmecocystus* colonies indicate that both of the preceding

---

290 | C. J. Lumsden and B. Hölldobler, "Ritualized combat and intercolony communication in ants," *Journal of Theoretical Biology* 100(1): 81–98 (1983).

assessment techniques are involved in intercolony communication. The data suggest, in particular, that immature and hence small colonies rely on the "caste-polling" technique, which enables them to determine quickly whether or not the opponent is a mature colony. When confronted with the evidence of a large colony, small colonies immediately retreat into the nest and close the nest entrance. This tactic enables them to prevent stronger opponent colonies from mounting a raid.

In the case of the head-counting method, film and video recordings have revealed that it is not the entire tournamenting worker force that does the "counting." A small group of "reconnaissance ants" move through the tournament and gather their information. These ants are of smaller body size and their encounter times are short; they linger with opponents only 1 to 3 seconds, the same period as with their own nestmates. Their routes through the tournament fields are also considerably longer than those of the displaying ants. Individuals of the reconnaissance group recruit reinforcements from the home nest by laying chemical trails with secretions of the rectal bladder, augmented by a rapid jerking display at the nest. These signals excite nestmates, which then follow the recruiting ant to the tournament site. Circumstantial evidence indicates that the recruiter, while on her way back to the tournament site, also discharges formic acid from the poison gland. This pheromone appears to be powerfully stimulating to nestmates. Those that remain in the tournament thereafter are on average larger in size. Further, their fat bodies and ovaries and their external wear and tear indicate that they are more advanced in age.

In summary, the *Myrmecocystus* colonies communicate to neighboring colonies their resource-holding potential by summoning cohorts of large display ants to tournament sites. Colonies unable to match the challenge retreat and forage in other directions or wait inside the nest until the dominant neighboring colony is inactive. The workers of dominant colonies often stay home when the soil is dry and termites, the main food source, are scarce. Adjusting to this "activity shadow," foragers of smaller colonies find the means to survive. They swarm out when the field clears and commence harvesting whatever food they can find and retrieve.

The territorial tournaments, one of the most sophisticated forms of animal social behavior, involve communication both within and between colonies. By means of chemical trails and motor displays, nestmates are summoned to the tournament site, and during encounters and confrontations with other ants, they use colony-specific chemical cues for recognition of nestmates and opponents.

The emergent property of the superorganism, comprising division of labor and communication, is the extended phenotype of the ant colony's collective membership. The territorial strategy is part of the behavioral phenotype of this superorganism. The tournamenting ants are for the *Myrmecocystus* superorganism what the antlers are for a deer. They announce their resource-holding potential.

Mathematical game theory models confirm common sense and the intuitive rules of military strategy: where fights endanger future survival and reproductive success, individuals should monitor the RHP of their opponent and the value of the resource and defend the resource if they are able to, but withdraw without escalation if they are likely to lose an ensuing fight. For animal species that live in social groups, contests are not necessarily between individuals; they are between groups that compete as units for the resource. This is certainly true of many social mammals, including some primates and social carnivores.[291] For example, groups of female lions adjust their agonistic behavior according to the number of individuals both in their own group and in the opposing group, and they assess the group size of their opponents by roaring contests.[292] The major difference between this example and that of *Myrmecocystus* colonies is that in the lion prides, each group member has full reproductive potential, whereas in *Myrmecocystus*, the tournamenting ants are sterile individuals that never reproduce. The ants function as somatic external display organs with which the superorganism advertises its RHP, and the reconnaissance ants are the sensory organs of the superorganism; they gather information at the tournament site and communicate it to the colony.

The ritualized territorial interactions of the honey ants are strikingly similar to the "nothing fights" that anthropologists describe in human tribes of New Guinea. Males of opposing groups assemble on tournament sites, where they display their weapons, shout insults at their enemies, and count. As long as both parties are of equal strength, the status quo is maintained. But if one party is outnumbered on the tournament site, the conflict escalates, and the weaker group may be raided. In both the tournamenting ants and the nothing fights of humans, opposing parties

---

291 | J. Grinnell and K. McComb, "Roaring and social communication in African lions: the limitations imposed by listeners," *Animal Behaviour* 62(1): 93–98 (2001); and C. Lazaro-Perea, "Intergroup interactions in wild common marmosets, *Callithrix jacchus*: territorial defense and assessment of neighbours," *Animal Behaviour* 62(1): 11–21 (2001).

292 | K. McComb, C. Packer, and A. Pusey, "Roaring and numerical assessment in contests between groups of female lions, *Panthera leo*," *Animal Behaviour* 47(2): 379–387 (1994).

communicate their strengths; and depending on their assessment, the fight might escalate to a raid or they may just continue to show off their strengths and hold the status quo.

CONCLUSION

The insect colony can be visualized as an information network, the pathways of which determine, without exception, all of the unique social qualities of its species. But as we have repeatedly stressed in this book, it is not the abstract conception of information alone that matters (see especially Chapters, 1, 3, and 5). Nor is it broad generalizations about decision rules, algorithms, kinship, distributed intelligence, context-dependent thresholds, amplification, phenotypic plasticity, and other elements of self-organization and emergence that we and others have developed.[293] These concepts, when skeletonized, easily lend themselves to mathematical modeling, and they can be of significant heuristic value and explanatory power. But on occasion they can also bedazzle and lead empirical and advanced theory onto sterile paths. More important now for future discovery and understanding are the fine details of natural history.

---

293 | The principles of self-organization and emergence with special reference to communication systems of social insects have been developed and reviewed over the last several decades; see, for example, E. O. Wilson, *The Insect Societies* (Cambridge, MA: The Belknap Press of Harvard University Press, 1971); B. Hölldobler and E. O. Wilson, *The Ants* (Cambridge, MA: The Belknap Press of Harvard University Press, 1990); R. F. A. Moritz and E. E. Southwick, *Bees as Superorganisms: An Evolutionary Reality* (New York: Springer-Verlag, 1992); T. D. Seeley, *The Wisdom of the Hive: The Social Physiology of Honey Bee Colonies* (Cambridge, MA: Harvard University Press, 1995); E. Bonabeau, G. Theraulaz, J.-L. Deneubourg, S. Aron, and S. Camazine, "Self-organization in social insects," *Trends in Ecology and Evolution* 12(5): 188–193 (1997); C. Detrain, J. L. Deneubourg, and J. M. Pasteels, eds., *Information Processing in Social Insects* (Basel: Birkhäuser Verlag, 1999); T. D. Seeley and S. C. Buhrman, "Group decision making in swarms of honey bees," *Behavioral Ecology and Sociobiology* 45(1): 19–31 (1999); S. Camazine, J.-L. Deneubourg, N. R. Franks, J. Sneyd, G. Theraulaz, and E. Bonabeau, *Self-organization in Biological Systems* (Princeton, NJ: Princeton University Press, 2001); C. Anderson and D. W. McShea, "Individual *versus* social complexity, with particular reference to ant colonies," *Biological Reviews of the Cambridge Philosophical Society* 76(2): 211–237 (2001); C. Anderson, G. Theraulaz, and J. L. Deneubourg, "Self-assemblages in insect societies," *Insectes Sociaux* 49(2): 99–110 (2002); D. E. J. Blazis et al., "The limits to self-organization in biological systems," a collection of 11 reviews and essays by different authors published as the proceedings of a workshop sponsored by the Center for Advanced Studies in the Space Life Sciences at the Marine Biological Laboratory, 11–13 May 2001, in *The Biological Bulletin* 202(3): 245–313 (2002); and J. H. Fewell, "Social insect networks," *Science* 301: 1867–1870 (2003).

Natural history matters fundamentally. Because the strong force in social evolution is ecological selection at the colony level, all the details of colony life history for each species in turn are important. Surprising exceptions and idiosyncrasies abound and demand ad hoc explanations. New phenomena routinely come to light, and they will always open doors to new kinds of research. Simple models will also routinely fall, and middle-level theory will reign, by its nature more complex and adaptable to large amounts of particular detail. To put the matter as simply as possible, we have only just arrived at the boundary between insect behavioral ecology and sociobiology. Many surprising discoveries and productive new initiatives in theory and empirical research await us.

The reason our research is at this still early stage of exploration is that the primary determinant of social behavior is genetic fitness of the colony. More precisely, what matters is the product of colony survivorship times colony reproduction summed in each time interval over all time intervals of maximum colony longevity. Direct genetic fitness of the individual is not the primary determinant; what counts is the maximization of inclusive fitness. The demographies and sociobiological traits of the colony members are shaped not by each individual's direct genetic fitness but by the summed effects of all their performances on the fitness of their colony. Collective wisdom, to use Thomas D. Seeley's felicitous phrase,[294] arises from poorly informed masses, whose interactions are shaped by natural selection among competing colony genotypes.

Each eusocial insect species has adapted to certain of the immense number of environments available to that species across evolutionary time. In these particularities, and in the patterns and principles they exemplify, lies a vast, still largely unexplored domain.

---

294 | T. D. Seeley, "Decision making in superorganisms: how collective wisdom arises from the poorly informed masses," in G. Gigerenzer and R. Selten, eds., *Bounded Rationality: The Adaptive Toolbox* (Report of the 84th Dahlem Workshop, 14–19 March 1999, Berlin) (Cambridge, MA: The MIT Press, 2001), pp. 249–261.

|| Plate 43. A huntress of the Australian dawn ant *Prionomyrmex* (formerly *Nothomyrmecia*) carries a captured wasp down a tree trunk at night.

7

THE RISE OF THE ANTS

No biological phenomenon can be fully understood without attention to its evolutionary history. From this perspective, we will now present a combined phylogenetic and ecological explanation of the immense diversity of the ants. This endeavor is favored by a strong fossil record, of which some of the most critical parts have only recently been discovered and analyzed.[1]

The contemporary ants and other social insects that swarm over the land are products of a long, cumulative process of ecological evolution that began 425 million years ago. From the first invasion of the land—the "greening of a strip of equatorial coastline," in Conrad Labandeira's felicitous phrase—animals have coevolved creating ever more complex ecosystems.[2] Arthropods, including primitive wingless insects, arose in the new environment by the Silurian period, no later than 400 million years ago and only a few tens of millions of years before plants began invading the land.[3] During Carboniferous times, between 300 and 360 million years ago, winged insects also made their appearance. A profusion of ecological specialists, including early versions of mayflies, dragonflies, cockroaches, and orthopterans, many of which were adapted to Earth's first forests, dominated the terrestrial habitats in both biomass and species diversity. With the great extinction spasm that ended the Permian period 248 million years ago came the end of 5 of the existing 31 orders. Among the most striking, paleodictyopterans, diaphanopterodeans, and

---

1 | An abbreviated version of this chapter was published as an article by E. O. Wilson and B. Hölldobler, "The rise of the ants: a phylogenetic and ecological explanation," *Proceedings of the National Academy of Sciences USA* 102(21): 7411–7414 (2005).

2 | C. C. Labandeira, "The history of associations between plants and animals," in C. M. Herrera and O. Pellmyr, eds., *Plant-Animal Interactions: An Evolutionary Approach* (Malden, MA: Blackwell Science, 2002), pp. 26–74 and supplementary information in an appendix on pp. 248–261.

3 | M. S. Engel and D. A. Grimaldi, "New light shed on the oldest insect," *Nature* 427: 627–630 (2004).

megasecopterans no longer graced the air. Mayflies, dragonflies, beetles, and cockroaches inherited the altered world.

On into the Mesozoic era, the surviving insects resumed a steady climb in diversity. By the middle of the Jurassic period, 180 million years ago, as many taxonomic families (into which orders are divided by hierarchical classification) existed as before the end-of-Permian crash.[4] About 90 to 110 million years later, in mid-Cretaceous times, the angiosperms (flowering plants) began a major diversification that was to lead to their replacing the gymnosperms as the dominant floristic elements. Some insect groups, exploiting the diverse new sources of food, shelter, and habitats the plants provided, experienced a parallel expansion and radiation. The angiosperms, in turn, came to depend on the insects for much of their pollination, seed dispersal, and nutrient cycling. This coevolution of the two dominant terrestrial groups has continued to the present time, despite a significant interruption by mass extinctions at the end of the Cretaceous period 65.5 million years ago.[5] Today we live in a land environment filled with flowering plants, nematodes, spiders, mites, and six ecological keystone insect groups: termites, hemipterans ("true" bugs), phytophagan beetles, cyclorrhaphan flies, glossatan moths, and hymenopterans (wasps, bees, and ants).[6]

THE ORIGIN OF ANTS

Ants arose during the Cretaceous period, and their history thereafter spanned well over 100 million years to the present time (Figure 7-1). The earliest known fossils fall into two groups. The first group consists of the anatomically most primitive ants known, the extinct Mesozoic subfamily Sphecomyrminae, and, if ranked a subfamily instead of its own family, Armaniinae. The second group, of common origin with the sphecomyrmines and armaniines, comprise primitive members of the subfamilies Aneuritinae, Ponerinae, and Formicinae.

---

4 | C. C. Labandeira and J. J. Sepkoski, Jr., "Insect diversity in the fossil record," *Science* 261: 310–315 (1993).

5 | C. C. Labandeira, K. R. Johnson, and P. Wilf, "Impact of the terminal Cretaceous event on plant-insect associations," *Proceedings of the National Academy of Sciences USA* 99(4): 2061–2066 (2002).

6 | D. Grimaldi, "Mesozoic radiations of the insects and origins of the modern fauna," *Proceedings of the Twenty-first International Congress of Entomology* 1: xix–xxvii (2000).

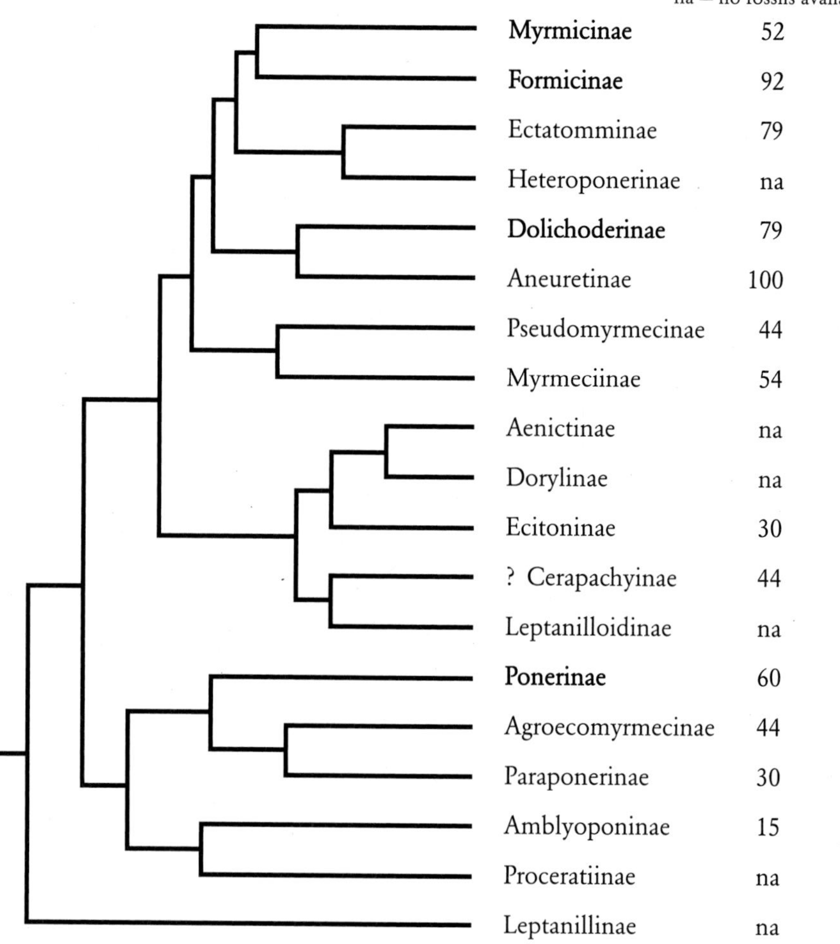

FIGURE 7-1. The phylogeny of the living ants, based on nucleic acid sequencing and representing 19 of the 20 subfamilies recognized in 2006. Not shown is the subfamily Aenictogitoninae, an obscure African subfamily known only from males and recently disclosed by Brady et al. (reference below) to be closely related to the subfamily Dorylinae. Three subfamilies not shown, namely Armaniinae, Brownimyrmeciinae, and Sphecomyrminae, are known only from Cretaceous fossils. Another, Formiciinae (not to be confused with Formicinae) lived in the mid-Cretaceous. The "big four" subfamilies, dominant at the present time in diversity and abundance, are highlighted in bold. The maximum diversification of the ants at the generic level, which generated many present-day genera, occurred in the late Cretaceous period, thence through Paleocene and early Eocene times, when flowering plants also radiated and rose to dominance. The dates of the earliest known fossils are given. The estimated time of the origin of the modern ants is between 140 and 168 million years ago. Based on C. S. Moreau, C. D. Bell, R. Vila, S. B. Archibald, and N. E. Pierce, "Phylogeny of the ants: diversification in the age of the angiosperms," *Science* 312: 101–104 (2006). A closely similar phylogeny was independently and simultaneously adduced by S. G. Brady, T. R. Schultz, B. L. Fisher, and P. S. Ward, "Evaluating alternative hypotheses for the early evolution and diversification of ants," *Proceedings of the National Academy of Sciences USA* 103(48): 18172–18177 (2006).

The workers of *Sphecomyrma*, the best-known sphecomyrmine genus, are a mosaic of ant and wasp traits—hence the derivation of the scientific name given them, "wasp ants." Small, relatively slender in form, they resemble no living species. They are nevertheless undoubtedly ants, with the wingless thorax and separated waist segment (abdominal segment II) diagnostic of the modern members of Formicidae. At the rear corners of the thorax are a pair of metapleural glands, which in modern ants secrete antibiotics and are also regarded as diagnostic of ants. The mandibles, however, are typically wasplike: narrow, each with two teeth in the manner of many wasps, they fold in repose tightly against the front of the head. The antennae, meanwhile, are intermediate in form between those of ants and wasps. In the four decades following the discovery of the first specimens in 1967, other sphecomyrmines have been found in deposits of amber (fossilized resin) in localities in Asia, Siberia, and North America, together embracing much of the northern supercontinent of Laurasia.

Unlike the ants of later, Paleocene deposits, the sphecomyrmines were evidently rare. Among the many thousands of insect specimens examined to date in Late Cretaceous amber of New Jersey, Canada, and Burma, only about a dozen specimens have enough features preserved to be called definitively sphecomyrmine. Of the biology of these ancient ants nothing can be said, except that they lived in most tropical or subtropical forests with rich floras and insect faunas. The specimens from New Jersey were trapped in what are thought to be sequoia or metasequoia resin, and those from Burma in the resin of metasequoia. The sphecomyrmines evidently did not survive the Cretaceous period. The exact timing and cause of their extinction in the final 10 or 20 million years of that era remain unknown.[7]

---

7 | The principal contributions to knowledge of Sphecomyrminae include E. O. Wilson, F. M. Carpenter, and W. L. Brown, "The first Mesozoic ants," *Science* 157: 1038–1040 (1967); E. O. Wilson, "Ants from the Cretaceous and Eocene amber of North America," *Psyche* (Cambridge, MA) 92(2–3): 205–216 (1985); E. O. Wilson, "The earliest known ants: an analysis of the Cretaceous species and an inference concerning their social organization," *Paleobiology* 13(1): 44–53 (1987); G. M. Dlussky, "A new family of Upper Cretaceous Hymenoptera: an 'intermediate link' between the ants and the scolioids," *Paleontological Journal* 17(3): 63–76 (1983); D. A. Grimaldi, D. Agosti, and J. M. Carpenter, "New and rediscovered primitive ants (Hymenoptera: Formicidae) in Cretaceous amber from New Jersey, and their phylogenetic relationships," *American Museum Novitates* No. 3208, 43 pp. (1997); D. Agosti, D. Grimaldi, and J. M. Carpenter, "Oldest known ant fossils discovered," *Nature* 391: 447 (1998); D. A. Grimaldi, M. S. Engel, and P. C. Nascimbene, "Fossiliferous Cretaceous amber from Myanmar (Burma): its rediscovery, biotic diversity, and paleontological significance," *American Museum Novitates* No. 3361, 71 pp. (2002); and M. S. Engel and D. A. Grimaldi, "Primitive new ants in Cretaceous amber from Myanmar, New Jersey, and Canada (Hymenoptera: Formicidae)," *American Museum Novitates* No. 3485, 23 pp. (2005).

THE EARLY RADIATION OF THE ANTS

Early in the history of the sphecomyrmines, a wider radiation beyond the stem genus *Sphecomyrma* occurred. In the newly explored Jersey amber fauna, a primitive species has been found, *Brownimyrmecia clavata*, that resembles *Sphecomyrma* overall but departs from it fundamentally in possession of thin, toothless mandibles that cross one another. Also present, and typical of the subfamily Ponerinae, is a girdling constriction of the gaster (posterior major segment of the body). The single specimen of *Brownimyrmecia* known is in fact intermediate between *Sphecomyrma* and primitive modern ponerines, particularly *Amblyopone* and related genera, and consequently has been placed by one systematist in a subfamily of its own, Brownimyrmeciinae.[8] A second species of ponerine (*Canapone dentata*) has been found in Canadian amber, of later Cretaceous provenance. A likely aneuretine (*Cananeuretus occidentalis*) was turned up in the same deposit.[9] Of equal significance is a formicine ant (*Kyromyrma neffi*), antecedent to a modern subfamily anatomically more advanced than Ponerinae.[10] Present in the Burmese amber, older even than the North American and Siberian *Sphecomyrma* (approximately 100 million years) and in fact accompanied by *Sphecomyrma*, are four other kinds of ants: an undescribed ponerine genus; *Haidomyrmex cerberus*, which has unique L-shaped mandibles and may be either sphecomyrmine or ponerine; *Burmomyrma*, likely an aneuretine; and *Myanmyrma*, a primitive myrmeciine or myrmeciine-sphecomyrmicine intermediate.[11]

What may be the oldest certifiable European ant fossil is *Gerontoformica cretacica*, from the Upper Abian (Lower Cretaceous) amber of France, dated to about 100 million years ago.[12] Because of imperfect preservation, it cannot be placed with

---

8 | B. Bolton, "Synopsis and classification of Formicidae," *Memoirs of the American Entomological Institute* 71: 1–370 (2003).

9 | M. S. Engel and D. A. Grimaldi, "Primitive new ants in Cretaceous amber from Myanmar, New Jersey, and Canada," *American Museum Novitates* No. 3485, 23 pp. (2005).

10 | D. A. Grimaldi and D. Agosti, "A formicine in New Jersey Cretaceous amber (Hymenoptera: Formicidae) and early evolution of the ants," *Proceedings of the National Academy of Sciences USA* 97(25): 13678–13683 (2000).

11 | M. S. Engel and D. A. Grimaldi, "Primitive new ants in Cretaceous amber from Myanmar, New Jersey, and Canada," *American Museum Novitates* No. 3485, 23 pp. (2005).

12 | A. Nel, G. Perrault, V. Perrichot, and D. Néraudeau, "The oldest ant in the Lower Cretaceous amber of Charente-Maritime (SW France) (Insecta: Hymenoptera: Formicidae)," *Geologica Acta* 2(1): 23–29 (2004).

reference to fossil or existing subfamilies, but evidently contains a mix of primitive ant traits (two-toothed mandible as in sphecomyrmines, clypeal denticles as in some primitive ponerines) and more advanced ant traits (well-developed, high petiolar node and long antennal scapes).

In still other very early deposits have been found additional products of the initial ant radiation, which appear to be members of or precursors of the bulldog ant subfamily Myrmeciinae, including the living "dawn ant" *Prionomyrmex* (formerly *Nothomyrmecia*) *macrops*. The subfamily is today limited to Australia, with one endangered species of *Myrmecia* on New Caledonia. A possible myrmeciine precursor, *Cariridris bipetidata*, has been described from a single, poorly preserved specimen from the Santana Formation of Brazil, approximately 110 million years old. What appear to us to fall within Myrmeciinae or close by, although placed variously in Ponerinae and Myrmicinae by describers, are ten rock fossils of ants (put in the new genera *Orapia*, *Afropone*, and *Afromyrma*) from the Late Cretaceous, about 90 million years old, found in the Orapa deposits of Botswana.[13] A wide variety of myrmeciines, including seven genera, have been identified in Early Eocene deposits of the U.S.-Canadian Pacific Northwest.[14] Several later myrmeciines of Paleogene age, *Ameghinoia* and *Polanskiella* of Argentina, *Archimyrmex* of the U.S. Green River Formation, and two species of *Prionomyrmex* (formerly *Nothomyrmecia*) recorded from the Baltic amber, bear witness to the spread of the subfamily around the world. However, it is reasonable, for the time being, to suppose that Myrmeciinae diverged from the sphecomyrmine stem and spread extensively by Late Cretaceous or Early Eocene times.[15]

By Paleocene times, as evidenced in ten specimens of ants from the presumed Paleocene amber of Sakhalin, dolichoderines and aneuretines (sister group of the dolichoderines, with one contemporary but unfortunately endangered species, *Aneuretus simoni*, surviving in Sri Lanka), which had made their appearance as early

---

13 | G. M. Dlussky, D. J. Brothers, and A. P. Rasnitsyn, "The first Late Cretaceous ants (Hymenoptera: Formicidae) from southern Africa, with comments on the origin of the Myrmicinae," *Insect Systematics and Evolution* 35(1): 1–13 (2004).

14 | S. Bruce Archibald, personal communication (2005).

15 | C. Baroni Urbani, "Rediscovery of the Baltic amber ant genus *Prionomyrmex* (Hymenoptera, Formicidae) and its taxonomic consequences," *Eclogae Geologicae Helvetiae* 93(3): 471–480 (2000); and P. S. Ward and S. G. Brady, "Phylogeny and biogeography of the ant subfamily Myrmeciinae (Hymenoptera: Formicidae)," *Invertebrate Systematics* 17(3): 361–386 (2003).

as Burmese amber times (100 million years ago), now flourished alongside ponerines and formicines.[16]

THE CENOZOIC RADIATION

By Early to Middle Eocene times, the diversification of the major groups of ants was in full swing, as revealed by the recently described Fushun amber fauna from northeastern China, about 50 million years in age or somewhat younger (see Figure 7-1).[17] Among some 20 identifiable specimens of workers and queens are represented a variety of primitive ponerines and myrmicines, as well as primitive members of the "formicoid" complex. The latter comprise formicines and an assortment of what are either primitive dolichoderines or aneuretines, or both. There is considerable anatomical variation in the shape of the bodies, heads, and mandibles of these forms and even of antennal segmentation numbers, indicating that an early radiation of more modern ants was under way. Still, most also bear traits shared with their presumed sphecomyrmine or sphecomyrmine-like ancestors, including short mandibles with small numbers of teeth, circular or ovoid head shapes, and relatively unmodified mesosomas (that is, lacking the spines, angulations, or extensive sclerite fusion common in modern faunas). Several sedimentary compression fossils of Early Eocene provenance, recently discovered in British Columbia, also represent relatively primitive ponerines.[18]

The next glimpse into the history of ants has come from a set of three specimens found in a later, Middle Eocene amber deposit at Malvern, Arkansas.[19] The amber is extremely "dirty"—crowded with extraneous debris—and the search for enclosed ant fossils is prohibitively laborious and slow. Nevertheless, and providentially, the specimens represent among them three of the dominant subfamilies in

---

16 | G. M. Dlussky, "Ants from (Paleocene?) Sakhalin amber," *Paleontological Journal* 22(1): 50–61 (1988).

17 | Y.-C. Hong, *Amber Insects of China,* 2 volumes (Beijing: Beijing Scientific and Technological Publishing House, first volume; Henan Scientific and Technological Publishing House, second volume, 2000), in Chinese. The amber is from the Xilutiàn Coal Mine of Fushun and the Guchengzi Formation, stratigraphically the equivalent of the European Ypresian stage.

18 | Studied by Bruce Archibald and cited here with permission.

19 | More precisely, the amber is from the lower Claiborne stratum (lower Middle Eocene): W. B. Saunders, R. H. Mapes, F. M. Carpenter, and W. C. Elsik, "Fossiliferous amber from the Eocene (Claiborne) of the Gulf Coastal Plain," *Geological Society of America Bulletin* 85(6): 979–984 (1974).

the present-day fauna: clearly demarcated Myrmicinae, Dolichoderinae, and Formicinae. Modern in aspect, with a representation of the contemporary dolichoderine genus *Iridomyrmex*, they reveal that key elements of what was to endure as the modern adaptive radiation were in place.[20]

That the radiation had not only begun but was full blown by the end of the Eocene epoch is suggested by the Baltic amber fossils, which are Middle Eocene in age. In examining 10,988 specimens, Gustav Mayr in the 1860s and William Morton Wheeler in 1914 distinguished between them no fewer than 92 genera.[21] The fauna as a whole had a distinctly modern aspect, despite its great age of approximately 45 million years. The most abundant specimens belong to the dolichoderine genus *Iridomyrmex*, and the second most abundant to the formicine genus *Lasius*. *Iridomyrmex* and several of its satellite genera are still numerically dominant and rich in species across southeast Asia, Australia, Melanesia, and tropical America. *Lasius* remains one of the several most abundant and species-rich genera in the cooler regions of North America and Eurasia. The same evidence of ecological success is present in other high-ranking ant genera that have persisted since Baltic amber times, including *Camponotus*, *Formica*, *Myrmica*, *Oecophylla*, *Ponera*, and *Technomyrmex*. An odd addition was the gigantic ants of the subfamily Formiciinae (possibly a distant relative of the subfamily Formicinae), known only from queens and males of Eocene age.

Not surprisingly, the richly documented and younger amber of the Dominican Republic, which is Early Miocene in age (roughly 20 million years old), has an even more modern composition. Of 38 genera and well-defined subgenera identified to date, 34 have survived somewhere in the New World tropics to the present, although all of the species studied thus far are extinct.[22] Of the surviving genera and subgenera, at least 22 persist on Hispaniola (Dominican Republic and Haiti).

---

20 | E. O. Wilson, "Ants from the Cretaceous and Eocene amber of North America," *Psyche* (Cambridge, MA) 92: 205–216 (1985).

21 | G. L. Mayr, "Die Ameisen des baltischen Bernsteins," *Beiträge zur Naturkunde Preussens herausgegeben von der Königlichen Physikalisch-Ökonomischen Gesellschaft zu Königsberg* 1: 1–102 (1868); and W. M. Wheeler, "The ants of the Baltic amber," *Schriften der Physikalisch-Ökonomischen Gesellschaft zu Königsberg* 55: 1–142 (1914). The earlier discoveries by Gustav Mayr are reanalyzed in Wheeler's review.

22 | E. O. Wilson, "Invasion and extinction in the West Indian ant fauna: evidence from the Dominican amber," *Science* 229: 265–267 (1985); E. O. Wilson, "The biogeography of the West Indian ants (Hymenoptera: Formicidae)," in J. K. Liebherr, ed., *Zoogeography of Caribbean Insects* (Ithaca, NY: Comstock Publishing Associates of Cornell University Press, 1988), pp. 214–230; and C. Baroni Urbani and E. O. Wilson, "The fossil members of the ant tribe Leptomyrmecini (Hymenoptera: Formicidae)," *Psyche* (Cambridge, MA) 94: 1–8 (1987).

Fifteen genera and subgenera have colonized the island since amber times, restoring the number to 37.

What was the fate of the first evolutionary radiation of the ants? The subfamily Sphecomyrminae evidently vanished by the end of the Mesozoic. The myrmeciines retreated to Australia and New Caledonia, and the aneuretines to Sri Lanka. But the myrmicines, formicines, dolichoderines, and ponerines not only flourished, but spread worldwide as dominant insect groups. The history of the ponerines is of special interest. Today there are six ponerine tribes (Amblyoponini, Ectatommini, Platythyreini, Ponerini, Thaumatomyrmecini, and Typhlomyrmecini), comprising 42 genera and more than 1,300 species. Together they are the most variable overall in anatomical characteristics and patterns of colony organization of all the subfamilies.[23]

As part of a comprehensive overview of ant classification, Barry Bolton has recently split Ponerinae into seven subfamilies (Ponerinae, Amblyoponinae, Ectatomminae, Heteroponerinae, Paraponerinae, Proceratiinae, and the fossil Brownimyrmeciinae).[24] Still, there is no reason as yet to doubt that the assemblage as a whole represents a diversification from a single Mesozoic ancestor.

THE PONERINE PARADOX

The subfamily Ponerinae, despite being, along with Formicinae and Myrmeciinae of Australia, the oldest documented phylogenetic assemblage, and despite having achieved prolific diversification and geographical breadth, oddly remains mostly primitive in its social organization.[25] In particular:

---

23 | C. Baroni Urbani, B. Bolton, and P. S. Ward, "The internal phylogeny of ants (Hymenoptera: Formicidae)," *Systematic Entomology* 17: 301–329 (1992); and C. Peeters, "Morphologically 'primitive' ants: comparative review of social characters, and the importance of queen-worker dimorphism," in J. C. Choe and B. J. Crespi, eds., *The Evolution of Social Behavior in Insects and Arachnids* (New York: Cambridge University Press, 1997), pp. 372–391.

24 | B. Bolton, "Synopsis and classification of Formicidae," *Memoirs of the American Entomological Institute* 71: 1–370 (2003).

25 | As reviewed in Chapter 8 of this book and in the synthesis by C. Peeters, "Morphologically 'primitive' ants: comparative review of social characters, and the importance of queen-worker dimorphism," in J. C. Choe and B. J. Crespi, eds., *The Evolution of Social Behavior in Insects and Arachnids* (New York: Cambridge University Press, 1997), pp. 372–391.

- Queens and workers of ponerine species are much closer in size than are the castes in the "higher" ant subfamilies Myrmicinae, Dolichoderinae, Formicinae, and army ant subfamilies. An exception is the ponerine *Pachycondyla* (formerly *Brachyponera*) *lutea*, of Australia.
- Ponerine queens have relatively low fertility, seldom producing more than 5 eggs each day. At the upper end of the fertility side in other subfamilies, queens of monogyne fire ant colonies (*Solenopsis invicta*) can lay 150 eggs per hour, and those of African driver ants lay into the millions per month, the highest of any insect species recorded.
- Corresponding to the low degree of queen fecundity, colony sizes of ponerines are small. In most cases, they comprise, according to species, between 20 and 200 workers and one to several reproductives. Exceptions include the legionary species of *Leptogenys*, such as *Leptogenys ocellifera* and *Leptogenys purpurea*, of tropical Asia and Melanesia, whose colonies contain thousands or even tens of thousands of workers.
- Young queens as a rule start new colonies independently, but not claustrally. That is, they leave their natal nest, mate, construct a nest of their own or else find a preformed cavity, then forage outside the nest to obtain at least part of the food with which they rear the first brood of workers. In contrast, the queens of many (but not all) higher subfamilies are fully claustral, remaining in the new nest permanently while feeding their young entirely on nutrients metabolized from their own wing muscles and fat.[26] An exception is, again, *Pachycondyla lutea*, of Australia, the queens of which are at least capable of fully claustral colony founding.
- The workers of many ponerine species forage alone and do not use odor trails or other pheromone signals to recruit nestmates to food sources they encounter. A substantial number of other ponerine species do recruit in this manner, and a few, such as the army-ant-like members of the genus *Onychomyrmex*, of Australia, the

---

26 | Ponerine-like semiclaustral founding has also been reported widely in myrmicines (*Acromyrmex*, *Atta* and other attines, *Strumigenys* and other dacetines, *Manica*, *Myrmica*, *Messor*, and *Pogonomyrmex*) and formicines (*Cataglyphis*, *Polyrhachis*). In at least some of these genera, it may be a secondarily evolved trait, as argued by M. J. F. Brown and S. Bonhoeffer, "On the evolution of claustral colony founding in ants," *Evolutionary Ecology Research* 5: 305–313 (2003); for a more recent review and discussion, see R. A. Johnson, "Capital and income breeding and the evolution of colony founding strategies in ants," *Insectes Sociaux* 53(3): 316–322 (2006).

legionary *Leptogenys*, of Asia and Melanesia, and the termite-raiding *Pachycondyla fochi* (formerly *Megaponera foetens*), of Africa, use chemical communication to coordinate group foraging. Within *Leptogenys*, group foraging has evolved on multiple occasions independently as a specialized adaptation to ecological conditions.[27]

- True oral trophallaxis, the exchange of regurgitated food among adults and between workers and larvae, is widespread in the higher subfamilies Myrmicinae, Dolichoderinae, and Formicinae. It is evidently much less so in Ponerinae and has so far been observed only in the closely related genera *Ponera* and *Hypoponera*.[28] Three other properties of ponerine biology may have contributed to the lack of trophallaxis in at least the more primitive genera: the reliance on insect prey, rendering liquid food exchange unnecessary; the fusion of the upper and lower sclerites of the fourth abdominal segment, which limits the expansion of the abdomen and within it the crop, the principal storage organ of liquid food; and finally, the small size of the colonies, which permits the easier sharing of each prey item among a large fraction of the colony.[29] Those ponerine species that do collect liquid food, such as the sugary secretions of extrafloral nectaries, carry the droplets between the mandibles, forming what has been called a "social bucket."[30]

This, then, is the ponerine paradox: a group that is globally successful yet socially primitive. The puzzle might be partially resolved if the more advanced ant

---

27 | C. Baroni Urbani, "The diversity and evolution of recruitment behaviour in ants, with a discussion of the usefulness of parsimony criteria in the reconstruction of evolutionary histories," *Insectes Sociaux* 40(3): 233–260 (1993); and V. Witte and U. Maschwitz, "Coordination of raiding and emigration in the ponerine army ant *Leptogenys distinguenda* (Hymenoptera: Formicidae: Ponerinae): a signal analysis," *Journal of Insect Behavior* 15(2): 195–217 (2002).

28 | Y. Hashimoto, K. Yamauchi, and E. Hasegawa, "Unique habits of stomodeal trophallaxis in the ponerine ant *Hypoponera* sp.," *Insectes Sociaux* 42(2): 137–144 (1995); and J. Liebig, J. Heinze, and B. Hölldobler, "Trophallaxis and aggression in the ponerine ant, *Ponera coarctata*: implications for the evolution of liquid food exchange in the Hymenoptera," *Ethology* 103(9): 707–722 (1997).

29 | C. Peeters, "Morphologically 'primitive' ants: comparative review of social characters, and the importance of queen-worker dimorphism," in J. C. Choe and B. J. Crespi, eds., *The Evolution of Social Behavior in Insects and Arachnids* (New York: Cambridge University Press, 1997), pp. 372–391.

30 | B. Hölldobler, "Liquid food transmission and antennation signals in ponerine ants," *Israel Journal of Entomology* 19(2): 89–99 (1985).

subfamilies can be shown to be derived from a ponerine stock; or put in the more exact language of cladistics, the paradox is partly soluble if the ponerines prove to be polyphyletic (of multiple origins), with the higher subfamilies sister groups to various contemporary "ponerine" lines. (It is a similar circumstance that allows us to speak of modern birds as the "last surviving dinosaurs.")

But even if such proves to be the case (contrary to the opinion of systematists who consider the ponerines basically monophyletic), there likely remain diverse modern subgroups, such as the very large and global tribe Ponerini, that are monophyletic. So the paradox would not be truly resolved.

The full and definitive solution to the ponerine paradox requires more information from paleontology and ecology than we now have. In the interim, and in the hope of stimulating research in the critical areas, we offer the following combined phylogenetic and ecological *dynastic-succession hypothesis* consistent with existing knowledge but in important aspects still speculative:

> *Possibly toward the end of Cretaceous times, but more likely during Paleocene and Early Eocene times following the end-of-Mesozoic extinction spasm, the ponerines underwent most of the adaptive radiation in taxonomic tribes and adaptive types that persist to the present time, an event to be described in Chapter 8.*

The fossil record of this interval, spanning some 30 million years, is still poorly represented for ants, preventing a more exact placement of the ponerine radiation in geological history. To continue:

> *During this primary expansion, the ponerines became entrenched worldwide as arthropod predators, especially in warm-temperate and tropical moist forests. They also evolved to favor ground and leaf-litter sites. In effect, they preempted this array of opportunities, partially blocking the sites from the later, otherwise more successful radiations of the advanced subfamilies Dolichoderinae and Formicinae. Of the "big four" in present-day diversity and geographical range, only Myrmicinae rivaled Ponerinae in invasion of the forest floor predator niches. The myrmicine radiation either coincided with the peak of the ponerine expansion or followed closely behind it.*

The ponerines are not all exclusively predators. *Odontomachus troglodytes*, for example, also attends coccids and aphids for their sugary excreta. Nevertheless, the diets of the best-known ponerines consist primarily of fresh insect and other arthropod prey, supplemented by the scavenging of arthropods newly killed by other causes. Many ponerine species are moreover specialized predators. Among the many examples are species of *Centromyrmex*, specialized on termites; *Leptogenys*, on isopods; *Myopias*, variously according to species on millipedes and beetles; and *Thaumatomyrmex*, on soft-bodied polyxenid millipedes.[31]

Ants living as huntresses have small colonies, a necessity imposed by the relative paucity of their food, especially when they specialize on particular kinds of prey. Predatory ponerines and members of other poneromorph subfamilies, not coincidentally, also tend to have low population densities. Most *Amblyopone* and *Proceratium* ants, for example, are famously scarce, while ants of the genus *Thaumatomyrmex* may be the rarest ants in the world (just finding a specimen is newsworthy among myrmecologists).

Predation as a way of life and small colony size in turn render other social traits simple and hence "primitive." Foraging tends to be solitary, recruitment and alarm elementary, worker subcastes few to none, workers prepared to reproduce on their own in the absence of the queen, and brood care relatively unorganized.

The ponerines (and other poneromorphs) do well as predators and also as occupants of nest sites on forest floors in warm regions around the world. The favored sites are small spaces in the many dimensions of the litter, including the interiors of rotting logs and stumps, tree limbs and twigs lying on the ground, clusters of dead leaves, masses of bryophytes, and the root systems of living trees, shrubs, and herbaceous plants. The colonies nest not only in cavities within these materials but also in spaces beneath them, as well as in chambers and galleries excavated in the soil.

The ponerines hold their own in this complex and nutrient-rich environment. In 110 samples analyzed by Philip S. Ward from forested localities around the world and containing 29,942 specimens, ponerines composed 22.2 percent of the species and 12.4 percent of the specimens, compared with 10.6 percent of the

---

31 | Full summaries of known ponerine diets are provided by B. Hölldobler and E. O. Wilson, *The Ants* (Cambridge, MA: The Belknap Press of Harvard University Press, 1990); and C. Peeters, "Morphologically 'primitive' ants: comparative review of social characters, and the importance of queen-worker dimorphism," in J. C. Choe and B. J. Crespi, eds., *The Evolution of Social Behavior in Insects and Arachnids* (New York: Cambridge University Press, 1997), pp. 372–391. See also M. B. Dijkstra and J. J. Boomsma, "*Gnamptogenys hartmani* Wheeler (Ponerinae: Ectatommini): an agro-predator of *Trachymyrmex* and *Sericomyrmex* fungus-growing ants," *Naturwissenschaften* 90(12): 568–571 (2003).

species and 12.9 percent of the specimens for the formicines and a relatively paltry 1.1 percent of the species and 0.5 percent of the specimens for the dolichoderines.[32] But these three subfamilies of the "big four" were dwarfed by the myrmicines, which made up 65.2 percent of the species and 73.7 percent of the specimens. Clearly, the myrmicines, many of which (such as members of the tribes Dacetini and Basicerotini) have habits similar to those of the ponerines, are the rulers of the world's forest litter.

Thus, the ponerine ants are also prevalent in the litter environment of most parts of the world, especially in the tropics and subtropics. At least one species of *Hypoponera* was present in 75.5 percent of Ward's 110 samples, matched only by the hyperdiverse and very abundant myrmicine genus *Pheidole. Pachycondyla*, at 43.6 percent, and *Anochetus*, at 25.5 percent, ranked sixth and tenth, respectively. Myrmicines dominated in general representation, with the highest-ranking *Pheidole*, *Strumigenys*, and *Solenopsis* followed at a distance by the highest-ranking formicine genera *Paratrechina* (53.6 percent) and *Brachymyrmex* (25.5 percent) and next only at a very great distance by the dolichoderines, possessing no genus in the top 40.

The ground litter of the world's angiosperm forests, and especially the tropical forests, is the habitat with the highest density and species diversity of ants. Because all of the subfamilies of ants since their origins in the mid-Cretaceous, save the subfamilies Sphecomyrminae and Formiciinae (the latter containing giant ants and not to be confused with the subfamily Formicinae), have living representatives, and most of the genera have living representatives as well since the Late Eocene, it is reasonable to suppose that the tropical forest litter has always had the same role. This habitat is reasonably interpreted to be the headquarters of ant evolution, from which major ant groups have spread into other habitats or, in a great many cases, failed to do so.

The picture changes radically for the other poneromorphs—for example, away from the tropical and warm-temperate forests. Except in Australia, they are notably scarce in cool-temperate forests, deserts, and arid grasslands.

---

32 | P. S. Ward, "Broad-scale patterns of diversity in leaf litter ant communities," in D. Agosti, J. D. Majer, L. E. Alonso, and T. R. Schultz, eds., *Ants: Standard Methods for Measuring and Monitoring Biodiversity* (Washington, DC: Smithsonian Institution Press, 2000), pp. 99–121; the pattern broadly duplicates the earlier but less extensive, and subjective survey by E. O. Wilson, "Which are the most prevalent ant genera?" *Studia Entomologica* 19(1–4): 187–200 (1976).

THE TROPICAL ARBOREAL ANTS

The picture changes across a few vertical meters in the tropical forests and thence up into the forest canopy. In the Amazonian forest, for example, subfamily dominance is flipped nearly upside down. There the formicines and dolichoderines have risen sharply in numbers relative to the myrmicines, while the ponerines have dropped to very low levels. In the 1987 canopy species series identified by E. O. Wilson from forest types within the Tambopata Reserved Zone of Peru,[33] the rank order for the seven most common genera were *Crematogaster* (Myrmicinae, 23.4 percent), *Camponotus* (Formicinae, 23.3 percent), *Azteca* (Dolichoderinae, 7.8 percent), *Dolichoderus* (Dolichoderinae, 5.8 percent), *Pseudomyrmex* (Pseudomyrmecinae, 4.9 percent), *Solenopsis* (formerly *Diplorhoptrum*, Myrmicinae, 4.7 percent), and *Cephalotes* (formerly *Zacryptocerus*, Myrmicinae, 3.9 percent). All of the genera of Ponerinae collectively composed only 4.0 percent of the series. Diversity also tipped away from the myrmicines and ponerines in comparison with the ground and litter fauna. Only 50 species of myrmicines and 10 of ponerines were identified, versus 16 of dolichoderines and 38 of formicines.

The great majority of the Amazonian canopy ant species, more than 90 percent, for example, of the Tambopata fauna, appear to be specialized for arboreal life. This is especially true of the four numerically dominant species, *Dolichoderus debilis*, *Camponotus femoratus*, *Crematogaster parabiotica*, and *Solenopsis parabiotica*. These ants are nest site specialists: they occupy and help sow "ant gardens," clusters of orchids, gesneriads, and other epiphytic plants with which they live in mutualistic symbiosis. Tropical arboreal ants nest in hollow stems and the abandoned burrows of wood-boring beetles. They occupy a diversity of myrmecodomatia, swollen cavities created by the plants themselves in mutualistic exchange for the protection the ants give them against herbivores.

Most importantly, the tropical arboreal ants as a whole are so abundant and compose such a large part of the animal biomass as to be inviable if they lived exclusively as predators and scavengers. There are not enough herbivorous insects available as immediate sources of protein to support them. This additional paradox now appears solved: it turns out that a large fraction of the ant populations of the

---

33 | E. O. Wilson, "The arboreal ant fauna of Peruvian Amazon forests: a first assessment," *Biotropica* 19(3): 245–251 (1987). The collections were made with canopy insecticidal fogging by T. L. Erwin.

canopy are not primarily predators but "cryptic herbivores," subsisting on the liquid exudates of hemipterous sap-feeding insects, including especially scale insects and treehoppers.[34] They further tend many of these insects as a kind of cattle, protecting them from parasites and predators. In tropical Asia (especially Borneo), several *Dolichoderus* species practice a remarkable symbiosis with mealybugs of the pseudococcid tribe Allomyrmococcini. These ants live as nomadic herders, culturing the mealybugs in their bivouacs and transporting them to freshly sprouted parts of host plants for "grazing." When nearby grazing sites have been depleted, the whole colony emigrates to new pastures, where the ants construct new bivouacs. The workers carry the mealybugs and their own brood to the new sites, guarding them fiercely along the way. *Dolichoderus* ants are strictly arboreal and live almost exclusively on the plant secretions passed on to them by the mealybugs.[35]

There is evidence that some arboreal ants pick up additional nutrients from pollen, fungal spores, and hyphae. Members of at least one myrmicine genus, *Cephalotes*, feed on nitrogen-rich bird droppings. The same is true for at least some species of *Camponotus*, including the gigantic *Camponotus gigas* of southeastern Asia. One of the functions of endosymbiotic bacteria in the midgut epithelium of *Camponotus* is to break down ammonia, which is a powerful cell poison. The bacterium (*Blochmania floridanus*) produces urease, which recycles the ammonia back to carbon dioxide and water, allowing the safe ingestion of bird feces into the digestive system of the ant hosts.

Around the world, in all major habitats, a large fraction of the species of dolichoderines and formicines have invested their biology heavily in symbioses with hemipterous insects. Myrmicines have done so to lesser degree, and ponerines almost never. While quantitative data are lacking, these disparities are commonly

---

34 | J. E. Tobin, "A Neotropical rainforest canopy, ant community: some ecological considerations," in C. R. Huxley and D. F. Cutler, eds., *Ant-Plant Interactions* (New York: Oxford University Press, 1991), pp. 536–538; J. E. Tobin, "Ants as primary consumers: diet and abundance in the Formicidae," in J. H. Hunt and C. A. Nalepa, eds., *Nourishment and Evolution in Insect Societies* (Boulder, CO: Westview Press, 1994), pp. 279–307; D. W. Davidson, S. C. Cook, R. R. Snelling, and T. H. Chua, "Explaining the abundance of ants in lowland tropical rainforest canopies," *Science* 300: 969–972 (2003); and J. H. Hunt, "Cryptic herbivores of the rainforest canopy," *Science* 300: 916–917 (2003).

35 | M. Dill, D. J. Williams, and U. Maschwitz, "Herdsmen ants and their mealybug partners," *Abhandlungen der Senckenbergischen Naturforschenden Gesellschaft Frankfurt am Main* 557: 1–373 (2002). See also C. A. Brühl, G. Gunsalam, and K. E. Linsenmair, "Stratification of ants (Hymenoptera, Formicidae) in a primary rain forest in Sabah, Borneo," *Journal of Tropical Ecology* 14(3): 285–297 (1998); and A. Floren, A. Biun, and K. E. Linsenmair, "Arboreal ants as key predators in tropical lowland rainforest trees," *Oecologia* 131(1): 137–144 (2002).

observed in natural history studies, and it seems clear that the dolichoderines and formicines have benefited in the growth of their biomass and diversity through coevolution with symbiotic hemipterans.

THE DYNASTIC-SUCCESSION HYPOTHESIS

The Cretaceous/Paleogene, or end-of-Mesozoic, extinction event, 65 million years ago, may have terminated the sphecomyrmine ants, but anatomically primitive ponerines soldiered on. During the reassembly and continued expansion of the flowering plants, which replaced much of the old gymnosperm flora worldwide, forest litter became more complex. (In all respects, especially structural but also chemical and microclimatic, the litter of angiosperms is much better suited for ant colonies than that of gymnosperms.) Insects inhabiting the litter, ground, and vegetation of the forests and savannas grew increasingly diverse and abundant. In Paleocene and Early Eocene times, from 65 to 50 million years ago, the ponerines experienced an adaptive radiation within this favorable period, and some of those genera have survived to the present time.

During and perhaps more precisely toward the end of the ponerine expansion, probably no later than the Early Eocene, the myrmicines began their own radiation. They became formidable competitors of the ponerines for both prey and nest sites. In time, they equaled and then surpassed the ponerines in biomass and diversity. Many also added seeds and elaiosomes to their diets, with the important addition of oils and carbohydrates. At least partly as a result, they were able to expand more effectively into deserts and dry grasslands.

Most importantly, some of the myrmicines added symbioses with hemipterans (members of the Order Hemiptera) to their repertoire—largely scale insects and treehoppers in tropical and warm-temperate vegetation, with aphids more commonly in cool-temperate vegetation, and mealybugs underground everywhere. Similar symbioses were contracted with the caterpillars of honeydew-secreting butterflies. In the New World, one line (Attini) acquired the ability to raise symbiotic fungi for food, expanding diversity and biomass dramatically.

Dolichoderines and formicines also diversified, perhaps with the myrmicines but probably later, in Early to Middle Eocene times. They were less successful than the ponerines and myrmicines in the forest litter environment, having been preempted there by these two groups, but more successful at creating hemipteran symbioses. Moreover, they were able to penetrate environments less available to predators, including cool-temperate climates and tropical forest canopies. Their success is reflected in their high levels of abundance in amber (especially worker specimens) and rock fossils (winged specimens), as would be expected from a preponderance in arboreal habitats.

The breakout of the dolichoderines and formicines, and to some extent that of the myrmicines, was due to a change in diet. This shift in turn was aided by the rising dominance of angiosperms over much of the land environment, an expansion that began in the Cretaceous and culminated in the Paleocene and Eocene. It was furthered by the expansion of the honeydew-producing hemipterans and lepidopterans, groups also favored by the angiosperm dominance.

The ecological history of the ants through geological time, culminating in the profusion of complexly social creatures around us today, must be regarded as one of the great epics of evolution. Its unfolding, however, can still be seen only in fragments. Large gaps remain in the fossil record, especially across the critical period of major radiation that reached from the Late Cretaceous into the Paleogene. Of equal consequence, the life cycles and natural history of the vast majority of living species, which still bear the indelible stamp of this history, remain unexplored.

|| Plate 44. A worker of the Asian jumping ant, *Harpegnathos saltator*, attends worker pupae.

8

PONERINE ANTS: THE GREAT RADIATION

The poneromorph complex of ants, and particularly the majority subfamily Ponerinae within it, displays among its species the greatest diversity of any ant subfamily. This diversity is evident in its reproductive cycles, patterns of internal conflict, communication systems, and colony-level organization. Hence, it is the most diverse ant clade in social organization overall. Many of these species are relatively primitive in terms of superorganismic structure, but with respect to interindividual interactions, they rival in complexity the most advanced nonhuman vertebrates, including primates. Because a substantial array of information is essential in illustrating this important property of the subfamily Ponerinae, and because so much of it is relatively recent, we have chosen to review it here in considerable detail.

THE SOCIAL REGULATION OF REPRODUCTION

In hymenopteran species that lack morphologically specialized queens and workers, all females can reproduce. Yet most individuals remain infertile because of antagonistic interactions among nestmates that generate reproductive dominance hierarchies. As a result, one or several dominant females have active ovaries, while the ovaries of subordinates are reduced and inactive. The role of behavioral patterns and chemical communication in dominance interactions in social hymenopterans have been elucidated in *Polistes* and *Ropalidia* wasps and especially in ponerine ants.[1, 2, 3] Individual interactions are highly directed in these

1 | S. Turillazzi and M. J. West-Eberhard, *Natural History and Evolution of Paper-Wasps* (New York: Oxford University Press, 1996); and H. A. Downing and R. L. Jeanne, "Communication of status in the social wasp *Polistes fuscatus* (Hymenoptera: Vespidae)," *Zeitschrift für Tierpsychologie* 67(1–4): 78–96 (1985).

insects, suggesting the existence of specific recognition mechanisms.

Ant species as a whole typically form colonies comprising morphologically distinct queen and worker castes. Even so, most of the workers possess ovaries, albeit usually much smaller than those of queens. Under certain circumstances, workers are capable of laying viable eggs. Because workers of most ant species have a degenerate spermatheca or lack the organ entirely, they cannot mate and store sperm. Their eggs, therefore, remain unfertilized and usually develop into males.[4] In most cases, however, workers do not reproduce at all in the presence of fertile queens.[5] The subfamily Ponerinae is of extreme interest in comparative studies of the social regulation of reproduction in ants; within this one subfamily, we find some species with distinct worker and queen castes, while in most other ponerine species, the dimorphism is only minor or else a queen caste does not exist at all, leaving the workers morphologically endowed for full functional reproduction.[6]

---

2 | R. Gadagkar, *Social Biology of* Ropalidia marginata (Cambridge, MA: Harvard University Press, 2001); S. Premmath, A. Sinah, and R. Gadagkar, "Dominance relationship in the establishment of reproductive division of labour in a primitively eusocial wasp (*Ropalidia marginata*)," *Behavioral Ecology and Sociobiology* 39(2): 125–132 (1996).

3 | C. Peeters, "Morphologically 'primitive' ants: comparative review of social characters, and the importance of queen–worker dimorphism," in J. C. Choe and B. J. Crespi, eds., *The Evolution of Social Behavior in Insects and Arachnids* (New York: Cambridge University Press, 1997), pp. 372–391.

4 | A few ant species reproduce by thelytokous parthenogenesis, in which unfertilized eggs yield females, and males may be absent.

5 | B. Hölldobler and E. O. Wilson, *The Ants* (Cambridge, MA: The Belknap Press of Harvard University Press, 1990); and J. Heinze, "Reproductive conflict in insect societies," *Advances in the Study of Behavior* 34: 1–57 (2004).

6 | C. Peeters, "Monogyny and polygyny in ponerine ants with or without queens," in L. Keller, ed., *Queen Number and Sociality in Insects* (New York: Oxford University Press, 1993), pp. 234–261; C. Peeters and F. Ito, "Colony dispersal and the evolution of queen morphology in social Hymenoptera," *Annual Review of Entomology* 46: 601–630 (2001); and B. Gobin, F. Itö, C. Peeters, and J. Billen, "Queen-worker differences in spermatheca reservoir of phylogenetically basal ants," *Cell and Tissue Research* 326(1): 169–178 (2006).

HARPEGNATHOS: LIFE CYCLE OF A COLONIAL ARCHITECT

To survey this remarkable array within the ponerine ants, a good point of departure is *Harpegnathos saltator*, a species of striking appearance that has been studied extensively in the field in southern India and in the laboratory.[7] The size difference between the queen and worker caste is moderate, due mainly to the presence of wing muscles in the queen (Figure 8-1). Queens and workers do not differ greatly in their ovarian anatomy. Both castes have the same number of ovarioles (usually eight), although the queen's ovarioles are approximately twice as long, and the queen's fecundity is correspondingly about twice that of a fertile worker.[8] In addition, all workers possess a fully functional spermatheca, and according to colony, 0 to 70 percent have mated, rendering them fully capable of becoming reproductives.[9] Yet it would obviously not be beneficial for the colony's reproductive efficiency if all mated individuals reproduced. Consider, then, how the solution to this problem has been achieved through the social regulation of reproduction built into the colony life cycle of *Harpegnathos saltator*.[10]

Sexually reproducing workers are called gamergates. Besides their occurrence in *Harpegnathos*, they coexist with queens only in a few other ponerine ants. The virgin queens seem normal: they are winged; they depart from the mother nest to mate outside with foreign males; and after mating, they shed their wings and excavate an incipient nest chamber. During this latter period, they hunt insects and spiders, which they paralyze with the venom of their sting. In this way, they are able to store live prey in the nest chamber, on which their larvae can feed at leisure.[11]

---

7 | J. Liebig, "Eusociality, female caste specialization, and regulation of reproduction in the ponerine ant, *Harpegnathos saltator* Jerdon" (Ph.D. thesis, University of Würzburg, Germany, 1998) (Berlin: Wissenschaft und Technik Verlag, 1998).

8 | C. Peeters, J. Liebig, and B. Hölldobler, "Sexual reproduction by both queens and workers in the ponerine ant *Harpegnathos saltator*," *Insectes Sociaux* 47(4): 325–332 (2000).

9 | This phenomenon has been found in a number of ponerine ants; see C. Peeters, "Monogyny and polygyny in ponerine ants with or without queens," in L. Keller, ed., *Queen Number and Sociality in Insects* (New York: Oxford University Press, 1993), pp. 234–261.

10 | C. Peeters and B. Hölldobler, "Reproductive cooperation between queens and their mated workers: the complex life history of an ant with a valuable nest," *Proceedings of the National Academy of Sciences USA* 92(24): 10977–10979 (1995).

11 | U. Maschwitz, M. Hahn, and P. Schönegge, "Paralysis of prey in ponerine ants," *Naturwissenschaften* 66(4): 213–214 (1979).

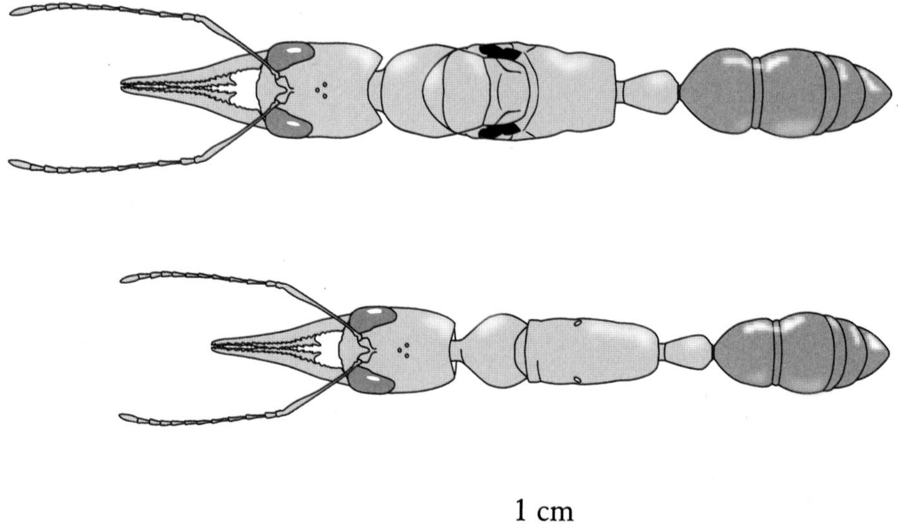

FIGURE 8-1. The female castes of the Asian ponerine *Harpegnathos saltator*, queen above and worker below. Based on C. Peeters, J. Liebig, and B. Hölldobler, "Sexual reproduction by both queens and workers in the ponerine ant *Harpegnathos saltator*," *Insectes Sociaux* 47(4): 325–332 (2000).

The incipient colony grows quite fast; at the end of the first year, 20 to 60 workers inhabit the nest.

With the emergence of her first adult daughters, the queen ceases foraging, but continues to lay eggs. With the number of workers increasing, the nest becomes more and more elaborate (Figure 8-2).[12] After 2 to 3 years, it has acquired an unusually complex architecture for that of a ponerine ant (Figure 8-3).[13] The inhabited part is built close to the ground surface. The uppermost chamber is protected by a thick vaulted roof, which is isolated from the surrounding soil by an empty space. As the colony grows, the vaulted roof is modified into a shell that encloses several superimposed chambers. Little openings, which may be encircled by molded flanges, are built in the upper region of the shell. The inside of the chamber is partly or completely lined with strips of empty cocoons. A refuse chamber is always found

12 | J. Liebig, "Eusociality, female caste specialization, and regulation of reproduction in the ponerine ant, *Harpegnathos saltator* Jerdon" (Ph.D. thesis, University of Würzburg, Germany, 1998) (Berlin: Wissenschaft und Technik Verlag, 1998).

13 | C. Peeters, B. Hölldobler, M. Moffett, and T. M. Musthak Ali, " 'Wall-papering' and elaborate nest architecture in the ponerine ant *Harpegnathos saltator*," *Insectes Sociaux* 41(2): 211–218 (1994).

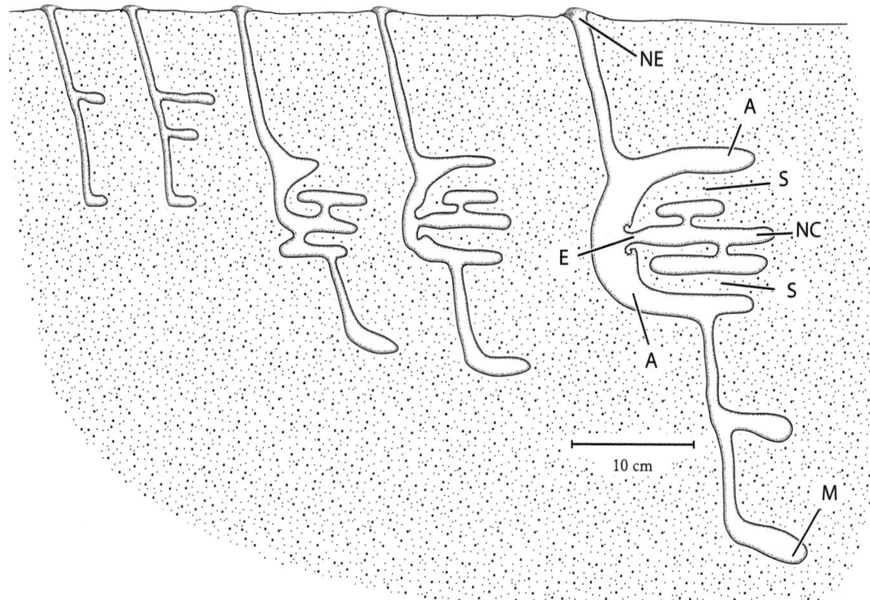

FIGURE 8-2. Growth of the nest of *Harpegnathos saltator*, based on excavations of nests in different developmental stages. A, atrium; E, entrance to the internal nest sphere; M, kitchen midden; NC, brood chambers; NE, nest entrance; S, nest sphere containing the brood chambers. *From left to right*: foundress nest; nest of 5 to 12 months, which consists of two floors; nest 1 to 1½ years old, which contains two brood chambers, but sphere and atrium are not yet complete; nest of 2 or more years; nest fully developed and shown in Figure 8-3. Based on J. Liebig, "Eusociality, female caste specialization, and regulation of reproduction in the ponerine ant, *Harpegnathos saltator* Jerdon" (Ph.D. thesis, University of Würzburg, Germany, 1998) (Berlin: Wissenschaft und Technik Verlag, 1998).

deeper than the inhabited chambers; it is filled with a moist, blackish brown mass of prey remains (crickets, moths, spiders, and other arthropods) and living commensal fly larvae (Milichiidae). The maggots eat the refuse, thereby preventing it from clogging up the chamber. The adult flies manage to travel from one nest to another by riding as hitchhikers on the back of homing *Harpegnathos* foragers (Figure 8-4). They then enter the refuse chambers and deposit their own eggs.

This strange architecture appears to be ideally suited to prevent flooding of the nest chambers. On the Indian subcontinent, where *Harpegnathos* occurs, there is a long dry period followed by intense monsoon rains, during which the large volumes of water saturate the soil. Waterlogged soil can destroy ordinary shallow ant nests, but the complex architecture of the *Harpegnathos* nests evidently plays a protective role. Although unfinished nests of young colonies most likely perish during the monsoon rains, the elaborate nest structures of adult colonies have a good chance

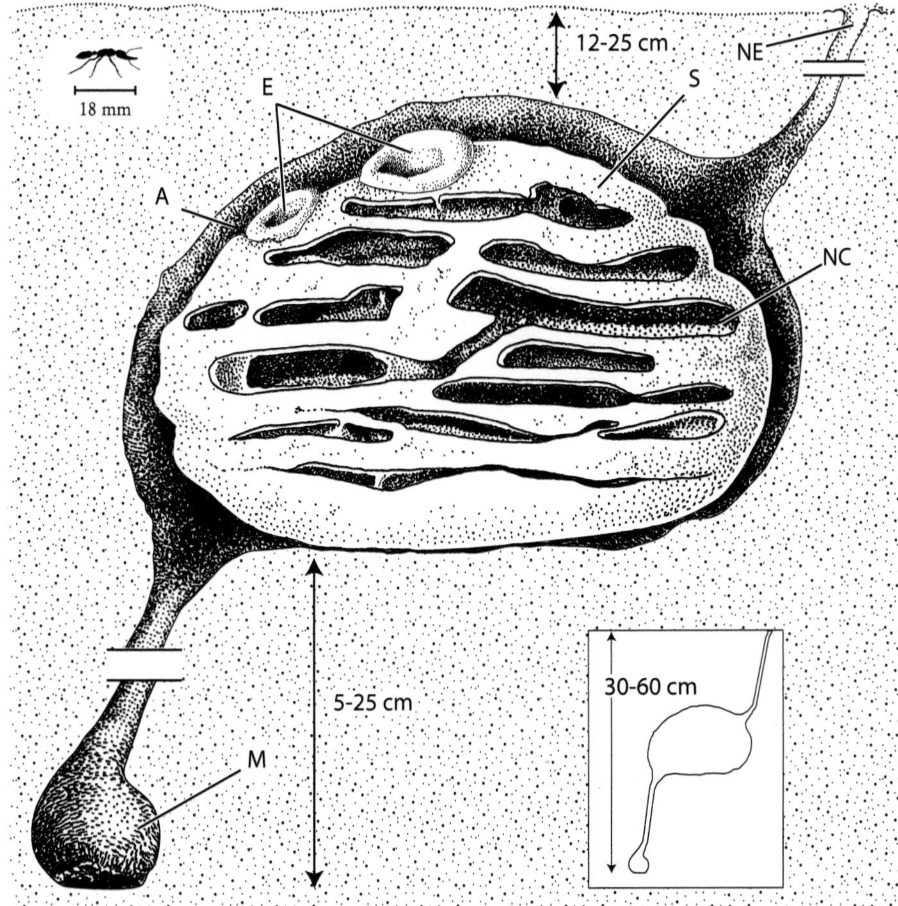

FIGURE 8-3. Mature nest of *Herpegnathos*. A, atrium; E, flange-shaped entrance to nest sphere; M, kitchen midden and flooding chamber; NC, nest and brood chambers; NE, nest entrance. The inside of the nest chambers are "wallpapered" with empty pupal cocoons and dry plant material. Based on C. Peeters, B. Hölldobler, M. Moffett, and T. M. Musthak Ali, "'Wall-papering' and elaborate nest architecture in the ponerine ant *Harpegnathos saltator*," *Insectes Sociaux* 41(2): 211–218 (1994).

of survival. *Harpegnathos* colonies thus create a very valuable and persistent microhabitat for themselves. In this way, they differ from most other ponerine species, which have simple nests and readily move from one nest site to another to meet emergencies.

The elaborate engineering of the nests has several consequences for the life cycle of *Harpegnathos*. The founding queen that survives the hazardous founding phase of her colony has a life span of 2 to 5 years. As long as she is fecund, she remains the only reproductive individual in the colony. But as her fecundity decreases, mated

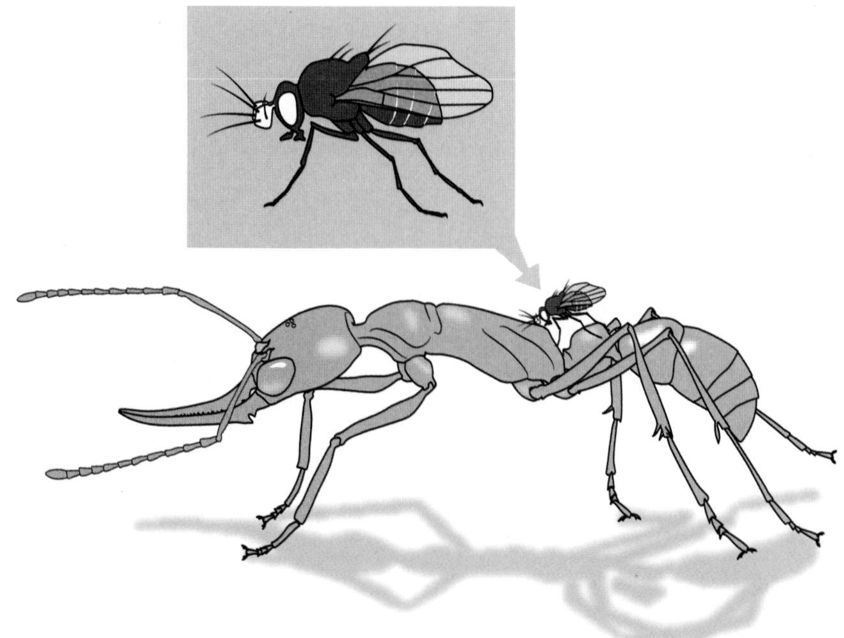

FIGURE 8-4. A milichiid fly hitches a ride on a foraging *Harpegnathos* worker. By this means, it will be transported back to the nest, where it can breed in the colony's refuse pile. Based on an unpublished original drawing by Malu Obermayer.

workers begin to battle for reproductive dominance. Finally, a group of gamergates (mated workers that reproduce) are established, and these become the top-ranked reproductive individuals. Many workers have mated with brothers inside the nest, but only a small group of those become active egg layers. Consequently, when the queen dies, the colony lives on. The foundress queen is succeeded by several gamergates, which themselves have a life span of 1 to 3 years. As soon as their fecundity wanes, they are replaced by other, younger workers, who have been monitoring the gamergates' fecundity and stand ready to challenge their reproductive rank.

Thanks to the rules of succession, gamergate-ruled colonies are potentially immortal. But how do the colonies propagate? Colony fission, which is common in other queenless ponerine ant species, does not occur in *Harpegnathos* colonies, because a specialized nest is needed for colony survival; any newly emigrated colony fragment with gamergates would lack an adequate nest for protection. In fact, no evidence of colony emigration in the field or laboratory has ever been found. Thus, a regular production of winged queens remains essential for the initiation of new colonies, and this is exactly what has been discovered in gamergate colonies: they continue every year to produce new winged queens, which leave the nest for mating and subsequent colony foundation (Plates 45 to 49).

Analyses of behavioral interactions within the *Harpegnathos* colonies have revealed intricate patterns of regulation of reproduction. As long as the foundress queen of each colony is present and fecund, only occasional antagonistic behavior occurs among workers. The queen appears to be the sole reproductive individual, and her offspring are reared by the workers. But as soon as the queen becomes senescent and suffers waning fecundity or dies, antagonistic tension among the workers appears and intensifies. After the queen's death, the monogynous society changes to a secondarily polygynous colony with gamergates, but only a fraction of the workers then become gamergates.

Consider next the role of individual behavior in reproductive regulation as well as its importance in the colony as a whole.[14] The occurrence and frequency of three agonistic interactions characterize the different levels of social stability in gamergate colonies. Each pattern has a distinct function. In *aggressive domination* (attack behavior), one ant stands over another ant and grasps it with the basal part of her mandibles, usually on the head or thorax (Figure 8-5). This is accompanied by vigorous downward jerks (usually fewer than 5, but sometimes more than 50). In most cases, the victim does not resist but instead crouches, and afterward she either remains immobile or runs away. This aggressive provocation is the most common type of antagonistic interaction. Gamergates direct such attacks at subordinate infertile workers, including those that have been mated.

The second form of agonistic behavior is termed *jump and hold. Harpegnathos* workers are able to leap forward several centimeters when capturing prey. In the context of policing interactions, a nestmate jumps 1 to 2 centimeters from a frontal position and grabs an opponent with the distal tip of her long mandibles. She then either proceeds with downward jerking movements or else immediately releases the other ant, which then crouches or walks away. In addition, an individual of lower rank often approaches an attacked nestmate from the side and grabs her at the thorax, neck, or petiole (Figure 8-6). The attacked ant is completely helpless and rarely escapes, despite offering strong resistance. She is often dragged around and held for hours.

---

14 | J. Liebig, "Eusociality, female caste specialization, and regulation of reproduction in the ponerine ant, *Harpegnathos saltator* Jerdon" (Ph.D. thesis, University of Würzburg, Germany, 1998) (Berlin: Wissenschaft und Technik Verlag, 1998).

PLATE 45. In the Indian jumping ant *Harpegnathos saltator*, the queen exhibits semiclaustral colony founding. *Above*: In the first phase of the founding process, the queen catches prey, which she paralyzes with her sting. The larvae feed on the prey progressively. *Below*: The first workers eclosing from the pupae are at the full size characteristic of large colonies, as opposed to dwarf workers (nanitics) found in evolutionarily more advanced ant species. (Photo: Jürgen Liebig.)

PLATE 46. As long as the *Harpegnathos* founding queen is fertile, she remains the sole reproductive in the colony.

Jump and hold behavior is directed mainly at workers in the process of becoming fertile, hence contenders for gamergate rank. The maneuver is clearly more a policing behavior than a dominance behavior, being launched mostly by low-ranking, infertile workers against nascent gamergates.[15]

The third agonistic behavior is the most intriguing: *antennal whipping* and *dueling*. An individual directs a rapid outburst of antennal lashes onto the head of her opponent. The latter ignores this challenge, or crouches down, or responds with antennal whipping of her own, thus engaging the challenger in a duel (Figure 8-7). The duels are quite complex. They begin when one ant lashes the other, then thrusts her body forward, causing the opponent to move backward for 5 to 10 millimeters.

---

15 | Such policing behavior, where infertile subordinates attack nascently fertile workers or destroy eggs laid by nestmates, is widespread in ants, particularly in ponerine ants. Among many reports, see T. Monnin and C. Peeters, Cannibalism of subordinates' eggs in the monogynous queenless ant *Dinoponera quadriceps*," *Naturwissenschaften* 84(11): 499–502 (1997); J. Liebig, C. Peeters, and B. Hölldobler, "Worker policing limits the number of reproductives in a ponerine ant," *Proceedings of the Royal Society of London B* 266: 1865–1870 (1999); B. Gobin, J. Billen, and C. Peeters, "Policing behaviour towards virgin egg layers in a polygynous ponerine ant," *Animal Behaviour* 58(5): 1117–1122 (1999); N. Kikuta and K. Tsuji, "Queen and worker policing in the monogynous and monandrous ant, *Diacamma* sp.," *Behavioral Ecology and Sociobiology* 46(3): 180–189 (1999); and P. D'Ettorre, J. Heinze, and F. L. W. Ratnieks, "Worker policing by egg eating in the ponerine ant *Pachycondyla inversa*," *Proceedings of the Royal Society of London B* 271: 1427–1434 (2004).

PLATE 47. *Above*: *Harpegnathos* workers, which have been marked individually in laboratory nests to track their behavior, frequently touch their antennae to the queen, apparently to monitor her fertility status. Below: A *Harpegnathos* male marked with a blue dot in a laboratory nest searches for freshly eclosed callow workers with whom it can mate.

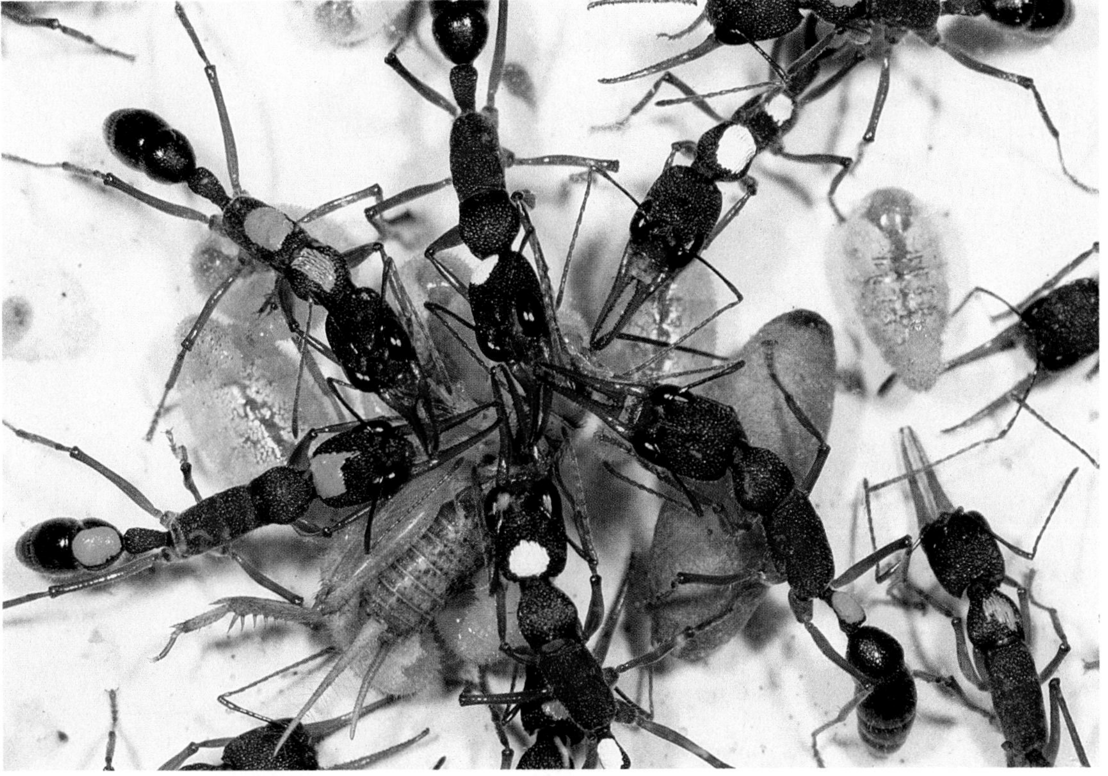

PLATE 48. *Above*: Although *Harpegnathos* workers amicably share captured prey, they engage in ritualized and escalated dominance fights when the queen's fertility wanes. *Below*: Competing ants (dotted white and green for recognition by the researcher) are shown in aggressive confrontation.

PLATE 49. *Above*: Dominant *Harpegnathos* workers become reproductives (gamergates), able to lay viable eggs. *Below*: Subordinate workers engage in brood care as well as nest maintenance and foraging.

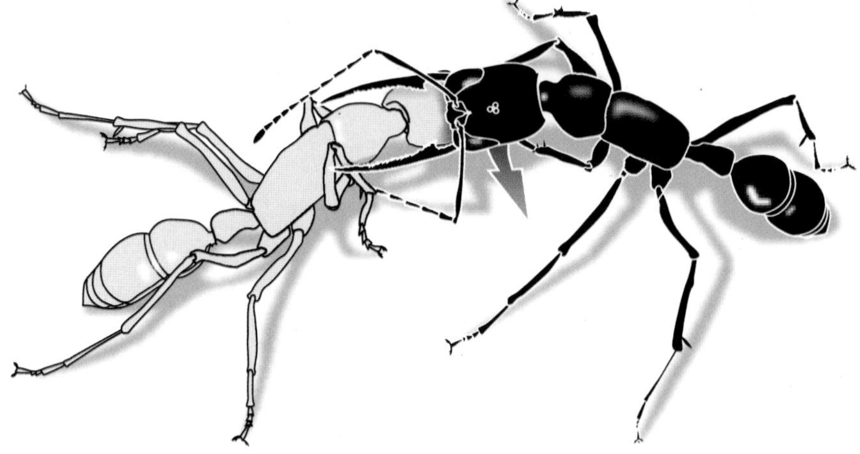

FIGURE 8-5. Aggressive domination by a *Harpegnathos* worker begins as it mounts its opponent, seizes the forepart of her body, and shakes it up and down. Based on photographs and videotapes provided by Jürgen Liebig.

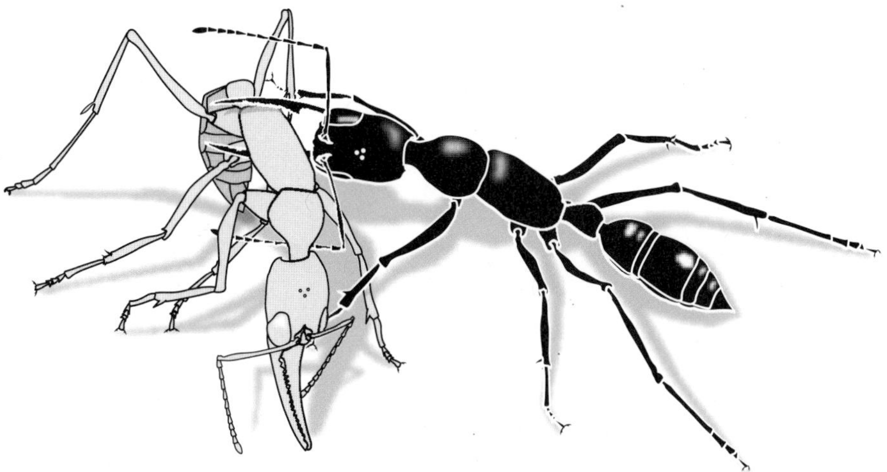

FIGURE 8-6. Policing of a *Harpegnathos* worker with partly developed ovaries. A nestmate (black) leaps to the side of an incipient fertile nestmate (gray) and seizes her by the thorax. Based on J. Liebig, C. Peeters, and B. Hölldobler, "Worker policing limits the number of reproductives in a ponerine ant," *Proceedings of the Royal Society of London B* 266: 1865–1870 (1999).

Then the entire process is reversed: the second ant now lashes the first ant and forces her to retreat. This odd pas de deux may be repeated up to 24 times, whereupon the two combatants simply walk away from each other. There is no obvious winner, and the whole performance appears to have been no more than a reaffirmation of social equality.

Such duels, in fact, occur most commonly between gamergates of equal rank.

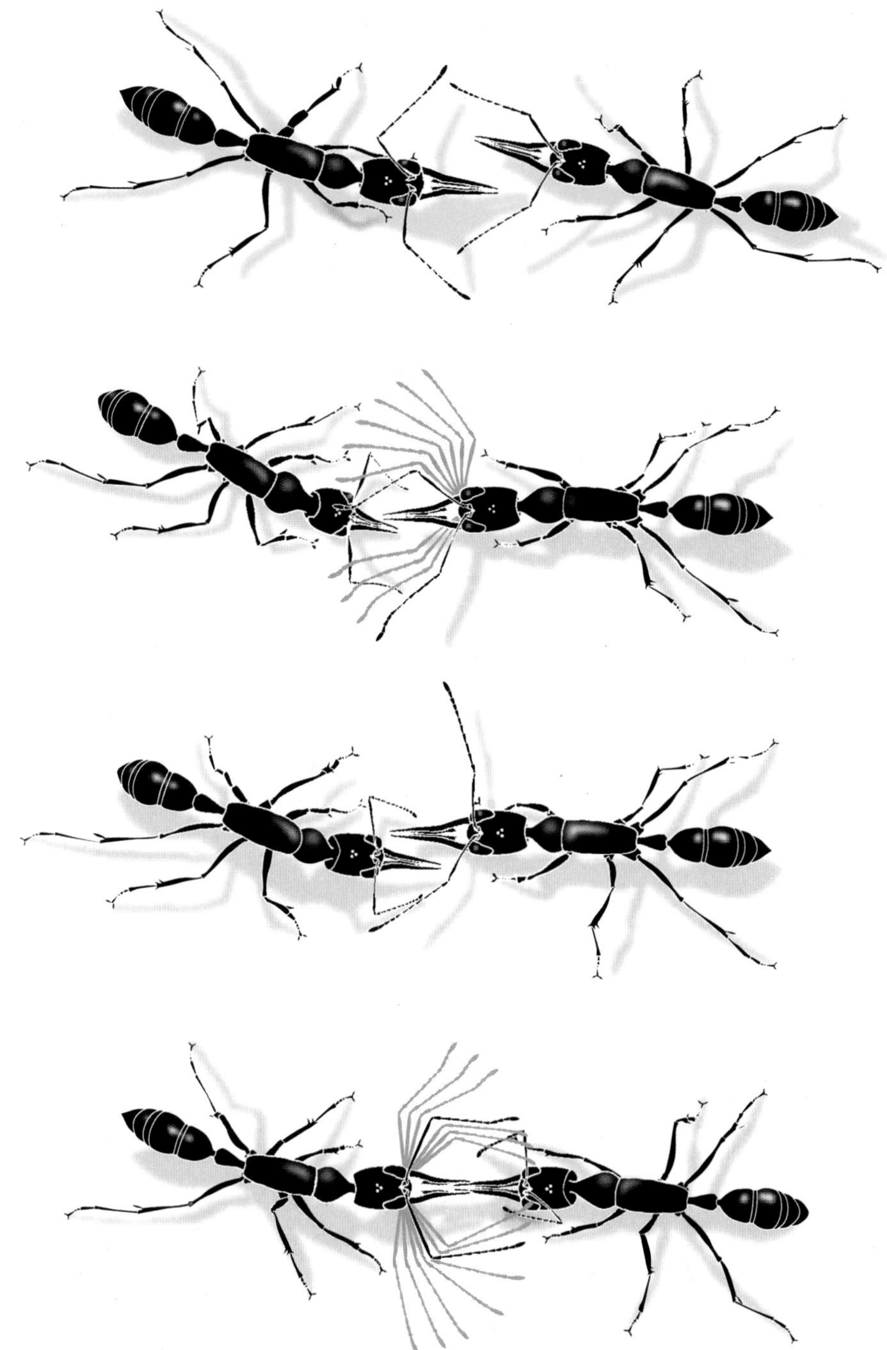

FIGURE 8-7. Antennal dueling between *Harpegnathos* workers. In the typical sequence depicted here (from top to bottom), the advancing worker lashes the backward-moving sister ant with her antennae. After the pair has moved about a body length, the procedure is reversed, and the lashed becomes the lasher. Based on videotapes provided by Jürgen Liebig.

Sometimes gamergates also initiate dueling with subordinate nascent fertile workers. The interactions sometimes escalate, with the gamergate jumping on and holding the subordinate worker. But subordinates also duel with one another, and other subordinates may jump and hold such dueling nestmates. Finally, young callow workers also engage in dueling with one another. The more assertive duelers have the greatest potential for later gamergate rank. Contests among subordinates always occur in bursts and stop only after the combatants have been assaulted and held repeatedly. In contrast, established gamergates engage in duels with each other only irregularly, at both a lower intensity and over more prolonged periods of time. The exchanges have no apparent consequence. Thus, the function of the duels can certainly not be attributed to an attempt of one gamergate to exclude the combatant from reproduction. To the contrary, the dueling among dominant gamergates may serve as a positive-feedback loop. Comparable procedures have been suggested by some writers as important for the molding of linear hierarchies in wasp societies.[16] Jürgen Liebig, who studied the agonistic interactions of *Harpegnathos saltator* in great detail, presents the following hypothesis.[17]

> Although linear hierarchies usually lead to the monopolization of reproduction by one individual, positive feedback loops may also work in polygynous societies. Imagine a linear hierarchy where the top ranking individual always dominates and thus receives positive feedback. As a consequence this individual should have the best developed ovaries. Subordinates are dominated by a varying number of higher ranking individuals, but also dominate a varying number of lower ranking nestmates. This will lead to a combination of positive and negative feedback on their status, which may result in an intermediate ovarian development. Lower ranking individuals are most often dominated, which inhibits ovarian development.

Harpegnathos workers have less reproductive potential than queens. Thus, to maintain the labor force needed for the colony to function, more than one worker must

---

16 | G. Theraulaz, E. Bonabeau, and J.-L. Deneubourg, "Self-organization of hierarchies in animal societies: the case of the primitively eusocial wasp *Polistes dominulus* Christ," *Journal of Theoretical Biology* 174(3): 313–323 (1995).

17 | J. Liebig, "Eusociality, female caste specialization, and regulation of reproduction in the ponerine ant, *Harpegnathos saltator* Jerdon" (Ph.D. thesis, University of Würzburg, Germany, 1998) (Berlin: Wissenschaft und Technik Verlag, 1998).

become a gamergate after the queen's death. According to Jürgen Liebig, one solution to this problem would be graduated egg laying corresponding to individual rank. But to maintain a finely tuned hierarchy would also require a large amount of aggression, with the result that colony productivity would be lessened. An alternative solution to this problem would be a symmetrical interaction among dominant individuals of equal rank in polygynous societies. They should be capable of aggressively interacting with other dominant individuals, with both participants receiving positive feedback. Such will occur if neither of the combatants is the loser in the interactions. Just this, in fact, is what happens in ritualized dueling among *Harpegnathos* gamergates.

Upon the severe decline or death of the foundress queen of the monogynous colony, several equally ranked gamergates assume the reproductive role. Instead of forming a linear hierarchy by dominating others, they engage in duels where no winner or loser can be determined, thus cementing their symmetrical interaction. The decrease in dueling activity can be expected to arise from a trade-off between the costs of the duels in terms of lost energy and the benefits in terms of received feedback. Nevertheless, every disturbance of the social stability in the colony instantly leads to an increase in dueling activity. Duels are most intense when workers try to become gamergates, but they do not commence until after the queen's decline or death or when senescent or deceased gamergates have to be replaced.[18]

It is important to remember that two behavioral mechanisms regulate reproduction in *Harpegnathos* societies. First, dominance interactions lead to the establishment of the reproductive oligarchy, consisting of a guild of gamergates; these individuals use dueling presumably to stimulate themselves and sustain their high reproductive rank. And second, policing subordinates prevent workers from moving prematurely to gamergate status.[19] Such individuals are often attacked and held down by infertile workers, an assault that has been demonstrated to inhibit fertility in precocious "hopeful reproductives." Policing behavior clearly maximizes colony efficiency and is favored by colony-level selection, whereas individual drive toward gamergate rank obviously is favored by individual-level direct selection. Colony efficiency can be expected to decrease if a certain ratio of reproductive individuals to infertile individuals

---

18 | J. Liebig, "Eusociality, female caste specialization, and regulation of reproduction in the ponerine ant, *Harpegnathos saltator* Jerdon" (Ph.D. thesis, University of Würzburg, Germany, 1998) (Berlin: Wissenschaft und Technik Verlag, 1998).

19 | J. Liebig, C. Peeters, and B. Hölldobler, "Worker policing limits the number of reproductives in a ponerine ant," *Proceedings of the Royal Society of London B* 266: 1865–1870 (1999).

is exceeded. A sufficient number of workers are needed to care for the brood, maintain the nest, and forage. A disproportionately high number of reproducing individuals would yield a surplus egg production and at the same time result in an insufficient worker force to raise the brood. Because gamergates do not share in worker tasks, such as foraging and brood care, more gamergates than needed for reproduction would be costly. Assuming that the costs diminish colony productivity, worker policing is selected as a means of maintaining colony efficiency.

Although the social organizations of queenright *Harpegnathos saltator* colonies, where the queen is also evidently singly mated, and polygynous gamergate colonies are very different, the social pattern of worker policing is similar. In queenright colonies, kin selection theory predicts that workers should prefer sons (life-for-life relatedness: $r = 0.5$) and nephews ($r = 0.375$) over their brothers ($r = 0.25$), but the situation differs for female offspring. Workers are more closely related to their sisters ($r = 0.75$) than to their inbred nieces ($r = 0.625$). The best relatedness–based policing strategy for workers would be to allow other worker nestmates to produce male offspring but prevent them from producing females. However, workers do not oviposit in queenright colonies, and in cases where there is premature ovarian development, workers police nestmates. One can argue that differential response to worker reproduction may easily lead to costly errors, such as incorrectly identifying the sex of eggs or first-instar larvae. On the other hand, it can also be argued that worker reproduction is detrimental to colony efficiency as long as the mother queen is sufficiently productive, and worker policing in queenright *Harpegnathos* colonies is, indeed, mainly due to colony-level (between-group) selection.[20]

In gamergate colonies of *Harpegnathos*, the relatedness pattern becomes even more complex due to the presence of several matrilines and partrilines. The limitation of the number of reproductives in a colony is most likely an important factor that favors worker policing behavior in *Harpegnathos saltator*. Without control of their ovarian activity, many mated workers could become gamergates. Thus, worker policing is most likely selected because it maintains colony efficiency, which is ultimately in the interest of every individual member of the colony. For example, without the continuous and labor-intensive maintenance of the elaborate nest structure, the whole colony would be doomed.

---

20 | J. Liebig, C. Peeters, and B. Hölldobler, "Worker policing limits the number of reproductives in a ponerine ant," *Proceedings of the Royal Society of London B* 266: 1865–1870 (1999).

To summarize to this point, worker oviposition in colonies of *Harpegnathos saltator* is regulated by highly directed aggressive interactions among nestmates, who can recognize different levels of ovarian activity. In fact, the regulation of the number of reproductives is only possible when the ovaries of the workers are still partly developed. As soon as their ovarian activity reaches a level similar to that of gamergates, infertile workers no longer differentiate between a newly ovipositing worker and a gamergate.[21] How, then, do workers recognize the condition of their nestmates?

Recent studies have documented the role of cuticular hydrocarbon (CHC) profiles as fertility indicators of individuals.[22] In *Harpegnathos saltator*, the differences in the profiles of both queens and workers are correlated with their physiological condition—in particular, the activity of their ovaries. Gamergates and reproductive queens can be clearly distinguished from nonreproductive individuals (Figure 8-8). Differences in age cannot explain these correlations, even though queens and gamergates of *Harpegnathos saltator* enjoy longer lives than their infertile worker nestmates. Moreover, the one relatively old gamergate that was assayed had a similar CHC profile to that of younger gamergates. There is also evidence that the CHC profile signaling a reproductive state reverts to the profile signaling the nonreproductive state after these individuals are policed and lose their rank. Thus, information about ovarian activity is encoded in the CHC profiles, which the ants use to assess each other's fertility. Further, ants can recognize the reproductive state of their nestmates solely on the basis of chemical information.[23, 24]

---

21 | J. Liebig, "Eusociality, female caste specialization, and regulation of reproduction in the ponerine ant, *Harpegnathos saltator* Jerdon" (Ph.D. thesis, University of Würzburg, Germany, 1998) (Berlin: Wissenschaft und Technik Verlag, 1998).

22 | C. Peeters, T. Monnin, and C. Malosse, "Cuticular hydrocarbons correlated with reproductive status in a queenless ant," *Proceedings of the Royal Society of London B* 266: 1323–1327 (1999); J. Liebig, C. Peeters, N. J. Oldham, C. Markstädter, and B. Hölldobler, "Are variations in cuticular hydrocarbons of queens and workers a reliable signal of fertility in the ant *Harpegnathos saltator*?" *Proceedings of the National Academy of Sciences USA* 97(8): 4124–4131 (2000); and V. Cuvillier-Hot, A. Lenoir, R. Crewe, C. Malosse, and C. Peeters, "Fertility signalling and reproductive skew in queenless ants," *Animal Behaviour* 68(5): 1209–1219 (2004).

23 | Physiological behavior analysis conducted on the Australian bulldog ant *Myrmecia gulosa* yielded unequivocal proof that the ants can in fact recognize differences in CHC profiles of conspecific individuals of different reproductive rank; see V. Dietemann, C. Peeters, J. Liebig, V. Thivet, and B. Hölldobler, "Cuticular hydrocarbons mediate discrimination of reproductives and nonreproductives in the ant *Myrmecia gulosa*," *Proceedings of the National Academy of Sciences USA* 100(18): 10341–10346 (2003).

24 | Furthermore, recent sensory studies on the ponerine ant *Pachycondyla inversa* demonstrate that the ants perceive the key compounds of the CHC fertility signal; see P. D'Ettorre, J. Heinze, C. Schulz, W. Francke, and

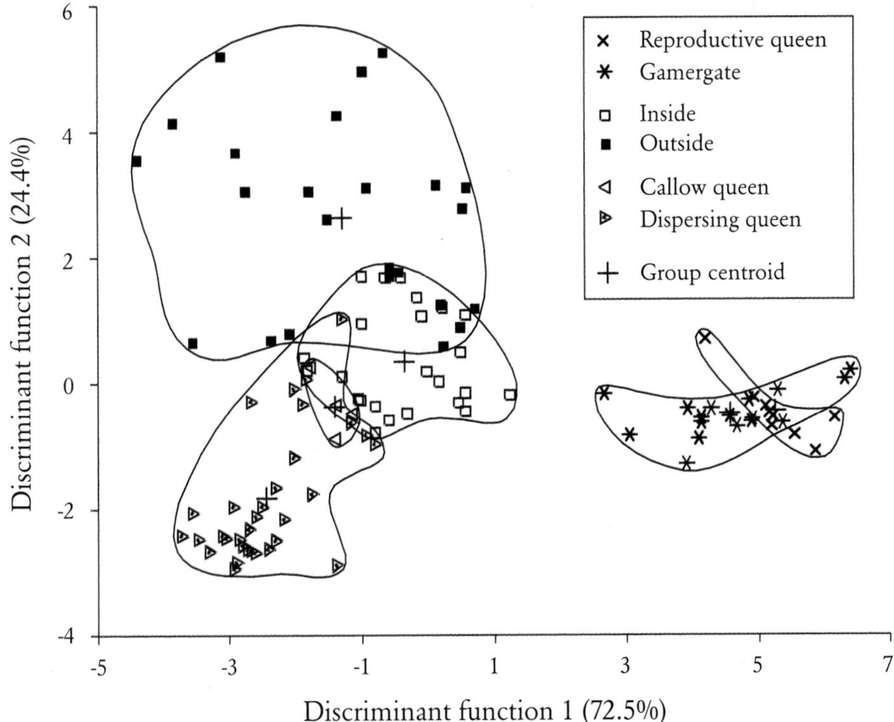

FIGURE 8-8. Discriminant analysis of the cuticular hydrocarbon (CHC) blends in *Harpegnathos saltator.* The analysis entails six groups of individuals (reproductive queens, gamergates, inside workers, outside workers, freshly eclosed callow virgin queens, and virgin queens ready for mating flight) and is based on 14 compounds. The encirclement of each group reflects that reproductive queens and gamergates are very similar in their hydrocarbon profiles, and these are clearly separated from those of virgin queens and workers. The hydrocarbon profiles of callow queens are significantly different from those of queens that depart from the nest for the mating flight and different from the profiles of workers; and the profiles of inside workers differ from those of outside workers. From J. Liebig, C. Peeters, N. J. Oldham, C. Markstädter, and B. Hölldobler, "Are variations in cuticular hydrocarbons of queen and workers a reliable signal of fertility in the ant *Harpegnathos saltator?*" *Proceedings of the National Academy of Sciences USA* 97(8): 4124–4131 (2000).

The most important feature of a fertility signal is its reliability, because workers benefit from helping the queen only if her productivity is sufficiently high. As her fertility drops, there is a point when workers will benefit more from their own reproduction. They can then be expected either to replace her or to start reproducing themselves. Because workers are more related to their own offspring than to that

---

M. Ayasse, "Does she smell like a queen? Chemoreception of a cuticular hydrocarbon signal in the ant *Pachycondyla inversa,*" *Journal of Experimental Biology* 207(7): 1085–1091 (2004).

of their sister nestmates, they should try to be among the reproductive individuals. It follows that workers should always carefully monitor the fertility and health of their mother in order to be prepared for reproductive competition once the queen dies. In *Harpegnathos saltator*, the function of a fertility signal is crucial, because queens and gamergates can also be replaced by other workers, which are ever ready as hopeful reproductives.

Cuticular hydrocarbon profiles have all the design features needed for a fertility signal. Cuticular hydrocarbons (CHCs) yield reliable information about ovarian developmental stage. There is only a small time lag between the change of ovarian activity and the change in the CHC profile, both for development of the reproductive profile and for its reversal. From circumstantial evidence, it appears that the CHC profile is a true fertility signal, selected in evolution to be a reliable indicator of the reproductive status of nestmates. In fact, the mutual control of egg-laying workers (worker policing[25]) in *Harpegnathos saltator* suggests that the chemical recognition of ovarian activity has also evolved by natural selection. Policing can be expected when worker reproduction imposes costs associated with reduced colony productivity. An individual worker may still benefit from her own reproduction in the presence of a queen or gamergates, even though her additional eggs lead to a marginal reduction of colony efficiency. However, she can succeed only if she is not detected and consequently policed by her nestmates. This is unlikely, and perhaps impossible, because she cannot conceal the modifications in her CHC profile.

The production of CHCs for the signaling of fertility entails energetic costs because a worker with incompletely developed ovaries is attacked and inhibited by nestmates in the presence of an egg-laying queen or gamergates. In contrast, the costs of the presumptive fertility signal to the colony is low, for the simple reason that neither a change of hydrocarbons nor their differential production is energetically expensive. The production of CHCs appears to be directly linked to fertility by the differential activity of certain enzymes and is thus part of the direct costs of reproduction.[26] The bearer of such chemicals, however, benefits if she is among the

---

25 | F. L. W. Ratnieks, "Reproductive harmony via mutual policing by workers in eusocial Hymenoptera," *American Naturalist* 132(2): 217–236 (1988).

26 | For more information on these physiological issues, see C. Schal, V. L. Sevala, H. P. Young, and J. A. S. Bachmann, "Sites of synthesis and transport pathways of insect hydrocarbons: cuticle and ovary as target tissues," *American Zoologist* 38: 382–393 (1998); Y. Fan, J. Chase, V. L. Sevala, and C. Schal, "Lipophorin-facilitated

established reproductives in the colony—that is, if she produces enough eggs to benefit the colony. It follows that such chemicals tuned to ovarian activity can be viewed as a pure fertility signal. It represents an honest index or a badge from the viewpoint of the colony as a whole.[27]

DINOPONERA: GIANT "WORKER QUEENS"

In contrast to *Harpegnathos saltator*, whose gamergate colonies are polygynous, the queenless ponerine ant *Dinoponera quadriceps* is monogynous. Thibaud Monnin and Christian Peeters have made detailed studies of the social organization and regulation of reproduction of the latter, Neotropical species, which is the world's largest ant.[28] All workers are morphologically equivalent, and reproduction is regulated by a near-linear dominance hierarchy involving from 5 to 10 workers among the approximately 40 to 140 workers composing the colony. Only one individual (the alpha worker) is mated and able to reproduce. The highest-ranking workers appear to recognize one another's status. The alpha worker uses several stereotyped displays to maintain dominance over other high-ranking nestmates (Figure 8-9). Among the most conspicuous is *blocking*, in which the aggressor stretches her antennae along either side of the head of the target worker. The latter then takes a more or less crouched position. If the target worker moves, the alpha often repositions herself in front of her and strikes her head with an antenna. An even more remarkable display is *gaster rubbing*, in which the dominant worker grabs one antenna of a target worker and rubs it against her gaster (the posteriormost part of the body) bent

---

hydrocarbon uptake by oocytes in the German cockroach, *Blattella germanica* (L.)," *Journal of Experimental Biology* 205(6): 781–790 (2002); and Y. Fan, L. Zurek, M. J. Dykstra, and C. Schal, "Hydrocarbon synthesis by enzymatically dissociated oenocytes of the abdominal integument of the German cockroach, *Blattella germanica*," *Naturwissenschaften* 90(3): 121–126 (2003).

27 | J. Liebig, "Eusociality, female caste specialization, and regulation of reproduction in the ponerine ant, *Harpegnathos saltator* Jerdon" (Ph.D. thesis, University of Würzburg, Germany, 1998) (Berlin: Wissenschaft und Technik Verlag, 1998); J. Liebig, C. Peeters, N. J. Oldham, C. Markstädter, and B. Hölldobler, "Are variations in cuticular hydrocarbons of queens and workers a reliable signal of fertility in the ant *Harpegnathos saltator*?" *Proceedings of the National Academy of Sciences USA* 97(8): 4124–4131 (2000).

28 | T. Monnin and C. Peeters, "Monogyny and regulation of worker mating in the queenless ant *Dinoponera quadriceps*," *Animal Behaviour* 55(2): 299–306 (1998); and T. Monnin and C. Peeters, "Dominance hierarchy and reproductive conflicts among subordinates in a monogynous queenless ant," *Behavioral Ecology* 10(3): 323–332 (1999).

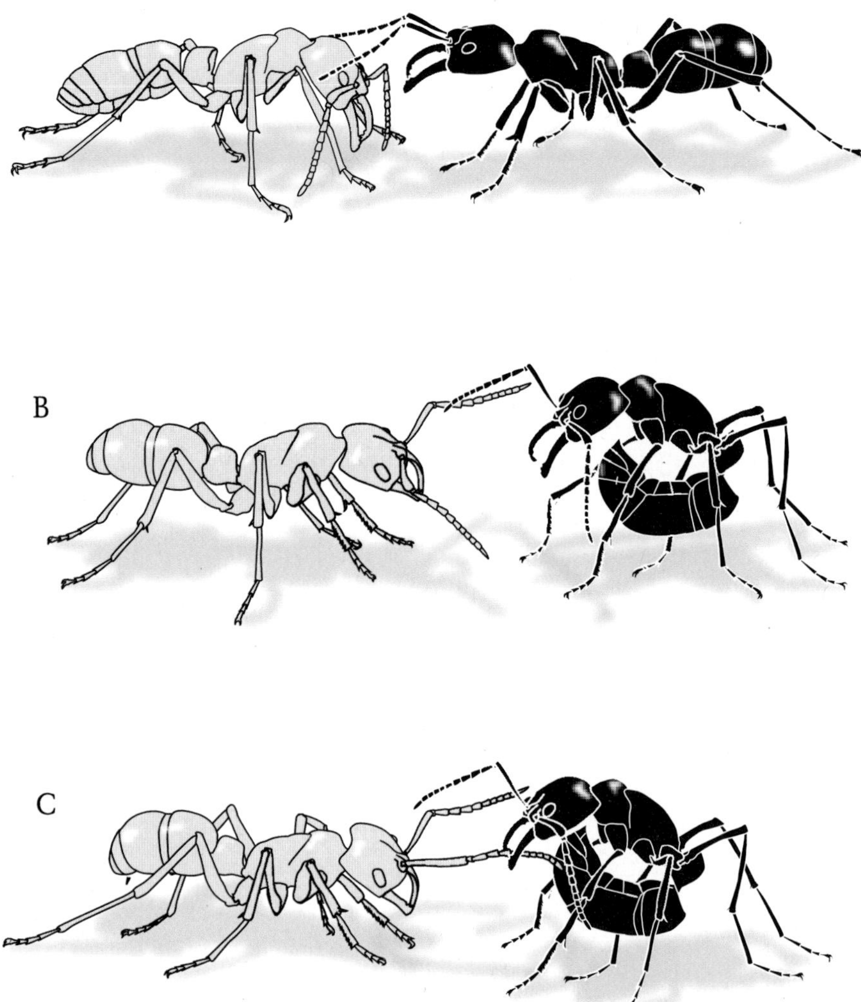

FIGURE 8-9. Dominance behavior by a worker of the giant South American ponerine ant *Dinoponera quadriceps*. *Above*: Blocking behavior. *Middle*: Gaster curling. *Below*: Gaster rubbing, whereby the dominant individual (black) grabs the antenna of the subordinate and rubs it on her gaster. Based on T. Monnin and C. Peeters, "Dominance hierarchy and reproductive conflicts among subordinates in a monogynous queenless ant," *Behavioral Ecology* 19(3): 323–332 (1999).

forward. A third display is *gaster curling*; the dominant individual bends her gaster forward between her legs, exposing its dorsal region to the subordinate. We have observed a similar case of such an aggressive display during dominance interaction in the African ponerine ant *Platythyrea cribrinoda* (Plate 50). The blocking display is most often performed by the alpha toward the beta individual.

PLATE 50. Dominance display by a worker of the ponerine ant *Platythyrea cribrinoda*. Note the unusual way workers of this species fold their forelegs behind their thorax. Note the black cocoons on which the ants are standing. In most ant species, the cocoon has a light brown color.

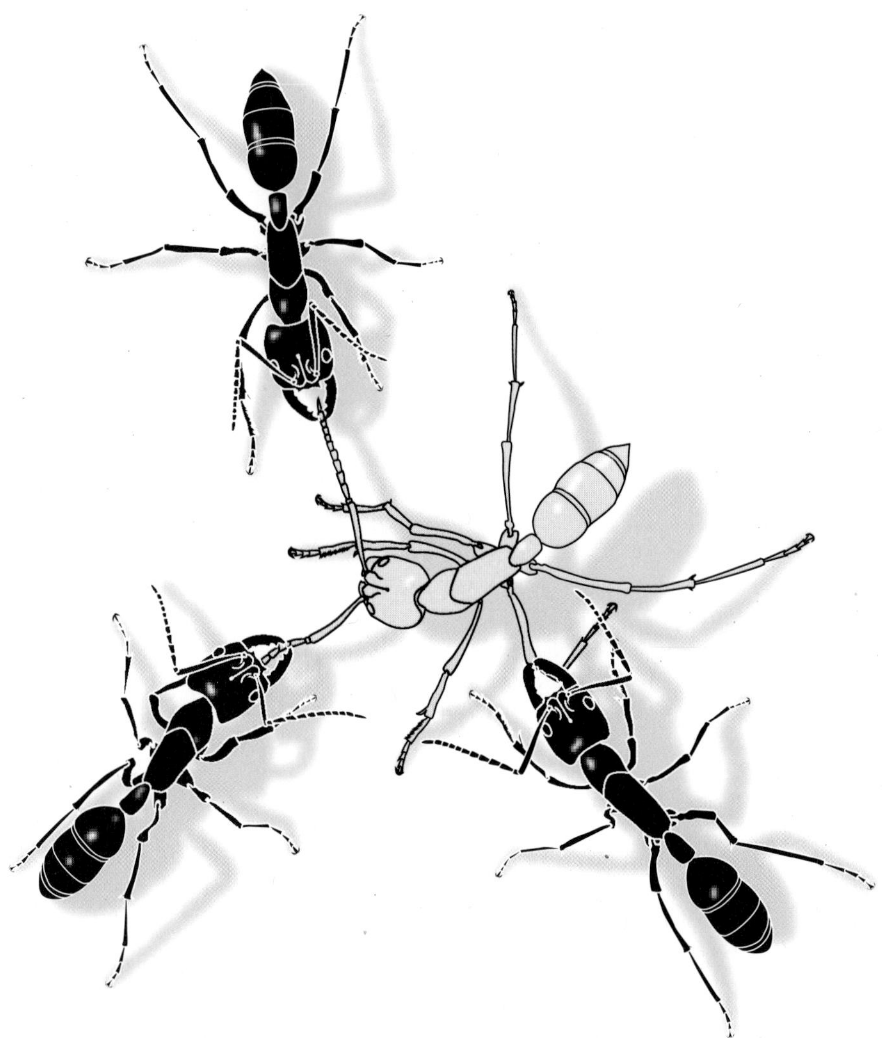

FIGURE 8-10. Policing in *Dinoponera quadriceps.* Workers (black) hold and spread-eagle a nestmate (gray) who is prematurely fertile. This immobilization negatively affects ovarian development in the policed individual. Based on T. Monnin and C. Peeters, "Dominance hierarchy and reproductive conflicts among subordinates in a monogynous queenless ant," *Behavioral Ecology* 10(3): 323–332 (1999).

The frequency of oviposition and number of eggs laid in *Dinoponera* colonies is correlated with rank, and only the alpha individual has mated. The alpha selectively destroys eggs laid by high-ranking subordinates. The likelihood that such a subordinate replaces a dead or declining alpha is also correlated with her previous rank. In contrast to high-ranking individuals, low-ranking subordinates are no longer able to become alphas. Nevertheless, they are involved in the regulation of reproduction;

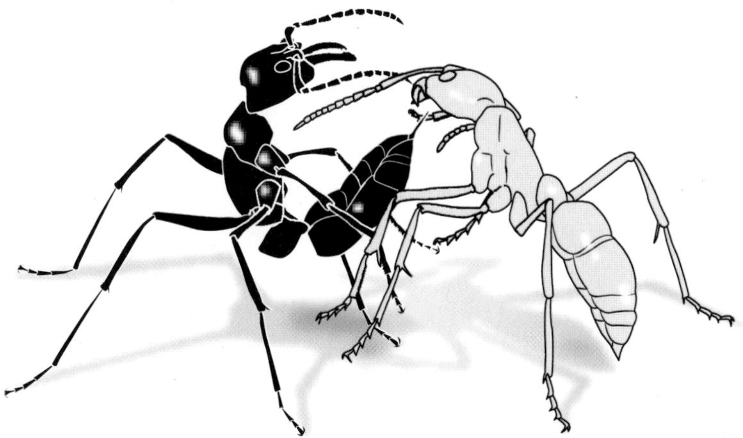

FIGURE 8-11. Another dominance interaction in *Dinoponera quadriceps* is the so-called "sting smearing." The dominant individual (gamergate, black) contaminates a subordinate nestmate that "pretends" a dominant position with secretion from the sting accessory gland (Dufour's gland). This marked pretender is now attacked and immobilized by low-ranking workers, as shown in Figure 8-10. Based on T. Monnin, F. L. W. Ratnieks, G. R. Jones, and R. Beard, "Pretender punishment induced by chemical signalling in a queenless ant," *Nature* 419: 61–64 (2002).

they sometimes immobilize high rankers, thus policing nestmates that might prematurely develop alpha traits (Figure 8-10).[29] In fact, the alpha and low-ranking subordinates commonly cooperate in this policing effort. If the alpha is seriously challenged by another high-ranking female, she chemically marks the pretender, who is subsequently grabbed, immobilized, and sometimes even killed by subordinates (Figure 8-11).[30] Under experimental conditions at least, the secretions used by the dominant gamergate, which originates in her Dufour's gland, induces immobilization of the contaminated ant significantly more often than do secretions from beta or low-ranking ants. Further, beta ants retain their high rank significantly more often following contamination with gamergate secretions than is the case with secretions from low-ranking nongamergates. The glands of gamergates have significantly more hydrocarbons than those of nongamergates, but not more than the glands of betas. Gamergate glands contain a chemical mixture with a higher proportion of high-molecular-mass hydrocarbons than do low-ranker glands, with beta glands

29 | T. Monnin and F. L. W. Ratnieks, "Policing in queenless ponerine ants," *Behavioral Ecology and Sociobiology* 50(2): 97–108 (2001).

30 | T. Monnin, F. L. W. Ratnieks, G. R. Jones, and R. Beard, "Pretender punishment induced by chemical signalling in a queenless ant," *Nature* 419: 61–65 (2002).

intermediate. These chemical data and bioassay results suggest that Dufour's gland secretions are a signal that induces policing, and that only the gamergate produces sufficient quantities or the correct composition of this particular glandular secretion to induce this form of aggressive behavior.

These fascinating observations parallel the behavior observed in queens of the myrmicine ant *Leptothorax gredleri*, which invade and then usurp conspecific nests by smearing resident queens with Dufour's gland secretions. The alien pheromones then induce resident workers to attack their own queen.[31] In this case, glandular secretions manipulate worker behavior in a nonadaptive way (the workers lose inclusive fitness); still, the operative behavioral mechanisms are very similar.

The pattern of policing in *Dinoponera quadriceps* is independent of relatedness. It does not matter whether the dominant gamergate is the mother of her nestmates (and hence their coefficient of relatedness to one another is 0.75), or whether she has more distant related relatives attending her. Thus, in *Dinoponera*, as in *Harpegnathos*, policing is best explained as the result of between-group (colony-level) selection. Premature replacement of a fully fertile alpha individual clearly is detrimental to colony efficiency, because such untimely transition at the top rank causes colony disruption.[32]

In *Dinoponera quadriceps*, as in *Harpegnathos saltator* (with queenright monogynous colonies that transform into polygynous gamergate colonies), the strictly monogynous form of reproduction is enforced by two distinct modes of agonistic interactions. One is dominance behavior that establishes high-rank positions or, in the case of *Dinoponera* colonies, a dominance hierarchy among the high-ranking individuals. The other mode of agonistic interaction is policing, a control practiced by low-ranking subordinates. Policing by subordinates ensures an efficient ratio of reproductives to nonreproductive nestmates in the colony, the case in *Harpegnathos*, or a limit of high-ranking individuals and even the monopoly of reproduction by the alpha ant, as in *Dinoponera* (Figure 8-12).

Such intricate interindividual behavioral exchanges require a communication

---

31 | J. Heinze, B. Oberstadt, J. Tentschert, B. Hölldobler, and H. J. Bestmann, "Colony specificity of Dufour gland secretions in a functionally monogynous ant," *Chemoecology* 8(4): 169–174 (1998).

32 | T. Monnin and C. Peeters, "Monogyny and regulation of worker mating in the queenless ant *Dinoponera quadriceps*," *Animal Behaviour* 55(2): 299–306 (1998); and T. Monnin and C. Peeters, "Dominance hierarchy and reproductive conflicts among subordinates in a monogynous queenless ant," *Behavioral Ecology* 10(3): 323–332 (1999). See also T. Monnin and F. L. W. Ratnieks, "Reproduction versus work in queenless ants: when to join a hierarchy of hopeful reproductives," *Behavioral Ecology and Sociobiology* 46(6): 413–422 (1999).

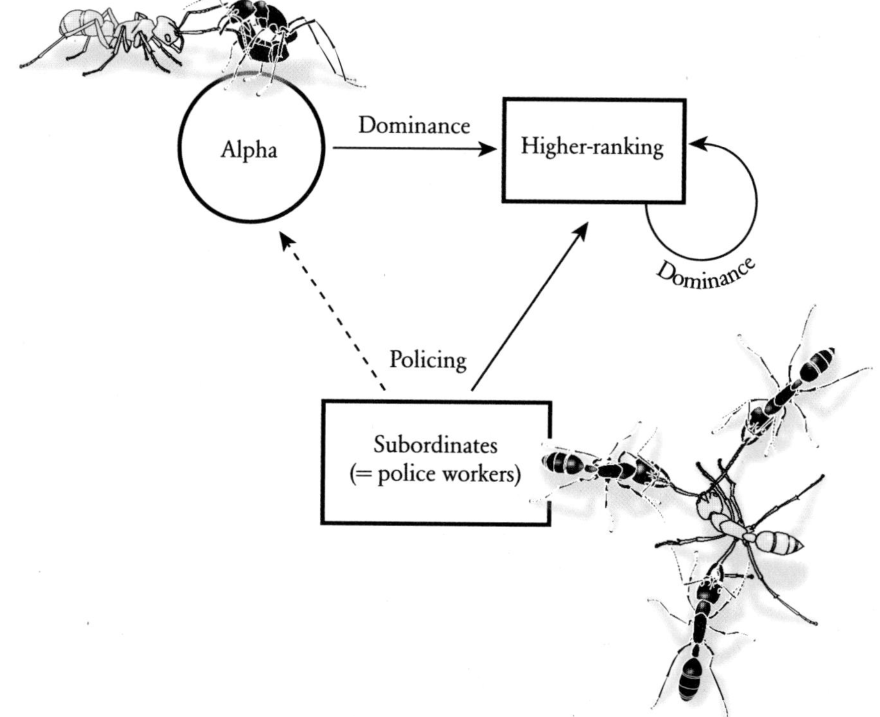

FIGURE 8-12. In *Dinoponera quadriceps*, a balance between female reproductives and the outwardly identical workers is maintained by individual dominance, rank, and policing. Based on an unpublished diagram provided by Christian Peeters.

system that signals rank and function within the colony, as illustrated so clearly in *Harpegnathos*. The use of Dufour's gland secretions by the *Dinoponera* gamergate to mark alpha pretenders and elicit policing behavior in nestmates against this individual is also such a system. However, it turns out that still other signals are used.

In *Dinoponera quadriceps,* dominance ranks are reflected clearly in the ovarian activity differences among alpha, beta, and other workers.[33, 34] Chemical analyses of CHCs demonstrate a strong correlation among workers between the relative proportion of the compound 9-hentriacontene (9-C_{31}) and egg-laying behavior.

---

33 | T. Monnin, C. Malosse, and C. Peeters, "Solid-phase microextraction and cuticular hydrocarbon differences related to reproductive activity in queenless ant *Dinoponera quadriceps*," *Journal of Chemical Ecology* 24(3): 473–490 (1998).

34 | C. Peeters, T. Monnin, and C. Malosse, "Cuticular hydrocarbons correlated with reproductive status in a queenless ant," *Proceedings of the Royal Society of London B* 266: 1323–1327 (1999).

This pattern has been confirmed by the increase in 9-C_{31} on the surface of the high-ranking worker that assumes alpha status following the experimental removal of the established alpha individual. Furthermore, 6 to 8 weeks are needed for the replacement alpha worker to exhibit a relative proportion of 9-C_{31} similar to that of the former alpha individual. Quite possibly, this interval corresponds to the time needed for the ovaries to become fully developed.

These data and other, similar findings on *Harpegnathos saltator* strongly suggest a relationship between ovarian activity and the production of long-chain hydrocarbons. In fact, such a relationship has also been found in some species of solitary insects, where the onset of oogenesis (vitellogenesis) is accompanied by the synthesis of cuticular hydrocarbon blends that serve as sex pheromones.[35]

Similar dominance interactions and reproductive hierarchies occur in other social hymenopterans, in particular *Polistes dominulus* and *Bombus hypnorum*. There the CHCs of the alpha individuals differ in the relative proportions of long-chain hydrocarbons.[36] What is unique in *Dinoponera quadriceps*, however, is the high level in the alpha ant of 9-C_{31}, as opposed to the other cases, where dominants are distinguished by a shift in their CHC profile to the long-chained assembly of CHCs. Yet as further investigations on *Dinoponera quadriceps* have shown, high levels of 9-C_{31} are not a prerequisite for procuring alpha rank. What counts is the ability to win physical contests; these, in turn, are a reliable correlate of the reproductive capability of participants.[37] *Dinoponera* betas compete aggressively with other high-ranked workers to replace the gamergate when the latter either dies or is experimentally removed. The new alpha performs more frequent gaster rubbing and blocking displays; these are directed at the new beta individual, even though such behaviors do not yet deliver the stimulus of high levels of 9-C_{31}. In other words, elevated levels of 9-C_{31} are not a prerequisite for becoming the new alpha individual. Once a worker

---

35 | See discussions in C. Peeters, T. Monnin, and C. Malosse, "Cuticular hydrocarbons correlated with reproductive status in a queenless ant," *Proceedings of the Royal Society of London B* 266: 1323–1327 (1999).

36 | A. Bonavita-Cougourdan, G. Theraulaz, A.-G. Bagnères, M. Roux, M. Pratte, E. Provost, and J.-L. Clément, "Cuticular hydrocarbons, social organization and ovarian development in a polistine wasp: *Polistes dominulus* Christ," *Comparative Biochemistry and Physiology B* 100(4): 667–680 (1991); and M. Ayasse, T. Marlovits, J. Tengö, T. Taghizadeh, and W. Francke, "Are there pheromonal dominance signals in the bumblebee *Bombus hypnorum* L. (Hymenoptera, Apidae)?" *Apidologie* 26(3): 163–180 (1995). See also H. A. Downing and R. L. Jeanne, "Communication of status in the social wasp *Polistes fuscatus* (Hymenoptera:Vespidae)," *Zeitschrift für Tierpsychologie* 67(1–4): 78–96 (1985).

37 | M. J. West-Eberhard, "Sexual Selection, social competition, and evolution," *Proceedings of the American Philosophical Society* 123(4): 222–234 (1979).

has established herself in the alpha position, she can be mated; and with the onset of vitellogenesis that follows, she then exhibits increasing levels of 9-C_{31} in her cuticular hydrocarbon blends. Thus, 9-C_{31} serves as a fertility signal embedded within the cuticular hydrocarbon profile of the alpha ant. Such fertility signals are "honest signals," insofar as they are physiologically linked to ovarian development. They play a key role in agonistic interactions in the colony, especially in the mediation of policing behavior by infertile nestmates. Such interactions that lead to the establishment of dominance hierarchies nevertheless depend exclusively on aggressive ability, and the signaling of fertility status is secondary. Still, both competence of aggression and information about fertility are ultimately essential for social regulation of reproduction in *Dinoponera quadriceps* colonies.[38]

The significance of agonistic behavioral interactions becomes especially apparent in *Dinoponera* colonies that have lost their gamergate due to death or to fission during colony propagation. In such orphaned colonies, aggressive interactions among some of the unmated nestmates lead to dominance hierarchies and the behavioral establishment of a top-ranking alpha worker. The virgin alpha individual then leaves the nest at night and mates nearby with foreign males, whom she apparently attracts with chemical signals.[39] The full nature of sexual calling in *Dinoponera quadriceps* is not yet known, but in at least a few other ponerine queenless ants, including *Rhytidoponera metallica*, workers ready to mate display a specific calling behavior, including the release of sex pheromones from the pygidial gland or other exocrine glands.[40] However they manage it, *Dinoponera quadriceps* males readily recognize virgin alpha workers as opposed to other workers. Thibaud Monnin and Christian Peeters give this account of the stereotyped mating behavior: "Immediately after touching the alpha with his antennae, a male changed his behavior radically: he followed her while vibrating his antennae and boxing her head and antennae, and tried to mount her. When the alpha was willing, intromission occurred very soon after the first antennal contact. The alpha soon went back into the nest dragging the male with her. Then she bent her abdomen forward, placing the male in front of her, and cut off the end of his abdomen within 1–2 min." Parts of the male's

---

38 | C. Peeters, T. Monnin, and C. Malosse, "Cuticular hydrocarbons correlated with reproductive status in a queenless ant," *Proceedings of the Royal Society of London B* 266: 1323–1327 (1999).

39 | T. Monnin and C. Peeters, "Monogyny and regulation of worker mating in the queenless ant *Dinoponera quadriceps*," *Animal Behaviour* 55(2): 299–306 (1998).

40 | B. Hölldobler and C. P. Haskins, "Sexual calling in primitive ants," *Science* 195: 793–794 (1977).

sexual apparatus remained attached to the alpha worker's genital tract for 15 to 73 minutes, until she removed it in entirety. When another male was placed with a freshly mated female, he did not try to remove the genital remains of the first male. This anatomical plug thus ensured a single insemination. Once mated, moreover, gamergates do not leave the nest again. Even when gamergates were experimentally brought in contact with males, no additional mating attempts ever ensued.

QUEENS, WORKERS, GAMERGATES IN PERMUTATIONS

The mating patterns of polygynous gamergate societies, unlike those of the monogynous queenless *Dinoponera quadriceps*, are not regulated by dominance interactions. For example, all young *Harpegnathos saltator* workers are sexually attractive and as many as 70 percent mate with male nestmates. A group of mated workers are then established as reproductives by dominance interactions, with the majority of the mated workers remaining sterile. A similar, though not identical, procedure has been discovered in *Pachycondyla tridentata.* Colonies of this Malaysian ponerine ant occur with queens or without them. More than 80 percent of the workers can be mated, regardless of the presence or absence of queens. As a result, queens and workers compete equally for reproduction. Although several ants lay eggs, continuous dominance interactions occur among colony members. A small number, not necessarily including a queen, occupy the top tier.[41]

The coexistence of queens and gamergates in colonies of the same species population have also been reported in two species of the ponerine genus *Gnamptogenys.* In neither do queens and gamergates appear to coexist in the same colony. In *Gnamptogenys striatula*, from northeastern Brazil, a small number of functional queens can be found in the same colony, unaccompanied by gamergates. However, several days after the queens have been experimentally removed, some workers exhibit a typical sexual calling behavior inside the laboratory nest. Other workers search for foreign males placed into the foraging arena and transport them to the nest chambers, where they mate with the workers that are sexually advertising themselves by the release of pheromones. Behavioral observations suggest that the

41 | K. Sommer, B. Hölldobler, and K. Jessen, "The unusual social organization of the ant *Pachycondyla tridentata* (Formicidae, Ponerinae)," *Journal of Ethology* 12(2): 175–185 (1994).

gamergates originate from the guild of nurse workers, presumably young virgin workers. Both queenright colonies and queenless gamergate colonies of *Gnamptogenys striatula* are polygynous. How the number of reproductive individuals is regulated remains unknown.[42]

The regulation of reproductives has been successfully investigated in *Gnamptogenys menadensis*, from South Sulawesi (Indonesia). Only a minority of colonies in the population have been found with multiple queens. In most colonies, multiple workers were found instead to have mated and reproduced. Although the ovaries of many virgin workers in both cases were active, they produced specialized trophic eggs that were then fed to larvae. However, when the gamergates were removed from a colony experimentally, some of the virgin workers switched to laying male eggs. When gamergates were reintroduced, sterile workers attacked and immobilized the newly fertile workers, often causing them to die. In such experiments, gamergates themselves never showed aggression toward the new egg layers. Worker policing alone directed at reproducing workers ensured that virgin workers laid only trophic eggs in the presence of gamergates or queens.[43]

In yet another variation in the vast array of schemes, a totally polygynous gamergate system with no dominance interactions whatsoever has been discovered in the African ponerine species *Pachycondyla* (formerly *Ophthalmopone*) *berthoudi*, where nearly all young workers are available for reproduction during the mating season. Foreign males enter the nests and seek out the young ants for mating. All mated workers oviposit; there is no indication that egg production is in any way controlled by dominance interaction like that seen in *Harpegnathos saltator*, *Pachycondyla tridentata*, or species of *Gnamptogenys*.[44] How excessive colony reproduction and loss of labor are controlled—and presumably such must be the case—is unknown.

Colony propagation in queenless gamergate societies occurs by colony fission. In polygynous systems, the newly separated colony fractions have an excellent chance to include mated workers. In monogynous societies, such as *Dinoponera quadriceps*,

---

42 | R. Blatrix and P. Jaisson, "Optional gamergates in the queenright ponerine ant *Gnamptogenys striatula* Mayr," *Insectes Sociaux* 47(2): 193–197 (2000).

43 | B. Gobin, J. Billen, and C. Peeters, "Policing behaviour towards virgin egg layers in a polygynous ponerine ant," *Animal Behaviour* 58(5): 1117–1122 (1999); and B. Gobin, J. Billen, and C. Peeters, "Dominance interactions regulate worker mating in the polygynous ponerine ant *Gnampogenys menadensis*," *Ethology* 107(6): 495–508 (2001).

44 | C. Peeters and R. Crewe, "Worker reproduction in the ponerine ant *Ophthalmopone berthoudi*: an alternative form of eusocial organization," *Behavioral Ecology and Sociobiology* 18(1): 29–37 (1985).

the new colony that ends up without a gamergate relies on new dominance interactions to determine the new alpha, who then mates and assumes her role. In other polygynous gamergate societies, such as older *Harpegnathos saltator* colonies, propagation depends exclusively on winged queens, who are produced annually and leave the nest for mating and subsequent colony founding. Finally, in species with coexisting gamergates and queens, such as *Pachycondyla tridentata*, some queens may leave the nest, mate, and found colonies independently, while others stay in the nest and mate like workers with male nestmates. An interesting sidelight has been discovered in laboratory colonies: the males' wings are often bitten off, presumably by workers who in this way prevent the males' departure and secure their own insemination (Plate 51).[45] Colonies may also reproduce by fission, especially if they are lacking alate queens. Variation in dispersal strategies is clearly due to ecological constraints. Alternative modes of colony reproduction within the species or even within a colony in a species such as *Pachycondyla tridentata* may serve different purposes: colony founding by dispersing queens is needed for long-range dispersal, to respond, for example, to unfavorable environmental conditions, whereas colony fission with gamergates or other wingless reproductives in tow lead to local dispersal of the colony offspring. Changing ecological conditions may favor one or the other mode of colony propagation in ponerine ants in any given season.

DIACAMMA: REGULATING REPRODUCTION BY MUTILATION

Standing out even in the strange world of ponerine biology cycles is a remarkable form of reproductive regulation found in the queenless ponerine genus *Diacamma*. In most species of *Diacamma* studied thus far, worker mating and reproductive dominance are regulated by a unique agonistic relationship among nestmates. All workers eclose from their cocoons with a pair of small, club-like thoracic appendages called gemmae (Figure 8-13). Each colony consists of up to 300 workers, but only one individual retains its gemmae intact; in all others, the gemmae have been amputated. The individual that retains her gemmae functions as the gamergate of

---

45 | K. Sommer and B. Hölldobler, "Coexistence and dominance among queens and mated workers in the ant *Pachycondyla tridentata*," *Naturwissenschaften* 79(10): 470–472 (1992).

PLATE 51. In the ponerine species *Pachycondyla tridentata*, queens and gamergates coexist in the same colony. *Above*: Behavioral recordings of marked individuals have demonstrated that either queens or gamergates can occupy the highest rank in the reproductive hierarchy. *Below*: Workers often cut off wings of some of the freshly eclosed males. This mutilation prevents them from leaving the nest on the wing and forces them to mate instead with their sister workers.

the colony. She and her mutilated nestmates clip off the gemmae of callow workers soon after the latter emerge. When deprived of their gemmae, workers lose their mating and aggressive drives and become nonreproductive helpers.[46]

The gemmae are covered with sensory hairs that probably act as mechanoreceptors. The sensory afferents arising from these hairs feature widely distributed collaterals invading all three thoracic ganglia as well as subesophageal and second abdominal ganglia.[47] Amputation of the gemmae evidently causes physiological and morphological changes in the central nervous system that direct the transition from aggressive to timid behavior. The gemmae are also richly endowed with exocrine cells.[48] The function of these glands is not known, but it is likely that they secrete chemical pheromones inducing the mutilation process.[49]

When a gamergate is absent in a colony, one of the callow workers, usually the first individual emerging after the colony has been orphaned, is not mutilated. She becomes instead the behaviorally dominant colony member, and all sisters that subsequently eclose from their pupal cocoons are relieved of their gemmae. However, the dominant worker can function as a gamergate only if she has mated. After 7 to 9 days, she leaves the nest, and when outside commences a peculiar calling behavior. Lowering her head and thorax and arching her gaster, she rubs the tibiae of her hind legs along the pleural and dorsal surfaces of her gaster (see Figure 8-13), an action that releases a sexual calling pheromone. Experiments have demonstrated that the most attractive compounds to the male come from

---

46 | Y. Fukumoto, T. Abe, and A. Taki, "A novel form of colony organization in the 'queenless' ant *Diacamma rugosum*," *Physiology and Ecology Japan* 26(1–2): 55–61 (1989); and C. Peeters and S. Higashi, "Reproductive dominance controlled by mutilation in the queenless ant *Diacamma australe*," *Naturwissenschaften* 76(4): 177–180 (1989).

47 | W. Gronenberg and C. Peeters, "Central projections of the sensory hairs on the gemmae of the ant *Diacamma*: substrate for behavioural modulation?" *Cell and Tissue Research* 273(3): 401–415 (1993). Whether gemmae can be considered homologous to insect wings has been investigated by S. Baratte, C. Peeters, and J. S. Deutsch, "Testing homology with morphology, development and gene expression: sex-specific thoracic appendages of the ant *Diacamma*," *Evolution and Development* 8(5): 433–445 (2006).

48 | J. Billen and C. Peeters, "Fine structure of the gemma gland in the ant *Diacamma australe* (Hymenoptera, Formicidae)," *Belgian Journal of Zoology* 121(2): 203–210 (1991).

49 | K. Tsuji, C. Peeters, and B. Hölldobler, "Experimental investigation of the mechanism of reproductive differentiation in the queenless ant, *Diacamma* sp., from Japan," *Ethology* 104(8): 633–643 (1998); and K. Ramaswamy, C. Peeters, S. P. Yuvana, T. Varghese, H. D. Pradeep, V. Dietemann, V. Karpakakunjaram, M. Cobb, and R. Gadagkar, "Social mutilation in the ponerine ant *Diacamma*: cues originate in the victims," *Insectes Sociaux* 51(4): 410–413 (2004).

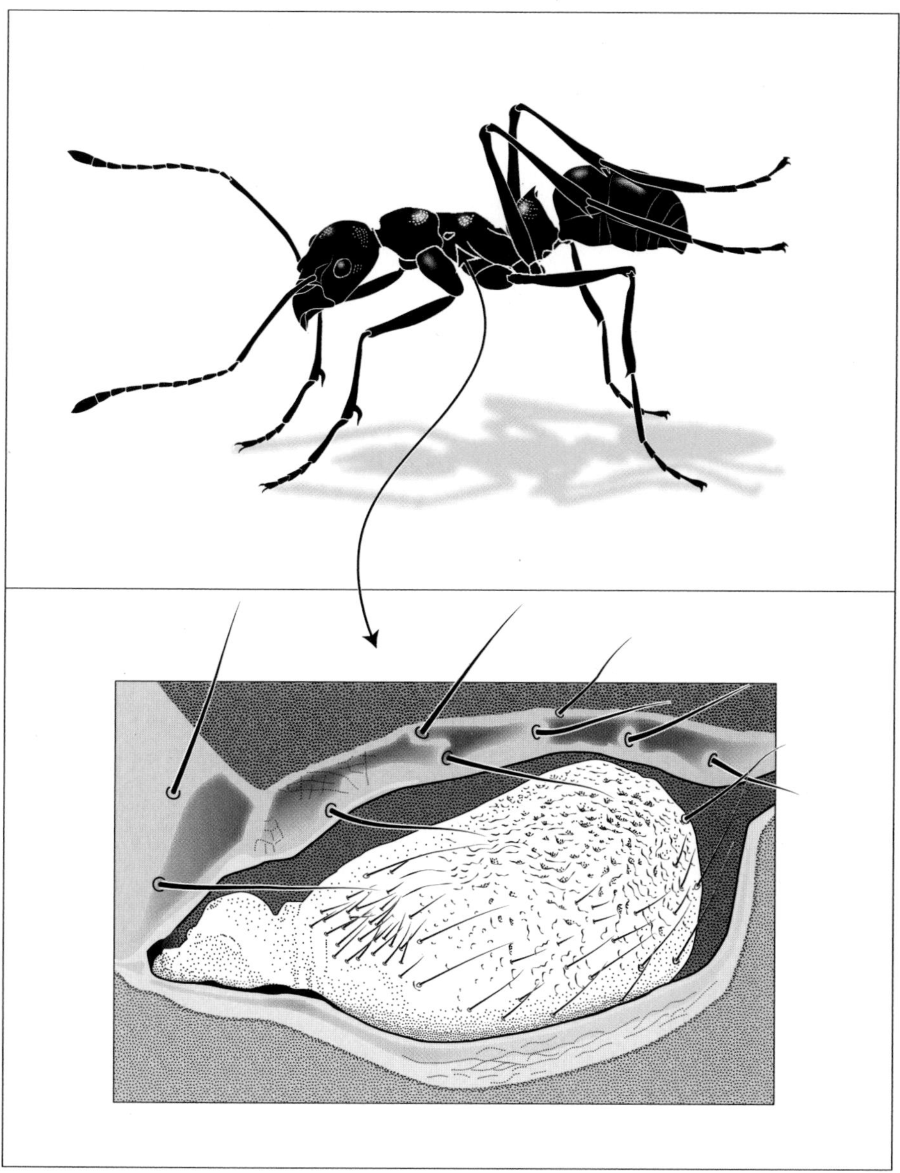

FIGURE 8-13. *Above*: A young worker of the ponerine genus *Diacamma* exhibits sexual calling behavior by rubbing the openings of the metatibial gland on the hind legs over the arched gaster. *Below*: The location of the gemmae (one gemma each on the left and right sides). A worker with this glandular organ serves as a reproductive female; when a worker loses its gemmae, it is committed to an existence as a nonreproductive. *Above*: Based on an original drawing by Malu Obermayer in B. Hölldobler, M. Obermayer, and C. Peeters, "Comparative study of the metatibial gland in ants (Hymenoptera, Formicidae)," *Zoomorphology* 116(4): 157–167 (1996). *Below*: Based on J. Billen and C. Peeters, "Fine structure of the gemma gland in the ant *Diacamma australe* (Hymenoptera, Formicidae)," *Belgian Journal of Zoology* 121(2): 203–210 (1991).

the metatibial glands.[50] When a *Diacamma* male encounters a calling female, he approaches her from behind, antennates her thorax and gaster, and if she runs away follows her closely. Eventually, he succeeds in mounting and copulating with her. As in the brutal practice of *Dinoponera*, the *Diacamma* female returns to the nest, dragging the motionless male in copula along. Nest workers then attack the male and bite off his head and thorax while leaving the gaster attached. There the severed body part remains for 1 or 2 days, until it is finally removed by nest workers or the gamergate herself.

Because only one individual in each colony retains her gemmae, and only females with gemmae exhibit sexual behavior and are mated, *Diacamma* societies are strictly monogynous and monandrous. Nevertheless, in the absence of a gamergate, workers without gemmae can become fertile and lay haploid eggs, something they rarely do in the presence of a gamergate. Behavioral analysis has revealed that only workers prevented from closely approaching the gamergate become aggressive and develop ovaries with fully mature oocytes. The gamergate signal evidently has very low volatility and is transmitted solely by direct physical contact.[51]

The *Diacamma* life cycle has implications for the genetic theory of social evolution. We should expect gamergate-worker conflict over male production to occur in such strictly monogynous and monandrous gamergate societies, because virgin workers have the same ovarian capacity as the gamergate. On the contrary, it turns out that workers in most cases respond to the gamergate's presence by voluntarily refraining from egg laying. Even when given the opportunity to avoid direct contact with the gamergate and thereby escape possible policing, the workers do not ordinarily lay eggs. In *Diacamma* colonies with a gamergate, mutilated virgin workers rarely try to lay eggs. And when it does happen, the gamergate often aggressively takes the egg from an ovipositing worker and eats it while the offender attempts to resist or escape. Only

---

50 | K. Nakate, K. Tsuji, B. Hölldobler, and A. Taki, "Sexual calling by workers using the metatibial glands in the ant, *Diacamma* sp., from Japan (Hymenoptera: Formicidae)," *Journal of Insect Behavior* 11(6): 869–877 (1998); and B. Hölldobler, M. Obermayer, and C. Peeters, "Comparative study of the metatibial gland in ants (Hymenoptera, Formicidae)," *Zoomorphology* 116(4): 157–167 (1996).

51 | K. Tsuji, K. Egashira, and B. Hölldobler, "Regulation of worker reproduction by direct physical contact in the ant *Diacamma* sp. from Japan," *Animal Behaviour* 58(2): 337–343 (1999); and S. Baratte, M. Cobb, and C. Peeters, "Reproductive conflicts and mutilation in queenless *Diacamma* ants," *Animal Behaviour* 72(2): 305–311 (2006).

when colonies are unusually large do a few worker-derived eggs escape gamergate policing and develop into males.[52]

For the most part, the *Diacamma* gamergate is able to signal her determination to police worker reproduction. If a gamergate's policing is effective and worker-derived eggs have little chance of surviving, self-regulation (self-policing) by workers in response to the gamergate signal would be favored by colony-level selection, because worker oviposition would be energetically wasteful and negatively affect colony reproductive efficiency. It has been argued that self-policing in workers can evolve by kin selection in monogynous and monandrous eusocial hymenopterans,[53] providing the production of males by workers reduces colony productivity by more than 20 percent.[54] However, even a much smaller cost of worker reproduction (4.4 percent reduction of colony productivity under some assumptions) is enough for colony-level selection to favor suppression of worker reproduction when workers mutually police oviposition.[55] Indeed, experimental evidence indicates that *Diacamma* workers police other workers' oviposition in the presence of a gamergate in the colony.[56]

In any case, the gamergate's odor appears to be an "honest signal" indicating her presence and reproductive dominance, to which most workers respond by self-policing. Although self-restraint of oviposition could be adaptive to most workers (because it enhances colony efficiency), worker oviposition occasionally occurs. This suggests that colony-level selection has not entirely eradicated all selfish incentives in workers and that some variation in the workers' response to the gamergate signal continues to exist. The patrolling behavior of the gamergate and the behavioral reactions of the workers when contacting the gamergate imply that the gamergate signal is of a chemical nature, most likely secretions with very limited volatility.

---

52 | N. Nakata and K. Tsuji, "The effect of colony size on conflict over male-production between gamergate and dominant workers in the ponerine ant *Diacamma* sp.," *Ethology Ecology & Evolution* 8(2): 147–156 (1996).

53 | In this case, kin selection and colony-level selection favor identical adaptive traits; see A. Bourke and N. Franks, *Social Evolution in Ants* (Princeton, NJ: Princeton University Press, 1995).

54 | B. J. Cole, "The social behavior of *Leptothorax allardycei* (Hymenoptera, Formicidae): time budgets and the evolution of worker reproduction," *Behavioral Ecology and Sociobiology* 18(3): 165–173 (1986).

55 | F. L. W. Ratnieks, "Reproductive harmony via mutual policing by workers in eusocial Hymenoptera," *American Naturalist* 132(2): 217–236 (1988).

56 | N. Kikuta and K. Tsuji, "Queen and worker policing in the monogynous and monandrous ant, *Diacamma* sp.," *Behavioral Ecology and Sociobiology* 46(3): 180–189 (1999).

Indeed, there is strong circumstantial evidence that such is the case: the cuticular hydrocarbons of nestmate workers vary in their proportions according to age and fertility, and workers that begin producing eggs develop a CHC profile that is different from the profile of workers that become foragers.[57]

Such a strict division of labor among one reproductive individual (gamergate) and sterile nestmates can even develop in orphaned *Diacamma* worker groups. Members of such groups start fighting about 1 week after their contact with the gamergate has been interrupted. A few days later, several workers start laying eggs, but after about 6 weeks, one worker establishes herself as the alpha individual and thereafter reigns as the sole egg layer.[58] An interesting relationship between reproductive dominance and juvenile hormone titer has been discovered in the *Diacamma* ants.[59, 60] Contrary to earlier findings in the fire ant *Solenopsis invicta* and wasp *Polistes gallicus*,[61, 62] juvenile hormone (JH) is not positively correlated with fecundity in *Diacamma*. Juvenile hormone is not detectable in gamergates and fertile workers, whereas juvenile hormone titer in nonreproductive workers increases with age. This is exactly the situation found in honeybees, in which queens and young workers have low titers, and foragers have high titers.[63]

When we now combine the pheromonal and endocrine evidence on *Diacamma*,

---

57 | V. Cuvillier-Hot, M. Cobb, C. Malosse, and C. Peeters, "Sex, age and ovarian activity affect cuticular hydrocarbons in *Diacamma ceylonense*, a queenless ant," *Journal of Insect Physiology* 47(4–5): 485–493 (2001); and V. Cuvillier-Hot, R. Gadagkar, C. Peeters, and M. Cobb, "Regulation of reproduction in a queenless ant: aggression, pheromones and reduction in conflict," *Proceedings of the Royal Society of London B* 269: 1295–1300 (2002).

58 | C. Peeters and K. Tsuji, "Reproductive conflict among ant workers in *Diacamma* sp. from Japan: dominance and oviposition in the absence of the gamergate," *Insectes Sociaux* 40(2): 119–136 (1993); and S. Baratte, M. Cobb, and C. Peeters, "Reproductive conflicts and mutilation in queenless *Diacamma* ants," *Animal Behaviour* 72(2): 305–311 (2006).

59 | K. Sommer, B. Hölldobler, and H. Rembold, "Behavioral and physiological aspects of reproductive control in a *Diacamma* species from Malaysia (Formicidae, Ponerinae)," *Ethology* 94(2): 162–170 (1993).

60 | These results have recently been confirmed in the queenless ponerine ant *Streblognathus*; see C. Brent, C. Peeters, V. Dietemann, R. Crewe, and E. Vargo, "Hormonal correlates of reproductive status in the queenless ponerine ant, *Streblognathus peetersi*," *Journal of Comparative Physiology A* 192: 315–320 (2006).

61 | J. F. Barker, "Neuroendocrine regulation of oocyte maturation in the imported fire ant *Solenopsis invicta*," *General and Comparative Endocrinology* 35(3): 234–237 (1978).

62 | P. F. Röseler, I. Röseler, A. Strambi, and R. Augier, "Influence of insect hormones on the establishment of dominance hierarchies among foundresses of the paper wasp, *Polistes gallicus*," *Behavioral Ecology and Sociobiology* 15(2): 133–142 (1984).

63 | G. E. Robinson, C. Strambi, A. Strambi, and Z.-Y. Huang, "Reproduction in worker honey bees is associated with low juvenile hormone titers and rates of biosynthesis," *General and Comparative Endocrinology* 87(3): 471–480 (1992).

it is reasonable to conclude that highly fertile *Diacamma* workers or gamergates have distinct CHC profiles, and these are physiologically coupled with the endocrine processes that regulate fertility.

STREBLOGNATHUS: DOMINANCE AND FERTILITY UNCOUPLED

Studies of the African ponerine *Streblognathus peetersi* by Virginie Cuvillier-Hot and her coworkers have documented the physiological mechanisms of dominance and fertility in queenless ponerine ants in even greater depth.[64, 65] The same negative correlation exists in this species between degree of fertility and juvenile hormone titer as in *Diacamma*, but dominance and fertility are uncoupled. When the dominant ant is treated with Pyriproxyfen, a juvenile hormone analog, the fertility of the alpha individual is markedly decreased. Although such treated alpha ants remain aggressive, they are attacked and immobilized by low-ranking workers. Interestingly, the treated individual is not challenged by the next highest ranking ants in the dominance hierarchy, but policed by low-ranking subordinates. While the alpha individual is immobilized, one of the high-ranking workers begins to exhibit dominance behavior—and it subsequently assumes the alpha position. During this process, the CHC profile of the treated ant undergoes modifications that are opposite to those of the high-ranking challenger ant. Such CHC profiles most likely signal fertility status, and although the treated alpha still exhibits the full repertoire of aggressive dominance, the police workers perceive the decline in fertility and react by immobilizing the alpha. These findings have impressively demonstrated the crucial role of sterile helpers in regulating reproduction. They underscore yet again the distinction between dominance interactions and policing as discrete agonistic behavioral mechanisms.[66, 67]

---

64 | V. Cuvillier-Hot, A. Lenoir, and C. Peeters, "Reproductive monopoly enforced by sterile police workers in a queenless ant," *Behavioral Ecology* 15(6): 970–975 (2004).

65 | V. Cuvillier-Hot, A. Lenoir, R. Crewe, C. Malosse, and C. Peeters, "Fertility signalling and reproductive skew in queenless ants," *Animal Behaviour* 68(5): 1209–1219 (2004).

66 | V. Cuvillier-Hot, A. Lenoir, and C. Peeters, "Reproductive monopoly enforced by sterile police workers in a queenless ant," *Behavioral Ecology* 15(6): 970–975 (2004).

67 | C. Brent, C. Peeters, V. Dietemann, R. Crewe, and E. Vargo, "Hormonal correlates of reproductive status in the queenless ponerine ant, *Streblognathus peetersi*," *Journal of Comparative Physiology A* 192(3): 315–320 (2006).

FIGURE 8-14. Dominance behavior in *Streblognathus peetersi.* Two agonistic acts are especially conspicuous. *Above*: In gaster curling, the dominant (black) worker bends her gaster forward and exposes the intersegmental membrane close to the gastral tip. During this confrontation, the dominant individual grabs the mandible at the base of the subordinate's (gray) antennae. *Below*: In the gaster rise posture, the dominant arches her gaster, again exposing the intersegmental membrane, and turns around and places her gaster in front of the subordinate. Based on V. Cuvillier-Hot, A. Lenoir, and C. Peeters, "Reproductive monopoly enforced by sterile police workers in a queenless ant," *Behavioral Ecology* 15(6): 970–975 (2004).

Each colony of *Streblognathus peetersi* comprises 30 to 130 workers, with only one individual occupying the alpha position in the hierarchy. This individual has mated and assumed the gamergate role. The dominance hierarchy is formed, as in other ponerine ants described earlier, by aggression. Among the several agonistic acts, two are especially conspicuous. In *gaster curling*, the dominant worker bends her gaster forward and exposes the intersegmental membrane close to the gastral tip. This has the look of uncapping an intersegmental gland of some kind, perhaps the pygidial gland itself, but no evidence has yet been found that true glandular secretions are involved. More likely, the dominant ant exposes intersegmental tissue loaded with a special blend of hydrocarbons. During the confrontation, the dominant ant grabs the mandibles or the base of the subordinate's antennae (Figure 8-14). In the *gaster rise* movement, the dominant arches her gaster, again with the intersegmental membranes at the distal region of the gaster exposed. She then turns around and places her gaster in front of the subordinate individual (see Figure 8-14).

For weeks after the alpha ant has established herself in the top position, she remains aggressive, but on a relatively low level—just enough, it appears, to confirm her top rank. In fact, *Streblognathus* colonies, like colonies of *Harpegnathos* and some other queenless ponerine ant species with dominance systems, go through phases of social stability and instability. During periods of calm, few acts of aggression occur and the alpha ant exhibits high fertility. During periods of instability, which occur soon after a dominance turnover has taken place, the new alpha individual has only slightly developed ovaries and uses aggression to enforce her position. A similar cycle has been observed in colonies of *Dinoponera quadriceps*.

To summarize to this point, aggression is the mechanism by which dominance hierarchies are established in the ponerine ants thus far studied, but the maintenance of reproductive monopoly is due to fertility signaling. Gamergate replacement is a very significant event in the life cycle of colonies, making colonies of queenless ants potentially immortal. It is also an important event in the life history of individual members of the colony, since nestmates compete from time to time to seize the reproductive role. The winners advertise their status. The alpha ant and her high-ranking nestmates in *Streblognathus peetersi* colonies are distinguished by their CHC profiles, "which encode sufficient and graded information to label not only egg layers but also workers with intermediate reproductive potential. . . . Moreover, a new alpha of *Streblognathus peetersi* can be detected by nestmates several days before the onset of her oviposition, which suggests that the CHC-profile reflects the hormonal state of a worker more than its current egg-laying rate."[68] Presumably, the ants use variation in CHC profiles to signal or detect reproductive capacity. Once the alpha individual is reproductively active, her chemical signals appear to replace most aggressive interactions as the mechanism for regulating reproduction in the *Streblognathus* colony.

Strong supporting evidence for the hypothesis that CHC profiles function as "honest" signals of social status and fertility has been provided by the measurement

---

68 | V. Cuvillier-Hot, A. Lenoir, R. Crewe, C. Malosse, and C. Peeters, "Fertility signalling and reproductive skew in queenless ants," *Animal Behaviour* 68(5): 1209–1219 (2004). Virginie Cuvillier-Hot and Alain Lenoir investigated the neurochemical basis of the behavioral plasticity observed in the queenless *Streblognathus* colonies in V. Cuvillier-Hot and A. Lenoir, "Biogenic amine levels, reproduction and social dominance in the queenless ant *Streblognathus peetersi*," *Naturwissenschaften* 93(3): 149–153 (2006).

of vitellogenin (Vg) levels in the blood (hemolymph) of different kinds of ants.[69] Vitellogenin is a yolk precursor that is synthesized in the fat body, secreted into the hemolymph, and incorporated in the oocytes to form vitellins, the major storage proteins in eggs. The relative amount of vitellogenin in the hemolymph is an excellent indicator of the readiness for egg production, regardless of whether the ant is currently in her oviposition phase.[70] Correlative analysis has demonstrated that the CHC profiles match both the social status and the levels of circulating vitellogenin in the *Streblognathus* workers. Workers with no more than trace levels of vitellogenin, or with none at all, exhibit the cuticular hydrocarbon blends identified in infertile foragers or freshly eclosed callows. Workers with higher vitellogenin level express CHC profiles close to those of egg layers or alpha individuals. Most importantly, "the more vitellogenin a worker had in her hemolymph, the further along the axis of fertility her cuticular profile occurred. The cuticular hydrocarbon signature thus provides reliable information not only about current fertility, but also about the reproductive potential of high rankers."[71]

GAMERGATES VERSUS ERGATOID QUEENS

The term *gamergate* (married worker) is applied to workers who mate and produce fertilized eggs.[72] William Morton Wheeler and his associate James Chapman were the first, in the early 1920s, to describe worker mating in *Diacamma rugosum*;[73] morphologically distinct queens had never before that time or since been found in this genus. Wheeler and Chapman correctly pointed out that reproductive workers are morphologically identical to nonreproductive workers and are different from ergatoid queens, a permanently wingless queen caste found in other kinds of ants. Nevertheless, as Christian Peeters has stressed, "the use of the term 'ergatoid' has

---

69 | V. Cuvillier-Hot, A. Lenoir, R. Crewe, C. Malosse, and C. Peeters, "Fertility signalling and reproductive skew in queenless ants," *Animal Behaviour* 68(5): 1209–1219 (2004).

70 | T. Martinez and D. E. Wheeler, "Effect of the queen, brood and worker caste on haemolymph vitellogenin titre in *Camponotus festinates* workers," *Journal of Insect Physiology* 37(5): 347–352 (1991).

71 | V. Cuvillier-Hot, A. Lenoir, R. Crewe, C. Malosse, and C. Peeters, "Fertility signalling and reproductive skew in queenless ants," *Animal Behaviour* 68(5): 1209–1219 (2004).

72 | C. Peeters and R. Crewe, "Insemination controls the reproductive division of labour in a ponerine ant," *Naturwissenschaften* 71(1): 50–51 (1984).

73 | W. M. Wheeler and J. Chapman, "The mating of *Diacamma*," *Psyche* (Cambridge, MA) 29: 203–211 (1922).

become confused, because it has been used indiscriminately in all ponerine species lacking alate queens."[74] For a clear distinction of the various reproductive modes in ants, it is important to use the term *caste* in its physical sense; that is, caste differences within a species occur when larvae follow particular developmental pathways that result in more than one specialized morphological trait. Caste in Peeters's strict sense refers to morphology and not to roles. Thus, reproducing mated workers (gamergates) should not be referred to as "queens."

The complete loss of the queen caste is an important evolutionary event that has occurred in multiple lines within several tribes of the subfamily Ponerinae. This convergence was relatively simple, because workers in many ponerine species have retained important reproductive ancestral traits, such as functional ovaries and a functional spermatheca. Worker and queen castes have not diverged as much as in most other ant subfamilies. In other words, they have barely passed the "point of no return." This is also true for Australian bulldog ants of the subfamily Myrmeciinae, although up to the present time, only one clearly documented case of gamergate reproduction has been reported,[75] in the myrmeciine *Myrmecia pyriformis*. Elsewhere, reproduction by gamergates has been documented in two species of the primitive myrmicine genus *Metapone* (subfamily Myrmicinae, not Myrmeciinae).[76]

In contrast to gamergates, which belong to the worker caste, ergatoid queens exhibit all the attributes of a distinct queen caste. They have evolved from alate queens in at least 12 ponerine genera and three cerapachyine genera. Ergatoid queens are also known from some species of the genus *Myrmecia* (subfamily Myrmeciinae), some myrmicine genera (subfamily Myrmicinae), one dolichoderine genus (*Leptomyrmex*), and one *Proformica* species (subfamily Formicinae).[77] Ergatoid queens are always wingless. Hence, their thorax does not need to accommodate wing muscles, and in this respect they are anatomically closer to workers than to alate queens. On the other hand, their thoracic sclerites remain distinct, their gaster

74 | C. Peeters, "Ergatoid queens and intercastes in ants: two distinct adult forms which look morphologically intermediate between workers and winged queens," *Insectes Sociaux* 38(1): 1–15 (1991).

75 | V. Dietemann, C. Peeters, and B. Hölldobler, "Gamergates in the Australian ant subfamily Myrmeciinae," *Naturwissenschaften* 91(9): 432–435 (2004).

76 | B. Hölldobler, J. Liebig, and G. D. Alpert, "Gamergates in the myrmicine genus *Metapone* (Hymenoptera: Formicidae)," *Naturwissenschaften* 89(7): 305–307 (2002).

77 | C. Peeters, "Ergatoid queens and intercastes in ants: two distinct adult forms which look morphologically intermediate between workers and winged queens," *Insectes Sociaux* 38(1): 1–15 (1991).

is larger than that of workers, and their internal anatomy is often strikingly different from that of workers.

PACHYCONDYLA FOCHI: MASS TERMITE RAIDERS

The role of the ergatoid queen can be seen in the large African ponerine *Pachycondyla fochi* (formerly *Megaponera foetens*).[78] *Pachycondyla fochi* is a specialized predator that conducts highly organized raids on termite nests and also frequently emigrates to new nesting sites. The colonies are relatively large, comprising from several hundred to more than 2,000 workers. The workers are polymorphic and exhibit a distinct division of labor correlated with their size. Mated workers are absent, and reproduction is carried out exclusively by a single ergatoid queen.[79] The queen is only slightly larger than a major worker. But she is easily recognizable by her larger gaster, which contains an ovary with 32 ovarioles, about twice the number of a major worker (12 to 15 ovarioles).[80] The queen usually holds her gaster slightly up in the air, often waving it faintly from side to side. In the nest and on the emigration trail, she is frequently surrounded by a court of workers. When the nest is disturbed, still more workers gather around the queen. Sometimes they form a multilayered retinue, pressing close together with their heads facing her yet only occasionally making direct physical contact (Plate 52). Many of these praetorian guards are majors, but some minors also participate. When laying eggs, the queen continues to hold her gaster slightly upward. The egg is discharged within less than a minute after the beginning of its emergence and is immediately taken by a minor worker and placed onto the egg pile.

The ergatoid queen is clearly the center of social activity in a *Pachycondyla fochi* colony. Her powerful attraction is moreover clearly based on chemical queen signals. The responses to these signals are so striking that scientists have paid special

---

78 | B. Hölldobler, C. Peeters, and M. Obermayer, "Exocrine glands and the attractiveness of the ergatoid queen in the ponerine ant *Megaponera foetens*," *Insectes Sociaux* 41(1): 63–72 (1994).

79 | M. H. Villet, "Division of labour in the Matabele ant *Megaponera foetens* (Fabr.) (Hymenoptera, Formicidae)," *Ethology Ecology & Evolution* 2(4): 397–417 (1990).

80 | C. Peeters, "Morphologically 'primitive' ants: comparative review of social characters, and the importance of queen–worker dimorphism," in J. C. Choe and B. J. Crespi, eds., *The Evolution of Social Behavior in Insects and Arachnids* (New York: Cambridge University Press, 1997), pp. 372–391.

PLATE 52. An ergatoid queen of the species *Pachycondyla fochi* surrounded by a large retinue of workers. The body of the queen is clothed in erect hairs, which are associated with glandular openings and likely glandular emissions.

attention to them.[81] Researchers divided a *Pachycondyla fochi* colony into several groups, only one of which contained the queen. After 8 to 10 days, one worker from the queenright group and another from a queenless group were introduced into another queenless group. In repeated tests, the behavior of the resident ants toward the introduced worker from the queenless group was neutral, but the introduced worker from the queenright group elicited strong attraction on the part of the resident ants, which repeatedly licked and carried her around. This remarkable attractiveness of ants from the queenright group lasted almost 3 hours in some cases. When, in another combination, workers from a queenless group were introduced into the queenright group, they were initially treated with hostility by the

81 | B. Hölldobler, C. Peeters, and M. Obermayer, "Exocrine glands and the attractiveness of the ergatoid queen in the ponerine ant *Megaponera foetens*," *Insectes Sociaux* 41(1): 63–72 (1994).

resident ants, which attacked them with mandible strikes. The introduced workers responded by fleeing or assuming a pupal posture by curling up and folding their appendages, after which they were often picked up by the attacking ant and carried to the refuse heap.

These observations strongly suggest that the *Pachycondyla fochi* queen produces attractive chemical signals that are transferred onto her nestmate workers. Workers that later lack the queen signal are discriminated against by their nestmates. Of course, it could also be that both the lack of the queen signal and the development of fertility signals in isolated workers elicit rejection by queenright nestmates. Indeed, several studies have shown that large workers in orphaned colonies oviposit on their own.

Despite their overall superficial resemblance to workers, ergatoid queens of *Pachycondyla fochi* generally possess striking differences in anatomical details. Numerous erect hairs cover the entire surface of her body, whereas the workers are nearly bare. Less visible but major differences exist in the cuticle and epidermal epithelium. In workers, the epidermis is a thin layer of collapsed cells; in the queen, it is a well-developed thick glandular epithelium with large nuclei and many vacuoles. The cuticle of the queen is moreover penetrated by dense networks of dermal gland ducts. The large erect setae are innervated and most likely function as mechanoreceptors. Many dermal gland ducts open around the pits of the base of the setae, suggesting that these long hairs also serve as dispensers of the secretions.

ERGATOID QUEENS AND ARMY ANTS

We agree with Christian Peeters that ergatoid queens should be defined as individuals that possess all the specialized attributes of the reproductive caste, despite their external resemblance to workers. Thus, the ergatoid queen is not, as previously often assumed, an "ergatogyne," a wingless female that is merely a morphological intermediate between workers and winged queens. Instead, it is a distinct queen caste that must have evolved from alate queens, and it should not be lumped together with gamergates in a single category.[82]

---

82 | B. Hölldobler and E. O. Wilson, *The Ants* (Cambridge, MA: The Belknap Press of Harvard University Press, 1990); and C. Peeters, "Monogyny and polygyny in ponerine ants with or without queens," in L. Keller, ed., *Queen Number and Sociality in Insects* (New York: Oxford University Press, 1993), pp. 234–261.

The adaptive significance of ergatoid queens remains unclear. In part, the caste might be a special adaptation in species that conduct group predation and have a migratory lifestyle, including colony propagation by fission. In this case, the queens do not fly forth singly but march out with a part of the worker force over the ground to new sites. In fact, many poneromorph species with ergatoid queens are group predators, and they frequently migrate to new hunting grounds and settle in ephemeral nests. The known examples are *Pachycondyla fochi*, along with species of *Onychomyrmex*, *Leptogenys*, and *Simopelta*. This is also true for the cerapachyine genera *Cerapachys* and *Sphinctomyrmex*, for the leptanilline genus *Leptanilla*, and of course for all species of the "true" army ants that compose the subfamilies Dorylinae and Ecitoninae. As in *Pachycondyla fochi*, the ergatoid queens of *Onychomyrmex*, *Leptogenys*, *Leptanilla*, and the ecitonine genera *Eciton* and *Neivamyrmex* are also endowed with special exocrine glands not present in the worker caste of their respective species.[83] Further, the ergatoid queens are very attractive to workers. We hypothesize that in ants with a nomadic lifestyle, the queen requires especially powerful attractants to signal her presence to her workers during emigrations. This is crucial in relatively large colonies, typically seen in species that practice group predation.[84] The rule is most notable for ponerine species specialized to hunt other social insects, including termites, wasps, and even other kinds of ants. Independent colony foundation by single queens would result in a very small worker force in beginning colonies; thus, the adaptive mode of colony propagation must be fission. Alate queens are useless, while ergatoid queens suffice. Because group predators frequently emigrate, their colonies remain widely dispersed, even in the absence of winged propagules, thus reducing competition.

Such ponerine societies with a single ergatoid queen provide excellent examples of group-level organization. The queen is the undisputed, highly attractive reproductive individual. In the presence of the queen, workers apparently remain reproductively sterile, although they may lay trophic eggs with no chance of further

---

83 | N. R. Franks and B. Hölldobler, "Sexual competition during colony reproduction in army ants," *Biological Journal of the Linnean Society* 30(3): 229–243 (1987); B. Hölldobler, J. M. Palmer, K. Masuko, and W. L. Brown, Jr., "New exocrine glands in the legionary ants of the genus *Leptanilla* (Hymenoptera, Formicidae, Leptanillinae)," *Zoomorphology* 108(5): 255–261 (1989); and B. Hölldobler and M. Obermayer, unpublished.

84 | B. Hölldobler and E. O. Wilson, *The Ants* (Cambridge, MA: The Belknap Press of Harvard University Press, 1990); and C. Peeters, "Monogyny and polygyny in ponerine ants with or without queens," in L. Keller, ed., *Queen Number and Sociality in Insects* (New York: Oxford University Press, 1993), pp. 234–261.

development. Workers do not forage solitarily, but instead prey in groups organized through highly efficient communication systems. Not the slightest evidence exists for hierarchical dominance structures among such workers. These ergatoid queen societies appear to have "overcome" intracolonial hierarchical friction and function as a fully annealed superorganism, with the queen as the reproductive unit and the hundreds of workers as her supporting somatic elements. By any measure of colony fitness, the queen is the most precious unit. With powerful attractants, she signals her presence to the entire colony, and by this signal, it is likely that she also informs workers about her personal fecundity and health. Perceiving this signal, workers "consent" to remain sterile for the sake of colony efficiency.[85]

PACHYCONDYLA: SOCIOBIOLOGICALLY THE MOST DIVERSE ANT GENUS

Harmonious societies are far from the rule in the ponerines at large, as abundantly documented in species with gamergate colonies, which are frequently torn by antagonistic interactions among nestmates. Comparable friction also occurs in ponerine species without gamergate reproduction. Consider, for example, the peculiar behavior of the two Neotropical ponerine species *Pachycondyla stigma* and *Pachycondyla apicalis*.

The reproductive activity and social organization of *Pachycondyla stigma* is closely linked with the variable rate at which individual colony members perform mutual antennal rubbings with nestmates. During these encounters, the ants swipe their antennae over the openings of a front tibial gland of the nestmates they encounter. The inseminated queen engages in such mutual rubbings at a much higher rate than noninseminated queens and workers. The reproductive dominance of the mated queen is further enhanced by extremely aggressive behavior of workers toward other egg-laying queens. When the mated queen is experimentally removed from the colony, however, mutual antennal rubbing among virgin queens and aggressive behavior increases markedly. Presumably, the tibial glands in the front legs of queens produce chemical signals that inform the nestmates about the queens' fertility and

85 | T. D. Seeley, "Queen substance dispersal by messenger workers in honeybee colonies," *Behavioral Ecology and Sociobiology* 5(4): 391–415 (1979); T. D. Seeley, *Honeybee Ecology* (Princeton, NJ: Princeton University Press, 1985); L. Keller and P. Nonacs, "The role of the queen pheromones in social insects: queen control or queen signal?" *Animal Behaviour* 45(4): 787–794 (1993).

reproductive dominance status. Subordinate queens or other nestmates that defiantly try to "sneak in" male-destined eggs are fiercely attacked by workers.[86]

A very different mode of reproductive control is practiced by *Pachycondyla apicalis*, which forms relatively small colonies of 80 to 100 workers and a single queen.[87] The queen is not attractive to the workers, and yet she maintains reproductive dominance without agonistic interactions with nestmate workers. The workers have active ovaries, and others possess rich deposits of yellow bodies, indicating previously active ovaries. In most cases, those with active ovaries lay nonviable trophic eggs that are immediately fed to the queen. However, if the queen is experimentally excluded from direct contact with her nestmate workers, some workers begin to produce viable eggs. This alone suggests that the presence of the fertile queen has an inhibitory effect on the workers' reproductive activity. Most likely, a queen pheromone is transmitted by direct contact with workers during the queen's walk around the nest. In some *Pachycondyla apicalis* colonies, a few workers nevertheless produce viable eggs, even in the presence of the queen.[88, 89] In such cases, as in queenless colonies, workers establish loose dominance orders through agonistic interactions, which can ultimately lead to a differential production of eggs by individual workers. Such aggressive behavior entails either overt physical attacks, during which the subordinate individual often exhibits a submissive posture (Figure 8-15), or the robbing and destruction of eggs laid by nestmates. The dominance pattern among workers is not stable; a change in the social status of individuals within the dominance order often occurs. When the queen is excluded from the colony, an increased number of workers participate in antagonistic interactions; most conspicuously, attacks on egg-laying workers become more frequent. The social status of individual workers apparently is closely correlated with their ovarian development: the top group in the dominance order has the best-developed ovaries and is most frequently seen attending the egg pile. There is evidence that dominance is age

---

86 | P. S. Oliveira, M. Obermayer, and B. Hölldobler, "Division of labor in the Neotropical ant, *Pachycondyla stigma* (Ponerinae), with special reference to mutual antennal rubbing between nestmates (Hymenoptera)," *Sociobiology* 31(1): 9–24 (1998).

87 | V. Dietemann and C. Peeters, "Queen influence on the shift from trophic to reproductive eggs laid by workers of the ponerine ant *Pachycondyla apicalis*," *Insectes Sociaux* 47(3): 223–228 (2000).

88 | V. Dietemann and C. Peeters, "Queen influence on the shift from trophic to reproductive eggs laid by workers of the ponerine ant *Pachycondyla apicalis*," *Insectes Sociaux* 47(3): 223–228 (2000).

89 | P. S. Oliveira and B. Hölldobler, "Dominance orders in the ponerine ant *Pachycondyla apicalis* (Hymenoptera, Formicidae)," *Behavioral Ecology and Sociobiology* 27(6): 385–393 (1990).

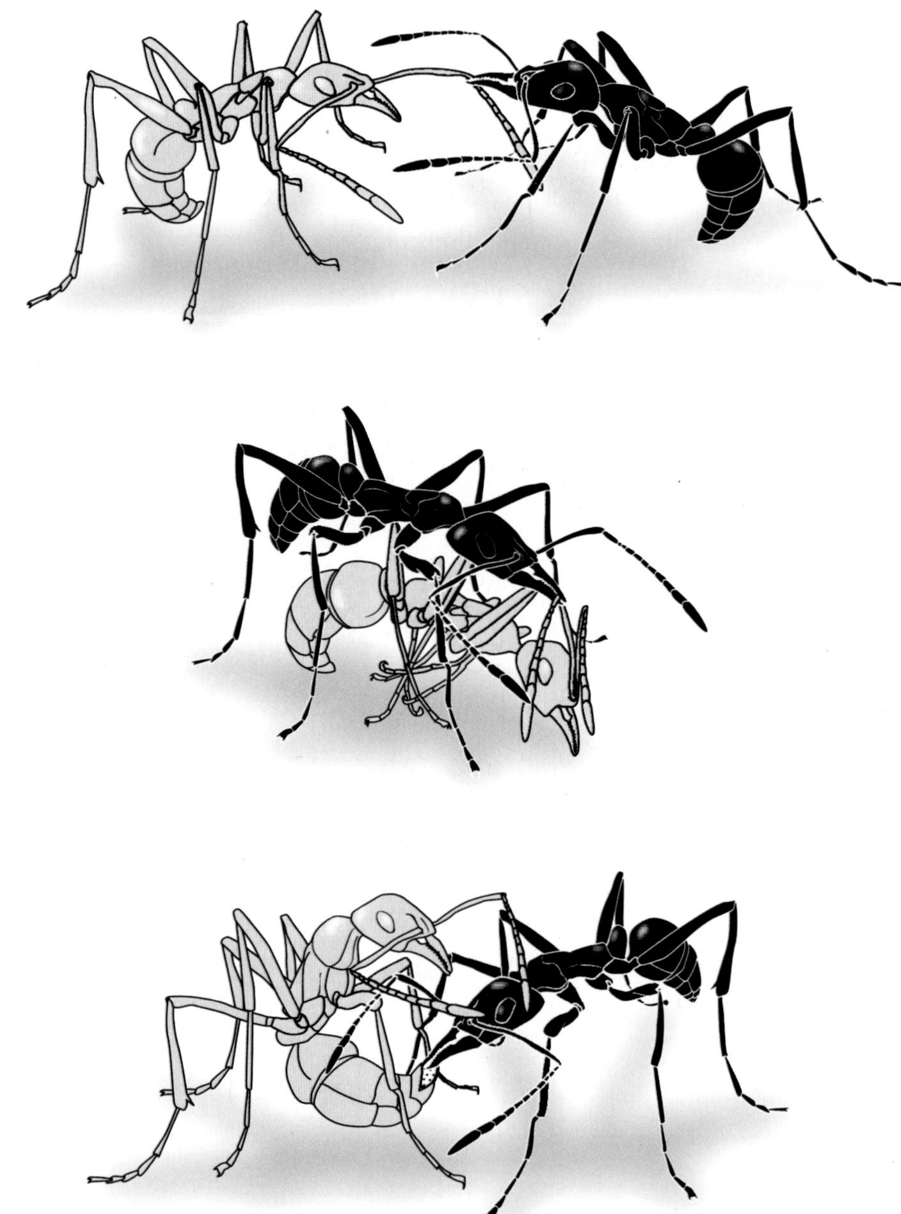

FIGURE 8-15. Dominance interactions between workers in the Neotropical ponerine ant species *Pachycondyla apicalis*. *Above*: A dominant individual (black) is pulling a nestmate by her antennae. *Middle*: In an escalation, the dominant continues to attack the subordinate individual. The latter assumes a pupal posture, with the appendages tightly folded to the body. *Below*: A dominant individual is pulling an egg out of the gaster of a nestmate. The egg eventually will be eaten by the aggressor. Based on an original illustration by Katherine Brown-Wing from P. S. Oliveira and B. Hölldobler, "Dominance orders in the ponerine ant *Pachycondyla apicalis* (Hymenoptera, Formicidae)," *Behavioral Ecology and Sociobiology* 27(6): 385–393 (1990).

dependent. Younger workers have better-developed ovaries, and they also challenge older workers in dominance interactions.[90]

Additional information concerning the behavioral physiology of reproductive dominance among workers has been obtained from comparisons of closely related *Pachycondyla* species in the *villosa* group, nearest to *inversa*, and possibly still taxonomically undescribed.[91] In colonies without a queen, the workers' rank order and their ovarian activity are highly correlated; moreover, the cuticular hydrocarbon blends of egg-laying workers are both quantitatively and qualitatively different from those of nonlaying workers. The reproductive workers' CHC blends resemble those of queens, although they do not completely match them.[92] Although top-ranking individuals are the most productive egg layers, several lower-ranking workers also lay considerable numbers of eggs. Many of the subordinates' eggs are eaten by high-ranking workers. Nevertheless, multilocus DNA fingerprinting has demonstrated unequivocally that low-ranking individuals occasionally succeed in producing adult males, even though the most dominant workers always have the highest reproductive success.[93, 94]

The situation, however, is very different in the presence of queens: workers kill all worker-laid eggs. Even those worker-laid eggs deposited in the egg pile of the queen will be detected by policing workers, because the characteristic hydrocarbon profiles of the queen-laid eggs are not transferred onto the worker-laid eggs by "cue scrambling" in queenright *Pachycondyla inversa* colonies.[95]

Returning to *Pachycondyla apicalis*, interesting questions are raised by the fact that

---

90 | Young individuals tend to become dominant over older nestmates, as documented in several ponerine species that form reproductive dominance orders; see S. Higashi, F. Ito, N. Sugiura, and K. Ohkawara, "Workers' age regulates the linear dominance hierarchy in the queenless ponerine ant, *Pachycondyla sublaevis* (Hymenoptera, Formicidae)," *Animal Behaviour* 47(1): 179–184 (1994); T. Monnin and C. Peeters, "Dominance hierarchy and reproductive conflicts among subordinates in a monogynous queenless ant," *Behavioral Ecology* 10(3): 323–332 (1999).

91 | K. Kolmer and J. Heinze, "Comparison between two species in the *Pachycondyla villosa* complex (Hymenoptera: Formicidae)," *Entomologica Basiliensia* 22: 219–222 (2000); and K. Kolmer, B. Hölldobler, and J. Heinze, "Colony and population structure in *Pachycondyla* cf. *inversa*, a ponerine ant with primary polygyny," *Ethology Ecology & Evolution* 14(2): 157–164 (2002).

92 | J. Heinze, B. Stengl, and M. F. Sledge, "Worker rank, reproductive status and cuticular hydrocarbon signature in the ant, *Pachycondyla* cf. *inversa*," *Behavioral Ecology and Sociobiology* 52(1): 59–65 (2002).

93 | J. Heinze, B. Trunzer, P. S. Oliveira, and B. Hölldobler, "Regulation of reproduction in the Neotropical ponerine ant, *Pachycondyla villosa*," *Journal of Insect Behavior* 9(3): 441–450 (1996).

94 | B. Trunzer, J. Heinze, and B. Hölldobler, "Social status and reproductive success in queenless ant colonies," *Behaviour* 136(9): 1093–1105 (1999).

95 | P. D'Ettorre, A. Tofilski, J. Heinze, and F. L. W. Ratnieks, "Non-transferable signals on ant queen eggs," *Naturwissenschaften* 93(3): 136–140 (2006).

some workers lay viable eggs and compete with one another for reproductive dominance, while others produce only trophic eggs. If *Pachycondyla apicalis* is monandrous (which is most likely the case because no nuptial flights occur in this species) and also monogynous (a fact confirmed by several independent studies[96, 97, 98]), any worker's genetic interest is to rear sons or nephews (relatedness coefficient 0.5 and 0.375, respectively) instead of brothers (relatedness 0.25). We should therefore expect reproductive conflict between the queen and her daughter workers.[99, 100] However, no such conflict between queen and workers has ever been observed. In many colonies, the queen's presence alone inhibits worker reproduction. In other colonies, some dominant workers succeed in producing viable eggs, and they actively suppress reproduction in their worker nestmates by attacking them or cannibalizing their viable eggs. But the queen does not seem to get involved in these aggressive interactions.[101, 102] It is as if such colonies are in a stage of incipient senescence, where the queen's productivity wanes. During the period of the queen's full fecundity, colony-level selection favors behavioral and physiological traits in workers that bring about reproductive restraint and the laying of trophic eggs to add to the queen's diet. Such behavior maximizes colony reproductive efficiency; enough viable eggs are produced and the labor force is maximized. On the other hand, once the queen's fecundity declines, young workers with the greatest reproductive potential begin to compete for reproductive dominance with their nestmates. Egg cannibalism and physical attack ensure that the most powerful individual lays male-destined eggs, and the subordinate workers rear this brood to adulthood. Obviously, at this stage, the force of individual direct

---

96 | V. Dietemann and C. Peeters, "Queen influence on the shift from trophic to reproductive eggs laid by workers of the ponerine ant *Pachycondyla apicalis*," *Insectes Sociaux* 47(3): 223–228 (2000).

97 | P. S. Oliveira and B. Hölldobler, "Dominance orders in the ponerine ant *Pachycondyla apicalis* (Hymenoptera, Formicidae)," *Behavioral Ecology and Sociobiology* 27(6): 385–393 (1990).

98 | D. Fresneau, "Biologie et comportement social d'une fourmi ponerine néotropicale (*Pachycondyla apicalis*)," Ph.D. thesis, Université Paris XIII, Villetaneuse.

99 | V. Dietemann and C. Peeters, "Queen influence on the shift from trophic to reproductive eggs laid by workers of the ponerine ant *Pachycondyla apicalis*," *Insectes Sociaux* 47(3): 223–228 (2000).

100 | F. L. W. Ratnieks, "Reproductive harmony via mutual policing by workers in eusocial Hymenoptera," *American Naturalist* 132(2): 217–236 (1988).

101 | V. Dietemann and C. Peeters, "Queen influence on the shift from trophic to reproductive eggs laid by workers of the ponerine ant *Pachycondyla apicalis*," *Insectes Sociaux* 47(3): 223–228 (2000).

102 | P. S. Oliveira and B. Hölldobler, "Dominance orders in the ponerine ant *Pachycondyla apicalis* (Hymenoptera, Formicidae)," *Behavioral Ecology and Sociobiology* 27(6): 385–393 (1990).

selection is stronger than the forces of either kin or colony-level selection. Individual competition is fierce, with the result that even dominant workers can be replaced by other young workers with greater reproductive potential.

We see in *Pachycondyla apicalis* many of the same evolutionary selection patterns previously encountered in *Harpegnathos saltator*, but with a significant difference. After the queen's demise, *Harpegnathos* colonies become potentially perpetual gamergate colonies, whereas the colonies of *Pachycondyla apicalis*, whose workers do not mate (even though they have spermathecae), die out after the founding queen's death. From that restriction alone, it is logically adaptive for the workers to begin producing males toward the end of their queen's reproductive tenure.[103]

Pachycondyla obscuricornis, a Neotropical ponerine species closely related to *Pachycondyla apicalis*, is almost identical to it in social structure.[104] Here, too, workers establish reproductive dominance orders once the fertile queen becomes senescent or dies, and many eggs laid by workers or virgin queens are destroyed by nestmates.[105] In one colony observed, however, half the workers were able to add their eggs to the colony pile. Success was enhanced by extra effort: while depositing their eggs, some of the ants shuffled them within the pile for 5 to 10 minutes and then often guarded them by sitting on the egg pile for up to 60 minutes after deposition.[106] The pile is constantly inspected by patrolling ants, and shuffling and guarding probably reduces the chance that a newly deposited egg will be detected and destroyed by other ants. In fact, it has recently been demonstrated that ant eggs, including those of *Pachycondyla* species, are marked with caste- or rank-specific blends of hydrocarbons, and egg recognition and destruction in ant colonies is based on these particular signals.[107] It seems likely that egg shuffling on

---

103 | V. Dietemann and C. Peeters, "Queen influence on the shift from trophic to reproductive eggs laid by workers of the ponerine ant *Pachycondyla apicalis*," *Insectes Sociaux* 47(3): 223–228 (2000).

104 | D. Fresneau, "Développement ovarien et statut social chez une fourmi primitive *Neoponera obscuricornis* Emery (Hym. Formicidae, Ponerinae)," *Insectes Sociaux* 31(4): 387–402 (1984).

105 | P. S. Oliveira and B. Hölldobler, "Agonistic interactions and reproductive dominance in *Pachycondyla obscuricornis* (Hymenoptera: Formicidae)," *Psyche* (Cambridge, MA) 98: 215–225 (1991).

106 | P. S. Oliveira and B. Hölldobler, "Agonistic interactions and reproductive dominance in *Pachycondyla obscuricornis* (Hymenoptera: Formicidae)," *Psyche* (Cambridge, MA) 98: 215–225 (1991).

107 | A. Endler, J. Liebig, T. Schmitt, J. E. Parker, G. R. Jones, P. Schreier, and B. Hölldobler, "Surface hydrocarbons of queen eggs regulate worker reproduction in a social insect," *Proceedings of the National Academy of Sciences USA* 101(9): 2945–2950 (2004); and P. D'Ettorre, J. Heinze, and F. L. W. Ratnieks, "Worker policing by egg eating in the ponerine ant *Pachycondyla inversa*," *Proceedings of the Royal Society of London B* 271: 1427–1434

the egg pile confounds the egg signals signifying origin and hence lowers the risk of an egg being detected. Egg shuffling was especially conspicuous in *Pachycondyla obscuricornis*, but it was also seen in *Pachycondyla unidentata* and to a lesser degree in *Pachycondyla apicalis*.

A very peculiar alternative reproductive system has been discovered in one colony of *Pachycondyla obscuricornis* collected in Brazil.[108] Three adults differed morphologically from the workers in its population: although wingless, their thorax was more or less like that of winged queens. The scutum and scutellum were separated by a suture, and the metanotum was larger than that of ordinary workers. Another 11 such individuals were found by opening cocoons. Most likely, these were intercastes as defined by Christian Peeters—forms truly intermediate in anatomy between workers and queens. They also differed from ergatoid queens in lacking distinctive queen caste features.[109]

Surprisingly, two of the *Pachycondyla obscuricornis* intercastes were mated and laid eggs. All of the workers were virgin, but some laid distinct trophic eggs, which were fed to the two reproductive intercastes. When the intercastes were experimentally removed, dominance interactions sprang up among the workers, some of whom then began to lay reproductive eggs. Intercastes with a reproductive function have not been reported elsewhere in *Pachycondyla*, although intercastes have been described in the ponerine species *Ponera pennsylvanicus*, *Hypoponera bondroiti*, and *H. eduardi*.[110]

The term *intercaste* describes a morphological state only and implies nothing about function. An important characteristic of intercastes is their morphological variability

---

(2004). Also, eggs laid by *Dinoponera quadriceps* alpha workers carry distinctly more of the specific alpha hydrocarbon 9-hentriacontene than eggs laid by lower-ranked workers, and alpha workers destroy subordinates' eggs; see T. Monnin and C. Peeters, "Cannibalism of subordinates' eggs in the monogynous queenless ant *Dinoponera quadriceps*," *Naturwissenschaften* 84(11): 499–502 (1997).

108 | O. Düssmann, C. Peeters, and B. Hölldobler, "Morphology and reproductive behaviour of intercastes in the ponerine ant *Pachycondyla obscuricornis*," *Insectes Sociaux* 43(4): 421–425 (1996).

109 | In most species, intercastes arise from accidental deviation from normal caste differentiation. Because in some ponerine species and a few species of other ant subfamilies intercastes exist as reproductives, Jürgen Heinze has questioned whether a distinction between "reproductive intercastes" and "ergatoid queens" is helpful. He has suggested instead a distinction between accidental deviations and normal caste differentiation involving wingless female reproductives in J. Heinze, "Intercastes, intermorphs, and ergatoid queens: who is who in ant reproduction?" *Insectes Sociaux* 45(2): 113–124 (1998).

110 | For review, see C. Peeters, "Ergatoid queens and intercastes in ants: two distinct adult forms which look morphologically intermediate between workers and winged queens," *Insectes Sociaux* 38(1): 1–15 (1991).

within a species and even within a colony. Thoracic structure, body size, and internal traits may vary across a continuum between the queen and worker extremes.

Intercastes are rare and aberrant in some species of ants, but are common and reproductively functional in others. Such evidently normal intercastes are found in some leptothoracine species, the dolichoderine *Technomyrmex albipes*, and the myrmicine *Myrmecina graminicola*.[111] Some of these reproductive intercastes are able to replace the founding queen, a substitution that has been associated with colony fission. We do not know if intercastes are produced in queenright colonies of *Pachycondyla obscuricornis* in addition to queenless ones. It is plausible, however, that reproduction through mated intercastes and colony fission is an adaptive alternative strategy of colony propagation in this species. It may even represent a stage along the evolutionary pathway toward reproduction through gamergates. Such is the case in the Australian *Pachycondyla sublaevis*, a species with exceptionally small colonies, rarely more than ten individuals, that form dominance hierarchies with a single gamergate at the top position.[112]

Overall, ponerine societies are characterized by patterns of strife and interindividual conflict in the course of fierce competition for reproductive dominance. But fighting to gain dominance is costly for the colony, and excessive, prolonged strife could be fatal. Recently, the first quantitative analysis of such costs has been carried out for a society of *Pachycondyla obscuricornis*.[113] Researchers found that aggressive behaviors within the society are reflected in colony-level carbon dioxide emissions. A colony as a whole consumes more energy during periods of fighting than during periods of social tranquility. The same study revealed another, quite remarkable fact. Just after the removal of the queen, the carbon dioxide emission of the colony dropped. The downturn coincided with a decrease in activity of individual workers, which occurred about 3 hours after the loss of the queen. Bruno Gobin and his coauthors, who conducted the analysis, state: "Removing the queen thus has a clear effect on worker behavior, apparently reducing their inclination to work for

---

111 | C. Peeters, "Ergatoid queens and intercastes in ants: two distinct adult forms which look morphologically intermediate between workers and winged queens," *Insectes Sociaux* 38(1): 1–15 (1991).

112 | C. Peeters, S. Higashi, and F. Ito, "Reproduction in ponerine ants without queens: monogyny and exceptionally small colonies in the Australian *Pachycondyla sublaevis*," *Ethology Ecology & Evolution* 3(2): 145–152 (1991); and F. Ito and S. Higashi, "A linear dominance hierarchy regulating reproduction and polyethism of the queenless ant *Pachycondyla sublaevis*," *Naturwissenschaften* 78(2): 80–82 (1991).

113 | B. Gobin, J. Heinze, M. Strätz, and F. Roces, "The energy cost of reproductive conflicts in the ant *Pachycondyla obscuricornis*," *Journal of Insect Physiology* 49(8): 747–752 (2003).

the colony. . . . A reduction in work performance is a cost to the colony in itself and might account for the observed reduction in energy consumption. If we correct our CO_2 emission data with the observed activity rates, we still find an overall correlation with aggression." They conclude: "In sum, fighting to gain dominance is thus costly to the colony, both in energetics, as the colony spends more energy overall, and in work performance, as workers carry out less work. These costs decrease once dominance relationships stabilize and egg laying starts."

A general principle of insect sociobiology is that aggression among workers in the queen's presence is relatively rare in ponerine species as well as in those of other ant subfamilies. If young workers in perennial queenright colonies were to strive to reach top rank in a reproductive worker hierarchy, this would constantly disrupt social stability and negatively affect colony productivity, thereby reducing the inclusive fitness of all the colony members. In general, worker reproduction and the formation of hierarchies among workers should be selected against by colony-level selection when the cost to colony productivity outweighs direct benefits.[114] Aggression among workers in the queen's presence is consequently rare in ants, except during the period when the queen is senescent and her fecundity markedly reduced. In such a situation, and especially when the queen dies, workers can be expected to maximize their fitness by attempting to raise sons.[115] The situation is different, however, when workers can mate and dominant gamergates can replace a queen (as in *Harpegnathos*) or a deceased gamergate. Aggression among workers in such societies should be expected across a wider range of costs. And indeed, aggression among ponerine workers in gamergate societies has been found to be more common than in queenright societies.[116]

Aggression and dominance in ponerine societies occur not only among workers,

---

114 | A. F. G. Bourke, "Colony size, social complexity and reproductive conflict in social insects," *Journal of Evolutionary Biology* 12(2): 245–257 (1999); and T. Monnin and F. L. W. Ratnieks, "Policing in queenless ponerine ants," *Behavioral Ecology and Sociobiology* 50(2): 97–108 (2001).

115 | A. F. G. Bourke, "Worker reproduction in the higher eusocial Hymenoptera," *The Quarterly Review of Biology* 63(3): 291–311 (1988).

116 | See previous sections, and also F. Ito and S. Higashi, "A linear dominance hierarchy regulating reproduction and polyethism of the queenless ant *Pachycondyla sublaevis*," *Naturwissenschaften* 78(2): 80–82 (1991); F. Ito, "Social organization in a primitive ponerine ant: queenless reproduction, dominance hierarchy and functional polygyny in *Amblyopone* sp. (*reclinata* group) (Hymenoptera: Formicidae: Ponerinae)," *Journal of Natural History* 27: 1315–1324 (1993); and T. Monnin and F. L. W. Ratnieks, "Reproduction versus work in queenless ants: when to join a hierarchy of hopeful reproductives?" *Behavioral Ecology and Sociobiology* 46(6): 413–422 (1999).

but also among fertile queens that occupy the same nest. In the Neotropical species of *Pachycondyla* close to *inversa* mentioned earlier, queens often found new colonies by pleometrosis, with several unrelated young mated queens living together in a single nest chamber.[117] In such associations, division of labor is strongly affected by aggressive interactions among the queens. Dominant individuals remain in the nest and tend the brood, while one or more of the subordinates accept the risky task of foraging. Egg-laying rates do not differ significantly between nestmate queens, but dominant individuals destroy and eat some of the eggs laid by subordinates.[118] The cuticular hydrocarbon blends obtained from live queens have revealed consistent differences between the CHC profiles of queens with high rank as opposed to those with low rank; in particular, only high-ranking queens possess considerable amounts of pentadecane and heptadecane.[119] Interestingly, in the *Pachycondyla* species close to *inversa*, pleometrosis leads to permanent primary polygyny, with unrelated queens still sharing the same nest when the colony reaches full size. In most of these mature colonies, queens contribute equally to the production of workers and sexuals. Egg cannibalism, common in founding associations in this species, does not occur in mature colonies.[120]

Optimal skew theory provides a framework for explaining social interactions between cofounding queens.[121] According to this theory, the behavior of cofounding unrelated queens depends on several parameters: relative fighting ability, productivity of the group in comparison to a solitary foundress, and extent of ecological constraints. Kerstin Kolmer and Jürgen Heinze have argued convincingly that "if ecological constraints are large, a queen with low fighting ability might still benefit from joint nesting, even if she is forced to specialize in dangerous tasks and some of

---

117 | B. Trunzer, J. Heinze, and B. Hölldobler, "Cooperative colony founding and experimental primary polygyny in the ponerine ant *Pachycondyla villosa*," *Insectes Sociaux* 45(3): 267–276 (1998).

118 | K. Kolmer and J. Heinze, "Rank orders and division of labour among unrelated cofounding ant queens," *Proceedings of the Royal Society of London B* 267: 1729–1734 (2000).

119 | J. Tentschert, K. Kolmer, B. Hölldobler, H.-J. Bestmann, J. H. C. Delabie, and J. Heinze, "Chemical profiles, division of labor and social status in *Pachycondyla* queens (Hymenoptera, Formicidae)," *Naturwissenschaften* 88(4): 175–178 (2001).

120 | J. Heinze, B. Trunzer, B. Hölldobler, and J. H. C. Delabie, "Reproductive skew and queen relatedness in an ant with primary polygyny," *Insectes Sociaux* 48(2): 149–153 (2001).

121 | H. K. Reeve and F. L. W. Ratnieks, "Queen-queen conflicts in polygynous societies: mutual tolerance and reproductive skew," in L. Keller, ed., *Queen Number and Sociality in Insects* (New York: Oxford University Press, 1993), pp. 45–85.

her eggs are destroyed by the dominant. . . . What might tip the scale very strongly towards pleometrosis in *Pachycondyla* cf. *inversa* is that, in striking contrast to most other cooperatively founding ants, the queen number in founding associations is not regulated to monogyny after eclosion of workers." As expected from optimal skew theory, reproduction is quite evenly partitioned among unrelated queens in polygynous societies of this species of *Pachycondyla*. It obviously pays for a subordinate queen to take over the foraging task only if she does not risk being expelled from the nest once the colony has raised workers.[122]

In contrast to founding groups, coexisting queens in mature colonies of *Pachycondyla* species close to *inversa* do not seem to exhibit aggressive behavior toward each other, and as already mentioned, there is no indication of significant reproductive skew. A different situation exists in the polygynous ponerine ant *Odontomachus chelifer*, of southeastern Brazil.[123] One colony collected contained 13 dealated queens, 27 alate females, 5 males, and approximately 130 workers. Studies revealed intricate agonistic interactions among the dealate queens. Behavioral domination entailed an array of stereotyped displays that occasionally escalated from vigorous antennation bouts to full mandibular strikes. In extreme situations, a dominant queen grasped the subordinate opponent and lifted her off the ground, while the subordinate assumed a pupal posture (Figure 8-16). Behavior during domination contests and the rank positions inferred from them correlated well with individual egg production and ovarian development. Highly ranked queens possessed better-developed ovaries, had all been inseminated, and laid more eggs. Although they also contested among themselves, with one individual holding the top position, their aggressive actions were also aimed at alate virgin females. Some of the latter had in fact developed ovaries and were laying eggs. In general, aggression toward egg-laying queens and the instant destruction of newly laid eggs were common, and conspicuous aggression was practiced by the dominant individuals. Several small egg piles were scattered among different nest chambers, each of which were guarded by individual queens. Sometimes a newly laid egg was destroyed by an inspecting queen

---

122 | K. Kolmer and J. Heinze, "Rank orders and division of labour among unrelated cofounding ant queens," *Proceedings of the Royal Society of London B* 267: 1729–1734 (2000).

123 | F. N. S. Medeiros, L. E. Lopes, P. R. S. Moutinho, P. S. Oliveira, and B. Hölldobler, "Functional polygyny, agonistic interactions and reproductive dominance in the Neotropical ant *Odontomachus chelifer* (Hymenoptera, Formicidae, Ponerinae)," *Ethology* 91(2): 134–146 (1992).

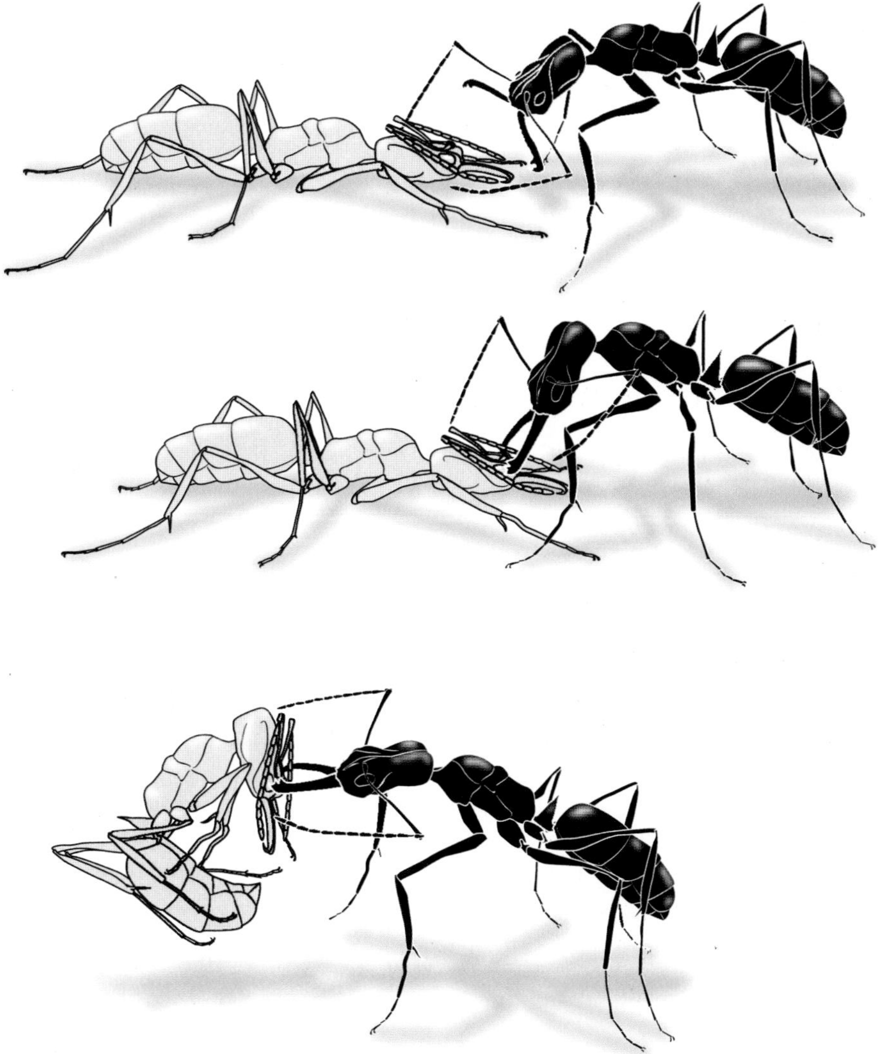

FIGURE 8-16. Dominance behavior between nestmate queens of the tropical American predatory ant *Odontomachus chelifer*. *Above*: The dominant queen (black) threatens with open mandibles while her sister crouches in submission. *Middle*: She escalates the conflict, grasping the head of her sister and then lifting her off the ground. *Below*: When handled this way, the sister ant indicates submission by pulling her legs into the posture of an immature pupa. Based on an original drawing by Katherine Brown-Wing in B. Hölldobler and E. O. Wilson, *Journey to the Ants* (Cambridge, MA: The Belknap Press of Harvard University Press, 1994).

if the egg pile containing it was not attended. During these aggressive contests, the workers remained relatively neutral.

The data on *Odontomachus chelifer* clearly reveals a dominance hierarchy among the queens, but the control exerted by the highest-ranked individuals does not entirely inhibit oviposition by lower-ranked ones. Because we lack information concerning ecological constraints, mode of colony foundation, and relatedness patterns among the queens in this species of *Odontomachus*, it is not possible to provide evolutionary scenarios that explain the origin of its peculiar social organization.[124]

PLATYTHYREA PUNCTATA: EXTREME PLASTICITY IN REPRODUCTION

We have now reviewed the bewildering diversity of social organizations and reproductive modes found in ponerine species, so far as we know. Researchers have confirmed that physical castes, especially those concerned with reproduction, have progressed only slightly from one species to another. As a consequence, there is a high degree of evolutionary plasticity in the reproductive systems and within-colony interactions. This phenomenon is illustrated perhaps most impressively in the extreme complexity of reproductive tactics in the Neotropical ponerine ant *Platythyrea punctata*.[125] In addition to reproduction by gamergates (mated workers) and also occasional mated queens, thelytokous parthenogenesis is very common. That is, unmated workers and intercastes (true morphological intermediates between workers and queen) produce diploid offspring from unfertilized eggs. It remains unknown whether unmated dealate queens, which also occasionally occur in colonies, are also capable of thelytokous reproduction. Furthermore, as in most other hymenopteran species, at least some unmated individuals are capable of producing males from unfertilized eggs by arrhenotokous parthenogenesis.

The ability to produce diploid female offspring by thelytokous parthenogenesis

---

124 | For a comprehensive review of queen interactions in polygynous ants, see J. Heinze, "Queen-queen interactions in polygynous ants," in L. Keller, ed., *Queen Number and Sociality in Insects* (New York: Oxford University Press, 1993), pp. 334–361; and J. Heinze, "Reproductive conflict in insect societies," *Advances in the Study of Behavior* 34: 1–57 (2004).

125 | K. Schilder, J. Heinze, and B. Hölldobler, "Colony structure and reproduction in the thelytokous parthenogenetic ant *Platythyrea punctata* (F. Smith) (Hymenoptera, Formicidae)," *Insectes Sociaux* 46(2): 150–158 (1999).

is, however, rare among the hymenopterans. In addition to *Platythyrea punctata*, obligate or facultative thelytokous parthenogenesis has been confirmed in only three other ant species: the myrmicine *Pristomyrmex pungens*, the cerapachyine *Cerapachys biroi*, and the formicine *Cataglyphis cursor*.[126, 127, 128]

Thelytokous reproduction has been seen as an evolutionary dead end that may trap genotypes adapted to a specialized ecological niche. Such systems presumably lack the ability to rapidly generate genotype diversity and hence provide adaptive variants to be selected as environmental conditions change. However, rates of evolution under unisexual conditions may be of similar magnitude, and parthenogens may, in theory at least, perform as well as obligately bisexual species under similar environmental conditions, provided mutation rates are elevated and environmental sensitivity is depressed to a sufficient degree.[129] The origin of a parthenogenetic lineage may therefore be a dynamic evolutionary process, based on alterations in the life history patterns and environmental conditions that favor unisexual populations. One such condition is adaptation to environments where locating a mate is difficult due to low colony density. Another is adaptation to environments where there is a high risk of colony fragmentation and hence loss of the queen. And finally, parthenogenesis may arise when it is favorable for small worker groups to found new colonies, particularly in environments where an exceptional colonizing ability proves advantageous. Although thelytoky limits genetic recombination, long-term selection will allow the persistence of thelytokous clonal populations that collectively possess broad tolerance of abiotic environmental variation.[130]

Short episodes of sexual reproduction can be expected, allowing for increased rates of phenotypic evolution and the generation of new clonal lineages. Indeed, whereas a rarity of males in *Platythyrea punctata* has been noted in both the field

---

126 | K. Tsuji, "Obligate parthenogenesis and reproductive division of labor in the Japanese queenless ant *Pristomyrmex pungens*: comparison of intranidal and extranidal workers," *Behavioral Ecology and Sociobiology* 23(4): 247–255 (1988).

127 | K. Tsuji and K. Yamauchi, "Production of females by parthenogenesis in the ant, *Cerapachys biroi*," *Insectes Sociaux* 42(3): 333–336 (1995).

128 | H. Cagniant, "La parthénogénèse thélytoque et arrhénotoque chez la fourmi *Cataglyphis cursor* Fonsc. (Hym. Form.) cycle biologique en élevage des colonies avec reine et des colonies sans reine," *Insectes Sociaux* 26(1): 51–60 (1979).

129 | W. Gabriel and G. P. Wagner, "Parthenogenetic populations can remain stable in spite of high mutation rate and random drift," *Naturwissenschaften* 75(4): 204–205 (1988).

130 | M. Lynch, "Destabilizing hybridization, general-purpose genotypes and geographic parthenogenesis," *The Quarterly Review of Biology* 59(3): 257–290 (1984).

and laboratory, they can still function: 2 out of 824 females from field colonies sampled were found to be inseminated.[131] And the data collected thus far from multiple field localities strongly suggest that parthenogenesis is indeed the predominant mode of reproduction in all sampled populations of this species. This mode of reproduction is therefore not confined to small, marginal subpopulations where thelytoky is generally expected to be advantageous in nature, but instead is general for the species.

Although all the workers in a *Platythyrea punctata* colony are equally capable of laying eggs, reproduction is monopolized by one worker or, occasionally, a pair of workers. Other workers perform tasks necessary for nest maintenance, foraging, and brood rearing. Such reproductive division of labor is regulated by antagonistic interactions very similar to those that operate among workers, gamergates, or queens of other ponerines.[132] This may seem inconsistent at first, because *Platythyrea punctata* colonies are clones; no genetic conflict exists between members of the colony and neither policing nor reproductive competition are to be expected. However, uncontrolled reproduction by all workers can, as we have stressed in our overall presentation of ponerine biology, seriously disrupt colony efficiency. Recent analyses have in fact demonstrated exactly that causal phenomenon. Worker policing through aggressive attacks by nestmates against additional reproducing workers keeps the number of *Platythyrea punctata* reproductives low and thereby group efficiency in brood rearing significantly enhanced.[133] In fact, Jürgen Heinze and his coworkers showed that rearing of brood is most efficient if only one egg layer is present. In addition, the scientists found that only a small group of high-ranking ants engage in nestmate policing through aggression and egg eating and thus "prevent destructive selfishness."

---

131 | K. Schilder, J. Heinze, and B. Hölldobler, "Colony structure and reproduction in the thelytokous parthenogenetic ant *Platythyrea punctata* (F. Smith) (Hymenoptera, Formicidae)," *Insectes Sociaux* 46(2): 150–158 (1999).

132 | J. Heinze and B. Hölldobler, "Thelytokous parthenogenesis and dominance hierarchies in the ponerine ant, *Platythyrea punctata*," *Naturwissenschaften* 82(1): 40–41 (1995).

133 | A. Hartmann, J. Wantia, J. A. Torres, and J. Heinze, "Worker policing without genetic conflicts in a clonal ant," *Proceedings of the National Academy of Sciences USA* 100(22): 12836–12840 (2003).

AGGRESSION AND DOMINANCE: ORIGIN AND LOSS

Aggression among nestmates and the establishment of reproductive dominance are typically restricted to species with small mature societies. This is true not only for ponerine species, but also for other ants, including species of the myrmicine genera *Leptothorax* and *Temnothorax*.[134] Such societies consist of only a few dozen or, rarely, several hundred individuals, of which only a small proportion engage in aggressive interactions. In species with mature colonies containing hundreds or thousands of potential contenders, physical aggression is probably no longer suitable to regulate reproduction because of ineffectiveness and high cost to the individuals.[135] In addition, physical antagonism occurs most frequently in species whose colonies have a high average ratio of reproductives to nonreproductives. Such is the case when queen-worker dimorphism is limited and workers have an elevated reproductive potential. Aggression is hence common in permanently queenless ponerine ants, where caste specialization is only feebly developed and all workers are potentially capable of mating and of laying fertilized eggs. To demonstrate this principle, let us return once more to *Harpegnathos saltator*, where workers exhibit an extreme degree of reproductive resilience.

HARPEGNATHOS: RESILIENCE IN REPRODUCTIVE BEHAVIOR

Although *Harpegnathos* workers never start colonies on their own in nature, laboratory experiments have proved that they retain the behavioral repertoire for colony founding. In one experiment illustrating this capacity,[136] several kinds of colony fragments were set up in the laboratory. Some fragments comprised only a single infertile worker, others 3 infertile workers, and still others 3 gamergates. Food was

---

134 | J. Heinze, B. Hölldobler, and C. Peeters, "Conflict and cooperation in ant societies," *Naturwissenschaften* 81(11): 489–497 (1994).

135 | H. K. Reeve and F. L. W. Ratnieks, "Queen-queen conflicts in polygynous societies: mutual tolerance and reproductive skew," in L. Keller, ed., *Queen Number and Sociality in Insects* (New York: Oxford University Press, 1993), pp. 45–85.

136 | J. Liebig, B. Hölldobler, and C. Peeters, "Are ant workers capable of colony foundation?" *Naturwissenschaften* 85(3): 133–135 (1998).

supplied ad libitum. Although 20 to 40 percent of the workers in these experimental arrangements died before they produced adult offspring, successful founding groups produced between 2 and 14 workers within 160 days and continued to grow. Two founding groups consisting of 1 and 3 originally infertile workers, respectively, reached sizes of 110 and 167 workers after only 1 year. Most remarkable was the case in which an originally infertile single worker produced 2 males within the first 23 weeks, suggesting that she was not mated. Thirteen weeks later, her first daughter eclosed, indicating that she had mated with her son. In a second case, a single gamergate produced 6 workers before she died. These workers produced males, and after several weeks new workers eclosed, indicating again that sons can mate with their mothers.

Harpegnathos workers clearly retain the behavioral repertoire for independent colony founding, although this behavior is probably rarely expressed in nature. The only scenario in which such behavioral resilience may be adaptive is the accidental destruction of significant parts of the elaborate nest that protects colonies against enemies and flooding. While this may be a relatively uncommon event, surviving workers could continue the colony or even start a new one, either alone or with nestmates and brood. Alternatively, the retention of the ancestral trait of independent colony foundation by workers can be explained simply by evolutionary inertia, since they have evolved from species with totipotent females.[137]

COLONY SIZE AS AN ECOLOGICAL ADAPTATION

Ponerine societies that feature internal reproductive competition, dominance structures, and policing are characterized by two additional traits: poorly developed division of labor among workers and simple communication systems. Foragers are usually solitary huntresses and engage in no recruitment communication. Most of those species build relatively simple nests in soil or rotting wood, with a few exceptions, such as *Harpegnathos saltator*. The colonies have relatively small nest populations. Their demography appears tailored to fit narrow niches where they can

---

137 | J. Liebig, B. Hölldobler, and C. Peeters, "Are ant workers capable of colony foundation?" *Naturwissenschaften* 85(3): 133–135 (1998).

sustain themselves as specialized predators on particular arthropod prey. Some of them also sporadically collect nectar from extrafloral nectaries and in rare instances even solicit honeydew from homopteran insects (see Plate 26).[138]

A few ponerine species have overcome the intrinsic limitations of their colony size caused by hierarchical social organizations and internal competition and evolved to highly efficient group predators with sophisticated mass recruitment communication based on chemical and motor signals.[139] These ants often are also specialist predators of other social insects or large single prey, or else they are generalists and sweep the floor, capturing every arthropod they can catch and pin down, in the manner of driver ants. Such colonies have abandoned hierarchical organization; they assign reproductive monopoly to the queen. Worker sterility is physiologically regulated by queen-derived signals, in turn allowing the vast expansion of worker population size. The Asian species *Leptogenys distinguenda* has colonies of as many as 30,000 workers.[140]

Another unusual exception to the ponerine rule, not connected with group predation, characterizes the African stink ant *Pachycondyla tarsata* (formerly *Paltothyreus tarsatus*). This species is widely distributed throughout Africa south of the Sahara. The ants are hunters and scavengers, and although the workers forage individually, they recruit nestmates with chemical trail signals when retrieving large or abundant prey items.[141]

An extensive field study in East and West Africa has yielded unexpected information concerning nest and colony size of this species.[142] Each of 14 colonies excavated

---

138 | B. Hölldobler and E. O. Wilson, *The Ants* (Cambridge, MA: The Belknap Press of Harvard University Press, 1990).

139 | B. Hölldobler, "Multimodal signals in ant communication," *Journal of Comparative Physiology A* 184(2): 129–141 (1999).

140 | U. Maschwitz, S. Steghaus-Kovaç, R. Gaube, and H. Hänel, "A South East Asian ponerine ant of the genus *Leptogenys* (Hym., Form.) with army ant life habits," *Behavioral Ecology and Sociobiology* 24(5): 305–319 (1989).

141 | B. Hölldobler, "Communication during foraging and nest-relocation in the African stink ant, *Paltothyreus tarsatus* Fabr. (Hymenoptera, Formicidae, Ponerinae)," *Zeitschrift für Tierpsychologie* 65(1): 40–52 (1984); E. Janssen, B. Hölldobler, and H. J. Bestmann, "A trail pheromone component of the African stink ant, *Pachycondyla* (*Paltothyreus*) *tarsata* Fabricius (Hymenoptera: Formicidae: Ponerinae)," *Chemoecology* 9(1): 9–11 (1999); and B. Hölldobler and E. O. Wilson, *The Ants* (Cambridge, MA: The Belknap Press of Harvard University Press, 1990).

142 | U. Braun, C. Peeters, and B. Hölldobler, "The giant nests of the African stink ant *Paltothyreus tarsatus* (Formicidae, Ponerinae)," *Biotropica* 26(3): 308–311 (1994); and U. Braun, B. Hölldobler, and C. Peeters, "Colonial reproduction and large queen-worker dimorphism in the African stink ant *Pachycondyla tarsata*," unpublished.

in the Shimba Hills Reserve of Kenya contained a single queen in the company of 157 to 2,444 workers. The smaller colonies were most likely younger ones, a status also reflected in the size of their nests and the number of brood. These data have been supplemented by others obtained in Ivory Coast national parks. Excavations of 41 nests in three different habitats yielded mature colonies with an average of more than 7,000 workers, close to 6,000 winged queens, more than 2,000 worker pupae, more than 10,000 larvae, and in excess of 1,000 eggs. A curious geographical variation was revealed: whereas all colonies excavated in savanna were monogynous with maximum worker populations of about 4,000, all colonies in gallery forest of the Comoé National Park and rain forest of the Tai National Park in Côte d'Ivoire, West Africa, possessed 2 to 9 queens per colony. The largest colonies were found in the gallery forest of the Comoé National Park. Since 2 of the 60 excavated founding colonies had 2 founding queens each, it is possible that polygynous colonies of *Pachycondyla tarsata* originate from the associations of 2 or more queens. Colony founding appears to be semiclaustral, since founding queens were observed foraging.

The nests of *Pachycondyla tarsata* are gigantic. The core area can be recognized by crater-shaped nest entrances and conspicuous mounds of soil and refuse piles spread over a large area (Figure 8-17). A majority of the nest chambers are located 30 to 150 centimenters below the surface. An extensive system of shallow tunnels 5 to 10 centimeters underground lead from the central nest chambers to foraging grounds up to 60 meters away. The tunnels are connected to the surface through several vertical, irregularly spaced ducts through which foragers can leave the tunnels to enter the surface foraging grounds. The subterranean foraging tunnel system of one nest excavated covered an astounding area of about 1,200 square meters. In addition, a single colony can occupy two or more nest centers spaced several meters from each other and connected by subterranean tunnels about 50 to 70 centimeters underground. Such enormous, decentralized territories, when crisscrossed by subterranean trunk route tunnels, allow workers to forage throughout the interior of their large colony territory while reducing the loss of foragers through predation.

FIGURE 8-17. *Above*: Core area of a nest of the African stink ant *Pachycondyla tarsata* (formerly *Paltothyreus tarsatus*), showing the funnel-shaped entrances surrounded by mounds of soil and refuse piles. Part of the underground tunnels leading to the foraging grounds are also indicated, together with exit holes. *Below*: Vertical section through the central chambers and connecting galleries. This drawing is based on numerous field sketches and photographs taken during the excavation. From U. Braun, C. Peeters, and B. Hölldobler, "The giant nests of the African stink ant *Paltothyreus tarsatus* (Formicidae, Ponerinae)," *Biotropica* 26(3): 308–311 (1994).

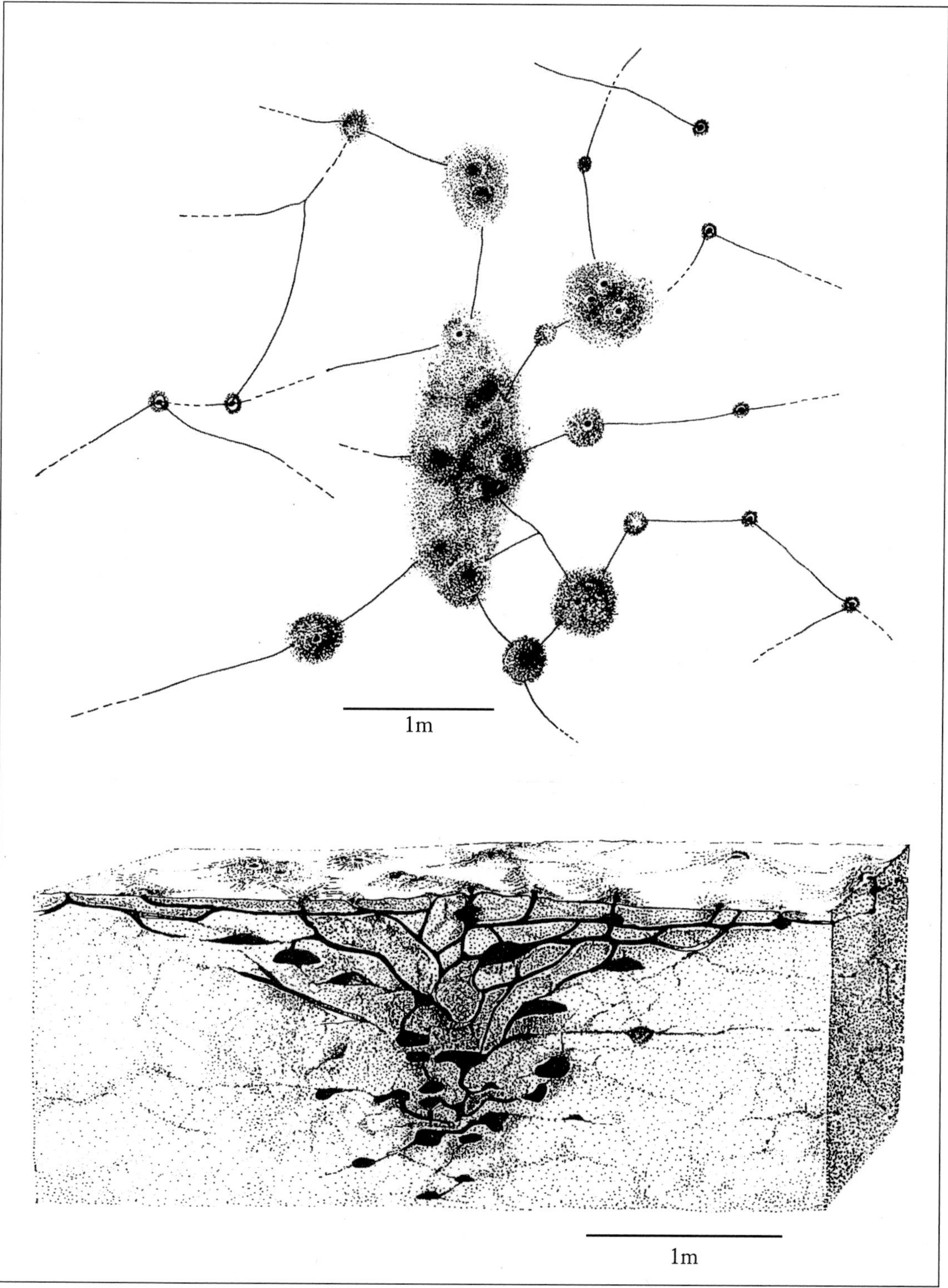
1m
1m

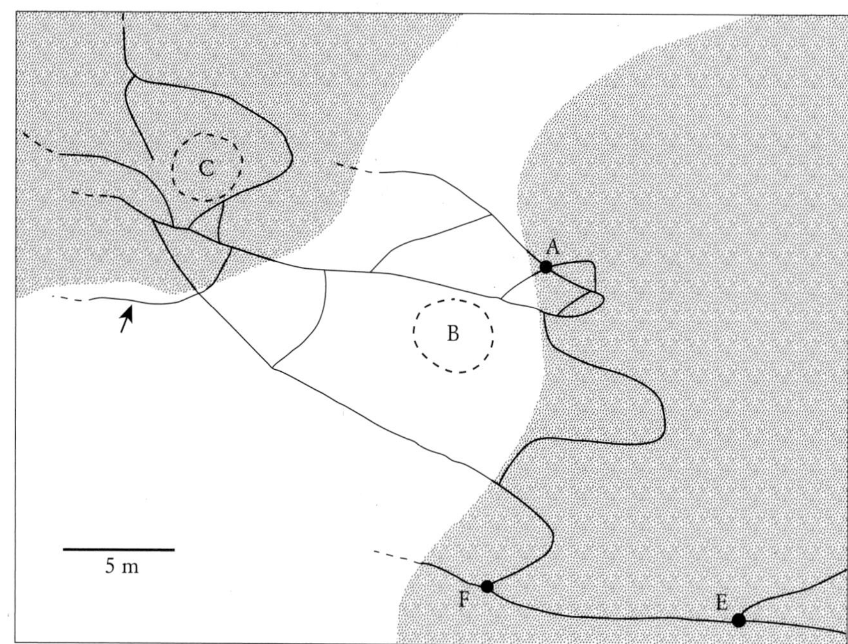

FIGURE 8-18. The subterranean tunnel system of a nest of the African stink ant *Pachycondyla tarsata* (formerly *Paltothyreus tarsatus*). Only shallow tunnels are indicated (5 to 10 centimeters below the surface). The arrow indicates the spot where excavation began. A, E, and F are the locations of deep chambers where workers and brood were collected. Many workers were also found aboveground around B and C. Stippled area indicates forest and thicket vegetation; clear area represents grassland. From U. Braun, C. Peeters, and B. Hölldobler, "The giant nests of the African stink ant *Paltothyreus tarsatus* (Formicidae, Ponerinae)," *Biotropica* 26(3): 308–311 (1994).

It also enables more precise orientation through the often restricted and confusing lighting conditions of forests (Figure 8-18).[143]

Dissections of *Pachycondyla tarsata* workers from field and laboratory colonies revealed that workers are sterile in the presence of a queen. However, when a group of approximately 500 workers were experimentally separated from their queen, workers began egg laying about 15 weeks later. This response suggests that *Pachycondyla tarsata* worker fertility, as in many other ant species, is physiologically inhibited by queen signals. But this creates a problem. *Pachycondyla tarsata* colonies are very large, and one must assume that the queen pheromones are spread to workers throughout the expansive nest, but how this feat is achieved remains unknown. One possibility is that queen-laid eggs, labeled with a chemical queen signature, are

143 | B. Hölldobler, "Canopy orientation: a new kind of orientation in ants," *Science* 210: 86–88 (1980).

transported to most nest chambers, so that workers are kept informed of the queen's presence and degree of fertility. Such mode of transmission of queen signals has in fact been recently documented in another ponerine as well as in the formicine ant *Camponotus floridanus*. Circumstantial evidence suggests that this mechanism of queen signaling may also be employed in species of *Oecophylla* weaver ants.[144]

More than likely, queens of *Pachycondyla tarsata* are singly mated. Although virgin queens have been reported to call with wafted pheromones from the ground,[145] massive nuptial flights were observed to take place in the Comoé National Park after the second heavy rainfall that followed the dry season. During the flights, which occurred in the early morning, hundreds of males were observed flying around treetops about 25 meters above the ground. Somewhat later, hundreds of virgin queens flew into these male aggregations. Copulation was initiated on the wing. Not long thereafter, clusters of copulating pairs began falling to the ground. Up to ten males competed to gain access to each female, but only one succeeded. As soon as the copulating pair separated, the queen flew away from the male aggregation, having mated only once.[146]

Here, again, we should expect that workers are selected to produce sons and rear nephews, to which they are more closely related than to brothers. Because they have functional ovaries and can lay viable eggs, and because the nests are so extensive that many of the workers rarely have direct physical contact with the queen, they can escape policing by the queen. Further, no sign of worker-worker policing has ever been detected. Yet workers still refrain from reproduction in queenright colonies. We conclude yet again that worker sterility in the presence of the queen and without any signs of active policing is favored by colony-level selection. However, with the removal of the queen, the situation changes. Under this condition, workers become fertile, and agonistic behavior among them commences. This response is best explained by individual direct selection. A monogynous *Pachycondyla tarsata* colony that has lost its queen is doomed unless it adopts a replacement queen

---

144 | A. Endler, J. Liebig, T. Schmitt, J. E. Parker, G. R. Jones, P. Schreier, and B. Hölldobler, "Surface hydrocarbons of queen eggs regulate worker reproduction in a social insect," *Proceedings of the National Academy of Sciences USA* 101(9): 2945–2950 (2004); and P. D'Ettorre, J. Heinze, and F. L. W. Ratnieks, "Worker policing by egg eating in the ponerine ant *Pachycondyla inversa*," *Proceedings of the Royal Society of London B* 271: 1427–1434 (2004).

145 | M. Villet, R. Crewe, and H. Robertson, "Mating behavior and dispersal in *Paltothyreus tarsatus* Fabr. (Hymenoptera: Formicidae)," *Journal of Insect Behavior* 2(3): 413–417 (1989).

146 | U. Braun, B. Hölldobler, and C. Peeters, unpublished observation.

from outside. In such a situation, individual-level selection "overpowers" colony-level selection. Older workers, whose ovaries have degenerated and can no longer be activated, rear nephews instead. This particular response can be explained as an outcome of kin selection.

There remains one additional feature of *Pachycondyla tarsata* colonies of considerable sociobiological interest. The nest structures of the colonies represent a tremendous investment for their members. In polygynous or oligogynous societies, which regularly occur in the Tai rain forest and occasionally also in the gallery forest, the precious "real estate" is maintained for multiple generations. Unfortunately, we do not as yet know how coexistence of up to nine queens in rain forest colonies comes about—or its genetic consequences.

PACHYCONDYLA: HYPERDIVERSITY SUMMARIZED

Let us summarize the astonishing diversity of social organizations thus far encountered in the ponerine genus *Pachycondyla* alone. There are first the relatively small monogynous societies of *Pachycondyla apicalis* and *Pachycondyla obscuricornis*, with dominance interactions and policing among workers. Next are gamergate societies of the queenless *Pachycondyla sublaevis*, which are based on a "reversed gerontocracy," whereby young individuals mate and replace older individuals in the dominance order of reproduction. Then an utterly different organization is encountered in the large colonies of the group-raiding *Pachycondyla fochi*, where the reproductive monopoly of one very attractive ergatoid queen is unchallenged by workers. And finally, providing further contrast, are the gigantic colonies of *Pachycondyla tarsata*. No other ant genus is known that displays such an enormous diversity in social organization as *Pachycondyla*.

|| Plate 53. Two *Atta sexdens* media workers cooperate to cut a live twig of a plant

9

THE ATTINE LEAFCUTTERS: THE ULTIMATE SUPERORGANISMS

Because they possess one of the most complex communication systems known in animals, the most elaborate caste systems, air-conditioned nest architecture, and populations into the millions, leafcutter ants deserve recognition as Earth's ultimate superorganisms.

THE ATTINE BREAKTHROUGH

Both human civilization and the evolution of extreme insect superorganisms were attained by agriculture, a form of mutualistic symbiosis of animals with plants or fungi. Human agriculture, which originated about 10,000 years ago, was a major cultural transition that catapulted our species from a hunter-gatherer lifestyle to a technological and increasingly urban existence, accompanied by an enormous expansion of population. Humanity thereby turned itself into a geophysical force and began to alter the environment of the entire planetary surface.

Approximately 50 to 60 million years before this momentous shift, some social insects had already made the evolutionary transition from a hunter-gatherer existence to agriculture. In particular, macrotermitine termites in the Old World and attine ants in the New World invented the culturing of fungi, which then became an essential part of their diet. The most advanced agricultural insect societies, like their human counterparts, rose to ecological dominance. The trend is especially marked in the leafcutter ants.[1]

---

1 | R. Wirth, H. Herz, R. J. Ryel, W. Beyschlag, and B. Hölldobler, *Herbivory of Leaf-Cutting Ants: A Case Study on* Atta colombica *in the Tropical Rainforest of Panama* (New York: Springer-Verlag, 2003).

Whereas most fungus-growing attine ants, comprising the anatomically and behaviorally "primitive" species, gather and process rotting leaf fragments and dead organic material on which they grow their specific fungi, the evolutionary invention of cutting and harvesting live plant material opened up a huge new nutritional niche for the species of *Acromyrmex* and *Atta.* As in human history, the innovation propelled further evolutionary development.[2]

The tribe Attini is a morphologically very distinctive group limited to the New World. Most of its 13 genera and approximately 220 species occur in tropical regions of Mexico and Central and South America. Some species occur in the southern portions of the United States, and several are adapted to arid habitats in the southwestern states. One species, *Trachymyrmex septentrionalis*, ranges north to the pine barrens of New Jersey, while in the opposite direction, several species of *Acromyrmex* penetrate to the cold-temperature deserts of central Argentina.[3]

The attine ants, all of which are fungus growers, are overall a monophyletic group. The two genera of the so-called leafcutter or leaf-cutting ants, *Acromyrmex* and *Atta*, are combined with three additional genera in the derived, monophyletic group of the "higher attines." The remaining eight genera are assembled in the "lower attines," a paraphyletic assemblage of basal lineages.[4, 5]

Most cultivated fungi belong to the basidiomycete family Lepiotaceae (Agaricales: Basidiomycota), with the great majority belonging to two genera, *Leucoagaricus* and *Leucocoprinus* (Leucocoprineae).[6, 7] Ulrich Mueller and his coworkers

---

2 | For an inspiring comparison of the convergent appearances of agriculture in human and ant societies, we refer to T. R. Schultz, U. G. Mueller, C. R. Currie, and S. A. Rehner, "Reciprocal illumination: a comparison of agriculture in humans and in fungus-growing ants," in F. E. Vega and M. Blackwell, eds., *Insect-Fungal Associations: Ecology and Evolution* (New York: Oxford University Press, 2005), pp. 149–190.

3 | B. Hölldobler and E. O. Wilson, *The Ants* (Cambridge, MA: The Belknap Press of Harvard University Press, 1990).

4 | T. R. Schultz and R. Meier, "A phylogenetic analysis of the fungus-growing ants (Hymenoptera: Formicidae: Attini) based on morphological characters of the larvae," *Systematic Entomology* 20(4): 337–370 (1995); and T. R. Schultz and S. G. Brady, "Major evolutionary transitions in ant agriculture," *Proceedings of the National Academy of Sciences USA* 105(14): 5435–5440 (2008).

5 | U. G. Mueller, S. A. Rehner, and T. R. Schultz, "The evolution of agriculture in ants," *Science* 281: 2034–2038 (1998).

6 | I. H. Chapela, S. A. Rehner, T. R. Schultz, and U. G. Mueller, "Evolutionary history of the symbiosis between fungus-growing ants and their fungi," *Science* 266: 1691–1694 (1994).

7 | G. Hinkle, J. K. Wetterer, T. R. Schultz, and M. L. Sogin, "Phylogeny of the attine ant fungi based on analysis of small subunit ribosomal RNA gene sequences," *Science* 266: 1695–1697 (1994).

reason that because "most basal attine lineages cultivate leucocoprineous mutualists, attine fungi culture likely originated with the cultivation of leucocoprineous fungi."[8] The majority of fungi of the lower attines is a polyphyletic mix within the family Lepiotaceae, representing two clades.[9] There are a couple of remarkable exceptions to this fidelity: some species of the attine genus *Apterostigma* have secondarily changed to non-lepiotaceous fungi that belong to the family Tricholomataceae.[10] In addition, a small group of lower attine ants cultivate yeast, and in a unicellular phase. Contrary to previous assumptions, this is not an ancestral state of fungus growing. A cladistic analysis of yeast-culturing attine ants (*Cyphomyrmex rimosus* group) revealed that this clade is not basal but actually derived within the lower attine ants.[11, 12] Finally, all higher attine ants cultivate a derived monophyletic group of fungi in the tribe Leucocoprineae.

It has long been assumed that the fungal transmission is strictly vertical, that is, a transfer of fungal cultivars from parent nests to offspring nests. This would imply that the clonally propagated fungal lineages evolved in parallel with the lineages of the ant mutualists over millions of years. However, at least some of the lower attines propagate cultivars that were recently domesticated from free-living populations of Lepiotaceae.[13] While the "higher attines" are still thought to propagate ancient clones several million years old,[14] how ancient these clones really are remains uncertain. In fact, patterns of lateral transfer of fungal cultivars have been demonstrated in species of the lower attine genus *Cyphomyrmex*. Laboratory colonies deprived of their fungus garden regained cultivars either by joining a neighboring colony,

---

8 | U. G. Mueller, T. R. Schultz, C. R. Currie, R. M. M. Adams, and D. Malloch, "The origin of the attine ant-fungus mutualism," *The Quarterly Review of Biology* 76(2): 169–197 (2001).

9 | T. R. Schultz and R. Meier, "A phylogenetic analysis of the fungus-growing ants (Hymenoptera, Formicidae, Attini) based on morphological characters of the larvae," *Systematic Entomology* 20(4): 337–370 (1995).

10 | U. G. Mueller, T. R. Schultz, C. R. Currie, R. M. M. Adams, and D. Malloch, "The origin of the attine ant-fungus mutualism," *The Quarterly Review of Biology* 76(2): 169–197 (2001); and P. Villesen, U. G. Mueller, T. R. Schultz, R. M. M. Adams, and A. C. Bouck, "Evolution of ant-cultivar specialization and cultivar switching in *Apterostigma* fungus-growing ants," *Evolution* 58(10): 2252–2263 (2004).

11 | J. K. Wetterer, T. R Schultz, and R. Meier, "Phylogeny of fungus-growing ants (tribe Attini) based on mtDNA sequence and morphology," *Molecular Phylogenetics and Evolution* 9(1): 42–47 (1998).

12 | S. L. Price, T. Murakami, U. G. Mueller, T. R. Schultz, and C. R. Currie, "Recent findings in the fungus-growing ants: evolution, ecology, and behavior of a complex microbial symbiosis," in T. Kikuchi, N. Azuma, and S. Higashi, eds., *Genes, Behavior and Evolution of Social Insects* (Sapporo, Japan: Hokkaido University Press, 2003), pp. 255–280.

13 | U. G. Mueller, S. A. Rehner, and T. R. Schultz, "The evolution of agriculture in ants," *Science* 281: 2034–2038 (1998).

14 | I. H. Chapela, S. A. Rehner, T. R. Schultz, and U. G. Mueller, "Evolutionary history of the symbiosis between fungus-growing ants and their fungi," *Science* 266: 1691–1694 (1994).

stealing a neighbor's garden, or invading such a garden. As we will show later, pathogens can devastate gardens of attine colonies under natural conditions. Joining, stealing, or usurping neighbors' gardens is probably an important adaptation to counter garden loss, which otherwise would be a fatal catastrophe for the afflicted colony.[15] Similar patterns of occasional lateral transfer of fungus material have been demonstrated between two sympatric *Acromyrmex* species.[16]

In addition, recent genetic evidence contradicts the once widely held perceptions of obligate clonality in the fungal symbionts of leafcutter ants. This research documents "long-lasting horizontal transmission of symbionts between leafcutter taxa on mainland Central and South America and those endemic to Cuba." This suggests that the coevolution of leafcutter ants and their fungal symbionts is not reciprocal. The researchers propose that "a single widespread and sexual fungal symbiont species is engaged in multiple interactions with divergent ant lineages."[17]

THE ASCENT OF THE LEAFCUTTERS

Most fungus-growing ant species exist at a level still far from the highest evolutionary grade of superorganismic organization. Lower attines live mostly in relatively small colonies of fewer than 100 to 1,000 individuals, and their nests are inconspicuous and harbor relatively small fungus gardens. Species of the lower attines do not cut and use leaves as the main substrate for their symbiotic fungus, but rather gather a large variety of dead vegetable matter, including bits of leaves, plant seeds, and fruits, as well as insect feces and corpses.[18] Their social organization is relatively simple, with at most only minor polymorphism in minor worker size. It stands in sharp contrast to the leaf-cutting ants of the genera *Acromyrmex* (24 species, 35 subspecies) and *Atta* (15 species). At the extreme, the mature societies of certain *Atta* species are made up of millions of workers inhabiting huge subterranean nest

15 | R. M. M. Adams, U. G. Mueller, A. K. Holloway, A. M. Green, and J. Narozniak, "Garden sharing and garden stealing in fungus-growing ants," *Naturwissenschaften* 87(11): 491–493 (2000).

16 | A. N. M. Bot, S. A. Rehner, and J. J. Boomsma, "Partial incompatibility between ants and symbiotic fungi in two species of *Acromyrmex* leaf-cutting ants," *Evolution* 55(10): 1980–1991 (2001).

17 | A. S. Mikheyev, U. G. Mueller, and P. Abbot, "Cryptic sex and many-to-one co-evolution in the fungus-growing ant symbiosis," *Proceedings of the National Academy of Sciences USA* 103(28): 10702–10706 (2006).

18 | I. R. Leal and P. S. Oliveira, "Foraging ecology of attine ants in a Neotropical savanna: seasonal use of fungal substrate in the cerrado vegetation of Brazil," *Insectes Sociaux* 47(4): 376–382 (2000).

structures with hundreds of interconnected fungus garden chambers. The *Atta* societies represent a benchmark for the spectacular lifestyles "invented" by the ants in the course of their more than 120 million years of evolutionary history. In the following pages, we will focus mainly on this particular genus. The principal life history traits of the various species are very similar across *Atta*, permitting us to present a general picture of their natural history.

The leafcutter ants are of immense importance in tropical and subtropical ecosystems, and they are also major herbivorous pests in cultivated fields through much of Central and South America.[19] For example, in a recent long-term study in the Panamanian rain forest, Rainer Wirth and his colleagues determined that mature *Atta colombica* colonies harvest between 85 and 470 kilograms (dry weight) total plant biomass per colony per year; this corresponds to a harvested leaf area of 835 to 4550 square meters per year.[20] Such harvesting and processing of enormous amounts of plant material, which is needed for culturing the symbiotic fungus, is only possible by means of cooperation and division of labor among thousands of individuals.

THE *ATTA* LIFE CYCLE

Each of the gigantic *Atta* colonies usually consists of only one queen, the exclusive reproductive individual, and hundreds of thousands or even millions of sterile workers of different sizes and shapes (Plates 54 and 55). Each year, mature colonies produce young reproductive females and males, the alates, which depart from their mother colonies on nuptial (mating) flights. The flights of all *Atta* colonies belonging to the same species and living in the same habitat appear to be synchronized. In *Atta sexdens*, of South America, for example, they take place in the afternoon at any time from the end of October to the middle of December, while in *Atta texana*,

19 | For reviews, see J. M. Cherrett, "History of the leaf-cutting ant problem," in C. S. Lofgren and R. K. Vander Meer, *Fire Ants and Leaf-Cutting Ants: Biology and Management* (Boulder, CO: Westview Press, 1986), pp. 10–17; and B. Hölldobler and E. O. Wilson, *The Ants* (Cambridge, MA: The Belknap Press of Harvard University Press, 1990).

20 | R. Wirth, H. Herz, R. J. Ryel, W. Beyschlag, and B. Hölldobler, *Herbivory of Leaf-Cutting Ants: A Case Study on* Atta colombica *in the Tropical Rainforest of Panama* (New York: Springer-Verlag, 2003); H. Herz, W. Beyschlag, and B. Hölldobler, "Assessing herbivory rates of leaf-cutting ant (*Atta colombica*) colonies through short-term refuse deposition counts," *Biotropica* 39(4): 476–481 (2007); and H. Herz, W. Beyschlag, and B. Hölldobler, "Herbivory rate of leaf-cutting ants in tropical moist forest in Panama at the population and ecosystem scales," *Biotropica* 39(4): 482–488 (2007).

PLATE 54. A queen of an incipient colony of *Atta vollenweideri.*

of the southern United States, they occur at night. Mating itself occurs high in the air, and since many colonies conduct their nuptial flights during the same time of the day, the probability of outbreeding is high. Although mating has never been observed in nature, it has been estimated from sperm counts in the spermatheca of newly mated *Atta sexdens* queens that each queen is inseminated by three to eight males.[21] Such polyandry was later confirmed in studies employing DNA analyses. For example, in *Atta colombica*, the average number of fathers per colony is a bit below three. Due to variation in shared paternity, the effective paternity frequency in this species is only two.[22] The range of fathers per *Atta sexdens* colony is between one and five.[23] The mating frequency of *Acromyrmex* queens ranges between one

21 | W. E. Kerr, "Tendências evolutivas na reprodução dos himenópteros sociais," *Arquivos do Museu Nacional* (Rio de Janeiro) 52: 115–116 (1962).

22 | E. J. Fjerdingstad, J. J. Boomsma, and P. Thorén, "Multiple paternity in the leafcutter ant *Atta colombica*—a microsatellite DNA study," *Heredity* 80(1): 118–126 (1998).

23 | E. J. Fjerdingstad and J. J. Boomsma, "Queen mating frequency and relatedness in young *Atta sexdens* colonies," *Insectes Sociaux* 47(4): 354–356 (2000).

and ten males.[24] In contrast, queens of *Sericomyrmex* and *Trachymyrmex* and those of the lower attine genera appear to be all singly mated.[25]

The biological significance of multiple paternity in leafcutter ants is not entirely clear. Obviously, multiple matings of the queen decreases the average relatedness among the workers in the colony. It has been argued that an increase in genetic diversity may confer an advantage of colony fitness—for instance, in regard to disease resistance.[26, 27] This might be of particular importance in those fungus-growing ant colonies that exhibit vast expansions and exist for many years. Such a body of organic matter underground and such large numbers of ants are very susceptible to parasites and pathogens. An increase in genetically determined defense and resistance mechanisms in the ants obviously is beneficial for colony survival. An enhancement of genetic diversity in fungus-growing ants is especially important in attine ants that cultivate a clonal fungus lineage on which they have been dependent perhaps for millions of years. A long duration of this kind can be expected to lower genetic diversity in the gardens and thus render the cultivar more susceptible to disease, which in turn would require enhanced sanitary defenses in the ants.[28, 29]

Two studies with honeybees provide convincing evidence in support of the hypothesis that multiple mating of the queen improves the colony's vitality and

---

24 | J. J. Boomsma, E. J. Fjerdingstad, and J. Frydenberg, "Multiple paternity, relatedness and genetic diversity in *Acromyrmex* leaf-cutter ants," *Proceedings of the Royal Society of London B* 266: 249–254 (1999).

25 | T. Murakami, S. Higashi, and D. Windsor, "Mating frequency, colony size, polyethism and sex ratio in fungus growing ants (Attini)," *Behavioral Ecology and Sociobiology* 48(4): 276–284 (2000).

26 | W. D. Hamilton, "Kinship, recognition, disease, and intelligence: constraints of social evolution," in Y. Itô, J. L. Brown, and J. Kikkawa, eds., *Animal Societies: Theories and Facts* (Tokyo: Japan Scientific Societies Press, 1987), pp. 81–102.

27 | P. W. Sherman, T. D. Seeley, and H. K. Reeve, "Parasites, pathogens, and polyandry in social Hymenoptera," *American Naturalist* 131(4): 602–610 (1988).

28 | R. M. M. Adams, U. G. Mueller, A. K. Holloway, A. M. Green, and J. Narozniak, "Garden sharing and garden stealing in fungus-growing ants," *Naturwissenschaften* 87(11): 491–493 (2000).

29 | The most convincing evidence in support of the "disease resistance hypothesis" was recently published by W. O. H. Hughes and J. J. Boomsma, "Genetic diversity and disease resistance in leaf-cutting ant societies," *Evolution* 58(6): 1251–1260 (2004). This paper also presents a thorough review of these topics.

PLATE 55. *Above*: A queen of an established colony of *Atta cephalotes*. This caste in leafcutter ants always resides in the middle of the fungus garden; there she is covered by worker ants, which groom and protect her. *Below*: A fungus garden of the leafcutter ant *Atta sexdens*. As in all leafcutter ants, the fungus is fed with leaf fragments brought into the nest by foragers.

resistance to disease in insect societies. In one study, brood nest temperatures in genetically diverse colonies were found to be more stable than in genetically uniform colonies.[30] Even more important, Thomas Seeley and David Tarpy were able to demonstrate that polyandry improves the colony's resistance to disease.[31] Experimental honeybee colonies were inoculated with spores of the bacterium *Paenibacillus larvae*, which causes the highly virulent disease called American foulbrood. Colonies headed by a multiply inseminated queen had markedly lower disease intensity and higher colony strength relative to colonies headed by a singly inseminated queen.

An alternative hypothesis to explain multiple matings is that a queen requires a large lifetime supply of sperm. Leafcutter colonies are usually extremely populous and have a life span of 10 to 15 years or even longer. During her lifetime, the queen produces 150 million to 200 million female offspring (female alates and workers). She stores approximately 200 million to 320 million sperm cells in her spermatheca.[32] It can be argued that to obtain an optimal store of sperm that will last for more than a decade, a queen has to mate with multiple males. It has been documented that multiple mating, as expected, does increase the amount of sperm in the spermatheca in *Atta* queens.[33]

Finally, a third competing hypothesis states that high within-colony genetic diversity has a positive effect on worker task efficiency and thus enhances genetic disposition for the development of morphological subcastes in *Atta* colonies.[34] Such partial hardwiring of labor division would be favored by colony-level selection.

Each of these three hypotheses is supported by circumstantial and nonexclusive

---

30 | J. C. Jones, M. R. Myerscough, S. Graham, and B. P. Oldroyd, "Honey bee nest thermoregulation: diversity promotes stability," *Science* 305: 402–404 (2004).

31 | T. D. Seeley and D. R. Tarpy, "Queen promiscuity lowers disease within honeybee colonies," *Proceedings of the Royal Society of London B* 274: 67–72 (2007).

32 | W. E. Kerr, "Tendências evolutivas na reprodução dos himenópteros sociais," *Arquivos do Museu Nacional* (Rio de Janeiro) 52: 115–116 (1962).

33 | E. J. Fjerdingstad and J. J. Boomsma, "Multiple mating increases the sperm stores of *Atta colombica* leafcutter ant queens," *Behavioral Ecology and Sociobiology* 42(4): 257–261 (1998). Similar observations were reported for the African honeybee, *Apis mellifera capensis*, in F. B. Kraus, P. Naumann, J. van Draagh, and R. F. A. Moritz, "Sperm limitation and the evolution of extreme polyandry in honeybees (*Apis mellifera* L.)," *Behavioral Ecology and Sociobiology* 55(5): 494–501 (2004).

34 | R. H. Crozier and R. E. Page, "On being the right size: male contributions and multiple mating in social Hymenoptera," *Behavioral Ecology and Sociobiology* 18(2): 105–115 (1985); and R. E. Page Jr., "Sperm utilization in social insects," *Annual Review of Entomology* 31: 297–320 (1986). Also suggested is that polyandry reduces the occurrence of diploid males in honeybee colonies; see D. R. Tarpy and R. E. Page Jr., "Sex determination and the evolution of polyandry in honey bees (*Apis mellifera*)," *Behavioral Ecology and Sociobiology* 52(2): 143–150 (2002).

evidence, and in fact within-colony genetic diversity may have multiple adaptive significance (see also Chapter 2).

After the mating flight, all males die. The sole function of male ants is to provide sperm, which are stored and kept alive for many years in the spermatheca of the queens. Thus, the life span of male ants (which develop from unfertilized eggs and are therefore haploid) is very short. However, because of the long preservation time of sperm in the queen's internal "sperm bank," males can become fathers many years after they have died. Mortality is also very high for the young queens, especially during the mating flight and immediately after, when the queens attempt to start new colonies. Out of 13,300 newly founded colonies of *Atta capiguara* in Brazil followed during one study, only 12 were alive 3 months later. From a start of 3,558 incipient *Atta sexdens* colonies, only 90 (2.5 percent) were alive after 3 months. In another study, only 10 percent of *Atta cephalotes* colonies survived the first few months after colony foundation.[35]

Before departing on her mating flight, each *Atta* queen packs a small wad of mycelia of the symbiotic fungus into her infrabuccal pocket (cibarium), a cavity located beneath the opening of the esophagus. Following the nuptial flight, the queen casts off her wings and excavates a nest chamber in the soil. This incipient nest consists of a narrow entrance gallery that descends 20 to 30 centimeters to a single chamber about 6 centimeters in length (Figure 9-1). The queen now spits out the mycelial wad, which then serves as an inoculum to start a new fungus garden. By the third day, fresh mycelia have begun to grow, and the queen has laid three to six eggs.[36] At the end of the first month, the brood, now consisting of eggs, larvae, and perhaps pupae, is embedded in the center of a mat of proliferating fungus. During this initial phase of colony foundation, the queen cultivates the fungus garden herself, mainly by fertilizing the garden with fecal liquid. The queen consumes 90 percent of the eggs she lays. When the first larvae hatch, they are also fed with eggs. Apparently, the queen does not feed on the initial fungus culture, which is very fragile. If the queen fails to build up a healthy fungus garden, the whole

---

35 | H. G. Fowler, V. Pereira-da-Silva, L. C. Forti, and N. B. Saes, "Population dynamics of leaf-cutting ants: a brief review," in C. S. Lofgren and R. K. Vander Meer, eds., *Fire Ants and Leaf-Cutting Ants: Biology and Management* (Boulder, CO: Westview Press, 1986), pp. 123–145.

36 | M. Autuori, "La fondation des sociétés chez les fourmis champignonnistes du genre '*Atta*' (Hym. Formicidae)," in M. Autuori, M.-P. Bénassy, J. Benoit, R. Courrier, Ed.-Ph. Deleurance, M. Fontaine, K. von Frisch, R. Gesell, P.-P. Grassé, J. B. S. Haldane, Mrs. Haldane-Spurway, H. Hediger, M. Klein, O. Koehler, D. Lehrman, K. Lorenz, D. Morris, H. Piéron, C. P. Richter, R. Ruyer, T. C. Schneirla, and G. Viaud, *L'Instinct dans le Comportement des Animaux et de l'Homme* (Paris: Massone et Cie Éditeurs, 1956), pp. 77–104.

FIGURE 9-1. Colony founding in the leaf-cutting ant *Atta*. *A*: A queen in her first chamber with the beginning fungus garden. *B*: The queen manures the garden by pulling a hyphal clump free and applying an anal drop to it. *C*: Three stages in the growth of the garden and of the first ant brood, which occur simultaneously. Based on an original drawing by Turid Hölldobler-Forsyth in E. O. Wilson, *The Insect Societies* (Cambridge, MA: The Belknap Press of Harvard University Press, 1971); which in turn was based on J. Hüber, "Über die Koloniengründung bei *Atta sexdens*," *Biologisches Centralblatt* 256(18): 609–619 (1905); and M. Autuori, "La fondation des sociétés chez les fourmis champignonnistes du genre *'Atta'* (Hym. Formicidae)," in M. Autuori et al., eds., *L'Instinct dans le Comportement des Animaux et de l'Homme* (Paris: Masson, 1956), pp. 77–104.

colony-founding process is doomed. Instead, the queen subsists entirely on her own fat-body reserves and by catabolizing her now useless wing muscles.

When the first workers eclose, they begin to feed on the fungus, and they take over the fungus culture activities. The egg-laying rate of the queen now increases. Not all her eggs are viable; some are large trophic eggs formed in the oviduct by the fusion of two or more distinct but malformed eggs. These are given by workers to developing larvae. After a week or so, the young workers open the clogged nest entrance and start foraging in the immediate vicinity. They collect bits of leaves, which they add to the substrate of the fungus culture. By this time, the queen has ceased attending to the brood and fungus garden and has become an "egg-laying machine," a role she will keep for the rest of her long life. The workers have assumed all "somatic" duties of the colony: foraging, caring for the fungus garden, raising the brood, extending the nest structures, and defending the colony against predators and competitors.[37]

As the fresh leaves and plant cuttings are brought into the nest, they are cut into smaller and smaller pieces and treated with the ants' fecal liquid before being inserted into the garden substratum. The ants subsequently pluck tufts of mycelia from other parts of the garden and plant them on newly formed portions of the substratum. The inoculum proliferates swiftly thereafter: the transplanted mycelia grow as much as 13 micrometers per hour (Plates 56 to 60).

The fungus cultivated by *Atta* and *Acromyrmex* species produces hyphal tip swellings, called gongylidia, which form into densely packed clusters called staphylae. These aggregates are easily plucked by the ants and eaten or fed to the larvae. The structures are rich in lipids and carbohydrates, while the hyphae are richer in proteins.[38] When given a choice during feeding experiments, *Atta* workers prefer staphylae over hyphae. In addition, they live longer when feeding on staphylae rather than hyphae.[39] Thus, the staphylae appear to possess the best balanced blend of nutritional components.

---

37 | B. Hölldobler and E. O. Wilson, *The Ants* (Cambridge, MA: The Belknap Press of Harvard University Press, 1990).

38 | M. Bass and J. M. Cherrett, "Fungal hyphae as a source of nutrients for the leaf-cutting ant *Atta sexdens*," *Physiological Entomology* 20(1): 1–6 (1995).

39 | For a review, see J. M. Cherrett, R. J. Powell, and D. J. Stradling, "The mutualism between leaf-cutting ants and their fungus," in N. Wilding, N. M. Collins, P. M. Hammond, and J. F. Webber, eds., *Insect-Fungus Interactions* (New York: Academic Press, 1989), pp. 93–120. Also see U. G. Mueller, T. R. Schultz, C. R. Currie, R. M. M. Adams, and D. Malloch, "The origin of the attine ant-fungus mutualism," *The Quarterly Review of Biology* 76(2): 169–197 (2001).

PLATE 56. In the first step of vegetation processing and construction of a fungus garden, an *Atta sexdens* forager worker shears a leaf fragment from the tree at the harvesting site. Only one mandible functions as the "cutting knife," leaving the other to serve as the "pacemaker." The leafcutter ant also uses the foot of the foreleg on the side of the cutting mandible to pull the cut edge of the leaf fragment upward. This motion evidently increases the stiffness of the leaf and aids the cutting process. Note how the ant explores the leaf surface to which the cut is directed with her right antenna.

An important nutritional interdependency of the symbiotic fungus and leaf-cutting ants was revealed in a study of the metabolism of plant polysaccharides by the fungus of *Atta sexdens*. It is generally assumed that once the fungus degrades and assimilates cellulose, xylan, pectin, and starch, it is able to mediate the transference of carbon from plant material to the ants. This metabolic integration enables ants to exploit solid plant material not otherwise available to them. The integration primarily involves xylan and starch, both of which support rapid fungal growth. Cellulose, contrary to previous assumptions, seems to be less important because it is poorly degraded and assimilated by the fungus. Thus, if these biochemical analytical results

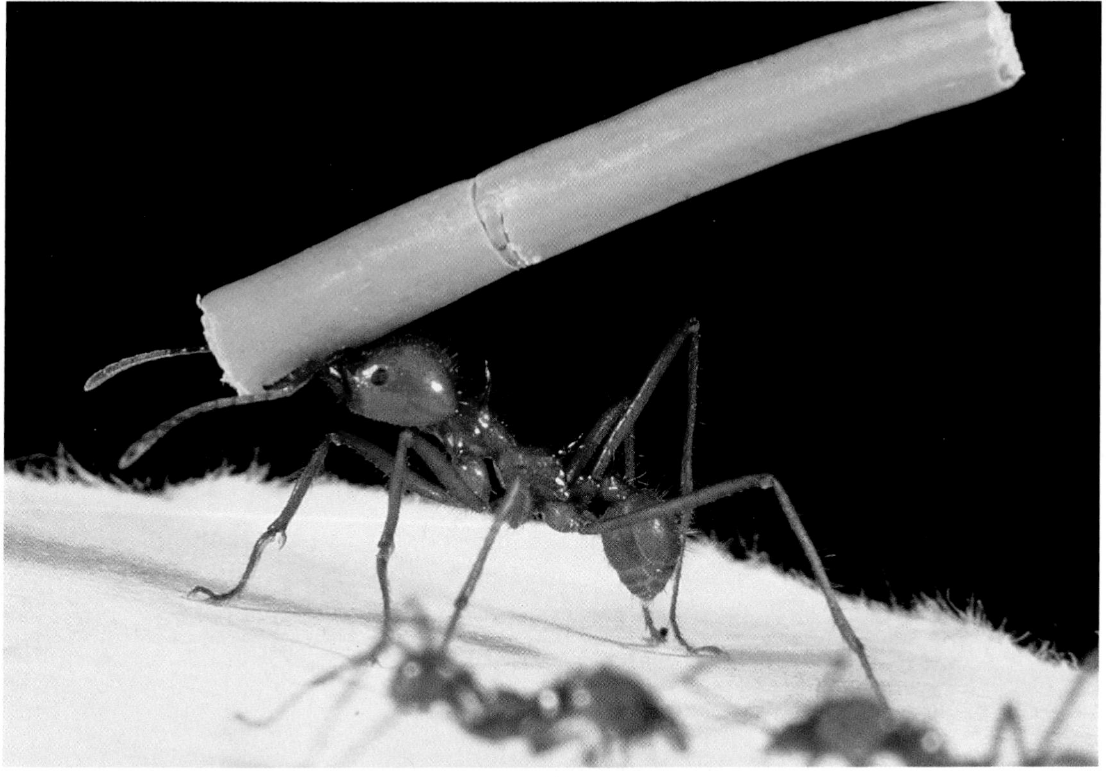

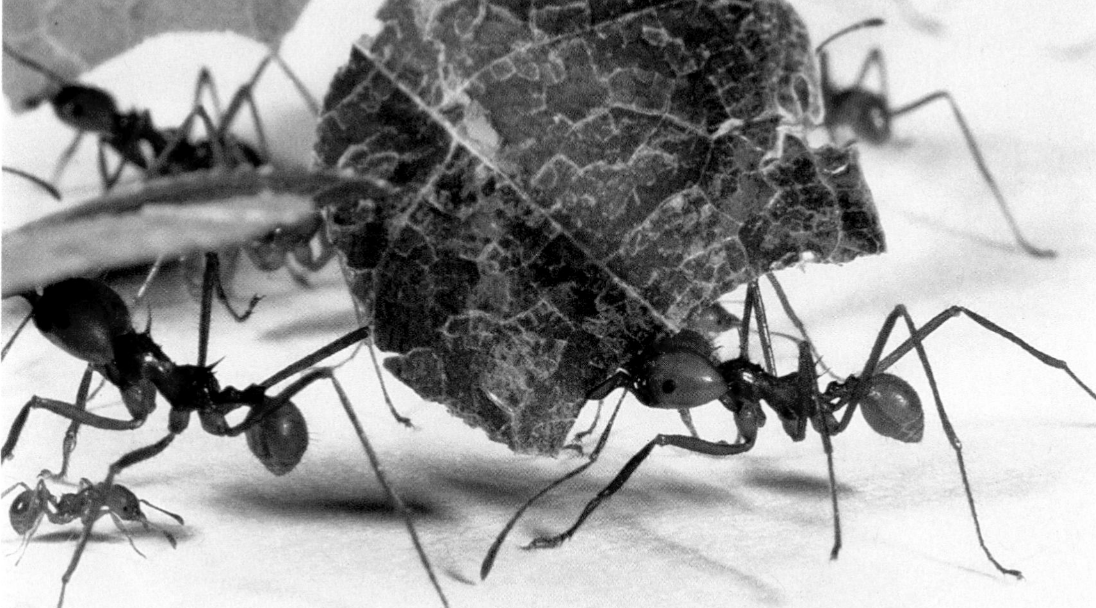

PLATE 57. Vegetable fragments harvested by *Atta* leafcutter ants are carried to the nest along pheromone-impregnated trunk routes.

PLATE 58. Often, members of the smallest worker subcaste (minims) ride on the transported vegetation during the harvesting operation. These hitchhikers protect the carrier ants from attack by parasitic phorid flies.

PLATE 59. The foraging columns of *Atta* species are typically crowded with leaf carriers. (Photo: Hubert Herz.)

drawn from laboratory cultures correctly reflect the fungal role in the symbiosis in nature, xylan and starch, not cellulose, are the main leaf polysaccharides contributing to ant nutrition.[40]

The main findings have been substantiated in recent studies in *Atta* and *Acromyrmex*, where it turns out that cellulose is not used as a main energy and carbon

---

40 | C. Gomes De Siqueira, M. Bacci Jr., F. C. Pagnocca, O. Correa Bueno, and M. J. A. Hebling, "Metabolism of plant polysaccharides by *Leucoagaricus gongylophorus*, the symbiotic fungus of the leaf-cutting ant *Atta sexdens* L.," *Applied and Environmental Microbiology* 64(12): 4820–4822 (1998).

source for the fungus-ant association. In fact, strong circumstantial evidence suggests that the fungus cannot degrade cellulose at all.[41]

In another study, worker extract has been reported to display high enzymatic activity on starch, maltose, sucrose, and a glycoside. Similar but higher enzymatic activity occurs in larval extract. In particular, the enzymes degrade sucrose, maltose, and laminarin, the latter a hemicellulose cell wall component of plants. Some variation in enzymatic activities occurs in the extract from symbiotic fungi of different *Acromyrmex* species. In the study, the fungal extract of *Acromyrmex subterraneus* was mostly active on laminarin, xylan, and cellulose, while the fungal extract of *Acromyrmex crassispinus* was most active on laminarin, starch, maltose, and sucrose.[42]

These results, in particular those regarding the degradation of laminarin and cellulose, seem to contradict earlier findings. That ant extracts can degrade the plant macromolecule laminarin is especially problematic. The difficulty may be resolved by the fact that fungal enzymes pass through the ant gut. Most likely, the enzymes detected in ant larval extract derived in part from the consumed fungus.

In any case, the fungus is not the only source of nutrition for leaf-cutting workers. In the laboratory at least, *Atta* and *Acromyrmex* workers feed directly on plant sap. The sap appears to be the "fuel" that provides the energy for the leaf cutters and harvest transporters. In fact, intake of sap appears to be crucial to the workers, because in laboratory experiments, only 5 percent of the energy requirements were met by ingestion of the contents of fungal staphylae.[43] In contrast, the larvae are able to subsist and grow entirely on the staphylae. The queen appears to obtain a substantial part of her food from trophic eggs laid by workers and fed to her at frequent intervals.

---

41 | A. B. Abril and E. H. Bucher, "Evidence that the fungus cultured by leaf-cutting ants does not metabolize cellulose," *Ecology Letters* 5(3): 325–328 (2002).

42 | P. D'Ettorre, P. Mora, V. Dibangou, C. Rouland, and C. Errard, "The role of the symbiotic fungus in the digestive metabolism of two species of fungus-growing ants," *Journal of Comparative Physiology B* 172(2): 169–176 (2002).

43 | R. J. Quinlan and J. M. Cherrett, "The role of fungus in the diet of the leaf-cutting ant *Atta cephalotes* (L.)," *Ecological Entomology* 4(2): 151–160 (1979).

PLATE 60. Even when not used, these long trunk routes are distinctly visible, being kept clear of vegetation by road worker ants. (Photo: Hubert Herz.)

THE *ATTA* CASTE SYSTEM

The growth of an incipient colony is slow in the first two years. During the next three years, it accelerates quickly and tapers off as the colony starts to produce winged males and queens. The ultimate colony size reached by *Atta* is enormous: the number of workers in a single colony has been estimated at 1 to 2.5 million in *Atta colombica*, 3.5 million in *Atta laevigata*, 5 to 8 million in *Atta sexdens*, and 4 to 7 million in *Atta vollenweideri*.[44]

Among the fungus-growing ants, only species of the two leaf-cutting genera, *Acromyrmex* and *Atta*, have highly polymorphic workers, with strong differences in size and anatomical proportion. This remarkable polymorphism is reflected in the complex division of labor exhibited within the colonies. A rich literature exists on various aspects of division of labor in *Atta* species. Most of the studies are in agreement concerning the major patterns that characterize division of labor in *Atta* colonies.[45] The following account is based on the labor system of *Atta cephalotes* and *Atta sexdens*.[46]

Atta leaf-cutting ants have a broad array of physical subcastes in the worker groups. In *Atta sexdens*, for example, the head width varies 8-fold and the dry weight 200-fold from the smallest minor workers to the huge major workers. However, developing colonies, started by a single queen, have a nearly uniform size frequency distributed across a relatively narrow head width range of 0.8 to 1.6 millimeters. The reason for this restriction, as detailed in Chapter 5, is an

---

44 | H. G. Fowler, V. Pereira-da-Silva, L. C. Forti, and N. B. Saes, "Population dynamics of leaf-cutting ants: a brief review," in C. S. Lofgren and R. K. Vander Meer, eds., *Fire Ants and Leaf-Cutting Ants: Biology and Management* (Boulder, CO: Westview Press, 1986), pp. 123–145.

45 | J. K. Wetterer, "Nourishment and evolution in fungus-growing ants and their fungi," in J. H. Hunt and C. A. Nalepa, eds., *Nourishment and Evolution in Insect Societies* (Boulder, CO: Westview Press, 1994), pp. 309–328.

46 | E. O. Wilson, "Caste and division of labor in leaf-cutter ants (Hymenoptera: Formicidae: *Atta*), I: The overall pattern in *A. sexdens*," *Behavioral Ecology and Sociobiology* 7(2): 143–156 (1980); E. O. Wilson, "Caste and division of labor in leaf-cutter ants (Hymenoptera: Formicidae: *Atta*), II: The ergonomic optimization of leaf cutting," *Behavioral Ecology and Sociobiology* 7(2): 157–165 (1980); E. O. Wilson, "Caste and division of labor in leaf-cutter ants (Hymenoptera: Formicidae: *Atta*), III: Ergonomic resiliency in foraging by *A. cephalotes*," *Behavioral Ecology and Sociobiology* 14(1): 47–54 (1983); E. O. Wilson, "Caste and division of labor in leaf-cutter ants (Hymenoptera: Formicidae: *Atta*), IV: Colony ontogeny of *A. cephalotes*," *Behavioral Ecology and Sociobiology* 14(1): 55–60 (1983).

experimentally demonstrated necessity: workers in the span of 0.8 to 1.0 millimeter are required as gardeners of the symbiotic fungus, whereas workers with a head width of 1.6 millimeters are the smallest that can cut vegetation of average toughness. The combined range (0.8 to 1.6 millimeters) also embraces the worker size groups most involved in brood care. Thus, the queen produces about the maximum number of individuals who together can perform all the essential colony tasks. As the colony continues growing, the worker size variation broadens in both directions, to head width 0.7 millimeter or slightly less at the lower end and to more than 5 millimeters at the upper end, while the frequency distribution becomes more sharply peaked and strongly skewed to the larger-size classes. This complex caste system reflects the division of labor in *Atta*, which is closely adapted to the collection and processing of fresh vegetation for fungal substrate and to the culturing of the fungus.

The *Atta* workers organize the gardening operation in the form of an assembly line. The most frequent size group among foragers, at the start of the line, consists of workers with a head width of 2.0 to 2.2 millimeters. At the end of the line, the care of the delicate fungal hyphae requires very small workers, a task filled within the nest by workers with a head width of predominantly 0.8 millimeter. The intervening steps in gardening are conducted by workers of graded intermediate size.

After the returning foragers (Figure 9-2, activity 1) drop the pieces of vegetation onto the floor of a nest chamber, the pieces are picked up by workers of slightly smaller size, who clip them into fragments about 1 to 2 millimeters across (activity 2). Within minutes, still smaller ants take over, crush and mold the fragments into moist pellets, add fecal droplets (activity 3), and carefully insert them into a mass of similar material (activity 4). Next, workers even smaller than those just described pluck loose strands of fungus from places of dense growth and plant them on the newly constructed surfaces (activity 5). Finally, the very smallest and most abundant workers patrol the beds of fungal strands, delicately probing them with their antennae, licking their surfaces, and plucking out spores and hyphae of alien species of mold (activity 6).

Superimposed on this division of labor, which is based on anatomical worker subcastes, is age polyethism: young workers of most subcastes perform tasks inside the nest, and older workers tend to be involved in tasks outside the nest. This distinction is strikingly illustrated by the smallest worker subcastes

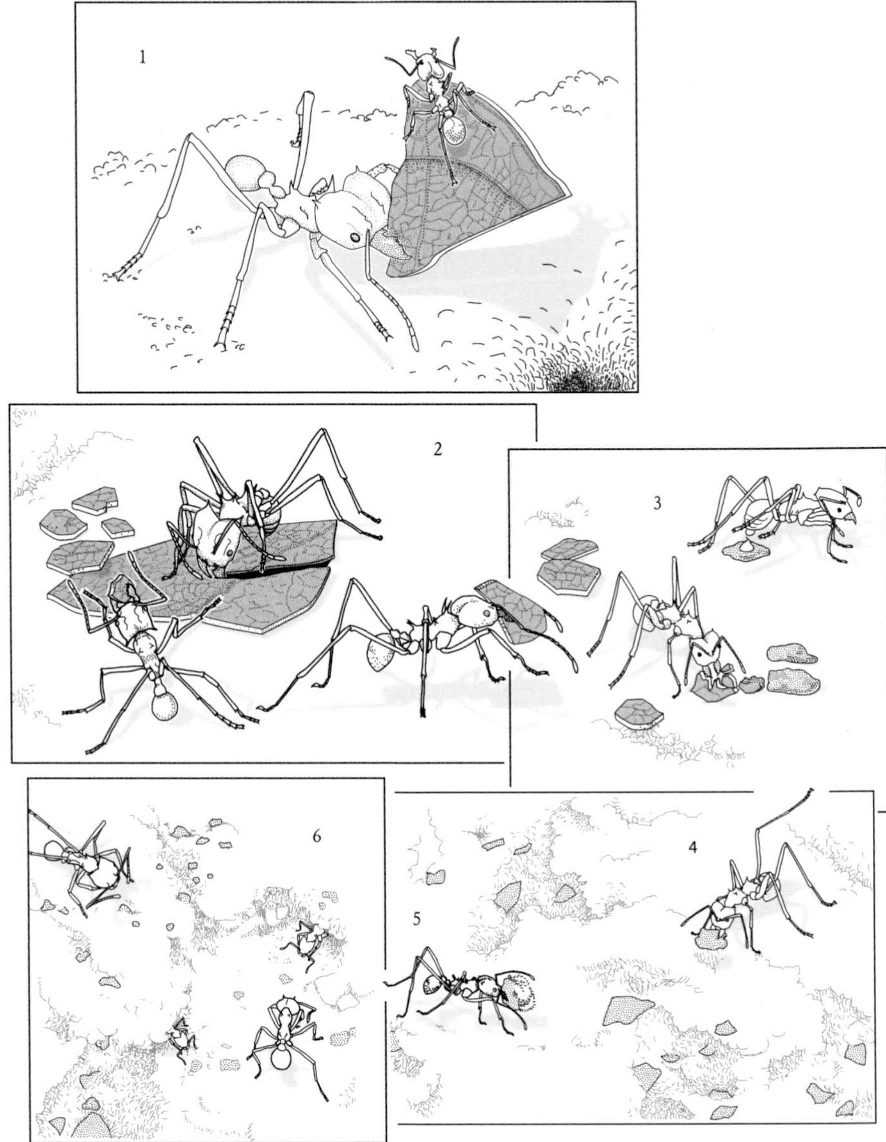

FIGURE 9-2. The "assembly line" by which colonies of *Atta cephalotes* create fungus gardens with fresh-cut leaves and other vegetation.

(the so-called minim workers), which inside the nest tend the fungus and small brood, but which can also be seen at the harvesting site, even though they are unable to cut and carry leaf fragments. Many of them do not walk back to the nest on their own, but ride ("hitchhike") on the leaf fragments being carried to the nest (see Plate 58). Most likely, these individuals are older minim workers

PLATE 61. A supersoldier of *Atta cephalotes*.

that defend the leaf carriers from attacks by parasitic phorid flies, which attempt to lay eggs on the ants' bodies.[47, 48]

Most size groups of *Atta* colonies engage in defense of the colony, but again, it is the older workers that most likely attack intruders and defend territories. At the same time, colony defense is organized to some extent according to worker size. There is, for example, a true soldier caste. These extremely large majors have sharp mandibles powered by massive adductor muscles (Plate 61). They are especially adept at repelling large enemies, especially vertebrates. The differential involvement of worker castes in colony defense has been well documented in a study of *Atta laevigata*. When the colony is threatened by a potential vertebrate predator, mostly the gigantic soldiers are recruited. However, when a colony has to defend its nest

---

47 | I. Ebil-Eibesfeldt and E. Eibl-Eibesfeldt, "Das Parasitenabwehren der Minima-Arbeiterinnen der Blattschneider-Ameise (*Atta cephalotes*)," *Zeitschrift für Tierpsychologie* 24(3): 278–281 (1967).

48 | D. H. Feener and K. A. G. Moss, "Defense against parasites by hitchhikers in leaf-cutting ants: a quantitative assessment," *Behavioral Ecology and Sociobiology* 26(1): 17–26 (1990).

or foraging area against conspecific or interspecific ant competitors, mainly smaller worker castes respond. These are more numerous and more suitable in territorial combat with enemy ants.[49] Similar results have been reported from studies of the grass-cutting ant, *Atta capiguara*. In this species, minor workers responded most readily to alarm pheromones experimentally released from the mandibular glands of *Atta capiguara* workers near the foraging trail. The response was strongest when the pheromone was released close to the trail.[50] Foragers transporting grass fragments did not respond at all. However, in *Atta colombica*, whose nests are raided by the ecitonine army ant *Nomamyrmex esenbeckii*, workers recruited mainly the majors (soldiers) as a specific defensive response to army ant attacks.[51]

HARVESTING VEGETATION

Variation in body size may also be important in the harvesting behavior of *Atta* workers.[52, 53] A leaf-cutting forager typically harvests leaf pieces the mass of which corresponds with her body size. This could be a result of the leaf-cutting behavior. During cutting, a worker usually anchors her hind legs on the leaf edge and slowly pivots around the body axis, pushing the cutting mandible through the leaf tissue (see Plate 56). In this way, the fragment size is correlated with the body size of the cutter. In other studies, however, no relationship between the length of the legs of *Atta cephalotes* ants and the cut curvature of the leaf fragment has been found.[54] Instead, the angle between head and thorax can be changed by the cutting ant, allowing it considerable flexibility. Hence, the fragment size is not a simple function of the legs acting as a pivot. It can also be argued that the leaf-

---

49 | M. E. A. Whitehouse and K. Jaffe, "Ant wars: combat strategies, territory and nest defence in the leaf-cutting ant *Atta laevigata*," *Animal Behaviour* 51(6): 1207–1217 (1996).

50 | W. O. H. Hughes and D. Goulson, "Polyethism and the importance of context in the alarm reaction of the grass-cutting ant, *Atta capiguara*," *Behavioral Ecology and Sociobiology* 49(6): 503–508 (2001).

51 | S. Powell and E. Clark, "Combat between large derived societies: a subterranean army ant established as a predator of mature leaf-cutting ant colonies," *Insectes Sociaux* 51(4): 342–351 (2004).

52 | C. M. Nichols-Orians and J. C. Schultz, "Leaf toughness affects leaf harvesting by the leaf-cutter ant, *Atta cephalotes* (L.) (Hymenoptera: Formicidae)," *Biotropica* 21(1): 80–83 (1989).

53 | For a review, see R. Wirth, H. Herz, R. J. Ryel, W. Beyschlag, and B. Hölldobler, *Herbivory of Leaf-Cutting Ants: A Case Study on* Atta colombica *in the Tropical Rainforest of Panama* (New York: Springer-Verlag, 2003).

54 | J. M. van Breda and D. J. Stradling, "Mechanisms affecting load size determination in *Atta cephalotes* (L.) (Hymenoptera, Formicidae)," *Insectes Sociaux* 41(4): 423–434 (1994).

cutting foragers do not directly assess fragment mass while cutting, but use leaf toughness as an indirect measure for adjusting the size of the cut fragments. Thus, while ants cannot cut larger pieces than their overall body size permits (unless they move along the cutting edge), they are able to change their posture in order to cut smaller leaf fragments.

During cutting, the two mandibles of *Atta* workers play different roles. While one mandible actively moves, the other remains almost fixed and serves as the cutting jaw. The steps in one full bite are as follows (Figure 9-3). The motile mandible is opened and anchored with its tip to the leaf tissue. The cutting mandible is not opened, but held steady. During the opening of the motile mandible, the cutting jaw is pushed against the leaf by lateral head movements. Next, the motile mandible is closed, pulling the cutting jaw further against the leaf and lengthening the incision. In this phase, the motile mandible also moves deeper into the leaf surface, thus preparing the way for the cutting jaw. As soon as both jaws meet, the cycle starts again. Thus, one jaw functions as "cutting knife," the other one as "pacemaker." But there is no "sidedness": either right or left jaw can function as cutting-knife, depending on the direction in which the leaf fragment is cut.

Leafcutters often stridulate while they are cutting. A number of workers cutting leaf fragments raise and lower their gasters in a motion identical to that performed by *Atta* workers when producing sound (see Figures 9-3 and 9-4). The sound comes from a stridulatory organ, composed of a cuticular file on the first gastric tergite and a scraper situated on the postpetiole. By rubbing the file against the scraper, the ants produce audible vibrations.[55, 56] The analysis of the temporal relation between mandible movements and stridulation, made by videotaping the cutting behavior and simultaneously recording the vibrational signals from the leaf surface with laser vibrometry, have revealed that stridulation occurs most often when the cutting mandible is moved through the plant tissue (see Figure 9-4). The stridulation generates complex vibrations of the mandibles, which give the mandibles some of the properties of a vibratome (the vibrating knife of a microtome). Indeed, when the cutting process was experimentally simulated, it turned out that the vibrating mandible reduces the force fluctuations

55 | H. Markl, "Stridulation in leaf-cutting ants," *Science* 149: 1392–1393 (1965).

56 | H. Markl, "Die Verständigung durch Stridulationssignale bei Blattschneiderameisen, II: Erzeugung und Eigenschaften der Signale," *Zeitschrift für vergleichende Physiologie* 60(2): 103–150 (1968).

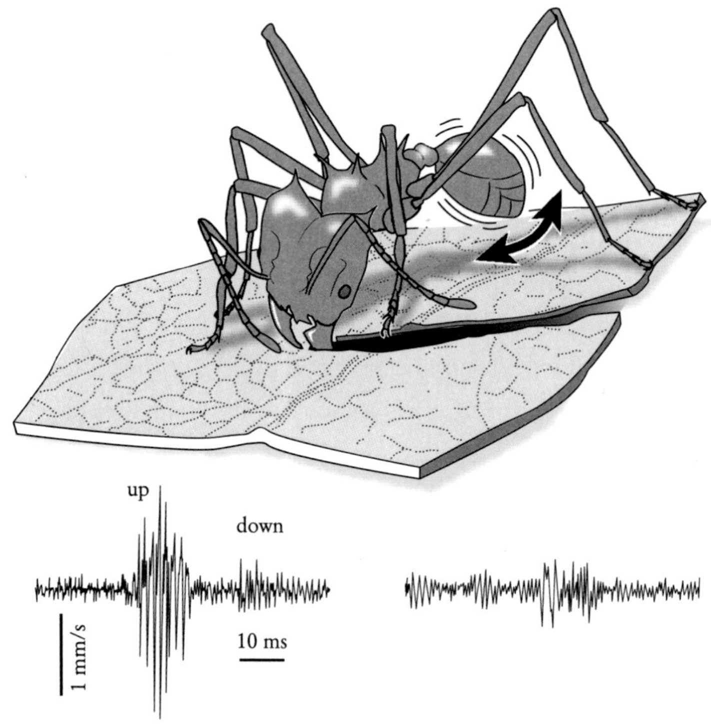

FIGURE 9-3. The cutting of a leaf by a leaf-cutting ant. The worker vibrates her abdomen, producing a stridulation sound, as she scissors her way along. The substrate-borne vibrations attract her nestmates from the vicinity and facilitate the cutting process. The stridulation signals were recorded by Laser-Doppler-Vibrometry (as velocity of the leaf's vibrations). *Lower left*: Substrate-borne vibrations transmitted mostly through the mandibles while cutting. *Lower right*: Vibrations transmitted onto the substrate through the legs when the mandibles do not touch the leaf. Based on F. Roces, J. Tautz, and B. Hölldobler, "Stridulation in leaf-cutting ants: short-range recruitment through plant-borne vibrations," *Naturwissenschaften* 80(11): 521–524 (1993).

that inevitably occur when material is being cut. Thus, stridulatory vibrations facilitate a smoother cut through tender leaf tissue.[57]

Cutting fragments out of leaves requires powerful mandible muscles. Accordingly, the mandibular muscles in *Atta* comprise more than 50 percent of their head capsule mass, or more than 25 percent of the entire body mass.[58] Leaf cutting is also an extraordinarily intense behavior energetically. The leaf-cutting metabolic rate, which has been determined in an extremely sensitive flow-through respirometry

57 | J. Tautz, F. Roces, and B. Hölldobler, "Use of a sound-based vibratome by leaf-cutting ants," *Science* 267: 84–87 (1995).

58 | F. Roces and J. R. B. Lighton, "Larger bites of leaf-cutting ants," *Nature* 373: 392–393 (1995).

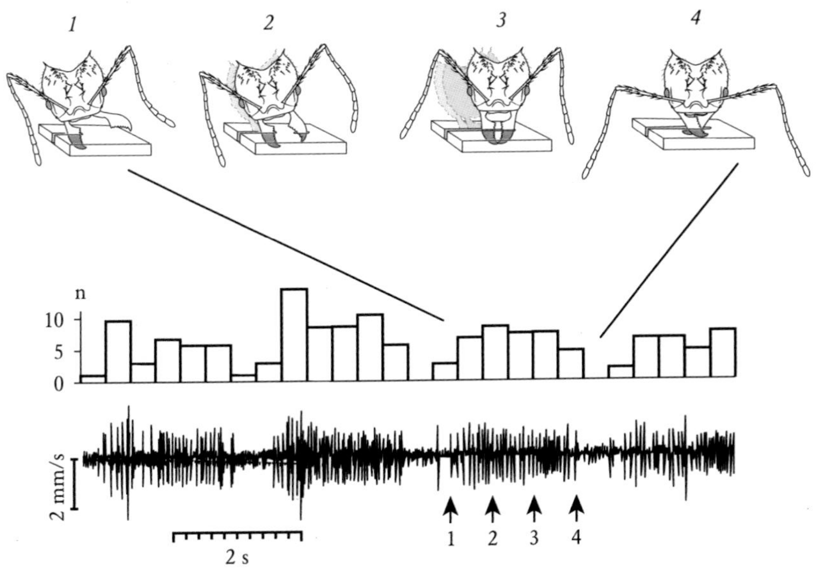

FIGURE 9-4. *Above*: Mandible and head movements during one cut into a tender leaf. *Below*: Stridulation during four bites. The histogram shows the number of chirps counted through a span of 400 milliseconds. The trace underneath depicts an original laser vibrometry of stridulation on a leaf. The arrows denote the temporal occurrence of the four cutting stages shown above. Based on J. Tautz, F. Roces, and B. Hölldobler, "Use of a sound-based vibratome by leaf-cutting ants," *Science* 267: 84–87 (1995).

system, is dramatically above both the standard rate and the postcutting locomotor metabolic rate. The aerobic scope of leaf cutting has been determined to be within the same range as that of flying insects, which are among the most metabolically active of all animals. The mandibular energetics of leaf cutting therefore probably plays an important role in the ants' load size selection and foraging efficiency at both the individual and colony levels.

During the past two decades, numerous papers have been published addressing the question of load size selection in leaf-cutting ants. It is beyond the scope of this presentation to review the many different and sometimes contradictory results obtained. Obviously, the parameters that affect load size are numerous. Although, as noted, there is a correspondence between the size of the leaf-cutting worker and the leaf fragment size (area) to be cut, fragment size is not always the best parameter to determine load size (mass). The reason is that the mass of a fragment depends on leaf mass per unit surface area as well as on leaf fragment volume. Foragers of *Atta cephalotes* and *Atta texana* do not tend to adjust leaf-cutting behavior as a function of leaf density. Instead, workers of different sizes tend to cut leaves of different densities.[59, 60] Similar patterns of forager polymorphism and resource matching have been found in other *Atta* species. On the other hand, several additional independent studies have revealed that the denser the leaf, the smaller the fragments.[61]

The mass of a leaf fragment being transported also affects the running speed of the carrier ant, and both parameters (load mass and retrieval time) affect the colony's rate of vegetable material intake.[62, 63] The slower speeds of the workers carrying heavier fragments may not negatively affect intake rates, because the yield per delivery is increased. But extended travel time, due to heavier loads, may have other detrimental consequences. For example, the transfer of information about the food resource to the colony can be delayed and therefore the speed and intensity of recruitment weakened.[64, 65]

Hence, short travel time appears to be an asset in the foraging system of leaf-cutting ants, thus favoring load sizes that have a minimal effect on the running speed of leaf carriers. In any case, individual maximization models, often usefully applied to solitary foraging animals, fail to explain fragment selection by *Atta* foragers.[66] In fact, it may well be the case that small loads are rate maximizing, but at the level of the colony rather than of the individual worker. Indeed, fragment size might be influenced by many factors, including size of the worker ant, energetic cost of cutting, density (mass) of the leaf, need to rapidly transfer foraging information to the colony, distance and quality of the harvesting site, and finally, "handling cost," which most likely increases with fragment size.

---

59 | J. K. Wetterer, "Forager polymorphism, size-matching, and load delivery in the leaf-cutting ant, *Atta cephalotes*," *Ecological Entomology* 19(1): 57–64 (1994).

60 | D. A. Waller, "The foraging ecology of *Atta texana* in Texas," in C. S. Lofgren and R. K. Vander Meer, eds., *Fire Ants and Leaf-Cutting Ants: Biology and Management* (Boulder, CO: Westview Press, 1986), pp. 146–158.

61 | See R. Wirth, H. Herz, R. J. Ryel, W. Beyschlag, and B. Hölldobler, *Herbivory of Leaf-Cutting Ants: A Case Study on* Atta colombica *in the Tropical Rainforest of Panama* (New York: Springer-Verlag, 2003).

62 | M. Burd, "Variable load size-ant size matching in leaf-cutting ant, *Atta colombica* (Hymenoptera: Formicidae)," *Journal of Insect Behavior* 8(5): 715–722 (1995).

63 | M. Burd, "Foraging performance by *Atta colombica*, a leaf-cutting ant," *American Naturalist* 148(4): 597–612 (1996).

64 | F. Roces and B. Hölldobler, "Leaf density and a trade-off between load size selection and recruitment behavior in the ant *Atta cephalotes*," *Oecologia* 97(1): 1–8 (1994).

65 | F. Roces and J. A. Núñez, "Information about food quality influences load-size selection in recruited leaf-cutting ants," *Animal Behaviour* 45(1): 135–143 (1993).

66 | For a detailed discussion of these issues, see M. Burd, "Server system and queuing models of leaf harvesting by leaf-cutting ants," *American Naturalist* 148(4): 613–629 (1996).

In almost every respect, size matters in the division of labor in *Atta* societies. Especially marked is the immense size and anatomical difference between workers and the gigantic queen, which reaches its extreme in the enormous ovaries of the queen when compared to the "degenerate" ovaries of workers. The queen is the sole reproductive individual in the society.

Additional labor specialization is achieved by programmed shifts with aging. At least three of the four physical castes of *Atta sexdens*, for example, pass through changes of behavior with aging. Although caste and division of labor in this and other *Atta* species are very complex in comparison with other ant systems, they are derived from surprisingly elementary processes of increased size variation, allometry, and alloethism. In fact, ant species in general and *Atta* species in particular have been remarkably restrained in the elaboration of their castes. They have relied on a single rule of deformation to create physical castes, which translates into a single allometric curve for any pair of specific dimensions, such as head width versus pronotal width. Hence, the *Atta* species have not evolved anywhere close to the conceivable limit. There are far more tasks than castes: by the first crude estimate, seven castes cover a total of 20 to 30 tasks. Furthermore, one can discern another important phenomenon in *Atta* species that constrains the elaboration of physical castes: polyethism has evolved further than polymorphism. In the course of evolution, *Atta* created its division of labor primarily by greatly expanding the size variation of the workers while adding a moderate amount of allometry and a relatively much greater amount of alloethism.[67]

Alloethism is the regular change in a particular category of behavior as a function of worker size. It stands in close relationship to the phenomenon of task partitioning: "A task can be said to be partitioned when it is split into two or more sequential stages so that material is passed from one worker to another."[68, 69] This phenomenon is well known to myrmecologists from a number of different species in various contexts. It includes "bucket-brigade" harvesting in leaf-cutting ants, in which leaves are cut by some workers and dropped to the ground for further

67 | B. Hölldobler and E. O. Wilson, *The Ants* (Cambridge, MA: The Belknap Press of Harvard University Press, 1990).

68 | C. Anderson and J. L. V. Jadin, "The adaptive benefit of leaf transfer in *Atta colombica*," *Insectes Sociaux* 48(4): 404–405 (2001).

69 | A. G. Hart and F. L. W. Ratnieks, "Leaf caching in the leafcutting ant *Atta colombica*: organizational shift, task partitioning and making the best of a bad job," *Animal Behaviour* 62(2): 227–234 (2001).

fragmentation. The material is then transported to the nest by other workers for varying distances along trunk trails, until the harvest material reaches the nest.[70] In some *Atta* species, such as *Atta colombica*, the leaf carriers establish one or more caches along the trail. In others, including *Atta vollenweideri*, the leaves are dropped haphazardly along the trail at variable distances. In grass-cutting ants (*Atta vollenweideri*), grass fragments are harvested and transported to the nest for distances of up to 150 meters along well-established trunk trails.[71, 72] Cutting and transporting of fragments are distinct activities, often performed by separate workers differing in body size. Because leaf cutting is a much more energetically intense activity than transport,[73] colonies can be expected to allocate larger workers to this task. This body size effect is less obvious when the harvest site is located very close to the nest; furthermore, no physical trail is present. In such situations, the cutter often carries the grass leaf fragment to the nest alone. However, on long foraging trails, the workers form transport chains composed of two to five carriers per grass fragment. As a rule, the first carriers cover only a short distance before dropping the fragments. Sometimes cutters participate in this first phase of harvest transportation but after dropping their load usually returned to the harvesting patch. The last carriers cover the longest distance. Furthermore, the probability of dropping the carried leaf fragment is independent of both worker size and load size.

What are the advantages of this kind of chain transport in *Atta vollenweideri*? Maximization of load transportation has been proposed for those leaf-cutting ant species that employ caches on the ground for transfer of harvested leaf fragments to the nest. However, the empirical data do not always support the assumptions.[74, 75, 76, 77]

---

70 | S. P. Hubbell, L. K. Johnson, E. Stanislav, B. Wilson, and H. Fowler, "Foraging by bucket-brigade in leaf-cutter ants," *Biotropica* 12(3): 210–213 (1980).

71 | J. Röschard and F. Roces, "The effect of load length, width and mass on transport rate in the grass-cutting ant *Atta vollenweideri*," *Oecologia* 131(2): 319–324 (2002).

72 | J. Röschard and F. Roces, "Cutters, carriers and transport chains: distance-dependent foraging strategies in the grass-cutting ant *Atta vollenweideri*," *Insectes Sociaux* 50(3): 237–244 (2003).

73 | F. Roces and J. R. B. Lighton, "Larger bites of leaf-cutting ants," *Nature* 373: 392–393 (1995).

74 | S. P. Hubbell, L. K. Johnson, E. Stanislav, B. Wilson, and H. Fowler, "Foraging by bucket-brigade in leaf-cutter ants," *Biotropica* 12(3): 210–213 (1980).

75 | H. G. Fowler and S. W. Robinson, "Foraging by *Atta sexdens* (Formicidae: Attini): seasonal patterns, caste and efficiency," *Ecological Entomology* 4(3): 239–247 (1979).

76 | C. Anderson and F. L. W. Ratnieks, "Task partitioning in insect societies, I: Effect of colony size on queuing delay and colony ergonomic efficiency," *American Naturalist* 154(5): 521–535 (1999).

77 | A. G. Hart and F. L. W. Ratnieks, "Leaf caching in the *Atta* leaf-cutting ant *Atta colombica*: organizational shift, task partitioning and making the best of a bad job," *Animal Behaviour* 62(2): 227–234 (2001).

Carl Anderson and his colleagues have discussed several advantages and disadvantages of a chain transport of harvest to nests, as in *Atta vollenweideri*, where the last carriers cover the longest distance.[78] The researchers argue that such task partitioning can be expected to enhance the work efficiency of individuals, because workers are more likely to become specialists when deployed sequentially. As a consequence, the colony's overall rate of resource retrieval should be higher. But again, the empirical data do not entirely support these theoretical considerations.

Finally, Jacqueline Röschard and Flavio Roces have proposed a second hypothesis: that the transport chains of *Atta vollenweideri* accelerate transfer of information about the plant species and food quality of the harvest.[79, 80] They argue that the dropping of fragments on the trail allows cutting workers to quickly return to their tasks. Furthermore, moving along short trail sections during foraging facilitates reinforcement of the trail pheromone markings, which in turn leads to a faster recruitment of a foraging force and subsequent monopolization of the harvesting site. In addition, fragments dropped on the trail may serve as information signals. For example, outgoing foragers could obtain information about the resources actually being harvested. If this "information transfer hypothesis" is correct, transport chains can be expected to occur more frequently under conditions in which the information is valuable, for instance, on discovery of high-quality resources or when the colony is harvest deprived.[81] Röschard and Roces have obtained evidence supporting this conjecture.[82]

For example, field experiments have revealed that when plant fragments of high quality are presented at selected foraging sites, transport chains occur more frequently, and independent of fragment sizes. In addition, high-quality fragments are transferred from one carrier to another after shorter transport travel. More chains and more segments in the chains are responses to an increase in the attractiveness of the load, allowing the first carriers to return quickly to the foraging site. These

78 | C. Anderson, J. J. Boomsma, and J. J. Bartholdi III, "Task partitioning in insect societies: bucket brigades," *Insectes Sociaux* 49(2): 171–180 (2002).

79 | J. Röschard and F. Roces, "The effect of load length, width and mass on transport rate in the grass-cutting ant *Atta vollenweideri*," *Oecologia* 131(2): 319–324 (2002).

80 | J. Röschard and F. Roces, "Cutters, carriers and transport chains: distance-dependent foraging strategies in the grass-cutting ant *Atta vollenweideri*," *Insectes Sociaux* 50(3): 237–244 (2003).

81 | F. Roces, "Individual complexity and self-organization in foraging by leaf-cutting ants," *The Biological Bulletin* 202(3): 306–313 (2002).

82 | F. Roces, personal communication.

results, as predicted, suggest that transport chains increase the information flow at the colony level. Additional data obtained with fragments of the same quality but different size do not support the hypothesis that transport chains enhance the economic load carriage at the individual level, as previously suggested.[83]

Overall, leaf quality is an important influence in recruitment and harvesting intensity in leaf-cutting ants. Its parameters include leaf tenderness, nutrient contents, and the presence and quantity of secondary plant chemicals. In one experiment, harvest preference in *Atta cephalotes* was tested by offering the ants fresh leaves of 49 woody plant species from a tropical deciduous forest in Costa Rica. Leaf protein content was positively correlated with the number of fragments cut, while secondary chemistry and nutrient availability interacted in determining the attractiveness of plant material to the ant foragers.[84] In another, related study, young, tender leaves of the tropical legume *Inga edulis* were found to be more loaded with secondary chemicals and contain fewer nutrients than mature leaves, but the latter are three times tougher and thus harder to cut. The investigators concluded that the quality of the colony's habitat likely determines whether a colony will harvest more of the less suitable leaves. These ants that locate and harvest from highly suitable host plants avoid *Inga edulis*, while those in poorer habitats accept the legume but, because of the toughness of this plant's older leaves, mostly harvest the otherwise less suitable young leaves.[85]

Several lines of evidence thus suggest that harvest preferences in *Atta* leafcutter colonies are determined by trade-offs between several parameters. Furthermore, they depend not only on particular leaf traits, but also on the properties of the ecosystem as a whole. Comparative bioassays focusing on a small set of parameters do not, in these ants, capture the complex multivariate picture of harvest selection.[86]

---

83 | C. Anderson, J. J. Boomsma, and J. J. Bartholdi III, "Task partitioning in insect societies: bucket brigades," *Insectes Sociaux* 49(2): 171–180 (2002).

84 | J. J. Howard, "Leaf-cutting and diet selection: relative influence of leaf chemistry and physical factors," *Ecology* 69(1): 250–260 (1988).

85 | C. M. Nichols-Orians and J. C. Schultz, "Interactions among leaf toughness, chemistry, and harvesting by attine ants," *Ecological Entomology* 15(3): 311–320 (1990).

86 | There exists a rich and sometimes contradicting literature on food plant selection in attine ants, which is partially reviewed in R. Wirth, H. Herz, R. J. Ryel, W. Beyschlag, and B. Hölldobler, *Herbivory of Leaf-Cutting Ants: A Case Study on* Atta colombica *in the Tropical Rainforest of Panama* (New York: Springer-Verlag, 2003).

COMMUNICATION IN *ATTA*

The highly organized cooperative foraging of the *Atta* leafcutters depends on information transfer and social communication. Much of this information transfer occurs on their harvesting routes. Leaf-cutting ants are famous for their extended and persistent foraging trails (see Plates 49 and 50). These durable routes are very obvious to the human eye. They lead masses of foragers to and from harvesting sites, which are either mostly the canopy of trees, a specialization of *Atta cephalotes*, *Atta colombica*, and *Atta sexdens*, or patches of savanna grass, the target, for example, of the grass-cutting *Atta vollenweideri*. Early behavioral experiments indicated that the foraging trails are chemically marked with secretions from the ants' poison gland sacs.[87] It has been suggested that this trail pheromone contains at least two functional components, one volatile, which serves as a recruitment signal, and the other much less volatile, which functions as a long-lasting orientation cue. Many chemical and behavioral details of the *Atta* poison gland contents and the response to them remain to be elucidated, but some important aspects with respect to pheromonal communication in foraging behavior have been analyzed.[88, 89]

The volatile recruitment component of some *Atta* species was the first ant trail pheromone whose chemical structure was identified.[90] This compound, methyl 4-methylpyrrole-2-carboxylate (MMPC) functions as a recruitment trail pheromone in all *Atta* species except *Atta sexdens*, whose main recruitment trail pheromone component is 3-ethyl-2,5-dimethylpyrazine (EDMP).[91] *Atta* workers in laboratory colonies readily respond to trails drawn with small amounts of these substances. They follow these trails through all the twists and turns chosen by the experimenter. As we noted in Chapter 6, the potency of MMPC is quite amazing: 1 milligram of this substance is theoretically sufficient to draw a trail to which

87 | J. C. Moser and M. S. Blum, "Trail marking substance of the Texas leaf-cutting ant: source and potency," *Science* 140: 1228 (1963).

88 | K. Jaffe and P. E. Howse, "The mass recruitment system of the leaf-cutting ant *Atta cephalotes* (L.)," *Animal Behaviour* 27(3): 930–939 (1979).

89 | B. Hölldobler and E. O. Wilson, *The Ants* (Cambridge, MA: The Belknap Press of Harvard University Press, 1990).

90 | J. H. Tumlinson, R. M. Silverstein, J. C. Moser, R. G. Brownlee, and J. M. Ruth, "Identification of the trail pheromone of a leaf-cutting ant, *Atta texana*," *Nature* 234: 348–349 (1971).

91 | J. H. Cross, R. C. Byler, U. Ravid, R. M. Silverstein, S. W. Robinson, P. M. Baker, J. S. De Oliveira, A. R. Jutsum, and J. M. Cherrett, "The major component of the trail pheromone of the leaf-cutting ant, *Atta sexdens rubropilosa* Forel: 3–ethyl-2,5-dimethylpyrazine," *Journal of Chemical Ecology* 5: 187–203 (1979).

foragers of *Atta texana* and *Atta cephalotes* will follow 3 times around Earth's circumference.[92] And that record has recently been broken in the case of *Atta vollenweideri*: 1 milligram of this trail pheromone would be enough to lay a trail 60 times around the planet, with approximately 50 percent of the foragers of the grass-cutting ant still following.[93]

The pheromonal markings of the long-distance foraging routes are continuously reinforced by the foragers. However, the fine tuning of their deposition and the recruitment that results depend on a number of parameters, including the quality of the food and the need of the colony's fungus for new vegetation.[94, 95] The trail pheromone and other markings also appear to affect the attractiveness of the food source, hence stimulating harvesting activities such as cutting and transport of leaf fragments away from the harvesting site.[96, 97] Trail pheromones are used to mark not only the major trunk routes, but also the tree branches and twigs frequented by the ants. As a result, foragers continuously perceive the chemical trail signal. Any additional signal that mediates short-range recruitment at the foraging site would be most effective if transmitted through a different sensory channel. In fact, just such a superimposed mechanical signal has been found.[98] Most harvesting by leaf-cutting ants as a whole occurs in the canopies of trees. Here one can often observe that collectives of ants cut fragments out of particular leaves until nothing is left except a few leaf veins, while other leaves nearby remain almost untouched. It appears that those leaves intensely frequented by foragers are more desirable than other leaves, perhaps because they are more tender or less loaded with secondary plant compounds. Ants are able to summon nestmate foragers to these higher-

---

92 | R. G. Riley, R. M. Silverstein, B. Carroll, and R. Carroll, "Methyl 4-methylpyrrole-2-carboxylate: a volatile trail pheromone from the leaf-cutting ant, *Atta cephalotes*," *Journal of Insect Physiology* 20(4): 651–654 (1974).

93 | C. J. Kleineidam, W. Rössler, B. Hölldobler, and F. Roces, "Perceptual differences in trail-following leaf-cutting ants relate to body size," *Journal of Insect Physiology* 53(12): 1233–1241 (2007).

94 | F. Roces and B. Hölldobler, "Leaf density and a trade-off between load-size selection and recruitment behavior in the ant *Atta cephalotes*," *Oecologia* 97(1): 1–8 (1994).

95 | C. M. Nichols-Orians and J. C. Schultz, "Interactions among leaf toughness, chemistry, and harvesting by attine ants," *Ecological Entomology* 15(3): 311–320 (1990).

96 | J. W. S. Bradshaw, P. E. Howse, and R. Baker, "A novel autostimulatory pheromone regulating transport of leaves in *Atta cephalotes*," *Animal Behaviour* 34(1): 234–240 (1986).

97 | B. Hölldobler and E. O. Wilson, "Nest area exploration and recognition in leafcutter ants (*Atta cephalotes*)," *Journal of Insect Physiology* 32(2): 143–150 (1986).

98 | F. Roces, J. Tautz, and B. Hölldobler, "Stridulation in leaf-cutting ants: short-range recruitment through plant-borne vibrations," *Naturwissenschaften* 80(11): 521–524 (1993).

quality leaves by employing special short-range recruitment signals. The process is as follows. A number of *Atta* workers cutting leaf fragments produce stridulatory sounds. By employing Laser-Doppler vibrometry, it has been possible to record the signals transmitted by the ants onto the leaf surface (see Figure 9-3). When ants were offered leaves of different quality, the proportion of workers that stridulated during cutting differed markedly. Significantly more ants stridulated when tender leaves instead of thick leaves were offered. When the quality of the two kinds of leaves was enhanced by a sugar coating, almost all workers that were cutting also stridulated, regardless of differences in the physical properties of the material being cut. These observations suggest that the production of stridulatory vibrations is affected by the quality of the leaves, and leaf-cutting foragers use the sound to communicate leaf quality to their nearby nestmates.

Ants do not respond to airborne components of the stridulation sound,[99] but they are highly sensitive to vibrations propagated through the substrate. *Atta* workers on their way to the foraging site were given a choice between a vibrating twig and a silent one. When given a choice, more *Atta* foragers respond to recruitment pheromones than to substrate-borne stridulatory vibrations without pheromones. But the effectiveness of the recruitment pheromone is significantly enhanced when it is combined with the vibrational signal. Under natural conditions, nearby workers respond to the stridulatory vibrations transmitted through the plant material by orienting toward the source of the vibrations and subsequently joining in leaf cutting.[100]

The response of the ants to stridulatory signals is context specific. Workers of *Atta sexdens* stridulate as an alarm signal during nest defense. Because stridulation also mechanically facilitates the leaf-cutting process, it is tempting to suppose that leaf cutting was the first function of stridulation and that its employment in communication is a derived trait in evolution. However, subsequent studies have provided circumstantial evidence that the reverse is true: facilitation of cutting by the

---

99 | H. Markl, "Stridulation in leaf-cutting ants," *Science* 149: 1392–1393 (1965); and H. Markl, "Die Verständigung durch Stridulationssignale bei Blattschneiderameisen, II: Erzeugung und Eigenschaften der Signale," *Zeitschrift für vergleichende Physiologie* 60(2): 103–150 (1968).

100 | F. Roces, J. Tautz, and B. Hölldobler, "Stridulation in leaf-cutting ants: short-range recruitment through plant-borne vibrations," *Naturwissenschaften* 80(11): 521–524 (1993); and B. Hölldobler and F. Roces, "The behavioral ecology of stridulatory communication in leaf-cutting ants," in L. A. Dugkatin, ed., *Model Systems in Behavioral Ecology: Integrating Conceptual, Theoretical, and Empirical Approaches* (Princeton, NJ: Princeton University Press, 2001), pp. 92–109.

vibrations is more likely an auxiliary benefit emerging from the communication process.[101]

Leaf-cutting ants also stridulate frequently during nest building, particularly while manipulating soil particles with their mandibles. The stridulatory vibrations might serve in close-range recruitment to summon help from nestmates. But it is also possible that the vibrations simultaneously enhance excavation by acting like a vibrating pneumatic drill.[102]

There is yet one more context in which leafcutter stridulation serves in communication. Minim workers, the smallest worker subcaste, often ride on the leaf fragments being carried by other ants to the nest (see Figure 9-2, activity 1, and Plate 58). These tiny guard ants defend the leaf carriers from attack by parasitic phorid flies that try to oviposit on their sisters' bodies. It has been shown that leaf carriers communicate to the hitchhikers their readiness to load up and walk home by using plant-borne stridulatory vibrations. The stridulatory vibrations produced by the carrier in this initial transport phase seem to attract the tiny minims, who mount the carrier and leaf fragment.[103]

Leaf-cutting workers conspicuously stridulate when prevented from moving freely, whether trapped in a partial cave-in or held by an enemy ant. The substrate-borne stridulatory vibrations elicit a close-range alarm effect in nestmates.[104] Ants are attracted by the signal, and they start digging in an attempt to free the nestmate, or they attack the enemy ant that grasps her. Under natural conditions, such rescue signaling is usually multimodal: the mechanical stridulation is an important synergist of alarm pheromones. However, in massive aggressive interactions, with dozens or hundreds of workers involved in a melee, as in territorial defense, pheromones combined with defensive secretions are much more important.

Atta and *Acromyrmex* workers, like those of most other ant species, produce alarm pheromones in their mandibular glands. In fact, the mandibular gland pheromone in *Atta sexdens* was one of the very first such pheromones chemically and

---

101 | F. Roces and B. Hölldobler, "Use of stridulation in foraging leaf-cutting ants: mechanical support during cutting or short-range recruitment signal?" *Behavioral Ecology and Sociobiology* 39(5): 293–299 (1996).

102 | F. Roces, personal communication.

103 | F. Roces and B. Hölldobler, "Vibrational communication between hitchhikers and foragers in leaf-cutting ants (*Atta cephalotes*)," *Behavioral Ecology and Sociobiology* 37(5): 297–302 (1995).

104 | H. Markl, "Stridulation in leaf-cutting ants," *Science* 149: 1392–1393 (1965); and H. Markl, "Die Verständigung durch Stridulationssignale bei Blattschneiderameisen, II: Erzeugung und Eigenschaften der Signale," *Zeitschrift für vergleichende Physiologie* 60(2): 103–150 (1968).

behaviorally characterized.[105] Adolf Butenandt and his collaborators identified citral as the main compound in the mandibular gland secretions. These authors noted the relatively large size of the mandibular glands in the largest worker subcastes (soldiers), and they estimated that the glands occupy one-fifth of the volume of the head capsule. In behavioral tests, they demonstrated that the secretions have an alarm and repellent function. Subsequent studies by other researchers came to different results: although they identified citral, geranial, neral, and many other compounds, the effective alarm pheromone was determined to be 4-methyl-3-heptanone.[106] Later studies revealed that smaller workers, which are primarily active inside the nest, have mandibular gland secretions containing largely 4-methyl-3-heptanone, whereas secretions of larger workers engaged primarily outside the nest contain mainly citral.[107] These interesting findings are in accordance with the previously discussed division of labor based on physical worker subcastes and strongly suggest a context-specific use and function of the mandibular gland secretions in *Atta* workers. Of course, it would also be of interest to investigate age-specific changes of mandibular gland secretions in the worker subcastes and to analyze worker reactions to these components in different locational and behavioral contexts.[108]

Of central significance for the operation of the superorganism is the communication between its queen, acting as its reproductive unit, and the workers, comprising its somatic units. In her long lifetime, spanning more than 10 years, the queen of a large *Atta* colony can produce as many as 150 million daughters, of which the vast majority are workers. Every year, several thousands of these females in mature colonies grow up not into workers but into alate queens, each able to mate and found a new colony on her own. In addition, several thousands of the queen's progeny develop every year from unfertilized eggs to become the short-lived males. It is through the young queens and males that the colony reproduces and propagates its

105 | A. Butenandt, B. Linzen, and M. Lindauer, "Über einen Duftstoff aus der Mandibeldrüse der Blattschneiderameise *Atta sexdens rubropilosa* Forel," *Archives D'Anatomie Microscopique et de Morphologie Expérimentale* 48(Supplement): 13–19 (1959).

106 | M. S. Blum, F. Padovani, and E. Amante, "Alkanones and terpenes in the mandibular glands of *Atta* species (Hymenoptera: Formicidae)," *Comparative Biochemistry and Physiology* 26: 291–299 (1968).

107 | R. R. Do Nascimento, E. D. Morgan, J. Billen, E. Schoeters, T. M. C. Della Lucia, and J. M. S. Bento, "Variation with caste of the mandibular gland secretion in the leaf-cutting ant *Atta sexdens rubropilosa*," *Journal of Chemical Ecology* 19(5): 907–918 (1993).

108 | For additional information on caste variation of pheromones in *Atta*, see W. O. H. Hughes, P. E. Howse, and D. Goulson, "Mandibular gland chemistry of grass-cutting ants: species, caste, and colony variation," *Journal of Chemical Ecology* 27(1): 109–124 (2001).

genes. Colonies that produce the largest number of such healthy reproductive forms have the best chance of being represented in the next generation. However, to produce such a large crop, the colony needs a huge worker force to secure and retrieve the resources needed to rear the energetically quite expensive sexual brood. It is not too much to say that the sole purpose of workers is to put into the world as many royal siblings as possible.

The gigantic *Atta* queen is surrounded by her daughter workers at all times. She is continuously groomed and fed, and she produces an enormous number of eggs. A rough calculation reveals that the mother of a mature colony lays on average about 20 eggs per minute, thus 28,800 per day and 10,512,000 eggs per year. In the presence of a fertile queen, nest workers as a rule lay only deformed trophic eggs, which are then fed to the queen. The production of viable eggs by workers in intact *Atta* colonies would negatively affect colony efficiency and would be a serious handicap in reproductive competition with other mature colonies in a population. Thus, we should expect that nest workers are continuously informed about their queen's presence and fertility. But how is such a queen-to-worker communication in the huge *Atta* colonies possible? We do not yet know, but we can make a reasonable guess based on other information.

The *Atta* queen appears to remain quite stationary in one of the central fungus garden chambers of the expansive nest structure, where she does nothing but eat mostly trophic eggs and produce reproductive eggs. Her eggs are then distributed by workers over the entire fungus garden. (Such dispersion is necessary, because if the eggs remained in the queen's chamber, she would be suffocated by the growing mass.) Might the scattering eggs themselves carry a cue of the queen's presence? It has recently been shown in the monogynous carpenter ant *Camponotus floridanus* that this form of communication does occur: queen-laid eggs are coated with a queen-specific blend of hydrocarbons that serves as a queen fertility signal.[109] Workers respond to the pheromone by refraining from laying viable eggs. The distribution of queen-laid eggs by the workers spreads the queen signal in the colony. It seems very likely that a similar method of queen signal transmission will be found in colonies of *Atta*.

---

109 | A. Endler, J. Liebig, T. Schmitt, J. E. Parker, G. R. Jones, P. Schreier, and B. Hölldobler, "Surface hydrocarbons of queen eggs regulate worker reproduction in a social insect," *Proceedings of the National Academy of Sciences USA* 101(9): 2945–2950 (2004); and A. Endler, J. Liebig, and B. Hölldobler, "Queen fertility, egg marking and colony size in the ant *Camponotus floridanus*," *Behavioral Ecology and Sociobiology* 59(4): 490–499 (2006).

THE ANT-FUNGUS MUTUALISM

Whenever two kinds of organisms live in close mutualistic symbiosis, as is the case in leaf-cutting ants and their fungus, we should expect communication between the two mutualists. The fungus may signal to its host ants its preference for particular vegetable substrates or the need for a change in diet to maintain nutritional diversity or even the presence of a harmful substrate. To date, only a few studies have examined the possibility of communication between the fungus and the host ants.

It is well established that selection of the leaf material harvested by the leaf-cutting ants is dependent on both the physical and chemical characteristics of the plant.[110] It is reasonable to suppose, therefore, that if plant material is loaded with secondary compounds harmful to the fungus, the workers will cease harvesting these particular plants. However, this reaction might not be immediate. Several hours may ensue before the foragers completely abandon this food source.[111] However, once this delayed "rejection," as it is called, sets in for a particular plant material, the ants continue to refuse it for days or even weeks. How, then, is the information that the harvest material is unsuitable for the fungus transmitted to the foragers?

In laboratory experiments with *Atta* and *Acromyrmex* colonies, P. Ridley and his collaborators demonstrated that the ants learn to reject plant material that contains chemicals harmful to the fungus. Although the foragers initially carried baits containing orange peel laced with cycloheximide, a fungicide, into the nest, they eventually stopped collecting the bait, and the rejection was maintained for many weeks. The test colonies also rejected orange peel not contaminated with the

---

110 | M. Littledyke and J. M. Cherrett, "Defence mechanisms in young and old leaves against cutting by the leaf-cutting ants *Atta cephalotes* (L.) and *Acromyrmex octospinosus* (Reich) (Hymenoptera: Formicidae)," *Bulletin of Entomological Research* 68(2): 263–271 (1978); S. P. Hubell, D. F. Wiemer, and A. Adejare, "An antifungal terpenoid defends a Neotropical tree (Hymenaea) against attack by fungus-growing ants (*Atta*)," *Oecologia* 60(3): 321–327 (1983); and J. J. Howard, "Leafcutting and diet selection: relative influence of leaf-chemistry and physical features," *Ecology* 69(1): 250–260 (1988). See a review in R. Wirth, H. Herz, R. J. Ryel, W. Beyschlag, and B. Hölldobler, *Herbivory of Leaf-Cutting Ants: A Case Study on* Atta colombica *in the Tropical Rainforest of Panama* (New York: Springer-Verlag, 2003).

111 | J. J. Knapp, P. E. Howse, and A. Kermarrec, "Factors controlling foraging patterns in the leaf-cutting ant *Acromyrmex octospinosus* (Reich)," in R. K. Vander Meer, K. Jaffe, and A. Cedeno, eds., *Applied Myrmecology: A World Perspective* (Boulder, CO: Westview Press, 1990), pp. 382–409; and H. L. Vasconcelos and H. G. Fowler, "Foraging and fungal substrate selection by leaf-cutting ants," in R. K. Vander Meer, K. Jaffe, and A. Cedeno, eds., *Applied Myrmecology: A World Perspective* (Boulder, CO: Westview Press, 1990), pp. 410–419.

fungicide substance. The researchers hypothesized that if the substrate causes toxic effects on the fungus, the fungus will produce a chemical signal that acts as a negative reinforcement to the ant servicing that particular fungus garden.[112] In a follow-up study, investigators attempted to trace the pathway of this putative fungal signal.[113] Their results suggest that a signal produced by the fungus does not affect the foragers directly; instead, nonforager workers have to have contact with the fungus for the rejection to occur. The results thus suggest that the information is transferred from the smaller fungus garden workers to the larger forager workers.

The hypothetical chemical signal produced by stressed fungal tissue has yet to be characterized. Meanwhile, R. D. North and his collaborators have proposed an alternative hypothesis: rejection occurs when ants detect fungal breakdown products from unhealthy or dead fungus. The workers then associate dead fungus with "orange flavor" and consequently reject all substrate containing orange.[114] At least this much is known: leaf-cutting ants learn to associate odor with food,[115] as indicated by the fact that workers who have experienced contaminated orange peel also then reject uncontaminated orange peel. Further evidence of the existence of associative learning is that leafcutter workers, if exposed in the nest to particular odors by incoming scouts, then tend to seek material with that odor during their own foraging excursions.[116] Still unanswered by the sick-fungus hypothesis is the means by which the garden workers perceive the health of the fungus and the signals by which they transmit this information to the foragers.[117] In a new

---

112 | P. Ridley, P. E. Howse, and C. W. Jackson, "Control of the behaviour of leaf-cutting ants by their 'symbiotic' fungus," *Experientia* 52(6): 631–635 (1996).

113 | R. D. North, C. W. Jackson, and P. E. Howse, "Communication between the fungus garden and workers of the leaf-cutting ant, *Atta sexdens rubropilosa*, regarding choice of substrate for the fungus," *Physiological Entomology* 24(2): 127–133 (1999).

114 | R. D. North, C. W. Jackson, and P. E. Howse, "Communication between the fungus garden and workers of the leaf-cutting ant, *Atta sexdens rubropilosa*, regarding choice of substrate for the fungus," *Physiological Entomology* 24(2): 127–133 (1999).

115 | F. Roces, "Olfactory conditioning during the recruitment process in a leaf-cutting ant," *Oecologia* 83(2): 261–262 (1990); F. Roces, "Odour learning and decision-making during food collection in the leaf-cutting ant *Acromyrmex lundi*," *Insectes Sociaux* 41(3): 235–239 (1994); and J. J. Howard, L. Henneman, G. Cronin, J. A. Fox, and G. Hormiga, "Conditioning of scouts and recruits during foraging by a leaf-cutting ant, *Atta colombica*," *Animal Behaviour* 52(2): 299–306 (1996).

116 | F. Roces, "Odour learning and decision-making during food collection in the leaf-cutting ant *Acromyrmex lundi*," *Insectes Sociaux* 41(3): 235–239 (1994).

117 | J. J. Howard, M. L. Henneman, G. Cronin, J. A. Fox, and G. Hormiga, "Conditioning of scouts and recruits during foraging by a leaf-cutting ant, *Atta colombica*," *Animal Behaviour* 52(2): 299–306 (1996).

study tailored to resemble more natural conditions, Hubert Herz and his collaborators manipulated leaf suitability for the fungus by infiltrating the plants with a fungicide (cyclohexidine) not detectable to the ants. The ants' delayed rejection behavior was specific toward the respective fungicide-treated plant species. The rejection began 10 hours after treated leaves were carried into the fungus garden, and it continued for at least 9 weeks. Rejection was also observed in naive ants after contact with the fungus garden containing treated leaves. However, acceptance resumed after 3 weeks when ants were "force-fed" on untreated leaves of the previously treated plant species. This shows again that ants get information from the fungus that a particular plant species is not good for the fungus. The ants identify this particular plant species and thereafter avoid it as harvest material. They will, however, resume harvesting this plant species when the fungus does not exhibit a negative reaction. This species-specific flexible reception of unsuitable substrate may be a mechanism to avoid provisioning the fungus garden with plants containing harmful compounds, as such plants occur in the highly diverse natural habitats of the leafcutter ant colonies.[118]

Yet another kind of fungal signaling exists in leaf-cutting ant colonies: the ants recognize their own symbiotic fungal strain and protect it against competing strains introduced from other colonies.[119] Experiments by Michael Poulsen and Jacobus Boomsma have recently revealed that the mechanism on which this discrimination behavior originates is in the fungus.[120] The researchers used fungus gardens from colonies of two sympatric species of Panamanian leaf-cutting ants, *Acromyrmex echinatior* and *Acromyrmex octospinosus*. The clonal fungi of both species belong to the same genetically diverse clade. The compatibility of fungi from different colonies was assessed by inoculating pairs of mycelia 1.5 centimeters apart on an agar medium. After 2 months, mycelial intercompatibility could be measured on a scale from fully compatible to total rejection. By this means it was demonstrated that upon contact, domesticated fungi actively reject mycelial fragments from foreign (even neighboring) colonies. The intensity of the rejection is proportional to the

118 | H. Herz, B. Hölldobler, and F. Roces, "Delayed rejection in a leaf-cutting ant after foraging on plants unsuitable for the symbiotic fungus," *Behavioral Ecology* 19(3): 575–582 (2008).

119 | A. N. M. Bot, S. A. Rehner, and J. J. Boomsma, "Partial incompatibility between ants and symbiotic fungi in two sympatric species of *Acromyrmex* leaf-cutting ants," *Evolution* 55(10): 1980–1991 (2001).

120 | M. Poulsen and J. J. Boomsma, "Mutualistic fungi control crop diversity in fungus-growing ants," *Science* 307: 741–744 (2005).

overall genetic differences between the symbionts. Incompatibility compounds were detected in the fungal strains; their chemical structure has yet to be determined.

All fungus-growing ants manure the fungal garden with their own feces. Amazingly, the fungal enzymes that biochemically break down plant material are preserved during passage through the ant gut. After the ant has eaten fungus, the enzymes accumulate with other fecal matter in the rectal bladder of the ants. Fecal droplets containing the recycled enzymes are then deposited on the freshly cut pieces of leaf-bearing mycelial inocula or directly onto older fungal growth. Fecal droplets from ants of a foreign colony cause the same incompatibility effect on mycelial growth as the direct introduction of a foreign fungus. The intensity of rejection by fungi toward fecal droplets from nonresident ants corresponded to the genetic distance between the inoculum and the resident fungus receiving it. Oddly, initial incompatibility is lost and changes to compatibility when ants are forced to feed on an incompatible symbiont for 10 days or longer. The ants' new fecal droplets then become incompatible with their original resident fungus. From these striking results, Poulsen and Boomsma concluded that the symbiotic fungus uses its host ants to carry the fungus-specific incompatibility signal to all parts of the vast fungus garden of the leafcutter ant colony. Their results suggest that the ants' manuring practice is the decisive factor that constrains colonies to rearing a single clone of symbiont. Obligate manuring with feces allows the resident fungus to control the genetic identity of new gardens in the nest, causing the removal of unrelated fungi before they contribute to ant feeding, and hence ensuring the production of compatible fecal droplets.[121]

Let us summarize to this point. Because hostile interactions between symbionts reduce overall productivity, the introduction of an alien fungus clone not only harms the resident fungus, but also diminishes the growth and productivity of the host ant colony. It is therefore in the interest of both the resident fungus and the host to avoid competing fungal strains. The purity of the resident fungal clone is maintained through the action of the fungal incompatibility compound contained in the ants' fecal droplets.[122]

---

121 | M. Poulsen and J. J. Boomsma, "Mutualistic fungi control crop diversity in fungus-growing ants," *Science* 307: 741–744 (2005).

122 | For a more detailed review and excellent discussion of issues concerning ant-fungus conflict, see also U. G. Mueller, "Ant versus fungus versus mutualism: ant-cultivar conflict and the deconstruction of the attine ant-fungus symbiosis," *American Naturalist* 160(Supplement): S67–S98 (2002).

All this symbiotic webwork beautifully illustrates how much the symbiotic fungus has evolved to be an intricate part of the leafcutter ant superorganism. Neither of the mutualistic partners would be able to exist alone. The ants' division of labor and much of their social behavior are shaped by the details of this symbiotic relationship. In turn, the productivity and clonal propagation of the fungus is entirely dependent on its host ants. Although there might be some evolutionary conflict of interest and some exploitative manipulation for fitness advantages on both sides of the relationship, each kind of organism must be evolutionarily adjusted to the other, or the colony dies.

HYGIENE IN THE SYMBIOSIS

Maintaining a high level of vitality and hygienic condition of the fungus gardens is crucial for the host ant colony to survive and reproduce. An adequate level is not easy to accomplish; for the fungus to flourish, the requisite subterranean growth chambers need high humidity and tropical temperatures. The ants keep their gardens clean by an impressive variety of hygienic techniques: they pluck out alien fungi; they inoculate the correct fungal mycelia onto fresh substrate; they fertilize the substrate with fecal droplets that contain incompatibility substances to repel alien strains of the host fungus species; they secrete antibiotics to depress competing fungi and microorganisms; and they produce growth hormones.[123]

In 1970, Ulrich Maschwitz and his collaborators made the pioneering discovery that antibiotic substances are produced in the metapleural glands of *Atta sexdens* workers.[124] They suggested that the compounds play different roles in the purification of the symbiotic fungus culture: phenylacetic acid suppresses bacterial growth; myrmicacin (hydroxydecanoic acid) inhibits the germination of spores of

---

123 | For a review, see B. Hölldobler and E. O. Wilson, *The Ants* (Cambridge, MA: The Belknap Press of Harvard University Press, 1990); and R. Wirth, H. Herz, R. J. Ryel, W. Beyschlag, and B. Hölldobler, *Herbivory of Leaf-Cutting Ants: A Case Study on* Atta colombica *in the Tropical Rainforest of Panama* (New York: Springer-Verlag, 2003).

124 | U. Maschwitz, K. Koob, and H. Schildknecht, "Ein Beitrag zur Funktion der Metathoracaldrüse der Ameisen," *Journal of Insect Physiology* 16(2): 387–404 (1970); and U. Maschwitz, "Vergleichende Untersuchungen zur Funktion der Ameisenmetathorakaldrüse," *Oecologia* 16(4): 303–310 (1974).

alien fungi; and indoleacetic acid, a plant hormone, stimulates mycelial growth.[125] Recently, a more comprehensive analysis of metapleural gland secretions of *Acromyrmex octospinosus* revealed 20 previously unrecognized compounds.[126] They span the whole range of carboxylic acids, from acetic acid to long-chain fatty acids, in addition to keto acids, alcohols, and lactones.

The metapleural glands of leaf-cutting ant workers are relatively large compared with those of other kinds of ants, and interestingly, that is particularly true in the smallest workers.[127, 128] The latter disproportion suggests that the allocation of resources to metapleural gland secretions is most important in the minor workers, which predominantly tend the fungus and care for the brood.

The earlier prevailing assumption that fungus-culturing ants maintain their fungus gardens in completely pure conditions has had to be revised with the later discovery that fungus gardens are often contaminated by bacteria, yeasts, and other kinds of fungi.[129] A more thorough and deeper search for pathogens and parasites in leaf-cutting ant colonies demonstrated that while the ants cannot prevent contamination, they are nonetheless able to hold the growth of the invading microorganisms and foreign fungi to very low levels. It has been suggested that the main countermeasure of the ants against parasitic fungi is to maintain the fungus cultures at an acidic pH of 5, optimal for the symbiotic fun-

---

125 | H. Schildknecht and K. Koob, "Plant bioregulators in the metathoracic glands of myrmicine ants," *Angewandte Chemie* 9(2): 173 (1970); and H. Schildknecht and K. Koob, "Myrmicacin, the first insect herbicide," *Angewandte Chemie* 10(2): 124–125 (1971).

126 | D. Ortius-Lechner, R. Maile, E. D. Morgan, and J. J. Boomsma, "Metapleural gland secretion of the leaf-cutter ant *Acromyrmex octospinosus*: new compounds and their functional significance," *Journal of Chemical Ecology* 26(7): 1667–1683 (2000).

127 | E. O. Wilson, "Caste and division of labor in leaf-cutter ants (Hymenoptera: Formicidae: *Atta*), I: The overall pattern in *A. sexdens*," *Behavioral Ecology and Sociobiology* 7(2): 143–156 (1980); E. O. Wilson, "Caste and division of labor in leaf-cutter ants (Hymenoptera: Formicidae: *Atta*), II: The ergonomic optimization of leaf cutting," *Behavioral Ecology and Sociobiology* 7(2): 157–165 (1980); E. O. Wilson, "Caste and division of labor in leaf-cutter ants (Hymenoptera: Formicidae: *Atta*), III: Ergonomic resiliency in foraging by *A. cephalotes*," *Behavioral Ecology and Sociobiology* 14(1): 47–54 (1983); E. O. Wilson, "Caste and division of labor in leaf-cutter ants (Hymenoptera: Formicidae: *Atta*), IV: Colony ontogeny of *A. cephalotes*," *Behavioral Ecology and Sociobiology* 14(1): 55–60 (1983).

128 | A. N. M. Bot, M. L. Obermayer, B. Hölldobler, and J. J. Boomsma, "Functional morphology of the metapleural gland in the leaf-cutting ant *Acromyrmex octospinosus*," *Insectes Sociaux* 48(1): 63–66 (2001).

129 | C. R. Currie, "Prevalence and impact of a virulent parasite on a tripartite mutualism," *Oecologia* 128: 99–106 (2001). For an excellent review, see C. R. Currie, "A community of ants, fungi, and bacteria: a multilateral approach to studying symbiosis," *Annual Review of Microbiology* 55: 357–380 (2001).

gus but detrimental for pathogenic invading fungi.[130] Supporting this hypothesis is the fact that the pH rises to 7 or 8 when the ants are removed, and within a few days, parasitic fungi and bacteria spread rapidly in the symbiotic fungus cultures. For that reason, it has been suggested that one of the main functions of the metapleural gland secretions of *Acromyrmex* and *Atta* workers is to reduce the pH of the leaf material brought into the colony from approximately 7 or 8 to 5. It's an added benefit that each of the acids present in the secretions also has antibiotic properties.[131]

Recently, striking new discoveries have been reported concerning the "agricultural pathology" of ant fungus gardens. By extensive isolation of nonmutualistic fungi from the gardens of attine ants, Cameron Currie and his collaborators found specialized garden parasites belonging to the microfungus genus *Escovopsis* (Ascomycota: anamorphic Hypocreales). These parasites are horizontally transmitted between leaf-cutting ant colonies. *Escovopsis* is highly virulent, able to devastate ant gardens and thus doom the entire ant colony. Most remarkably, the genus *Escovopsis* appears to specialize on fungal gardens of attine ants. It has not been isolated from any other habitat, and it is especially prevalent in *Atta* and *Acromyrmex* colonies.

Currie and coworkers explain the success of the parasite with the following argument. The increased prevalence of *Escovopsis* within the more derived attine genera suggests that the long clonal history of the leafcutter fungal cultivars, perhaps as long as 23 million years, makes them more susceptible to losing the "arms race" with parasites. By contrast, lower attines routinely acquire new fungal cultivars from free-living sexual populations, leading to a greater genetic diversity in the fungal mutualist population. This may account for the apparent lower susceptibility to parasitism of the less derived attine lineages.[132]

In opposition to this hypothesis, however, is recently obtained evidence of sexual recombination in fungal symbionts of leafcutter ant taxa. If that phenomenon

130 | R. J. Powell and D. J. Stradling, "Factors influencing the growth of *Attamyces bromatificus*, a symbiont of attine ants," *Transactions of the British Mycological Society* 87(2): 205–213 (1986).

131 | D. Ortius-Lechner, R. Maile, E. D. Morgan, and J. J. Boomsma, "Metapleural gland secretion of the leaf-cutter ant, *Acroomyrmex octospinosus*: new compounds and their functional significance," *Journal of Chemical Ecology* 26(7): 1667–1683 (2000).

132 | C. R. Currie, U. G. Mueller, and D. Malloch, "The agricultural pathology of ant fungus gardens," *Proceedings of the National Academy of Sciences USA* 96(14): 7998–8002 (1999).

proves widespread, it suggests that clonality and vertical transmission have not played the critical role in leafcutter symbiotic evolution.[133]

Whatever the cause of the virulence, the question remains: How do the fungus-growing *Atta* and *Acromyrmex* cope with this continuous deadly threat? Obviously, the successful maintenance of a healthy fungus garden involves a continuous struggle to control the *Escovopsis* incursions. Some deterring effect might come from the metapleural gland secretions. But the main weapon against *Escovopsis* appears to be a third mutualist associated with attine ants, an actinomycetous filamentous bacterium of the genus *Pseudonocardia*.[134] This symbiont produces antibiotics that strongly suppress the growth of *Escovopsis*.[135] *Pseudonocardia* bacteria are true, evolved mutualists; they inhabit regions of the ants' cuticle that appear to be specific to the ant genus. In *Acromyrmex*, for example, they are housed on the laterocervical plates of the propleura (Figure 9-5). In this region, the *Acromyrmex* ants possess morphological modifications, such as crypts lined with integumental protrusions in the form of tubercles. Numerous exocrine gland cells connect to the tubercles through cuticular channels. The mutualistic filamentous bacteria are housed inside the crypts. Such bacterium-harboring structures have been found so far only in fungus-growing ants. However, the form and location of these structures are highly variable within the phylogeny of attine ants.[136] It is likely that the secretions of the glandular crypts help maintain the actinomycete cultures.

The mutualist bacteria are transmitted vertically (from parent to offspring colonies) within the body of the founding queen, in the same fashion as the symbiotic fungus. The bacteria are not just adapted to fight the parasitic fungus. They also promote the growth of the symbiotic fungus in vitro. In extreme cases of infestation, the ant colony may be forced to escape from *Escovopsis* by nest

---

133 | A. S. Mikheyev, U. G. Mueller, and P. Abbot, "Cryptic sex and many-to-one coevolution in the fungus-growing ant symbiosis," *Proceedings of the National Academy of Sciences USA* 103(28): 10702–10706 (2006).

134 | Originally, this actinomycete was thought to be of the genus *Streptomyces* (Streptomycetaceae: Actinomycetes). This identification appears to be incorrect (R. Wirth, personal communication), and ongoing molecular phylogenetic analyses have revealed that the symbiotic bacterium belongs to the actinomycetous family Pseudonocardiaceae; C. R. Currie, personal communication; see also corrigendum in *Nature* 423: 461 (2003).

135 | C. R. Currie, J. A. Scott, R. C. Summerbell, and D. Malloch, "Fungus-growing ants use antibiotic-producing bacteria to control garden parasites," *Nature* 398: 701–704 (1999).

136 | C. R. Currie, M. Poulsen, J. Mendenhall, J. J. Boomsma, and J. Billen, "Coevolved crypts and exocrine glands support mutualistic bacteria in fungus-growing ants," *Science* 311: 81–83 (2006).

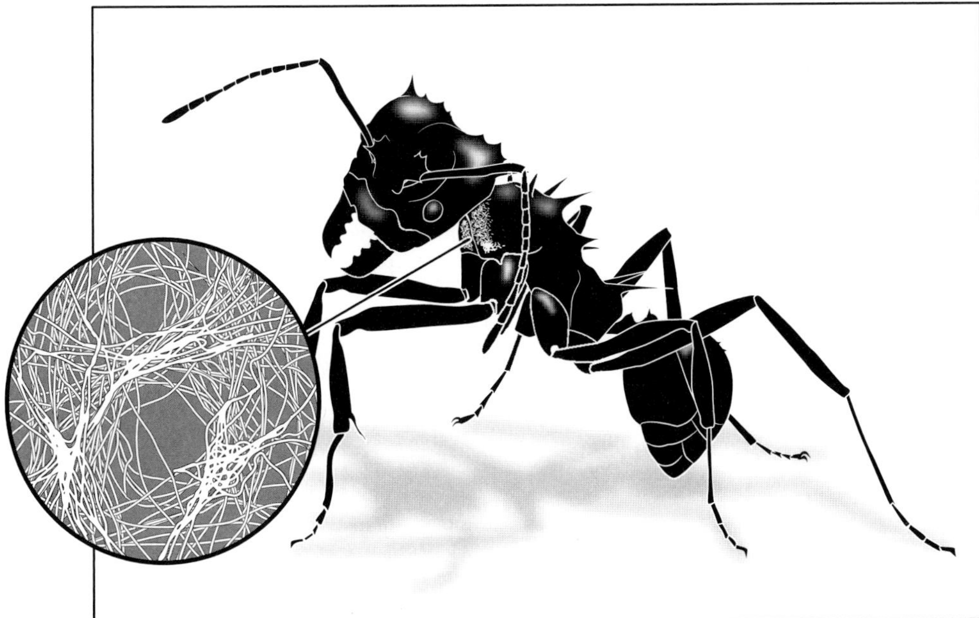

FIGURE 9-5. An actinomycete filamentous bacterium of the genus *Pseudonocardia* is a symbiont of leafcutter ants. It produces antibiotics that strongly suppress the growth of parasitic fungi. In the ant genus *Acromyrmex*, the symbiont is housed in the frontal region of the thorax (the laterocervical plates of the propleura), where it lives in special bacteria-harboring structures. Based on C. R. Currie, M. Poulsen, J. Mendenhall, J. J. Boomsma, and J. Billen, "Coevolved crypts and exocrine glands support mutualistic bacteria in fungus-growing ants," *Science* 311: 81–83 (2006).

emigration, carrying their bacteria with them, to continue the struggle in a new location.[137]

The uniqueness and tight fit in the relationship between attine ants and *Pseudonocardia* bespeak an ancient origin of the mutualism. Cameron Currie and his colleagues conclude overall: "Although the ant-fungus mutualism is often regarded as one of the most fascinating examples of highly evolved symbiosis, it is now clear that its complexity has been greatly underestimated. The attine symbiosis appears to be a co-evolutionary 'arms race' between the garden parasite, *Escovopsis*, on the one hand, and the tripartite association amongst the actinomycete, the ant host, and the fungal mutualist on the other" (Figure 9-6).[138]

---

137 | R. Wirth, H. Herz, R. J. Ryel, W. Beyschlag, and B. Hölldobler, *Herbivory of Leaf-Cutting Ants: A Case Study on* Atta colombica *in the Tropical Rainforest of Panama* (New York: Springer-Verlag, 2003).

138 | C. R. Currie, U. G. Mueller, and D. Malloch, "The agricultural pathology of ant fungus gardens," *Proceedings of the National Academy of Sciences USA* 96(14): 7998–8002 (1999).

WASTE MANAGEMENT

The depleted substrate left by the fungus forms a tremendous residue. Most *Atta* species build special refuse chambers in their nests to receive this waste, but *Atta colombica* has another solution: its colonies dispose of the material outside the nest. The refuse is loaded with secondary plant compounds and possibly parasitic fungal mycelia and other pathogens. The ants exhibit a strong avoidance of the waste once it is removed. Among many local people in Latin America, it has long been known that refuse from *Atta* nests can be used as a powerful repellent against ants. Experiments have shown that *Atta* refuse material scattered around young plants will protect them from *Atta* herbivory.[139, 140]

Atta colombica colonies usually manage their external waste removal by a form of task partitioning.[141] The refuse is taken from the nest and deposited on a cache along the trail to the dump. Other workers collect material from the cache and carry it to the main pile. The adaptive value of partitioning the task of waste removal may be the reduction of the spread of disease and parasites into the colony by segregating the garbage collectors inside the nest from the dump managers outside.[142] Ants exposed to waste material die at a higher rate, and waste is often infected by the fungal parasite *Escovopsis*. Waste management is mainly performed by older workers, who are destined soon to die anyway.[143, 144] The proneness of older workers to risk their lives is obviously an adaptive trait with respect to colony-level efficiency. This phenomenon of older workers taking greater risks is true for many ant species and in different contexts.[145]

---

139 | J. A. Zeh, A. D. Zeh, and D. W. Zeh, "Dump material as an effective small-scale deterrent to herbivory by *Atta cephalotes*," *Biotropica* 31(2): 368–371 (1999).

140 | C. R. Currie, J. A. Scott, R. C. Summerbell, and D. Malloch, "Fungus-growing ants use antibiotic-producing bacteria to control garden parasites," *Nature* 398: 701–704 (1999).

141 | C. Anderson and F. L. W. Ratnieks, "Task partitioning in insect societies: novel situations," *Insectes Sociaux* 47(2): 198–199 (2000).

142 | C. Anderson and F. L. W. Ratnieks, "Task partitioning in insect societies: novel situations," *Insectes Sociaux* 47(2): 198–199 (2000).

143 | See also A. N. M. Bot, C. R. Currie, A. G. Hart, and J. J. Boomsma, "Waste management in leaf-cutting ants," *Ethology Ecology & Evolution* 13(3): 225–237 (2001).

144 | A. G. Hart and F. L. W. Ratnieks, "Waste management in the leaf-cutting ant *Atta colombica*," *Behavioral Ecology* 13(2): 224–231 (2002).

145 | B. Hölldobler and E. O. Wilson, *The Ants* (Cambridge, MA: The Belknap Press of Harvard University Press, 1990).

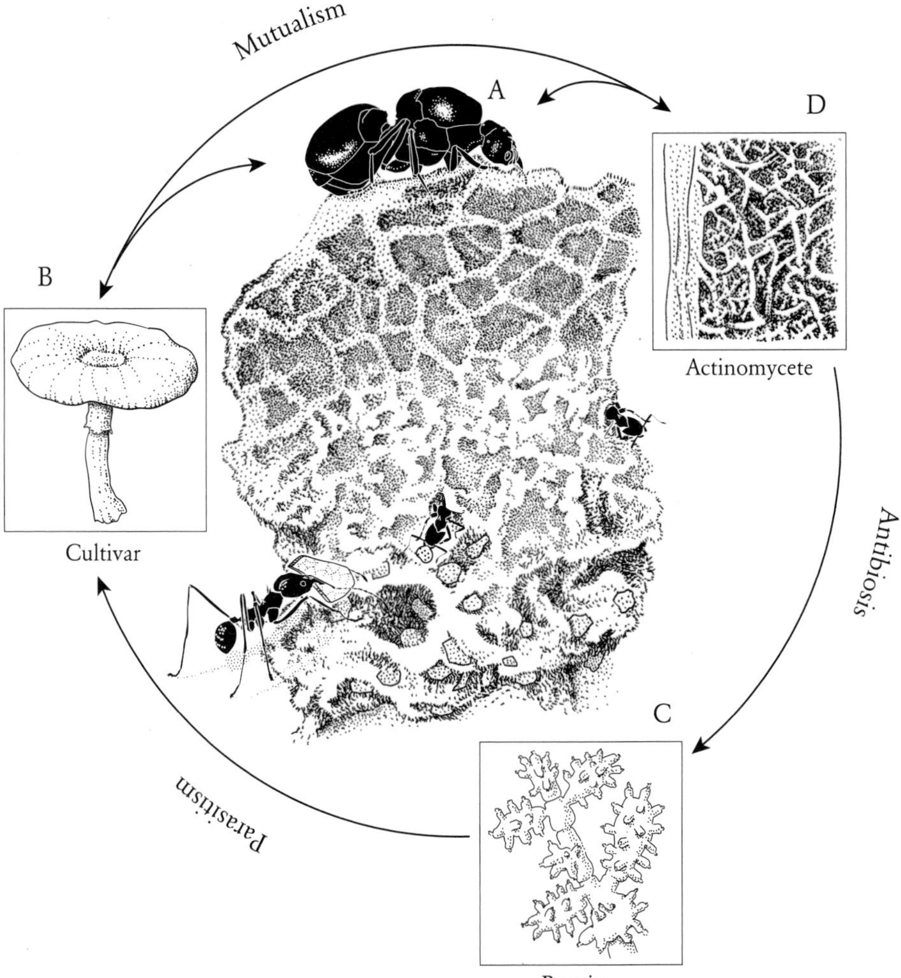

FIGURE 9-6. The quadripartite symbiosis in leafcutter ants. *A*: The queen is the reproductive unit of the leafcutter colony. *B*: Mushroom habitus of the free-living leucocoprineous fungi. *C*: The parasitic microfungus *Escovopsis*. *D*: The filamentous actinomycete *Pseudonocardia,* which grows on the cuticle of the ants and produces antibiotics that suppress the growth of *Escovopsis*. The arrows indicate interacting components. Based on an original illustration by Gara Gibson in C. R. Currie, "A community of ants, fungi, and bacteria: a multilateral approach to studying symbiosis," *Annual Review of Microbiology* 55: 357–380 (2001).

Occasionally, as noted, a fungus garden is struck by a massive invasion of *Escovopsis* or other pathogen, and the colony is forced to abandon its nest and fungus garden and emigrate to a new nest site. The colony then has to acquire a new fungal cultivar. Genetic data suggest that transfer of fungus from one colony to another can occur, and for the lower attine genus *Cyphomyrmex*, this phenomenon has been demonstrated experimentally.[146] In the laboratory, we have repeatedly "healed" declining gardens of *Atta cephalotes* colonies by introducing pieces of healthy gardens cultivated by other colonies. In this context, the reports of raid of incipient colonies by large colonies of *Atta sexdens rubropilosa*, with the transfer of brood and fungus material, are of special interest.[147] These observations were made with colonies cultured in the laboratory, and no such documentation exists as yet for *Atta* in the field. However, raids have been observed in nature between incipient colonies of *Acromyrmex versicolor.*[148] It seems likely that the stealing or usurpation of fungus gardens by older attine colonies whose own garden has been devastated does occur as a natural response to the loss of fungus gardens.[149]

AGROPREDATORS AND AGROPARASITES

The symbiotic fungus of attine colonies obviously is an attractive resource, not only for parasitic fungi that require the garden substrate and conspecific ant colonies that have lost their own fungus garden, but also for other ant species. Michiel Dijkstra and Jacobus Boomsma have described the predatory raids conducted by the ponerine ant *Gnamptogenys hartmani* in Panama. *Gnamptogenys* scouts that have discovered a fungus-growing ant nest of *Trachymyrmex* or *Sericomyrmex* species recruit raiding columns of nestmates that attack and usurp the nest of the assaulted colony. There is hardly any defense; the workers of the raided colony abscond their nest in panic. The raiders' colony then moves into the captured nest, where its workers and larvae consume the fungus and host brood. After this resource is depleted, the colony conquers the nest of another fungus-growing ant colony.[150]

---

146 | R. M. M. Adams, U. G. Mueller, A. K. Holloway, A. M. Green, and J. Narozniak, "Garden sharing and garden stealing in fungus-growing ants," *Naturwissenschaften* 87(11): 491–493 (2000).

147 | M. Autuori, "Contribuição para o conhecimento da saúva *(Atta* spp.—Hymenoptera—Formicidae), V: Número de formas aladas e redução dos sauveiros iniciais," *Arquivos do Instituto Biológico São Paulo* 19(22): 325–331 (1950).

148 | S. W. Rissing, G. B. Pollock, M. R. Higgins, R. H. Hagen, and D. R. Smith, "Foraging specialization without relatedness or dominance among co-founding ant queens," *Nature* 338: 420–422 (1989).

149 | In this context the new findings on long-distance horizontal transmission of fungal symbionts between leafcutter ants is of particular interest. See A. S. Mikheyev, U. G. Mueller, and P. Abbot, "Cryptic sex and many-to-one co-evolution in the fungus-growing ant symbiosis," *Proceedings of the National Academy of Sciences USA* 103(28): 10702–10706 (2006).

150 | M. B. Dijkstra and J. J. Boomsma, "*Gnamptogenys hartmani* Wheeler (Ponerinae: Ectatommini): an agropredator of *Trachymyrmex* and *Sericomyrmex* fungus-growing ants," *Naturwissenschaften* 90(12): 568–571 (2003).

This raiding behavior closely resembles that discovered in the myrmicine genus *Megalomyrmex*.[151, 152] These agropredators raid nests of the attine ant *Cyphomyrmex longiscapus*, and they consume the cultivated fungus and the attine brood after all the resident ants have been killed or expelled. Raiding *Megalomyrmex* species may represent an early evolutionary grade in the phylogenetic pathway toward a trophic and social parasitic cohabitation and usage of the fungus garden with the host ants, an adaptation described for *Megalomyrmex symmetochus*.[153]

Surprisingly, no agropredators are yet known to attack the leaf-cutting genera *Atta* and *Acromyrmex*. It might be that their nests are too large to be usurped, and their elaborate worker subcaste systems with specialized defenders make them resistant to such predation. However, at least two kinds of social parasitic ants live in *Acromyrmex* colonies. *Pseudoatta argentina*, which parasitizes colonies of *Acromyrmex lundi*, is a highly derived social parasite in which the worker caste has been lost. Another social parasite, *Acromyrmex insinuator*, apparently represents a less derived evolutionary grade of social parasitism. It retains a worker caste, and its morphology still closely resembles that of its host species, *Acromyrmex octospinosus*. These social parasites coexist intimately with their host ants and eat their fungus, but do not participate in the culturing efforts.[154] Although obviously an economic burden on the leaf-cutting ant colony, they do not go so far as to devastate the fungus garden.

LEAFCUTTER NESTS

The enormous number of worker ants and the huge fungus gardens of an *Atta* colony require a colossal nest capacity. One typical *Atta sexdens* nest, more than 6

---

151 | R. M. M. Adams, U. G. Mueller, A. K. Holloway, A. M. Green, and J. Narozniak, "Garden sharing and garden stealing in fungus-growing ants," *Naturwissenschaften* 87(11): 491–493 (2000); and R. M. M. Adams, U. G. Mueller, T. R. Schultz, and B. Norden, "Agro-predation: usurpation of attine fungus gardens by *Megalomyrmex* ants," *Naturwissenschaften* 87(12): 549–554 (2000).

152 | W. M. Wheeler, "A new guest-ant and other new Formicidae from Barro Colorado Island, Panama," *The Biological Bulletin* 49(1): 150–181 (1925).

153 | C. R. F. Brandão, "Systematic revision of the Neotropical ant genus *Megalomyrmex* Forel (Hymenoptera, Formicidae, Myrmicinae), with the description of thirteen new species," *Arquivos de Zoologia* (São Paulo) 31: 411–481 (1990).

154 | T. R. Schultz, D. Bekkevold, and J. J. Boomsma, "*Acromyrmex insinuator* new species: an incipient social parasite of fungus-growing ants," *Insectes Sociaux* 45(4): 457–471 (1998).

years old, contained 1,920 chambers, of which 238 were occupied by fungus garden and ants. The loose soil that had been brought out and piled on the ground by the ants during the excavation of their nest weighed approximately 40,000 kilograms (40 tons). Although nests of different *Atta* species have been excavated and reconstructed on paper by several authors,[155] the recent, quantitatively detailed work by Luiz Forti and his team in Brazil has delivered a breakthrough in our understanding of the megalopolis architecture of *Atta* colonies.[156]

The nest mound of the mature *Atta laevigata* colonies measured by these researchers varied from 26.1 to 67.2 square meters. In addition to a careful step-by-step excavation, the team perfected a method for making a mold of the nest interior. To obtain the mold, they poured liquid cement into the nest entrance. For one large nest, a mixture of 6,300 kilograms (6.3 tons) of cement and 8,200 liters of water was required, enough for a small human dwelling. After 2 to 3 weeks, the "petrified" nest structure was carefully excavated (Plates 62 to 65).[157] The number of nest chambers in the sample made by the Forti team ranged from 1,149 (smaller mature colony) to 7,864 (largest colony), both reaching as deep as 7 to 8 meters underground. Most of the chambers were located at a depth of 1 to 3 meters. In the very large nests, about 30 percent of the chambers were found below 4 meters, although several were empty. Some chambers contained declining fungus cultures, but many others housed flourishing fungus gardens complete with brood and ants. In addition, a number of chambers were filled with plant debris and degraded fungal material. Extensive subterranean foraging tunnels led into a central area of the highest concentration of fungus garden chambers. Smaller tunnels branched off from the main channels, and still smaller ramifications connected directly to individual fungus garden chambers. Most of the garden chambers had only one such small tunnel, with an opening located in the middle or near its base. The volume of the largest chambers ranged from about 25 to 51 liters, and that of the smallest ones from 0.03 to 0.06 liter.

---

155 | See B. Hölldobler and E. O. Wilson, *The Ants* (Cambridge, MA: The Belknap Press of Harvard University Press, 1990); colonies of *Acromyrmex* species are correspondingly smaller and less complex.

156 | A. A. Moreira, L. C. Forti, A. P. P. Andrade, M. A. C. Boaretto, and J. F. S. Lopes, "Nest architecture of *Atta laevigata* (F. Smith, 1858) (Hymenoptera: Formicidae)," *Studies on Neotropical Fauna and Environment* 39(2): 109–116 (2004); and A. A. Moreira, L. C. Forti, M. A. C. Boaretto, A. P. P. Andrade, J. F. S. Lopes, and V. M. Ramos, "External and internal structure of *Atta bisphaerica* Forel (Hymenoptera: Formicidae) nests," *Journal of Applied Entomology* 128(3): 204–211 (2004).

157 | L. C. Forti and F. Roces, personal communication.

PLATE 62. The mature nests of *Atta* species are gigantic. Shown here is a nest of *Atta vollenweideri* in Argentina. (Photo: Flavio Roces.)

The nests of all species of *Atta* possess a comparatively complex architecture, comprising many tunnels and chambers of variable sizes and shapes.[158] A particular feature common to nests of at least *Atta laevigata*, *Atta sexdens rubropilosa*, *Atta vollenweideri*, and *Atta bisphaerica* is the location of fungus garden chambers mainly below the loose soil mound and down to about 3 meters deep. The Brazilian researchers have suggested the following architectural rationale: "Probably the accumulation of loose soil over the fungus chambers has the purpose of thermal insulation, since *Atta laevigata* often build their nest in open habitats with the fungus chambers rather close to the surface. From the localization of the fungus chambers

158 | For a discussion, see A. A. Moreira, L. C. Forti, A. P. P. Andrade, M. A. C. Boaretto, and J. F. S. Lopes, "Nest architecture of *Atta laevigata* (F. Smith, 1858) (Hymenoptera: Formicidae)," *Studies on Neotropical Fauna and Environment* 39(2): 109–116 (2004); and A. A. Moreira, L. C. Forti, M. A. C. Boaretto, A. P. P. Andrade, J. F. S. Lopes, and V. M. Ramos, "External and internal structure of *Atta bisphaerica* Forel (Hymenoptera: Formicidae) nests," *Journal of Applied Entomology* 128(3): 204–211 (2004).

PLATE 63. A mature nest of *Atta laevigata* in Brazil was excavated after 6 tons of cement and 8,000 liters of water had been poured into the nest to preserve its structure in petrified form. (Photo: Wolfgang Thaler.)

we must conclude that the upper 3 m of soil offer the best microclimatic conditions for fungus growth."

An *Atta* nest thus houses the huge fungal body together with millions of workers and immature ants in a distributed but interconnected network of nest chambers. As the enormous biomass metabolizes, it produces large quantities of carbon dioxide, which can be fatal to the ant colony if the concentration becomes too high. *Atta* workers are equipped with very sensitive carbon dioxide receptors on their antennae, enabling them to measure carbon dioxide concentrations.[159]

Concentrations of carbon dioxide inside nests of *Atta vollenweideri* vary according to the size of the nests, to differences in the effectiveness of nest ventilation, and

159 | C. Kleineidam and J. Tautz, "Perception of carbon dioxide and other 'air-condition' parameters in the leaf-cutting ant *Atta cephalotes*," *Naturwissenschaften* 83(12): 566–568 (1996).

PLATE 64. Portions of the cement-filled subterranean tunnels, ducts, and fungus chambers of the *Atta laevigata* nest. (Photo: Wolfgang Thaler.)

finally to differences between small and large colonies. Small colonies tend to close their nest entrances during rain to protect the fungus garden from flooding. In such situations, carbon dioxide concentrations increase rapidly, and colony respiration rates are reduced. It appears that the ants' respiration remains unchanged, but the respiration of the symbiotic fungus is lowered. This subsidence, of course, negatively affects the growth rate of the fungus, and ultimately that of the colony, because the fungus is the main food source of the larvae. Young growing colonies are thus confronted with a trade-off: minimizing the risk of being flooded and drowned versus providing adequate gas exchange inside the nests.[160]

---

160 | C. Kleineidam and F. Roces, "Carbon dioxide concentrations and nest ventilation in nests of the leaf-cutting ant *Atta vollenweideri*," *Insectes Sociaux* 47(3): 241–248 (2000); and C. Kleineidam, R. Ernst, and F. Roces,

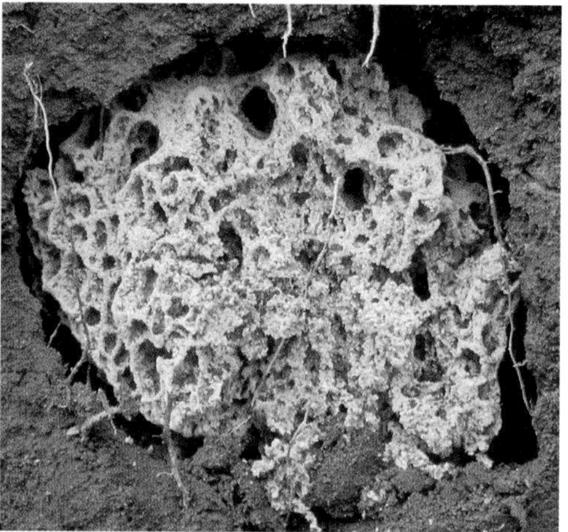

PLATE 65. *Left*: The ball-shaped structures connected by tunnels and ducts are "petrified" fungus chambers. *Right*: A live fungus chamber. (Photo: Wolfgang Thaler.)

In nests of mature colonies, on the other hand, provided with many nest openings and with deep chambers, gas exchange occurs without interruption, albeit with variable intensity. The openings in the central area of the nest mound are often shaped like turrets (see Plate 62). Investigators have found a markedly negative correlation between wind velocity above the nest mound and carbon dioxide

---

"Wind-induced ventilation of the giant nests of the leaf-cutting ant *Atta vollenweideri*," *Naturwissenschaften* 88(7): 301–305 (2001).

concentration inside the nest, with the wind on the surface most likely inducing an increased outflow of carbon dioxide–laden air. The decaying organic material in the refuse chambers often raises the ambient temperature within the nest. If the outside temperature is lower, warmer carbon dixoide–laden air from the nest chambers rises and flows out of the turrets, with cooler fresh air pulled back into the nest through other entrance tubes.[161]

It follows that both passive ventilation due to external wind velocity and thermal convection likely drive the gas exchange system in *Atta vollenweideri* nests and probably also nests of other *Atta* species. *Atta* nests located in open grasslands are more exposed to strong winds than nests located in forest habitats, such as those of *Atta cephalotes*. In the latter species, thermal convection is probably more important.

Leaf-cutting ants do not depend entirely on nest architecture to regulate nest ventilation. The gardening workers are also able to sense differences in relative humidity and establish fungus gardens in chambers with the highest humidity. When the chambers start to dry out, workers tear up and relocate the garden into chambers with higher humidity.[162] In fact, Martin Bollazzi and Flavio Roces report fascinating experiments with the leaf-cutting ant *Acromyrmex ambiguus*. In the laboratory, a colony was exposed to either dry or humid air flowing through the nest chambers. Circulation of dry air triggered an increase in building activity, and tunnels through which the dry air entered the nest were plugged at the inflow, but much less at the outflow sections. Inflow of humid air did not elicit much, if any, plugging behavior. The direction of airflow served as an environmental cue for spatial guidance of building activities, and control of the nest climate may be a major determinant in the design of nest structures.[163]

TRAILS AND TRUNK ROUTES

Yet another architectural feature of *Atta laevigata* nests are the extensive horizontal foraging tunnels, constructed about 40 to 50 centimeters below the ground

---

161 | F. Roces, personal communication.

162 | F. Roces and C. Kleineidam, "Humidity preference for fungus culturing by workers of the leaf-cutting ant *Atta sexdens rubropilosa*," *Insectes Sociaux* 47(4): 348–350 (2000).

163 | M. Bollazzi and F. Roces, "To build or not to build: circulating dry air organizes collective building for climate control in the leaf-cutting ant *Acromyrmex ambiguus*," *Animal Behaviour* 74(5): 1349–1355 (2007).

surface. In cross section they are elliptical, measuring 4 to 48 centimeters in width and 2 to 6 centimeters in height. In larger nests, the foraging tunnels are wider, but not necessarily higher.[164] They channel the masses of foragers arriving on the aboveground trunk routes. Such subterranean tunnels are also known from other *Atta* species, including *Atta vollenweideri*.[165, 166] These tunnels are occasionally more than 6 meters long. They in turn open out onto trunk routes connecting nest and harvesting areas, sometimes extending more than 250 meters from the nest. Because they are static and long-lived, these trunk routes are considered part of the nest architecture. In most cases, they are deeply engraved into the ground and conspicuous even to the most casual observer. They are retained for months or years. Even when abandoned for a time, they are commonly used again by the ants. Serving as the superhighways of the *Atta* colonies, they are continuously cleaned of invading vegetation and other obstacles by "road workers" (see Plate 60). The trunk route system enhances foraging efficiency by increasing foraging speed four- to tenfold compared with that on uncleared ground.[167] They are moreover part of the defended territories that serve to protect the colony's resources from competitors.[168]

Trail construction and maintenance add a considerable amount to the overall energy investment for resource acquisition. However, as documented in a study of *Atta colombica*, the energy expenditure is small relative to the energy yielded by their use, and their cost does not constrain their construction.[169] In addition, most *Atta* species appear to be trail centered. That is, their search for high-quality resources is restricted to the vicinity of the trunk routes.

In one study conducted by Rainer Wirth and his colleagues in Panama, the trunk trail system of *Atta colombica* colonies were monitored continuously over a

---

164 | A. A. Moreira, L. C. Forti, A. P. P. Andrade, M. A. C. Boaretto, and J. F. S. Lopes, "Nest architecture of *Atta laevigata* (F. Smith, 1858) (Hymenoptera: Formicidae)," *Studies on Neotropical Fauna and Environment* 39(2): 109–116 (2004).

165 | J. C. M. Jonkman, "The external and internal structure and growth of nests of the leaf-cutting ant *Atta vollenweideri* Forel, 1893 (Hym.: Formicidae)," *Zeitschrift für angewandte Entomologie* 89(2): 158–173 (1980).

166 | F. Roces, personal communication.

167 | L. L. Rockwood and S. P. Hubbell, "Host-plant selection, diet diversity, and optimal foraging in a tropical leafcutting ant," *Oecologia* 74(1): 55–61 (1987).

168 | H. G. Fowler and E. W. Stiles, "Conservative resource management by leaf-cutting ants? The role of foraging territories and trails, and environmental patchiness," *Sociobiology* 5(1): 25–41 (1980).

169 | J. J. Howard, "Costs of trail construction and maintenance in the leaf-cutting ant *Atta colombica*," *Behavioral Ecology and Sociobiology* 49(5): 348–356 (2001).

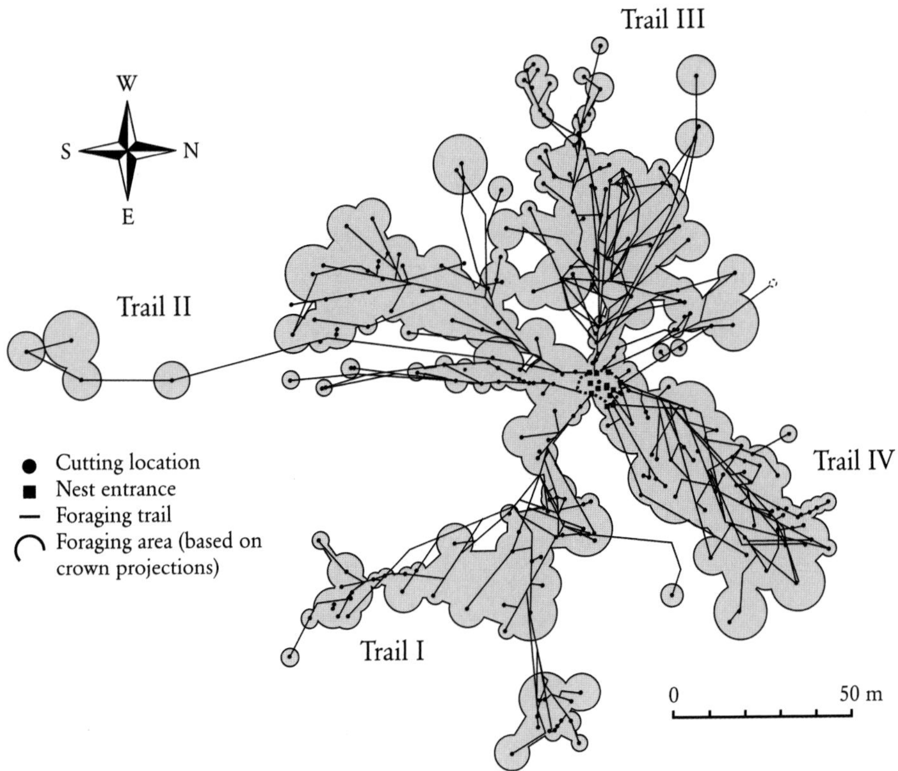

FIGURE 9-7. The trail system of one colony of the leaf-cutting ant *Atta colombica* covered an entire hectare, as illustrated here. The estimates of foraging area (gray) are based on regression estimates of crown projection areas. From R. Wirth, H. Herz, R. J. Ryel, W. Beyschlag, and B. Hölldobler, *Herbivory of Leaf-Cutting Ants: A Case Study on* Atta colombica *in the Tropical Rainforest of Panama* (New York: Springer-Verlag, 2003).

period of a full year. The data obtained were then used for an assessment of the true size of the foraging area. One representative nest had four major trunk routes, each of which opened out into numerous branches. The foraging areas of the four routes were 2,712, 2,597, 2,640, and 2,409 square meters. The total of the foraging area visited was thus estimated at an impressive 1.03 hectares (Figure 9-7).[170] This happens to be approximately the same "ecological footprint," or average amount of land utilized to sustain one person, in developing countries.

170 | R. Wirth, H. Herz, R. J. Ryel, W. Beyschlag, and B. Hölldobler, *Herbivory of Leaf-Cutting Ants: A Case Study on* Atta colombica *in the Tropical Rainforest of Panama* (New York: Springer-Verlag, 2003); see also C. Kost, E. G. de Oliveira Kost, T. A. Knoch, and R. Wirth, "Spatio-temporal permanence and plasticity of foraging trails in young and mature leaf-cutting ant colonies (*Atta* spp.)," *Journal of Tropical Ecology* 21(6): 677–688 (2005).

Comparative studies of the nest architecture of *Atta laevigata*, *Atta bisphaerica*, *Atta vollenweideri*, *Atta capiguara*, *Atta sexdens*, and *Atta texana* reveal much similarity but also species-specific differences.[171, 172, 173, 174] Nest structures are the product of innate collective behavior. They are, to use a metaphor coined by Richard Dawkins, the "extended phenotype" of each kind of superorganism in turn. No less than the anatomy and physiology of the ants themselves, architectural features of the nests are shaped by natural selection. They are an exemplary illustration of natural selection acting at the level of the entire society. Like any other group-level units of biological organization, the nests are, to quote Thomas D. Seeley, "elegant devices that nature has evolved for integrating thousands of insects into a higher order entity, one whose abilities transcend those of the individual."[175]

A half century after William Morton Wheeler's publication of this idea of an insect colony as a superorganism, scientists have revived the superorganism concept with an emphasis on colony-level adaptive demography and organization based on division of labor, visualizing the colony as a self-organized entity and the target of social selection. In particular, David S. Wilson and Elliott Sober have argued that insect colonies can be true superorganisms that are targeted by selection, provided colonies have differential group fitness and the variation in group fitness is caused by heritable variation. In addition, no reproductive competition within groups should exist; or at the very least, in their view, intragroup competition should be conspicuously lower than intergroup competition.[176]

If we were to accept this concept of a superorganism, with its codicil of little or no reproductive competition among nestmates, many of the poneromorph and myrmeciine societies we have described might not be considered true superorganisms,

---

171 | A. A. Moreira, L. C. Forti, A. P. P. Andrade, M. A. C. Boaretto, and J. F. S. Lopes, "Nest architecture of *Atta laevigata* (F. Smith, 1858) (Hymenoptera: Formicidae)," *Studies on Neotropical Fauna and Environment* 39(2): 109–116 (2004); and A. A. Moreira, L. C. Forti, M. A. C. Boaretto, A. P. P. Andrade, J. F. S. Lopes, and V. M. Ramos, "External and internal structure of *Atta bispherica* Forel (Hymenoptera: Formicidae) nests," *Journal of Applied Entomology* 128(3): 204–211 (2004).

172 | L. C. Forti and F. Roces, personal communication.

173 | N. A. Weber, *Gardening Ants: The Attines* (Philadelphia: American Philosophical Society, 1972).

174 | J. C. Moser, "Contents and structure of *Atta texana* nest in summer," *Annals of the Entomological Society of America* 56(3): 286–291 (1963).

175 | T. D. Seeley, *The Wisdom of the Hive: The Social Physiology of Honey Bee Colonies* (Cambridge, MA: Harvard University Press, 1995).

176 | D. S. Wilson and E. Sober, "Reviving the superorganism," *Journal of Theoretical Biology* 136(3): 337–356 (1989).

because intracolony reproductive competition is indeed conspicuously common. We share the view now generally held by other researchers that insect societies are dynamic, self-organized systems of different complexity subject to hierarchical levels of selection.[177] This view does not contradict that of Wilson and Sober, but appears to avoid its narrow and potentially confusing restriction. Nevertheless, the thousands of social insect species display among themselves almost every conceivable grade in the division of labor, from little more than competition among nestmates for reproductive status to highly complex systems of specialized subcastes. The level of this gradient at which the colony can be called a superorganism is subjective; it may be at the origin of eusociality (preferred by E.O.W.) or at a high level, beyond the "point of no return," in which within-colony competition for reproductive status is greatly reduced or absent (preferred by B.H.).

But whatever criteria may be adopted, there can be little doubt that the gigantic colonies of the *Atta* leafcutters, with their interlocking symbiont communities and extreme complexity and mechanisms of cohesiveness, deserve special attention as the greatest superorganisms discovered to the present time.

177 | T. D. Seeley, "Honey bee colonies are group-level adaptive units," *The American Naturalist* 150(Supplement): S22–S41 (1997); S. D. Mitchell, *Biological Complexity and Integrative Pluralism* (New York: Cambridge University Press, 2003); and R. E. Page and S. D. Mitchell, "The superorganism: new perspectives or tired metaphor?" *Trends in Ecology and Evolution* 8(7): 265–266 (1993).

|| Plate 66. Cast of a subterranean tunnel of a nest of the leafcutter ant *Atta laevigata*. The subterranean highway leads the ants to the fungus chambers left and right of the tunnel. (Photo: Wolfgang Thaler.)

10

NEST ARCHITECTURE AND HOUSE HUNTING

Animal-built structures can be considered external organs of their builders or, more precisely, part of the extended phenotype of the organisms that produce them.[1, 2] In social insects species-typical nest structures result from the collective actions of many individuals and thus represent extended phenotypes of the cooperating groups, the superorganisms, that build them.

THE ANALYSIS OF NEST ARCHITECTURE

Although a wealth of information exists about nest structures in termites, social bees, and wasps, the subterranean nest architecture of ants has remained largely unknown.[3] Only a few studies have provided quantitative descriptions—in particular, those of Walter Tschinkel, who has also produced three-dimensional casts of nest structures by pouring a thin slurry of dental plaster or molten aluminum into the nests.[4, 5, 6] In other studies, nests have been carefully excavated and the three-dimensional architecture then reconstructed. From these studies, together with those

---

1 | J. S. Turner, *The Extended Organism: The Physiology of Animal-Built Structures* (Cambridge, MA: Harvard University Press, 2000).

2 | R. Dawkins, *The Extended Phenotype: The Gene as the Unit of Selection* (San Francisco: W. H. Freeman, 1982).

3 | M. H. Hansell, *Animal Architecture and Building Behaviour* (New York: Longman, 1984).

4 | W. R. Tschinkel, "Subterranean ant nests: trace fossils past and future?" *Palaeo* 192: 321–333 (2003).

5 | W. R. Tschinkel, "The nest architecture of the Florida harvester ant, *Pogonomyrmex badius*," *Journal of Insect Science* 4(21), 19 pp. (2004).

6 | W. R. Tschinkel, "The nest architecture of the ant, *Camponotus socius*," *Journal of Insect Science* 5(9), 18 pp. (2005).

on *Atta* species described in the previous chapter, it has been established that ant nests not only display intricate structure but also frequently exhibit features that characterize particular species. Colonies of most poneromorph species occupy relatively simple nests (albeit with few notable exceptions; see Chapter 8). Colonies of many evolutionarily more derived species construct complex nests in wood or soil, or out of leaves and thatch, or from a paper-like carton manufactured by the ants. The nests of the leaf-cutting ant colonies exhibit by far the most complex nest architecture of all known ant species.

In colonies of leafcutter ants and most other advanced social insects, the distribution and activities of workers and brood within the nest, as well as on the aboveground nest terrain, are spatially structured. The nest design serves as a "template" for the sorting and distribution of worker task groups, a phenomenon also reported from other ant species.[7] Experimental studies have moreover revealed a remarkable social resilience of the spatial distribution among ant workers in both simple and complex nest structures, at least during limited time spans.[8, 9]

The techniques developed by Tschinkel have revealed beautiful examples of subterranean nest structures for different ant species (Plates 67 to 69). Although there is wide variation in shape and size, genus- or even species-typical features are easily recognizable. Tschinkel describes the study and its importance in the following account:

> As a subject of behavioral study, nest architecture offers an appealing feature that practically no other behaviors offer; namely, the nest is a perfect record of the collective digging effort of a colony, and once cast, is ready to study. By studying a series of casts of increasing size it is possible to describe the nest's growth and ontogeny, infer its species-typical characteristics, and bracket the range of variation. By doing this under different environments and soil types, possibly with transplanted

---

7 | W. R. Tschinkel, "The nest architecture of the Florida harvester ant, *Pogonomyrmex badius*," *Journal of Insect Science* 4(21), 19 pp. (2004).

8 | W. R. Tschinkel, "Sociometry and sociogenesis of colony-level attributes of the Florida harvester ant (Hymenoptera: Formicidae)," *Annals of the Entomological Society of America* 92(1): 80–89 (1999); and D. Cassill, W. R. Tschinkel, and S. B. Vinson, "Nest complexity, group size and brood rearing in the fire ant, *Solenopsis invicta*," *Insectes Sociaux* 49(2): 158–163 (2002).

9 | S. J. Backen, A. B. Sendova-Franks, and N. R. Franks, "Testing the limits of social resilience in ant colonies," *Behavioral Ecology and Sociobiology* 48(2): 125–131 (2000).

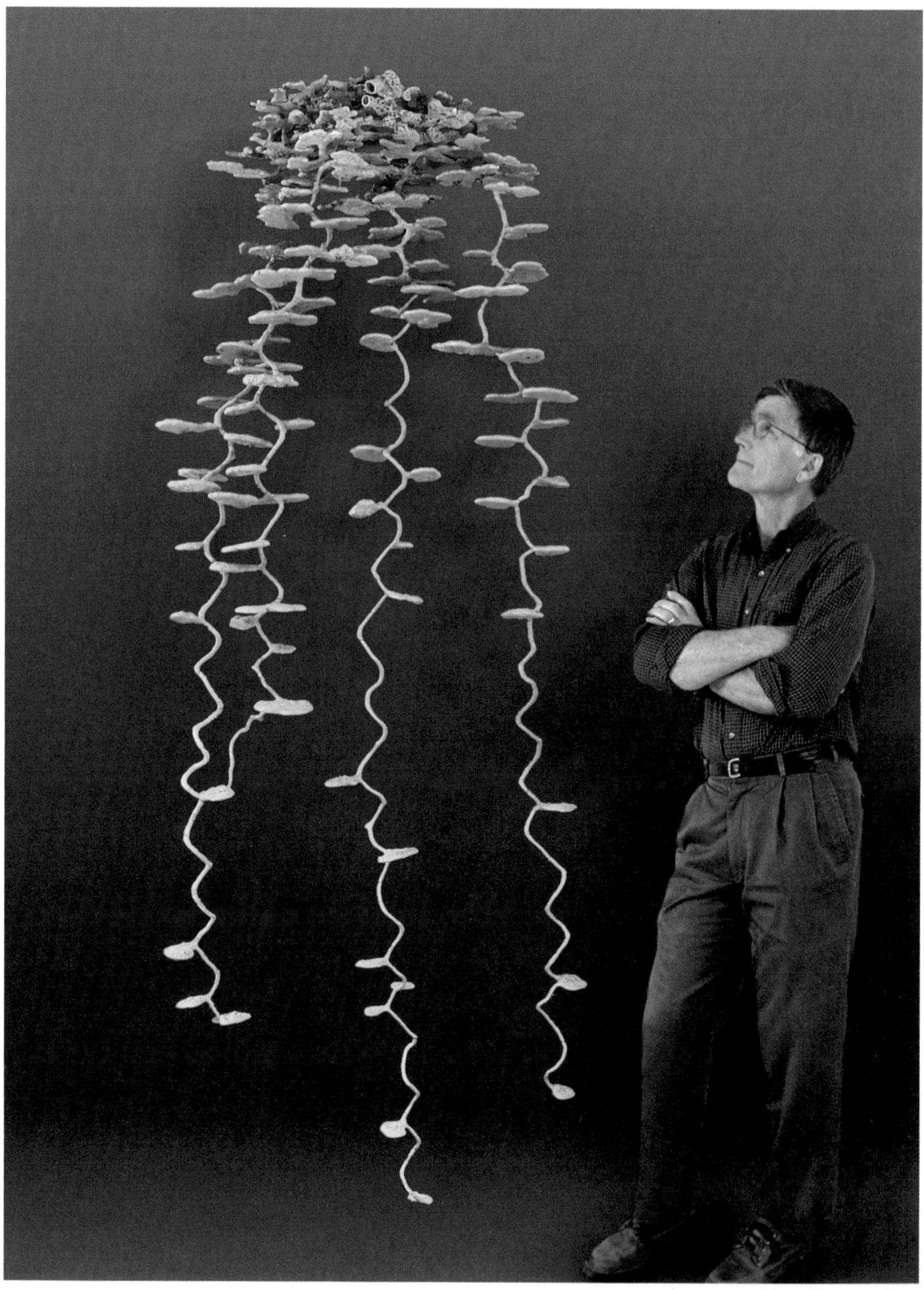

colonies, it is possible to tease out the variation that the environment imposes on the architecture. The current study is only a small, initial step toward creating a field of nest architecture studies, whose ultimate goal is an understanding of how the nest emerges from self-organizing behavior, what function it serves, how it varies within and between species, and how it evolves. In addition, these casts reveal something previously unseen. The study of nest architecture is thus a true exploration of a hidden world that holds unsuspected beauty, pattern and complexity.[10]

HOW ARCHITECTURE IS ACHIEVED

Some social insects can build complex structures complete with air-conditioning and fortified "castles." But unlike human construction, there is no architect, no blueprint, no global design that governs the course of construction. Instead, nest structure emerges through the self-organization of multiple workers interacting with each other and with the environment as they modify it. Theories of this process were originally developed in physics and chemistry to describe the emergence of macroscopic patterns out of processes of interactions defined at the microscopic level.[11] The basic concept has been extended to social insects—to model complex collective behaviors and thereby explain how social patterns and nest structures emerge from interactions between individuals guided by the same behavioral algorithms.[12] For example, Scott Camazine has shown how the characteristic order of brood, pollen, and honey cells on the comb of a honeybee colony emerges through remarkably simple behavioral dynamics.[13] The typical pattern on the comb consists of three distinct concentric regions; a central brood area,

---

10 | W. R. Tschinkel, "The nest architecture of the Florida harvester ant, *Pogonomyrmex badius*," *Journal of Insect Science* 4(21), 19 pp. (2004).

11 | E. Bonabeau, G. Theraulaz, J. L. Deneubourg, S. Aron, and S. Camazine, "Self-organization in social insects," *Trends in Ecology and Evolution* 12(5): 188–193 (1997).

12 | S. Camazine, J.-L. Deneubourg, N. R. Franks, J. Sneyd, G. Theraulaz, and E. Bonabeau, *Self-Organization in Biological Systems* (Princeton, NJ: Princeton University Press, 2001).

13 | S. Camazine, "Self-organizing pattern formation on the combs of honey bee colonies," *Behavioral Ecology and Sociobiology* 28(1): 61–76 (1991).

PLATE 67. The cast of a mature nest of the eastern harvester ant *Pogonomyrmex badius*. Walter Tschinkel, who made the cast and stands next to it, is 178 centimeters tall. (Photo: Charles F. Badland.)

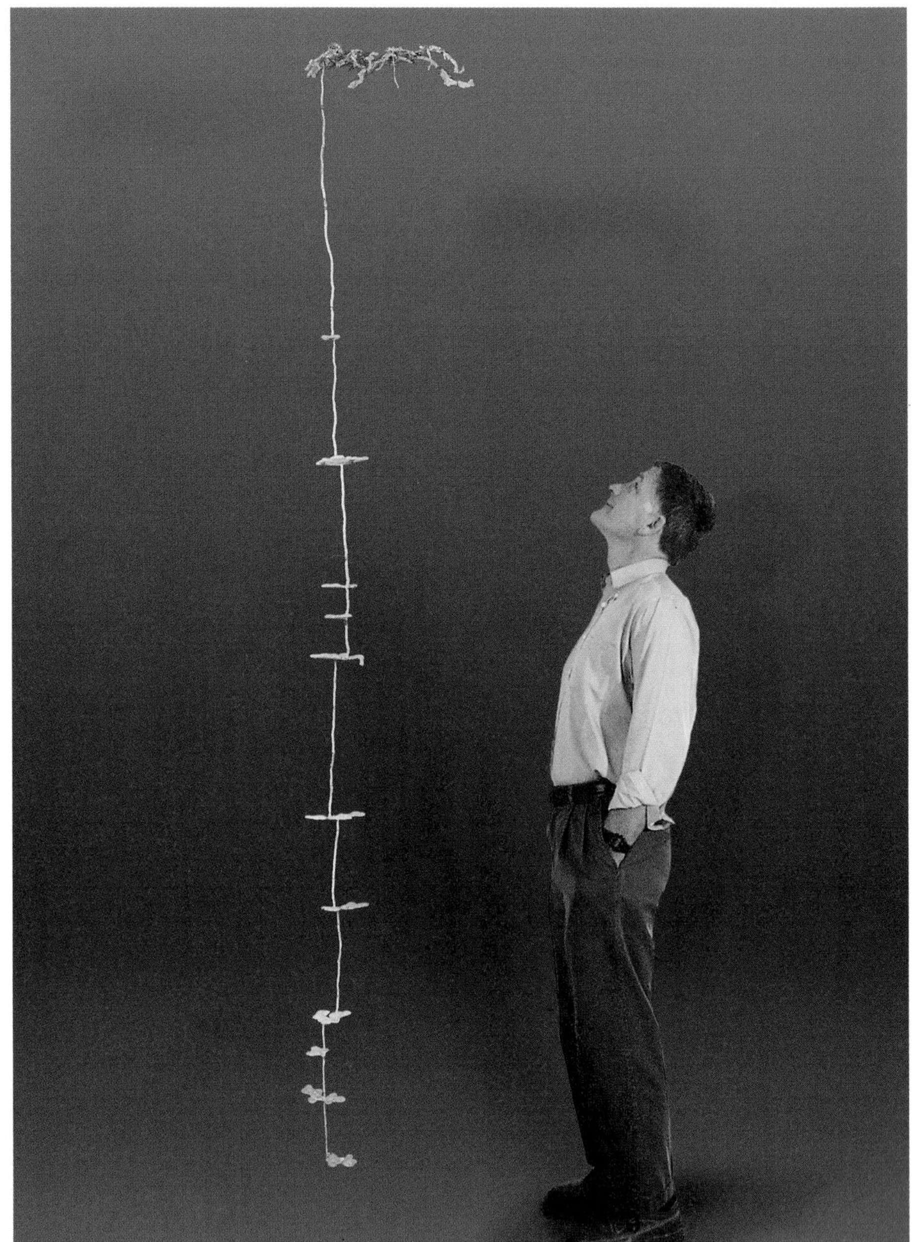

PLATE 68. The remarkable nest structure of the formicine "winter ant" *Prenolepis imparis*, is revealed here in a cast made by Walter Tschinkel. (Photo: Charles F. Badland.)

a surrounding zone of pollen cells, and a large peripheral region of honey cells (Plate 70). Based on a self-organization model and experimental observations, Camazine determined the following governing rules (algorithms) underlying the emergence of this pattern:

1 | The queen randomly patrols the combs and lays most eggs in the neighborhood of cells already occupied by brood.
2 | Honey and pollen are deposited randomly in available cells.
3 | Four times as much honey is brought to the hive as pollen.
4 | Typical removal-to-input ratios for honey and pollen are 0.6 and 0.95, respectively.
5 | Removal of honey and pollen is proportional to the number of cells containing brood.

Simulations based on these rules and experimental observations show how the formation of the concentric regions of brood, pollen, and honey emerge. Rules 1 and 5 ensure the growth of a central brood area, providing the first eggs are laid approximately at the center of the comb. Honey and pollen are initially randomly deposited, but in conformity to rules 3 and 4, pollen cells are more likely emptied and refilled with honey. This means that pollen cells located in the periphery will be replaced by honey cells, and the only cells

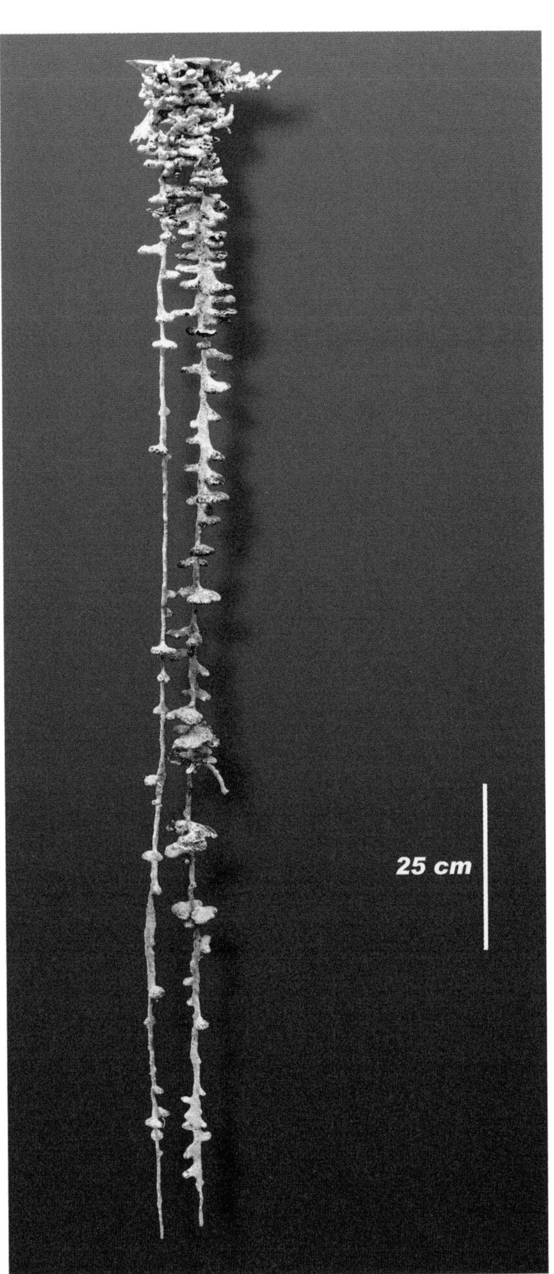

PLATE 69. Nest structure of *Pheidole morrisi*. (Cast and photo: Walter Tschinkel.)

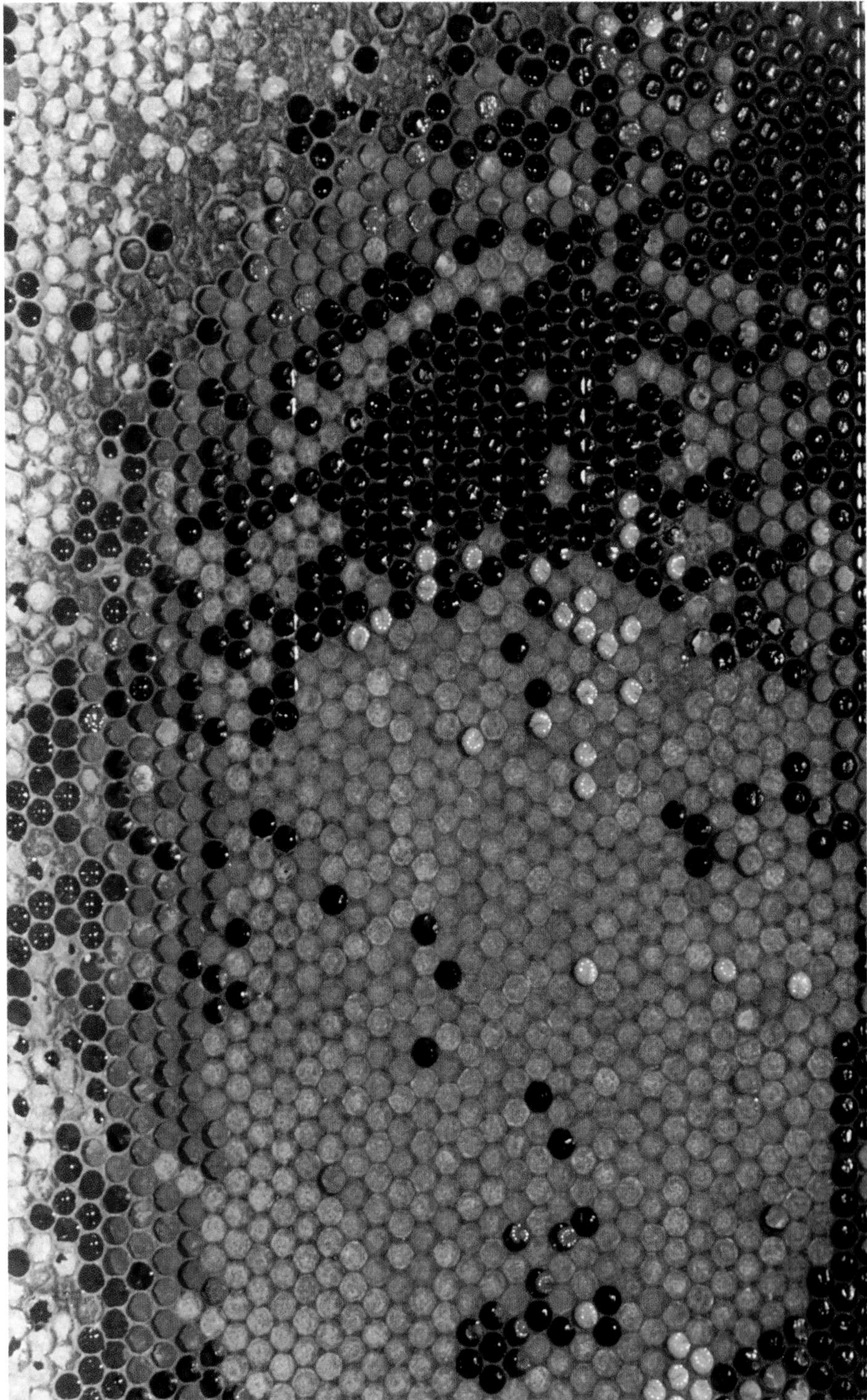

PLATE 70. The defining pattern of a honeybee comb consists of three more or less distinct concentric regions: a central brood area, a surrounding zone of pollen cells, and a large peripheral region of honey cells. (Photo: Marco Kleinhenz.)

available for pollen are those surrounding the brood area, because they have a higher turnover rate.[14]

Thus, the random pollen and nectar deposits throughout the comb and the preferential removal of honey and pollen near the brood cells play a crucial organizing role. They create a zone between the honey cells and the brood cells, which is then devoted to pollen storage. But as the amount of brood increases, the removal rate of pollen near brood increases too, so that more cells become available in which the queen can deposit eggs, thus leading to a further increase in pollen consumption. More honey and pollen cells are emptied, and this in turn increases the available storage space for new pollen to meet the nutritional needs of the developing larvae. Such positive feedback interacts with the simple behavioral rules of emptying and filling cells on the comb.

This study beautifully illustrates how colony-level patterns in nest organization can be created when bees employ simple rules in response to local cues. It shows "how global order can emerge from what appears to be local chaos—a concept that computer programmers have been cribbing from insect behavior for years to make networks run more efficiently."[15]

The prospects of this approach are very promising. Consider the precise arrangement of hexagonal comb cells in a honeybee nest. It turns out to arise from restraining physical parameters and simple behavioral rules, as recently disclosed through an analysis by Christian Pirk and his colleagues.[16] When these researchers filled the cells with polyester resin to obtain casts from the inner surface of cells, they obtained a surprising result. The sides of hardened resin block were hexagonal in shape, as expected; but the bottom did not consist of three rhomboids, as previously assumed, but instead had the shape of a flat spheroid. The appearance of rhomboids is an optical artifact. The structure of the combs results from wax as a thermoplastic building material that softens and hardens by increasing and decreasing temperatures. The wax flows around an array of closely packed cylinders that are, in fact, the builder bees themselves. Individual bees, by working side by side in adjacent circular tubes, extend the cell to its final length, and as they

14 | See also E. Bonabeau, G. Theraulaz, J.-L. Deneubourg, S. Aron, and S. Camazine, "Self-organization in social insects," *Trends in Ecology and Evolution* 12(5): 188–193 (1997).

15 | B. Shouse, "Getting the behavior of social insects to compute," *Science* 295: 2357 (2002).

16 | C. W. W. Pirk, H. R. Hepburn, S. E. Radloff, and J. Tautz, "Honeybee combs: construction through a liquid equilibrium process?" *Naturwissenschaften* 91(7): 350–353 (2004).

work, they warm up the wax. The round walls gradually expand against the adjacent cells until they end up hexagonal. The cell bases, however, are hemispherical from the outset of construction and never form three rhomboids. The bees themselves are heat engines as well as mechanics. It has been found that bees inside the cells may heat up to over 40°C, a temperature that optimizes the thermoplastic properties of the wax.

The analysis of pattern formation on the comb of honeybees is an example of an investigation in which mathematical modeling led to empirical experimentation, which in turn revealed the emergence of orderly structures through self-organization. Other such studies have addressed nest building and wall construction behavior in ants and termites,[17, 18, 19, 20] cell arrangements in paper wasp nests,[21] the bodily self-assemblages in insect societies,[22, 23] brood sorting in ants,[24] the generation of foraging columns and trunk routes,[25] and other structured mass body formations and biological properties that can emerge only from collective actions.

No species-specific blueprints exist that direct the building activities of various termites, wasps, bees, and ants; instead, a set of simple specific behavioral "rules" and feedback loops govern the collective efforts out of which the stereotypical nest structures emerge. The responses of individual colony members to key stimuli are genetically encoded and hence subject to natural selection. The study of

---

17 | N. R. Franks, A. Wilby, B. W. Silverman, and C. Tofts, "Self-organizing nest construction in ants: sophisticated building by blind bulldozing," *Animal Behaviour* 44(2): 357–375 (1992).

18 | J. L. Deneubourg and N. R. Franks, "Collective control without explicit coding: the case of communal nest excavation," *Journal of Insect Behavior* 8(4): 417–432 (1995).

19 | G. Theraulaz and E. Bonabeau, "Modelling the collective building of complex architectures in social insects with lattice swarms, "*Journal of Theoretical Biology* 177(4): 381–400 (1995).

20 | P. Rasse and J. L. Deneubourg, "Dynamics of nest excavation and nest size regulation of *Lasius niger* (Hymenoptera: Formicidae)," *Journal of Insect Behavior* 14(4): 433–449 (2001).

21 | I. Karsai and Z. Pénzes, "Optimality of cell arrangements and rules of thumb of cell initiation in *Polistes dominulus*: a modeling approach," *Behavioral Ecology* 11(4): 387–395 (2000).

22 | C. Anderson, G. Theraulaz, and J. L. Deneubourg, "Self-assemblages in insect societies," *Insectes Sociaux* 49(2): 99–110 (2002).

23 | A. Lioni and J.-L. Deneubourg, "Collective decision through self-assembling," *Naturwissenschaften* 91(5): 237–241 (2004).

24 | N. R. Franks and A. B. Sendova-Franks, "Brood sorting by ants: distributing the workload over the work-surface," *Behavioral Ecology and Sociobiology* 30(2): 109–123 (1992).

25 | N. R. Franks, N. Gomez, S. Goss, and J. L. Deneubourg, "The blind leading the blind in army ant raid patterns: testing a model of self-organization (Hymenoptera: Formicidae)," *Journal of Insect Behavior* 4(5): 583–607 (1991); and I. D. Couzin and N. R. Franks, "Self-organized lane formation and optimized traffic flow in army ants," *Proceedings of the Royal Society of London B* 270: 139–146 (2003).

architectural variation among species has further led to the understanding of the evolutionary adaptations of nest structures in social insects and the general rules and feedback responses that guide their construction.[26, 27, 28]

THE PROCESS OF STIGMERGY

Stigmergy, the pioneering concept of self-organization in nest architecture, was first proposed by the great French zoologist Pierre-Paul Grassé in 1959 in order to explain nest construction by the macrotermitine termites of Africa.[29] The term is derived from the Greek phrase meaning "incite to work." No direct interactions are required between the workers. Instead, as Grassé demonstrated, individual builders affect each other through the by-products of their activity. At the very start, a worker modifies its environment by placing a fecal pellet or some other material at a particular spot. The environmental alteration then becomes a new stimulus, able to induce in other workers a new response—for example, adding a second pellet or rearranging a pile of them. In stigmergic labor, it is the product of work previously accomplished rather than direct communication among nestmates that induces the insects to perform additional labor. Even if the workforce is constantly renewed, the nest structure already completed determines by its location, height, shape, and probably also its odor what further work will be done. Such stigmergic labor also enables social insects to build at multiple locations simultaneously, an important evolutionary step toward development of the most complex nest architecture.[30]

Of equal importance, stigmergy includes positive feedback. The most rapidly growing structures act as the strongest stimuli and as a consequence continue to

---

26 | J. S. Turner, *The Extended Organism: The Physiology of Animal-Built Structures* (Cambridge, MA: Harvard University Press, 2000).

27 | See, for example, R. L. Jeanne, "The adaptiveness of social wasp nest architecture," *The Quarterly Review of Biology* 50(3): 267–287 (1975).

28 | J. W. Wenzel, "Evolution of nest architecture," in K. G. Ross and R. W. Matthews, eds., *The Social Biology of Wasps* (Ithaca, NY: Cornell University Press, 1991), pp. 480–519.

29 | P.-P. Grassé, "La reconstruction du nid et les coordinations interindividuelles chez *Bellicositermes natalensis* et *Cubitermes* sp. la théorie de la stigmergie: essai d'interprétation du comportement des termites constructeurs," *Insectes Sociaux* 6(1): 41–84 (1959).

30 | I. Karsai and J. W. Wenzel, "Productivity, individual-level and colony-level flexibility, and organization of work as consequences of colony size," *Proceedings of the National Academy of Sciences USA* 95(15): 8665–8669 (1998).

grow more quickly until they reach a stimulus plateau, triggering negative feedback with a stimulus decline for further building. These guiding rules most likely change in different environmental contexts. Otherwise the architecture would be homogeneous, and major elements, such as subterraneous foraging tunnels and spherical fungus garden chambers, could not be constructed.

Stigmergic responses have proved to be major elements in nest construction by social insects generally. Their explicit identification has helped clarify the control of nest construction behavior in paper wasps,[31, 32] the communal nest excavation and self-organized nest construction in some ant species,[33, 34] and the baffling cooperative labor of the weaver ants (*Oecophylla longinoda* and *Oecophylla smaragdina*).

Consider, then, the weaver ants, dominant arboreal species in tropical Africa and Asia. Working in the treetops, they construct nests from green leaves that are bound together by sticky larval silk. To build these unique leaf-and-silk pavilions, it is necessary for groups of workers to pull leaves together simultaneously while others move the silk-secreting larvae back and forth between the edges of the leaves like animated shuttles (see Plates 17 and 19). How is such cooperation achieved?

The basic solution, discovered by John Sudd, involves a simple form of stigmergy.[35] Workers work independently in their first attempts to pull down or roll up leaves. When success is achieved by one or more of them at any part of a leaf, other workers in the vicinity abandon their own effort and join in. Thus, the environmental change achieved by successful workers acts as a stimulus for other workers to cooperate. That, in turn, enhances the stimulus, attracting even more workers to the work crew. They line up in a row and pull together. Alternatively, in cases where a gap longer than the ant's body remains to be closed, the workers form a living chain by seizing one another's petiole ("waist") and then pulling on the edge of a leaf as a single unit. When multiple rows of chains are aligned, the combined force is surprisingly powerful.

Recently, Arnaud Lioni and Jean-Louis Deneubourg developed a mathematical

---

31 | I. Karsai and Z. Pénzes, "Nest shapes in paper wasps: can the variability of forms be deduced from the same construction algorithm?" *Proceedings of the Royal Society of London B* 265: 1261–1268 (1998).

32 | I. Karsai, "Decentralized control of construction behavior in paper wasps: an overview of the stigmergy approach," *Artificial Life* 5(2): 117–136 (1999).

33 | N. R. Franks, A. Wilby, B. W. Silverman, and C. Tofts, "Self-organizing nest construction in ants: sophisticated building by blind bulldozing," *Animal Behaviour* 44(2): 357–375 (1992).

34 | J. L. Deneubourg and N. R. Franks, "Collective control without explicit coding: the case of communal nest excavation," *Journal of Insect Behavior* 8(4): 417–432 (1995).

35 | J. H. Sudd, *An Introduction to the Behaviour of Ants* (London: Edward Arnold, 1967).

model that explains the dynamics of chain formation in weaver ants. They confirmed that the bigger the chain, the greater the probability that others will join it, and the smaller it is, the lower the probability that individuals will leave.[36] All this, of course, depends on the interplay between the size of the worker group in the vicinity of the forming chain and the individual response to the number of nestmates in the chain. In fact, recruitment of nestmates to a nest construction site by employment of chemical trails and motor signals amplifies and stabilizes the collective choice.[37, 38]

The elaboration of Grassé's stigmergy concept was and continues to be very successful. However, as so often is the case in science, and to give due historical credit, the basic idea of stigmergy actually precedes Grassé. It was the perspicacious Pierre Huber who, in 1810, first devised the concept, but without coining a special term for it. Speaking of nest building in the ant *Formica fusca*, he said, "From these observations, and thousands like them, I am convinced that each ant acts independently of its companions. The first that hits upon an easy plan of execution immediately produces the outline of it; others only have to continue along the same lines, guided by an inspection of the first efforts."[39]

HOUSE HUNTING AND COLONY EMIGRATION

Many ant species, as well as wasps and bees, build their nests in preformed cavities in the soil or trees, as well as in stone cracks or hollow twigs and acorns. The size of these enclosed spaces is decisive for insects when they select a suitable nesting site. It has long been known that scouts of swarming honeybees are very particular in making this choice. The most thorough investigation of the behavioral process of nest site selection in honeybees has been that by Thomas Seeley.[40] He presented the bees

36 | A. Lioni and J.-L. Deneubourg, "Collective decision through self-assembling," *Naturwissenschaften* 91(5): 237–241 (2004).

37 | B. Hölldobler and E. O. Wilson, "The multiple recruitment systems of the African weaver ant *Oecophylla longinoda* (Latreille) (Hymenoptera: Formicidae)," *Behavioral Ecology and Sociobiology* 3(1): 19–60 (1978).

38 | A. Lioni and J.-L. Deneubourg, "Collective decision through self-assembling," *Naturwissenschaften* 91(5): 237–241 (2004).

39 | P. Huber, *Recherches sur les Moeurs des Fourmis Indigènes* (Paris: J. J. Paschoud, 1810).

40 | T. Seeley, "Measurement of nest cavity volume by the honey bee *(Apis mellifera)*," *Behavioral Ecology and Sociobiology* 2(2): 201–227 (1977); and T. D. Seeley and R. A. Morse, "Nest site selection by the honey bee, *Apis mellifera*," *Insectes Sociaux* 25(4): 323–337 (1978).

with a number of differently shaped and sized nest boxes, which served as substitutes for the tree cavities normally used by the bees. By examining the frequency with which wild swarms occupied the various nest boxes, he was able to demonstrate that scout bees measure all three dimensions and calculate volume directly from them. Recording the exploratory movements within the box, Seeley noted that the bees spend considerable time—40 minutes or so—walking over the walls of the cavity in a systematic fashion. It appeared that the bees made their assessment of the nest box volume by pacing off the interior. To test this hypothesis, Seeley offered the bees a nest box with rotating walls. When a scout began to explore the inner surface of the cavity, the wall could be rotated either with or against the bee's direction of motion. This was intended to give the scout the erroneous impression that the cavity was either larger or smaller than it really was. The data clearly demonstrated that a scout bee assesses linear dimensions by pacing them off in some way and that the "bee is constantly measuring the angle between the line struck off by her path and the line to a fixed point, perhaps the entrance to the nest. By constantly monitoring the changing relationship among these angles, and by remembering the distance moved at each angle, the bee could determine the relevant cross-sections of the space and 'calculate' the volume directly."[41]

The abilities of honeybee scouts appear astonishing at first glance. However, as Seeley points out, such integrative operations invoked by the "vector calculus hypothesis" are within the honeybees' range of capacities, as also exhibited by their behavior during waggle dance communication: "Numerous detour experiments . . . in which bees were forced to fly indirect, angular routes between hives and food sources have demonstrated that the bees can calculate the angle of dancing corresponding to the direct air line between hive and food source. This involves the integration of solar angle and length of different flight segments. . . . Bees forced to walk angular routes between their hive and food source can also perform the operations of measuring solar angles, quantifying movement distances and calculating the direct line between hive and food source."[42] Volume of prospective nest cavities

---

41 | T. Seeley, "Measurement of nest cavity volume by the honey bee (*Apis mellifera*)," *Behavioral Ecology and Sociobiology* 2(2): 201–227 (1977); and T. D. Seeley and R. A. Morse, "Nest site selection by the honey bee *Apis mellifera*," *Insectes Sociaux* 25(4): 323–337 (1978).

42 | Recently, a new algorithm was suggested for volume estimation by honeybees, but this has not yet tested experimentally; see N. R. Franks, and A. Dornhaus, "How might individual honeybees measure massive volumes?" *Proceedings of the Royal Society of London B* 270(Supplement): S181–S182 (2003).

is only one parameter for quality assessment. Other attributes include height, area, direction of the entrance, entrance position relative to the cavity floor, and qualities of location and presence of combs from a previous colony.[43, 44]

Once the scouts have discovered and explored new nesting sites, they return to their nestmates waiting for them with the mother queen in the swarm cluster. The cluster comprises bees that flew from the parental hive to establish a new colony, leaving behind a daughter queen and approximately half of the workers to perpetuate the old colony. The swarm cluster is only a preliminary assembly until a new dwelling place of appropriate size has been found. It has been known for over half a century that successful scouts return to the cluster and communicate the location and quality of the newly found nest sites by means of waggle dances (Figure 10-1).[45, 46] The parameters of the waggle dance containing directional and distance information are well understood.[47] The advertisement of the nest site's quality by the graded "endurance" and "liveliness" of the dance has also been well studied.[48] But there are hundreds of scouts searching for a new home, and many of them return and report different sites. How does the swarm cluster decide which one of the advertised new shelters to choose? It was Martin Lindauer who first showed, in the 1950s, that the dancing scout bees not only compete with one another but also influence one another.[49] He observed that scouts who had ceased to dance for one site were later recruited to another site and danced for it. However, the precise mechanism of the decision-making process during nest site selection in honeybee swarm clusters remained to be elucidated.

How is it possible that a honeybee swarm, composed of small-brained bees, can pursue the most sophisticated strategy of decision making? That was the question

---

43 | T. D. Seeley, *Honeybee Ecology: A Study of Adaptation in Social Life* (Princeton, NJ: Princeton University Press, 1985).

44 | P. C. Witherell, "A review of the scientific literature relating to honey bee bait hives and swarm attractants," *American Bee Journal* 125(12): 823–829 (1985).

45 | M. Lindauer, "Schwarmbienen auf Wohnungssuche," *Zeitschrift für vergleichende Physiologie* 37(4): 263–324 (1955).

46 | M. Lindauer, *Communication among Social Bees* (Cambridge, MA: Harvard University Press, 1961).

47 | K. v. Frisch, *The Dance Language and Orientation of Bees* (Cambridge, MA: The Belknap Press of Harvard University Press, 1967; 2nd printing, 1993).

48 | M. Lindauer, "Schwarmbienen auf Wohnungssuche," *Zeitschrift für vergleichende Physiologie* 37(4): 263–324 (1955).

49 | M. Lindauer, "Schwarmbienen auf Wohnungssuche," *Zeitschrift für vergleichende Physiologie* 37(4): 263–324 (1955); and M. Lindauer, *Communication among Social Bees* (Cambridge, MA: Harvard University Press, 1961).

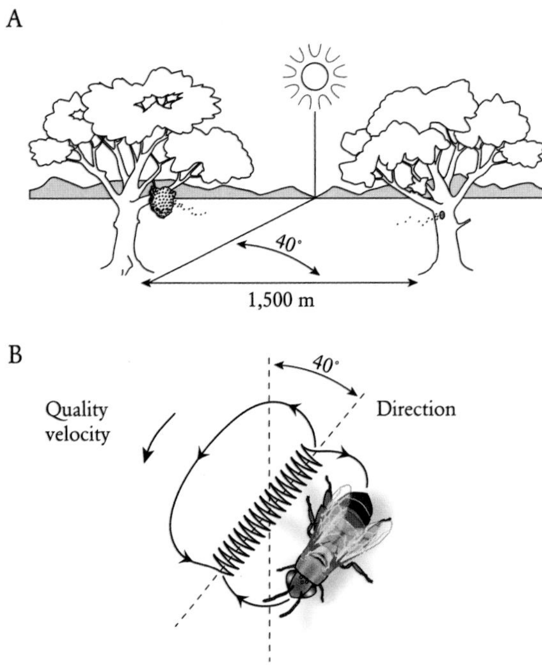

FIGURE 10-1. Recruitment to new nest sites in honeybees. *Above*: The colony finds a preliminary nesting site on a tree, where all the bees gather around the queen, forming a swarm cluster, and a scout discovers a suitable cavity inside a tree trunk 1,500 meters away from the swarm cluster. *Below*: The scout indicates the distance and direction of the newly discovered nest cavity, conducting a waggle dance correlated with the quality of the new nest site. Based on M. Lindauer, "Schwarmbienen auf Wohnungssuche," *Zeitschrift für vergleichende Physiologie* 37(4): 263–324 (1955); and T. D. Seeley and S. C. Buhrman, "Group decision making in swarms of honey bees," *Behavioral Ecology and Sociobiology* 45(1): 19–31 (1999).

that Thomas Seeley and Susannah Buhrman addressed when they began a new analysis of this remarkable phenomenon.[50, 51] Employing videotaping techniques, the researchers were able to obtain complete records of the dances of individual bees and hence track the interactions between the scouts and the swarm clusters.[52] From "reading" the dances, Seeley and Buhrman confirmed that scout bees initially find

---

50 | T. D. Seeley, and S. C. Buhrman, "Group decision making in swarms of honey bees," *Behavioral Ecology and Sociobiology* 45(1): 19–31 (1999).

51 | T. D. Seeley and S. C. Buhrman, "Nest-site selection in honey bees: how well do swarms implement the 'best-of-*N*' decision rule?" *Behavioral Ecology and Sociobiology* 49(5): 416–427 (2001).

52 | See also S. Camazine, P. K. Visscher, J. Finley, and R. S. Vetter, "House-hunting by honey bee swarms: collective decisions and individual behaviors," *Insectes Sociaux* 46(4): 348–360 (1999); and P. K. Visscher and S. Camazine, "Collective decisions and cognition in bees," *Nature* 397: 400 (1999).

nest sites in many directions and distances of up to several kilometers. Although the different scouts at first report many sites, they eventually switch to advertising only one site; and as a result, within an hour or so, the swarm departs and flies to the chosen site. Analyses of this process at the level of the individual bee have revealed that a bee advertising a better site produces waggle runs both for a longer time and at a higher rate, qualities that Martin Lindauer had earlier called "more lively" and "with greater endurance." Thus, it appears that the scout bees tune their dance behavior to site quality. Bees recruiting to a rather mediocre site are less successful in summoning nestmates than bees advertising better potential nest sites. Scouts visiting the high-quality site soon outnumber those visiting the medium-quality sites—with the ultimate result that the high-quality site is ultimately chosen.[53]

It appears that scout bees that have formerly danced for a certain site, but have ceased dancing, randomly monitor the dances of others without favoring particular dances—even those that advertise the site they had formerly reported but subsequently abandoned.[54] Relative rates of recruitment to different sites should be proportional to the relative strengths of the dances, reflected in the number of dance circuits produced for each of the different sites. By this measure, the site that receives the largest number of advertisements by its scouts is eventually chosen by the swarm.

Hence, the algorithm for consensus building among dancing bees that first dance for mediocre sites is to cease their dancing relatively quickly and so become available for recruitment to an alternate site of higher quality. Seeley and Buhrman conclude that scout bees are programmed "to gradually quit dancing" and that "this reduces the possibility of the decision-making process coming to a standstill with groups of unyielding dancers deadlocked over two or more sites."[55]

Choice experiments, in which swarms are offered a limited number of nest cavities of differing quality, have confirmed that the swarm is able to implement the "best-of-*N*" decision rule, meaning that the scouts adjust the recruitment rate to

---

53 | T. D. Seeley and S. C. Buhrman, "Nest-site selection in honey bees: how well do swarms implement the 'best-of-*N*' decision rule?" *Behavioral Ecololgy and Sociobiology* 49(5): 416–427 (2001).

54 | See also S. Camazine, P. K. Visscher, J. Finley, and R. S. Vetter, "House-hunting by honey bee swarms: collective decisions and individual behaviors," *Insectes Sociaux* 46(4): 348–360 (1999); and P. K. Visscher and S. Camazine, "Collective decisions and cognition in bees," *Nature* 397: 400 (1999).

55 | T. D. Seeley and S. C. Buhrman, "Group decision making in swarms of honey bees," *Behavioral Ecology and Sociobiology* 45(1): 19–31 (1999); and T. D. Seeley, "Consensus building during nest-site selection in honey bee swarms: the expiration of dissent," *Behavioral Ecology and Sociobiology* 53(6): 417–424 (2003).

a potential home site as a function of its quality: "The better the site, the stronger the recruitment, the speedier the buildup of bees, and the likelier the site will be chosen."[56, 57]

In holistic manner, using the algorithm that guides individual bees, the honeybee swarm evaluates each alternative nest site with respect to quality parameters and eventually arrives at a decision. Clearly, there is no supervisory bee that compiles all the individual evaluations and makes a decision for the colony. Instead, as Seeley and Buhrman point out, and consistent with the general theme of self-organization, decision making is a highly distributed process of friendly competition among the scout bees that identifies the best site. It is, in effect, a democracy. The evaluation of potential nest sites for the final selection of one particular site is a task distributed among many bees, each one exhibiting a different recruitment intensity to the particular site examined.

In fact, it is now understood that the decision-making process is not so much "consensus building" as "quorum sensing." That is, the choice among the various possible nesting cavities is a kind of race among different coalitions of scouts to see which one will first accumulate a threshold number of bees (quorum = 15 to 20 bees) at the site at any one time. Once this quorum is reached at one of the sites, the scouts there sense this and begin producing signals that stimulate the nonscouts to warm up for the flight to the site. That action in turn probably stimulates the scouts from other sites to quit their effort. Hence, a consensus among the dancing bees is generally reached, but it turns out that this is not what bees pay attention to in their decision making. Such quorum sensing is remarkably similar to the decision-making process during nest moving in *Leptothorax* and *Temnothorax* ants, to be described shortly. Both cases involve a "race to accumulate nestmates at a site."[58]

The cognitive effort of a single scout bee is obviously quite small relative to the global information processing performed by the entire swarm. The honeybee swarm

---

56 | T. D. Seeley and S. C. Buhrman, "Nest-site selection in honey bees: how well do swarms implement the 'best-of-*N*' decision rule?" *Behavioral Ecology and Sociobiology* 49(5): 416–427 (2001).

57 | For a mathematical treatment of this rule, see N. F. Britton, N. R. Franks, S. C. Pratt, and T. D. Seeley, "Deciding on a new home: how do honeybees agree?" *Proceedings of the Royal Society of London B* 269: 1383–1388 (2002).

58 | For a review and comparison of house hunting and information flow in honeybees and *Leptothorax* ants, see N. R. Franks, S. C. Pratt, E. B. Mallon, N. F. Britton, and D. J. T. Sumpter, "Information flow, opinion polling and collective intelligence in house-hunting social insects," *Philosophical Transactions of the Royal Society of London B* 357: 1567–1583 (2002).

is a higher-order cognitive entity. The way in which the swarm chooses its future home, along with other algorithm-guided, distributed processes, clearly qualifies the honeybee colony as a superorganism.[59, 60, 61]

The same principle is illustrated in the emigration of ant colonies. Emigration occurs when a colony propagates by budding or fission. It also occurs when a colony has grown too large for its current site and requires a more spacious home. Moreover, a colony may emigrate to avoid competitive pressures from neighboring colonies or to escape from parasites at the home site.

Colony emigrations in ants are initiated by scout ants who have discovered better nest sites. These individuals employ various recruitment mechanisms, and the communication signals employed are often specialized to serve colony emigration.[62]

Well-understood examples can be seen in the myrmicine ant genera *Leptothorax* and *Temnothorax*. Colonies of most species are quite small, rarely comprising more than 100 workers. They are accordingly able to occupy thin cracks in rocks, as well as hollow twigs or acorns. Because these nest sites are relatively unstable, the colonies frequently move to new homes. Field observations have revealed that even slight physical disturbances trigger nest emigrations. Michael Möglich presented the first detailed behavioral analysis of the social organization and underlying communication mechanisms during nest emigration in *Temnothorax* (formerly part of *Leptothorax*), utilizing *Temnothorax rugatulus*.[63] The process begins when a potential new nest site is inspected by the worker that discovered it. The scout returns to her colony. She may next turn around and revisit the new site and inspect it again, and she may do this repeatedly, as though "making up her mind," before she finally starts recruiting nestmates to the site. In this next step, she rapidly antennates

---

59 | T. D. Seeley and S. C. Buhrman, "Group decision making in swarms of honey bees," *Behavioral Ecology and Sociobiology* 45(1): 19–31 (1999).

60 | T. D. Seeley and S. C. Buhrman, "Nest-site selection in honey bees: how well do swarms implement the 'best-of-*N*' decision rule?" *Behavioral Ecology and Sociobiology* 49(5): 416–427 (2001); K. M. Passino, T. D. Seeley, and P. K. Visscher, "Swarm cognition in honey bees," *Behavioral Ecology and Sociobiology* 62(3): 401–414 (2008).

61 | See also S. Camazine, P. K. Visscher, J. Finley, and R. S. Vetter, "House-hunting by honey bee swarms: collective decisions and individual behaviors," *Insectes Sociaux* 46(4): 348–360 (1999); and P. K. Visscher and S. Camazine, "Collective decisions and cognition in bees," *Nature* 397: 400 (1999).

62 | B. Hölldobler and E. O. Wilson, *The Ants* (Cambridge, MA: The Belknap Press of Harvard University Press, 1990).

63 | M. Möglich, "Social organization of nest emigration in *Leptothorax* (Hym., Form.)," *Insectes Sociaux* 25(3): 205–225 (1978).

several nestmates, then turns around, arches her gaster upward, and, with her sting extruded, discharges volatile secretions from her poison gland. As soon as a nestmate attracted by this "tandem calling" touches the scout, tandem running commences (Figure 10-2).[64] As the nestmate runs close behind the scout, both mechanical and chemical signals are used to maintain contact between leader and follower. After the recruited ant inspects the new nest site, she may then become a recruiter herself, provided she has found the new site superior to the old dwelling. Individual scouts who have discovered a new nest site of notable promise strive to summon other potential scouts to it. This "recruitment of recruiters" occurs exclusively in the initial phase of nest emigration, and it serves to increase rapidly the number of scouts inspecting the site before a decision to emigrate is made by the colony as a whole.

The *Temnothorax* workers recruited through tandem running make their own decisions about the quality of the candidate nest site and whether or not to recruit nestmates. Individualists by ant standards, each appears to inspect the nest site as thoroughly as the first scout did, assessing location, distance, direction, difficulty of terrain, and conditions at the site. Further, being led in tandem is a good way to gather such information about the targeted site. Such recruitment of recruiters rapidly increases the number of scouts to the most desirable site, and this number by itself is probably important information influencing the decision of the colony as a whole whether to emigrate. Once a full exodus begins, the majority of the colony members, both adult and immature, are carried bodily (see Figure 10-2). Because former scouts also serve as transporters, the recruitment of recruiters jumpstarts the number of active carriers and thereby hastens the colony emigration.

Although the communication mechanisms and social organization of nest emigration in *Temnothorax rugatulus* colonies became known from Möglich's observations,[65] the parameters of the decision-making process in nest site selection and information flow within the colony remained unclear. These questions were then addressed by Nigel Franks and his colleagues. They chose *Temnothorax* (formerly *Leptothorax*) *albipennis* as the model system for their studies of nest site assessment, information flow, and decision making in an ant society. As in the honeybees, the question of interest is how the colony as a whole selects the best

---

64 | M. Möglich, U. Maschwitz, and B. Hölldobler, "Tandem calling: a new kind of signal in ant communication," *Science* 186: 1046–1047 (1974).

65 | M. Möglich, "Social organization of nest emigration in *Leptothorax* (Hym., Form.)," *Insectes Sociaux* 25(3): 205–225 (1978).

FIGURE 10-2. Recruitment behavior in *Leptothorax* and *Temnothorax* during nest emigration. *From top to bottom*: A recruiting worker assumes a calling position, extruding the sting on the tip of which appears a droplet of poison gland secretion. A nestmate approaches and touches the calling ant. Tandem running commences. The nestmate is led to the newly discovered nest site. Later in the emigration process, recruitment proceeds by adult transport. Based on M. Möglich, U. Maschwitz, and B. Hölldobler, "Tandem calling: a new kind of signal in ant communication," *Science* 186: 1046–1047 (1974).

nest site among competing alternatives. In other words, how well do ant colonies implement the "best-of-*N*" decision rule? Although the colonies of *Temnothorax* ants and honeybees are quite different in many ways, not least in population size, the researchers discovered some similarities in the algorithms the different societies follow when house hunting.[66]

66 | For a review and comparison of house hunting and information flow in honeybees and in *Temnothorax* and *Leptothorax* ants, see N. R. Franks, S. C. Pratt, E. B. Mallon, N. F. Britton, and D. J. T. Sumpter, "Information flow, opinion polling and collective intelligence in house-hunting social insects," *Philosophical Transactions of the Royal Society of London B* 357: 1567–1583 (2002).

The two kinds of insects, ants and bees, engage in different modes of communication. Scouts of *Temnothorax* colonies usually compare nest sites, but they communicate under a handicap; they cannot directly inform one another of their assessment of a potential dwelling place. The scouts recruit other nestmate scouts to particular nest sites, and each recruited ant makes its own on-the-spot assessment. The recruitment dance of honeybees, in contrast, contains information about both location and quality assessment delivered at a distance. Nestmates can therefore process such information directly from the scouts without visiting the nest site.

In their experiments on nest emigration, the researchers elicited scouting for new nest sites by destroying nest walls, enlarging the entrance hole, or disturbing the old nest site in other ways. The colony was simultaneously offered two alternative nests in the experimental arena.[67, 68] Each worker of the experimental colony had been marked with an individual color code so that the complete behavioral record of the entire colony could be videographed. These studies revealed that individual ants were indeed able to visit and compare two different sites and choose the better one. Although during the process most of the ants encountered only one site, they still contributed to the colony's decision, because their probability of initiating recruitment to a site depended on its quality.[69]

The transition in *Temnothorax albipennis* colonies from recruitment of recruiters by tandem running to the much faster recruitment by carrying nestmates, hence the emigration itself, is triggered by a population increase at the new site.[70] As Stephen Pratt and his colleagues discovered, once a quorum of nestmates is present at the new site, the scouts commence to carry workers, the queen, and the brood from the old nest to the new one. This quorum is a central requirement for a colony

---

67 | E. B. Mallon, S. C. Pratt, and N. R. Franks, "Individual and collective decision-making during nest site selection by the ant *Leptothorax albipennis*," *Behavioral Ecology and Sociobiology*, 50(4): 352–359 (2001).

68 | S. C. Pratt, E. B. Mallon, D. J. T. Sumpter, and N. R. Franks, "Quorum sensing, recruitment, and collective decision-making during colony emigration by the ant *Leptothorax albipennis*," *Behavioral Ecology and Sociobiology* 52(2): 117–127 (2002); and S. C. Pratt, D. J. T. Sumpter, E. B. Mallon, and N. R. Franks, "An agent-based model of collective nest choice by the ant *Temnothorax albipennis*," *Animal Behaviour* 70(5): 1023–1036 (2005).

69 | E. B. Mallon, S. C. Pratt, and N. R. Franks, "Individual and collective decision-making during nest site selection by the ant *Leptothorax albipennis*," *Behavioral Ecology and Sociobiology* 50(4): 352–359 (2001).

70 | S. C. Pratt, E. B. Mallon, D. J. T. Sumpter, and N. R. Franks, "Quorum sensing, recruitment, and collective decision-making during colony emigration by the ant *Leptothorax albipennis*," *Behavioral Ecology and Sociobiology* 52(2): 117–127 (2002); and S. C. Pratt, D. J. T. Sumpter, E. B. Mallon, and N. R. Franks, "An agent-based model of collective nest choice by the ant *Temnothorax albipennis*," *Animal Behaviour* 70(5): 1023–1036 (2005).

to choose the best available site, even when only relatively few scouts have compared sites directly. The reason is that recruiters to a given site launch rapid transport of the bulk of the colony only if enough active ants have been "convinced" of the worth of the site. The algorithm is another example of insect societies realizing adaptive colony-level behavioral decisions "from the decentralized interactions of relatively poorly informed insects, each combining her own limited direct information with indirect cues about the experience of her nestmates."[71]

The decision-making system is simple and efficient, but it is far from perfect. Colonies readily move from intact nests to another dwelling site whenever the latter is determined to be superior, especially when the old one has been disturbed. Thus, it can happen that the colony chooses an acceptable new nest site rather than the best one available. Nevertheless, abandoning an undisturbed intact nest of good quality for an even more suitable dwelling place apparently requires a larger number of scouts to agree on one site before starting emigration, making an optimum choice more likely.[72]

Among the several features that account for nest site quality, the most important is space. *Temnothorax* colonies habitually inhabit small cavities, and *Temnothorax albipennis,* in particular, prefers flat rock crevices. These cracks have to be of a certain size, not too large and not too small, for the entire colony to settle comfortably. Thus, one of the most important tasks for the scouts is to select nest sites of the correct shape and size. They face a problem similar to the one honeybee scouts must solve when choosing a suitable hive cavity for the waiting swarm.

How do the *Temnothorax* ants estimate the size of a potential nest site? Eamon Mallon and Nigel Franks have argued that *Temnothorax* scouts employ a rule of thumb called Buffon's needle algorithm.[73, 74] Adapting the statistical geometry of Comte Georges-Louis Leclerc de Buffon, an eighteenth-century French naturalist,

71 | S. C. Pratt, E. B. Mallon, D. J. T. Sumpter, and N. R. Franks, "Quorum sensing, recruitment, and collective decision-making during colony emigration by the ant *Leptothorax albipennis*," *Behavioral Ecology and Sociobiology* 52(2): 117–127 (2002); and S. C. Pratt, D. J. T. Sumpter, E. B. Mallon, and N. R. Franks, "An agent-based model of collective nest choice by the ant *Temnothorax albipennis*," *Animal Behaviour* 70(5): 1023–1036 (2005).

72 | A. Dornhaus, N. R. Franks, R. M. Hawkins, and H. N. S. Shere, "Ants move to improve: colonies of *Leptothorax albipennis* emigrate whenever they find a superior nest site," *Animal Behaviour* 67(5): 959–963 (2004).

73 | E. B. Mallon and N. R. Franks, "Ants estimate area using Buffon's needle," *Proceedings of the Royal Society of London B* 267: 765–770 (2000).

74 | S. T. Mugford, E. B. Mallon, and N. R. Franks, "The accuracy of Buffon's needle: a rule of thumb used by ants to estimate area," *Behavioral Ecology* 12(6): 655–658 (2001).

it is possible to estimate the area of a plane from the frequency of intersections between two sets of randomly scattered lines of known length. The researchers point to the fact that *Temnothorax albipennis* scouts, after walking through the entrance of a candidate site, exit and reenter it. During their first inspection, the scouts mark their route with a trail pheromone personal to themselves. (Earlier experimental studies by Ulrich Maschwitz and his colleagues on *Temnothorax affinis* had shown that workers can distinguish their own pheromone from that of nestmates.[75]) Mallon and Franks reasoned, from both theoretical and empirical evidence, that ants size up the nest sites by "counting" the number of times they cross their own first exploratory trail on their second inspection of the crevice. The smaller the area, the more likely they are to visit the same place twice. This remarkable hypothesis has not yet been tested experimentally, but the observation that the *Temnothorax* scouts always pace out a path of the same length on their first visit to the site strongly supports the assumption that the ants might indeed employ the Buffon's needle algorithm. The first move through the crevice provides the standard path that serves as a measuring stick that the ants use on return visits to complete their survey.[76]

During colony emigration, in monogynous ant species in which the phenomenon has been studied, the queen is dragged or escorted along in the middle or later part of the emigration process. Similar observations have been made in species of carpenter ants (*Camponotus*), weaver ants (*Oecophylla*), harvester ants (*Pogonomyrmex*, *Aphaenogaster*, *Pheidole*), and leaf-cutting ants (*Atta*), as well as in several species that have no stable nest sites but travel from bivouac to bivouac (for example, species of *Onychomyrmex* and *Leptogenys* and "true" army ants in *Eciton* and *Dorylus*). In those species where the queen moves on her own power, she is usually highly protected by a dense retinue of workers.[77] The pattern appears to be adaptive. The queen is the most vital unit of the colony, and it is therefore to be expected that she is urged along after a large bulk of the colony has already settled at the new site, but with another part of the colony still at the old nest. Such a pattern is the clear rule in *Temnothorax albipennis*, as proved by a statistical analysis of 32 colony

---

75 | U. Maschwitz, S. Lenz, and A. Buschinger, "Individual specific trails in the ant *Leptothorax affinis* (Formicidae: Myrmicinae)," *Experientia* 42(10): 1173–1174 (1986).

76 | S. T. Mugford, E. B. Mallon, and N. R. Franks, "The accuracy of Buffon's needle: a rule of thumb used by ants to estimate area," *Behavioral Ecology* 12(6): 655–658 (2001).

77 | B. Hölldobler and E. O. Wilson, *The Ants* (Cambridge, MA: The Belknap Press of Harvard University Press, 1990).

emigrations of this species. We may reasonably assume that the strategy is "favored by selection and is an aspect of colonies behaving as group-level adaptive units."[78]

We have chosen honeybees and *Temnothorax* ants to introduce collective decision making during house hunting because these are the cases best investigated. Their natural history properties make them ideal model systems to address several of the key questions concerning group-level decision making. However, colony emigration and nest site selection in many ant species appear to be even more complex, entailing major communicative and logistic problems that the ant colonies need to solve. In most cases, sociobiologists do not yet understand how the ants manage these tasks.

Most of the ant species that have been studied employ specific recruitment signals during nest emigration, and the phenomenon of recruitment of recruiters during the initial phase of nest emigration appears to be quite common. However, only a few species use the tandem running technique. In addition to those in the genus *Temnothorax*, the method has also been recorded in some ponerine species, including *Pachycondyla tesserinoda*, *Pachycondyla obscuricornis*, and several *Diacamma* species, as well as a few members of the formicine genus *Camponotus* and several *Polyrhachis* species (see Chapter 6).

The behavioral organization of tandem running allows only one ant to be recruited at a given time. As we discussed in detail in Chapter 6, the follower ant keeps close antennal contact with the leader ant. Whenever they lose each other, the leader halts while the follower moves in searching circles. Simple experiments have shown that tactile signals transmitted by the follower are sufficient to trigger leader behavior during tandem running. The signal patterns that bind the follower ant to the leader are considerably more complicated, but the basics are similar in all tandem running ants thus far investigated (see Chapter 6). In all cases, the follower ant is bound to chemical signals from the leader ant, which originate either from particular exocrine glands or from cuticular secretions on the gaster surface. Not only does the mechanical contact to the leader ant ensure a precise orientation during tandem following, but the follower ant is somehow able to move back from the target area to the nest on its own and subsequently lead another nestmate by tandem

78 | N. R. Franks and A. B. Sendova-Franks, "Queen transport during ant colony emigration: a group-level adaptive behavior," *Behavioral Ecology* 11(3): 315–318 (2000).

running to the target area.[79] Recently, this already well-known behavior, informing the naive nestmates by tandem running about the route to a new target area, has been referred to as a case of "teaching" in ants—the leader ant the "teacher" and the follower ant the "pupil."[80] This might be a charming metaphor, but it adds little, if anything, to our understanding of this fascinating recruitment behavior.

An example of a more complex communication is provided by *Camponotus sericeus* (see Figures 6-25 and 6-26), which incorporates the tandem running technique for recruitment to food sources and new nest sites. However, the behavior of the recruiting ant during nest emigration is markedly different from that used for recruitment to new food sources.[81] When facing a nestmate head-on, the scout jerks its body rapidly back and forth or else grasps the nestmate on the mandibles and pulls her forward. Then she turns 180° and presents her gaster to the nestmate. If the nestmate responds at this point by touching the scout's gaster or hind legs, tandem running starts. This behavior is very stereotyped and is regularly employed when nestmates are invited to follow to a new nest site. After inspection of the site, many of the initially recruited nestmates become recruiters themselves. As the emigration progresses, an increasing number of the recruited ants do not return to the old nest.

Meanwhile, as the emigration proceeds, colony members at the old nest unresponsive to the tandem running invitation are simply carried to the new nest site (see Figure 6-27). The behavioral sequence that initiates carrying behavior is almost identical to the one initiating tandem running, except that the recruiting ant keeps the firm grip of the nestmates' mandibles when turning around, thereby raising her slightly off the ground. When lifted, the nestmate assumes the posture of a pupa, folding her legs and antennae tightly to the body and rolling her gaster inward. In this compact posture, she is easily carried to the new nest by the scout. Males and

---

79 | B. Hölldobler, M. Möglich, and U. Maschwitz, "Communication by tandem running in the ant *Camponotus sericeus*," *Journal of Comparative Physiology A* 90(2): 105–127 (1974); M. Möglich, U. Maschwitz, and B. Hölldobler, "Tandem calling: a new kind of signal in ant communication," *Science* 186: 1046–1047 (1974); U. Maschwitz, B. Hölldobler, and M. Möglich, "Tandemlaufen als Rekrutierungsverhalten bei *Bothroponera tesserinoda* Forel (Formicidae: Ponerinae)," *Zeitschrift für Tierpsychologie* 35(2): 113–123 (1974); J. F. A. Traniello and B. Hölldobler, "Chemical communication during tandem running in *Pachycondyla obscuricornis* (Hymenoptera: Formicidae)," *Journal of Chemical Ecology* 10(5): 783–794 (1984); and U. Maschwitz, K. Jessen, and S. Knecht, "Tandem recruitment and trail laying in the ponerine ant *Diacamma rugosum*: signal analysis," *Ethology* 71(1): 30–41 (1986).

80 | N. R. Franks, and T. Richardson, "Teaching in tandem-running ants," *Nature* 439: 153 (2006).

81 | B. Hölldobler, M. Möglich, and U. Maschwitz, "Communication by tandem running in the ant *Camponotus sericeus*," *Journal of Comparative Physiology A* 90(2): 105–127 (1974).

winged virgin females are also sometimes carried, but the males are transported in a different posture, and the great majority of alate females are successfully led entirely by tandem running to the new nest (see also Chapter 6).

Quantitative sociograms made of small *Camponotus sericeus* colonies in the laboratory have revealed that nest moving is organized by a pronounced system of division of labor.[82] Only about 6 percent of the workers were found to be active recruiters in the great majority of all nest-moving events. Subsequent dissections of the entire worker forces revealed that the "movers" are a guild of older workers, with degenerate ovaries, whereas the ovaries of many of the ants being carried are relatively well developed. Older workers with degenerate ovaries are more generally involved in outside work, such as foraging and nest defense. However, only a small portion of these "outside workers" appear to function as mover specialists. If they are removed experimentally, they can be replaced by other workers with reduced ovaries. On the other hand, nest-bound workers, with well-developed ovaries and thus younger in age, seldom replace lost mover specialists. Similar results have been found in other formicine species, including the well-studied *Formica polyctena* and *Formica sanguinea*, although these species employ different recruitment behaviors.[83, 84, 85]

Nest moving in ant species that live in colonies of tens of thousands or even more individuals obviously present major logistic problems. How it is accomplished has been analyzed in the harvester ant species *Pogonomyrmex barbatus* and *Pogonomyrmex rugosus*.[86] The colonies live in complex subterranean nest structures, comprising elaborate interconnected tunnels and nest chambers. Before a colony can emigrate, a new nest has to be built complete enough to accommodate a major part of the nest population. Scout ants begin the process by recruiting nestmates to a suitable new site, usually 10 or more meters away, using trail pheromones laid from their poison glands. Soon after, the ants begin digging at the new site, and

---

82 | M. Möglich and B. Hölldobler, "Social carrying behavior and division of labor during nest moving in ants," *Psyche* (Cambridge, MA) 81: 219–236 (1974).

83 | M. Möglich and B. Hölldobler, "Social carrying behavior and division of labor during nest moving in ants," *Psyche* (Cambridge, MA) 81(2): 219–236 (1974).

84 | D. Otto, "Über die Arbeitsteilung im Staate von *Formica rufa rufopratensis* minor Gössw. und ihre verhaltensphysiologischen Grundlagen," *Wissenschaftliche Abhandlungen Deutsche Akademie der Landwirtschaftswissenschaften zu Berlin* 30: 1–169 (1958).

85 | G. Kneitz, "Saisonales Trageverhalten bei *Formica polyctena* Foerst. (Formicidae, Gen. *Formica*)," *Insectes Sociaux* 11(2): 105–130 (1964).

86 | B. Hölldobler, "Recruitment behavior, home range orientation and territoriality in harvester ants, *Pogonomyrmex*," *Behavioral Ecology and Sociobiology* 1(1): 3–44 (1976); and B. Hölldobler, unpublished observations.

with increasing worker traffic between the two sites, a trunk route is established that allows acceleration of the traffic. The chemical markings most likely include the shorter-lasting recruitment pheromones from the poison gland (mainly 3-ethyl-2,5-dimethylpyrazine), together with colony-specific blends of hydrocarbons from the Dufour's gland. The latter substances, which serve as longer-lasting orientation signals, are sporadically renewed by individual ants.[87, 88] After 1 to 2 weeks, the traffic between both sites grows still stronger, with workers and brood being carried to the new site (but also, perhaps perversely, in the reverse direction, to the old nest). Finally, the nest emigration gets fully under way as hundreds of workers, many bearing brood, adults, and harvested seeds, move to the new nest in large numbers. In a few cases, the queen has been seen during the second half of the emigration, walking under her own power and surrounded by a retinue of workers.

Although chemical trail communication is the main recruitment mechanism used in emigration by the *Pogonomyrmex* harvesters, recruitment by carrying adults also plays an important role. Social carrying behavior, which occurs in most other ant species as well, can occur in a number of contexts,[89] but is most frequently employed during emigration from one nest site to another. The behavioral patterns of adult transport in ants are very stereotyped and are often specific for individual taxonomic groups. While most formicine species carry nestmates held backward, most myrmicines and some ectatommines carry the transportee in parallel and curled over the transporter's own head. In the latter case, the transportee bends its gaster inward and folds its appendages tightly to its body (see Figure 10-2 and Plate 71). But there are a few exceptions: the harvester ants *Pogonomyrmex badius*, *Pogonomyrmex barbatus*, and *Pogonomyrmex rugosus* rarely employ the typical myrmicine mode, using instead a more elementary carrying technique in which the transportee is grasped at any part of the body, lifted up, and carried along. The individual thus handled folds her appendages to the body. This transporting behavior has been recorded in the anatomically primitive myrmeciines, the bulldog ants of

---

87 | B. Hölldobler, E. D. Morgan, N. J. Oldham, and J. Liebig, "Recruitment pheromone in the harvester ant genus *Pogonomyrmex*," *Journal of Insect Physiology* 47(4–5): 369–374 (2001).

88 | B. Hölldobler, E. D. Morgan, N. J. Oldham, J. Liebig, and Y. Liu, "Dufour gland secretion in the harvester ant genus *Pogonomyrmex*," *Chemoecology* 14(2): 101–106 (2004).

89 | For a review, see E. O. Wilson, *The Insect Societies* (Cambridge, MA: The Belknap Press of Harvard University Press, 1971); and M. Möglich and B. Hölldobler, "Social carrying behavior and division of labor during nest moving in ants," *Psyche* (Cambridge, MA) 81: 219–236 (1974).

PLATE 71. Carrying positions vary in ants according to species. *Above*: Most formicines transport nestmates held high and pointed backward as in *Camponotus perthiana*. *Below*: In contrast, most myrmicine and ectatommine species carry the transportee curled back over their own body, as in the *Ectatomma ruidum* ant shown here.

Australia, and some ponerines. It is remarkable that at least a few other species of *Pogonomyrmex*, including *Pogonomyrmex maricopa* and *Pogonomyrmex californicus*, use the stereotyped adult transport behavior typically employed by other myrmicine ants.[90]

Ant colonies that invest heavily in elaborate nest constructions can be expected to remain relatively sessile, refusing to abandon the high-quality "real estate" they have gradually built up. In fact, we have no evidence of emigration by mature colonies of honey ants (*Myrmecocystus* species), which construct numerous deep subterranean brood chambers and storage chambers, the latter housing the densely packed replete worker castes (honeypots; see Plates 15 and 31). The tenacity of the colonies persists even when desert badgers dig into the nests to raid the honeypot chambers. Similarly, emigrations are evidently never undertaken by *Lasius fuliginosus*, whose colonies live in hollow tree trunks filled with elaborate carton chambers. And yet another example is provided by the European carpenter ant *Camponotus herculeanus*, which lives in the solid trunks of trees carved out by the workers to create a complex network of tunnels and chambers.

Even so, there are ant species with large nests that do emigrate, albeit much less commonly than the majority of species with less investment in their real estate. The most remarkable examples are the leaf-cutting ant species of *Atta*. As described in the previous chapter, colonies of *Atta* species have the largest and most elaborate nest structures known in the ant world. Many *Atta* colonies stay in the same nest throughout their entire life span of 10 to 20 years. However, a few mature colonies have been observed to move to new sites, and over distances ranging from 33 to 258 meters. In a population of *Atta colombica* monitored on Barro Colorado Island, Panama, 25 percent of the colonies relocated during one year. The reason for this unexpectedly high number is not yet clear. The nests of the population were quite closely spaced, so that aggression between neighboring colonies could have been a contributing cause. Infections of the fungus gardens by parasites, such as the fungus *Escovopsis*, is also a likely factor.[91]

Relocating an entire *Atta* colony, comprising millions of workers, is by necessity

---

90 | M. Möglich and B. Hölldobler, "Social carrying behavior and division of labor during nest moving in ants," *Psyche* (Cambridge, MA) 81: 219–236 (1974).

91 | For a review, see R. Wirth, H. Herz, R. J. Reyel, W. Beyschlag, and B. Hölldobler, *Herbivory of Leaf-Cutting Ants: A Case Study on* Atta colombica *in the Tropical Rainforest of Panama* (New York: Springer-Verlag, 2003).

a complicated process. We do not know how new nest sites are selected or how the "construction crew" is attracted to the new site. It is clear that before emigration commences, the new nest must already be partly built. During this initial construction period, the transport of plant fragments to the new nest occurs. As the transport of brood, workers, and pieces of fungus grows more intense, excavation at the new nest is correspondingly increased. In the middle or latter part of the evacuation event, the queen travels across under her own power, surrounded by a large worker retinue. At the slightest disturbance, the workers push and drag her under the nearest leaf or other shelter. This escorted transport of the queen from the old to the new nest lasts a full day or even longer. Finally, even after the colony emigration to the new site is completed, there is still some two-way traffic along the trunk trail connecting the old with the new nest, a movement that is continued until the old nest is completely abandoned.[92] We have some understanding of the spectacular emigrations of *Atta* species, but still know almost nothing about the causes or the communication by which they are organized.

Overall, the diverse cooperative processes of colony-level emigration and nest construction are among the phenomena that most clearly reveal colony-level (between-group) selection and justify the characterization of insect societies as superorganisms.

---

92 | For a review, see R. Wirth, H. Herz, R. J. Reyel, W. Beyschlag, and B. Hölldobler, *Herbivory of Leaf-Cutting Ants: A Case Study on* Atta colombica *in the Tropical Rainforest of Panama* (New York: Springer-Verlag, 2003); and Hubert Herz, personal communication.

EPILOGUE

Our knowledge of the social insects, and the phenomenon of the superorganism they so beautifully display, has grown immensely during the past century. Yet we have only begun to explore this alien world.

About 14,000 species of ants are known, for example, but twice that number or more are likely to exist. Of those known, fewer than 100 can be said to have been well studied.

Where do we go from here? It is dangerous to predict the future of any field of study, but at least the following areas seem to promise rich discoveries: identifying and sequencing the alleles that were substituted at the evolutionary origins of eusociality and the point of no return; tracking the developmental processes prescribed by these major transitional alleles; tracking the developmental mechanisms that underlie the sociogenesis of superorganisms; pinpointing the ecological pressures that lead to effective between-group selection at the two key thresholds; and, through a greatly expanded scientific natural history, discovering new phenomena of multilevel selection that underlie the proximate phenomena of colonial life.

It is worth reflecting that René-Antoine Ferchault de Réaumur, when composing his *Mémoirs pour Servir à l'Histoire des Insectes* in 1737, could not have imagined what was to be reported by Pierre Huber in his 1810 *Recherches sur les Moeurs des Fourmis Indigènes*. Neither Réaumur nor Huber, in turn, could have dreamed of what was to be the content of Auguste Forel's *Le Fourmis de la Suisse* (1874) and, even more, William Morton Wheeler's magisterial two works, *Ants: Their Structure, Development, and Behavior* (1910) and *The Social Insects: Their Origins and Evolution* (1928). Yet Wheeler's thoughts came nowhere close to many of the later advances chronicled by the syntheses of the 1970s. And many of those have now been rendered outdated by the large population of later researchers reported in the preceding chapters.

And so it will be a half century from now. None of us now present can imagine the great advances certain to come. But of course, that is one reason we have future generations.

Finally, what is the significance of this knowledge for our species? In ants and other insects, we are privileged to see not only how complex societies evolved independently from those of humans (and in a different sensory modality—mostly chemosensory rather than audiovisual), but also, with increasing clarity, the relation between advanced social orders and the forces of natural selection that created and shaped them.

Rarity of occurrence and unusual preadaptations characterized the early species of *Homo*, as they did the early species of social insects. Both groups achieved spectacular ecological success and the exclusion of nonsocial species that competed with them. Both owe their success to cooperative behavior and division of labor practiced within groups. The evolution of both has been driven by group selection comprising competition and often direct conflict between groups. The residue of that evolutionary force affects us still in our irrational and destructive tribal wars.

There exists, however, a profound difference between us and them. Social insects are still ruled rigidly by instinct, and they will remain so forever. Humans have intelligence and swiftly evolving cultures. We have the potential for self-understanding and can find a way to curb our self-destructive conflicts. Further, for over 100 million years, the rigid instincts of the social insects have fitted them harmoniously into the living environment. Our intelligence has allowed us to control and destroy the global environment for short-term gain, the first time that was achieved by any species in the history of the planet. By coming to see more clearly who we are and how we came to be, our species might find better ways to live harmoniously, not only with one another but also with the rest of life.

ACKNOWLEDGMENTS

During the five years (2002–2007) we took to write *The Superorganism*, we accumulated a large debt to the many colleagues who advised and helped with research and composition.

It is appropriate to start with Kathleen M. Horton. Without her skills in library research, editing, complex manuscript production, and sustained hard work, this book probably would not have been finished.

We have benefited greatly from the encouragement and creative suggestions provided by our editor, Robert Weil. His advice has improved the text throughout. The book has also been notably enhanced by the outstanding work of Janet Greenblatt, the manuscript editor.

We are further grateful to those expert colleagues who read and commented on parts of the manuscript. They include James Costa, Jennifer Fewell, Kevin Foster, David Haig, Robert Page, David Queller, Kern Reeve, Gene Robinson, Flavio Roces, Thomas Seeley, Mary-Jane West-Eberhard, and David S. Wilson.

We have also benefited in many ways from colleagues who discussed with us research topics central to the book, including Gro Amdam, Kirk Anderson, Bruce Archibald, Jacobus Boomsma, Andrew F. G. Bourke, Cameron Currie, Annett Endler, Jürgen Gadau, David Grimaldi, Jürgen Heinze, James Hunt, Michael Kaspari, Jürgen Liebig, Timothy Linksvayer, Thibaud Monnin, Corrie Moreau, Christian Peeters, Stephen Pratt, Ted Schultz, Brian Smith, Philip Ward, Ming-Sheng Wang, Diana Wheeler, the graduate students of the Social Insects Research Group (SIRG) and the members of the Center for Social Dynamics and Complexity (CSDC), both of Arizona State University.

Others contributed photographic illustrations that have added greatly to the content and clarity of *The Superorganism*. They are Gro Amdam, Vincent Dietemann, Hubert Herz, James H. Hunt, Marco Kleinhenz, Flavio Roces,

Wolfgang Thaler, and Walter Tschinkel. All other photographs were taken by Bert Hölldobler.

The personal research of Bert Hölldobler reported in this book was supported by grants from the German Science Foundation, the National Science Foundation (USA), the National Geographic Society (USA), and Arizona State University. That of Edward O. Wilson was supported by the National Science Foundation (USA).

GLOSSARY

ACULEATE Pertaining to Aculeata, or stinging Hymenoptera, a group including the bees, ants, and many of the wasps.

AGGREGATION A group of individuals, composed of more than just a mated pair or a family, that have gathered in the same place but do not construct nests or rear offspring in a cooperative manner (as opposed to a colony). *See* Colony.

ALARM-DEFENSE SYSTEM Defensive behavior that also functions as an alarm-signaling device within the colony. Examples include the use by certain ant species of chemical defensive secretions that double as alarm pheromones.

ALLELE A particular form of a gene, distinguishable from other forms of the same gene.

ALLODAPINE A ceratine bee belonging to *Allodape* or one of a series of closely related genera, all of which are either eusocial or socially parasitic. Excluded from this informal taxonomic category is *Ceratina*, the only other major living genus of the tribe Ceratinini.

ALLOMETRY Any size relation between two body parts that can be expressed by $y = bx^a$, where a and b are fitted constants. In the special case of isometry, $a = 1$, and the relative proportions of the body parts therefore remain constant with change in total body size. In all other cases ($a \neq 1$), the relative proportions change as the total body size is varied.

ALPHA Referring to the highest-ranking individual within a dominance hierarchy.

ALTRUISM In evolutionary biology, behavior that lowers the genetic fitness of the altruist and increases the genetic fitness of others. *See* Genetic fitness.

AMBROSIA BEETLE A wood-boring scolytoid beetle that cultivates fungus ("ambrosia") for food.

ARACHNID A member of the class Arachnida, such as a spider, mite, or scorpion.

ARMY ANT Also called legionary ant. A member of an ant species that engages in both nomadic and group predatory behavior. In other words, the nest site is changed at relatively frequent intervals, in some cases daily, and the workers forage in groups.

ARTHROPOD Any member of the phylum Arthropoda, such as a crustacean, spider, millipede, centipede, or insect.

BINDING Factors or forces in evolution that promote cohesion and harmony in societies.

BIOMASS The weight of a set of plants, animals, or microorganisms. The set is chosen for convenience; it can be, for example, a colony of insects, a population of wolves, or an entire forest.

BIVOUAC The mass of army ant workers within which the queen and brood find refuge; also, the site where the mass is located.

BROOD The immature members of a colony collectively, including eggs, nymphs, larvae, and pupae. Eggs and pupae are sometimes not considered members of the society, but they are still referred to as part of the brood.

CASTE Broadly defined, as in ergonomic theory, any set of individuals of a particular morphological type or age-group, or both, that performs specialized labor in the colony. More narrowly defined, any set of individuals in a given colony that is both morphologically distinct and specialized in behavior.

CLADE A species or set of species representing a distinct branch in a phylogenetic tree and hence of single common ancestry.

CLONE A population of individuals all derived asexually from the same single parent.

COEFFICIENT OF RELATEDNESS Also known as the degree of relatedness, and symbolized by *r*, the fraction of genes identical by descent between two individuals.

COLLATERAL KIN Kin other than direct offspring and other lineal relatives such as parents or grandparents.

COLONY A group of individuals, other than a single mated pair, that constructs nests or rears offspring in a cooperative manner (as opposed to an aggregation). *See* Aggregation.

COLONY ODOR The odor found on the bodies of social insects that is peculiar to a given colony. By smelling the colony odor of another member of the same species, an insect is able to determine whether it is a nestmate or not. *See* nest odor.

COMMUNICATION Action on the part of one organism (or cell) that alters the probability pattern of behavior in another organism (or cell). Communication can be "manipulative," altering the behavior of the receiver to the advantage of the sender of the signal, or it can be to the advantage of both sender and receiver. The latter mode is called reciprocal communication and is most often observed in the social insects.

CUES A stimulus that conveys information but has not been "shaped" by natural selection to serve as a communication signal.

DARWINISM The theory of evolution by natural selection, as originally propounded by Charles Darwin. The modern version of this theory still recognizes natural selection as the central process, acting on levels of organization from genes to groups, and for this reason is often called neo-Darwinism.

DIMORPHISM In caste systems, the existence in the same colony of two different forms, including two size classes, not connected by intermediates.

DIPHASIC ALLOMETRY Polymorphism in which the allometric regression line, when plotted on a double logarithmic scale, "breaks" and consists of two segments of different slopes whose ends meet at an intermediate point.

DIPLODIPLOIDY The method of sex determination in which both male and female arise from a fertilized, diploid egg, and sex is determined by a difference in an allele or chromosome, or some other means. For comparison, *see* Haplodiploidy.

DIPLOID With reference to a cell or to an organism, having a chromosome complement consisting of two copies (called homologues) of each chromosome. A diploid cell or organism usually arises as the result of the union of two sex cells, each bearing just one copy of each chromosome. Thus, the two homologues in each chromosome pair in a diploid cell are of separate origin, one derived from the mother and the other from the father.

DISSOLUTIVE Referring to factors or forces in evolution that oppose cohesion and harmony in societies. Other authors speak of these as factors causing conflict that opposes cooperation.

DOMINANCE HIERARCHY Also called dominance order or peck order; the physical domination of some members of a group by other members in relatively orderly and long-lasting patterns. Except for the highest- and lowest-ranking individuals, a given member dominates one or more of its companions and is dominated in turn by one or more of the others. The hierarchy is initiated and sustained by hostile behavior, albeit sometimes of a subtle and indirect nature.

DOMINANCE ORDER Similar to a dominance hierarchy, but at times also having structures other than a strictly linear hierarchy.

DRIVER ANT African legionary ant belonging to the genus Anomma and, less frequently, other members of the tribe Dorylini.

DRONE A male social bee, especially a male honeybee or bumblebee.

ECLOSION Emergence of the adult (imago) from the pupa; less commonly, the hatching of an egg.

ECOLOGY The scientific study of the interaction of organisms with their environment, including both the physical environment and the other organisms that live in it.

ENTOMOLOGY The scientific study of insects.

ERGATOGYNE Any form morphologically intermediate between the worker and the queen; also called intercaste or intermorph.

ERGATOID QUEEN Wingless queen caste in ants, looking similar to the worker caste, but at close inspection clearly identified by external anatomical traits as a distinct queen caste.

ERGONOMICS The quantitative study of work, performance, and efficiency.

EUSOCIAL As it pertains to a group of individuals, displaying all of the following three traits: cooperation in caring for the young; reproductive division of labor, with more or less sterile individuals working on behalf of individuals engaged in reproduction; and overlap of at least two generations of life stages capable of contributing to colony labor. This is the formal equivalent of the expressions *advanced social* and *higher social*, which are commonly used but with less exact meaning.

EVOLUTION Any genetic change in organisms from generation to generation; or, more strictly, any change in gene frequencies within populations from generation to generation.

EVOLUTIONARY BIOLOGY The collective disciplines of biology that treat the evolutionary process and the characteristics of populations of organisms, including their ecology, behavior, and systematics.

EXOCRINE GLAND Any gland, such as the salivary gland, that secretes to the outside of the body or into the alimentary tract. Exocrine glands are the most common source of pheromones, the chemical substances used in communication by most kinds of animals.

FACILITATION *See* Social facilitation; *see also* Group effect.

FAMILY In sociobiology, parents and offspring, together with other kin who are closely associated with them. In taxonomy, the category below order and above genus—that is, a group of related, similar genera. Examples of taxonomic families include Formicidae, including all of the ants, and Felidae, including all of the cats.

FITNESS *See* Genetic fitness.

FREQUENCY CURVE A curve plotted on a graph to display a particular frequency distribution.

FREQUENCY DISTRIBUTION The array of numbers of individuals showing differing frequency (abundance) of some variable quantity; for example, the numbers of animals of different ages or the numbers of nests containing different numbers of young.

GAMERGATE A mated, egg-laying worker. The worker may function as the sole reproductive (monogyny) or one of several reproductives (polygyny) in the colony. Gamergates occur in some ponerine species; they have also been found in one myrmeciine and one myrmicine species. In some cases, morphological queens and gamergates exist in the same colony.

GASTER A special term occasionally applied to the metasoma, or terminal major body part, of ants and other aculeate hymenopterans.

GENE The basic unit of heredity.

GENETIC FITNESS The contribution to the next generation of one genotype in a population relative to the contributions of other genotypes. By definition, the process of natural selection eventually leads to the prevalence of those genotypes with the highest fitnesses.

GENOME The complete genetic constitution of an organism.

GENUS (plural, genera) A group of related, similar species. Examples include *Apis* (the four or more species of honeybees) and *Canis* (wolves, domestic dogs, and their close relatives).

GROOMING The licking of the body surfaces of nestmates. Self-grooming also occurs in ants, whereby individuals clean their own bodies both by licking and stroking with the legs.

GROUP EFFECT An alteration in behavior or physiology within a species brought about by signals or cues that are directed in neither space nor time. A simple example is social facilitation, in which there is an increase of an activity merely from the sight, smell, or sound (or other form of stimulation) coming from other individuals engaged in the same activity.

GROUP SELECTION More precisely, between-group selection; selection that operates on two or more members of a lineage group as a unit. Defined broadly, group selection includes interdemic selection. *See also* Interdemic selection.

HAPLODIPLOIDY The mode of sex determination in which males are derived from haploid eggs and females from diploid eggs.

HARVESTING ANTS Ant species that store seeds in their nests. Many taxonomic groups have developed this habit independently in evolution.

HOLOMETABOLOUS Undergoing a complete metamorphosis during development, with distinct larval, pupal, and adult stages. The hymenopterans, for example, are holometabolous.

HOME RANGE The area that an animal learns thoroughly and patrols regularly. The home range may or may not be defended; those portions that are defended constitute the territory. *See also* Territory.

HOMOPTERAN A member of, or pertaining to, the insect order Homoptera, which includes the aphids, jumping plant lice, treehoppers, spittlebugs, whiteflies, and related groups. In some recent classifications, Homoptera has been treated as a subgoup of the order Hemiptera.

HONEYBEE A member of the genus *Apis*. Unless qualified otherwise, a honeybee is a member of the domestic species *Apis mellifera*, and the term is further usually applied to the worker caste.

HONEYDEW A sugar-rich fluid derived from the phloem sap of plants and passed as excrement through the guts of sap-feeding aphids and other insects. Honeydew is a principal food of many kinds of ants.

HONEYPOT A container made by stingless bees or bumblebees from soft cerumen and used to store honey. Also, the special caste in some species of ants (for example, species of *Myrmecocystus* and *Prenolepis*) whose crop swells hugely as liquid storage receptacles for the colony as a whole.

HORNET A large wasp of the family Vespinae, particularly a member of the genus *Vespa* or (in the United States) the bald-faced hornet *Vespula (Dolichovespula) maculata*.

HYMENOPTERAN Pertaining to the insect order Hymenoptera; also, a member of the order, such as a wasp, bee, or ant.

IMAGINAL DISK A relatively undifferentiated tissue mass occurring in the body of a larva and destined to develop later into an adult organ.

INBREEDING The mating with close kin. The degree of inbreeding is measured by the fraction of genes that will be identical owing to common descent.

INCLUSIVE FITNESS *See* Kin selection.

INSECT SOCIETY In the strict sense, a colony of eusocial insects (ants, termites, eusocial wasps, or eusocial bees). In the broad sense adopted in this book, any group of presocial or eusocial insects.

INSTAR Any period between molts during the course of development.

INSTINCT Behavior that is highly stereotyped, more complex than the simplest reflexes, and usually directed at particular objects in the environment or other members of the same species and society. Learning may or may not be involved in the development of instinctive behavior; the important point is that the behavior develops toward a narrow, predictable end product.

INTERDEMIC SELECTION The selection of entire breeding populations (demes) as the basic unit. It is a form of group selection defined broadly. *See also* Group selection.

INVERTEBRATE An animal lacking a vertebral column. Invertebrates include a wide range of animals, from protozoans to insects and starfish.

ISOMETRY Growth of an individual that does not result in change in the relative size of different parts of the body.

KIN SELECTION The selection of genes due to one or more individuals favoring or disfavoring the survival and reproduction of relatives who possess the same genes by common descent. Kin selection theory is equivalent to inclusive fitness theory. In a narrow sense, kin selection means collateral (nondescendant) kin selection, entailing all relatives except direct offspring.

KINSHIP Parent and offspring or otherwise possession of a common ancestor in the not-too-distant past by two or more individuals. Kinship is measured precisely by the coefficient of relatedness. *See also* Coefficient of relatedness.

LABIUM In insects, the lower "lip," or lowermost mouthpart-bearing segment, located just below the mandibles and maxillae.

LABRUM The upper, hinged "lip" of an insect.

LARVA An immature stage that is radically different in form from the adult; characteristic of the holometabolous insects, including the hymenopterans. In the termites, the term is used in a special sense to designate an immature individual without any external trace of wing buds or soldier characteristics.

LEGIONARY ANT *See* Army ant.

LIFE CYCLE The entire span of the life of an organism (or of a society), from the moment it originates to the time it reproduces.

MAJOR WORKER A member of the largest worker subcaste, especially in ants. In ants, the subcaste is usually specialized for defense, and adults belonging to this subcaste are often referred to as soldiers. *See also* Media worker; Minor worker.

MASS COMMUNICATION The transfer of information among groups of individuals of a kind that cannot be transmitted from a single individual to another. Examples include the spatial organization of army ant raids, the regulation of numbers of worker ants on odor trails, and certain aspects of the thermoregulation of nests.

MASS PROVISIONING The act of storing all of the food required for the development of a larva at the time the egg is laid (as opposed to progressive provisioning). *See also* Progressive provisioning.

MATING FLIGHT *See* Nuptial flight.

MEDIA WORKER In polymorphic ant species containing three or more worker subcastes, an individual belonging to the medium-sized subcaste(s). *See also* Minor Worker; Major worker.

MINOR WORKER A member of the smallest worker subcaste, especially in ants; also called a minima. *See also* Nanitic worker; Media worker; Major worker.

MOLT In insects or other arthropods, the casting off of the outgrown skin or exoskeleton in the process of growth; also, the cast-off skin itself. The word is further used as a verb to designate the performance of the behavior.

MONANDRY The tendency of each female to mate with only one male.

MONOGYNY In animals generally, the tendency of each male to mate with only a single female (as opposed to polygyny). In social insects, the existence of only one functional queen in the colony. *See also* Polygyny.

MONOMORPHISM The existence within a species or colony of only a single worker subcaste.

MONOPHASIC ALLOMETRY Polymorphism in which the allometric regression line has a single slope, which is greater or lesser than one unit (of measurement).

MULTILEVEL SELECTION Selection that targets different levels of biological organization, in particular at the level of the organism and the level of the colony or some other group.

MUTATION In the broad sense, any discontinuous change in the genetic constitution of an organism. In the narrow sense, a "point mutation," a change along a very narrow portion of the nucleic acid sequence.

MYRMECOLOGY The scientific study of ants.

NANITIC WORKER A worker of extremely small size, a type usually limited to the first generation of workers produced by a nest-founding queen.

NATURAL SELECTION The differential contribution of offspring to the next generation by individuals of different genetic types but belonging to the same population. This is the basic mechanism proposed by Charles Darwin and is generally regarded today as the main guiding force in evolution.

NEST ODOR The distinctive odor of a nest, by which its inhabitants are able to distinguish the nest from those belonging to other colonies or at least from the surrounding environment. In some cases, the insects, (for example, honeybees and some ants) can orient toward the nest by means of the odor. The nest odor may be the same as the colony odor in some cases. The nest odor of honeybees is often referred to as the hive aura or hive odor. *See also* Colony odor.

NICHE The range of each environmental variable, such as temperature, humidity, and food items, within which a species can exist and reproduce. The preferred niche is the one in which the species performs best, and the realized niche is the one in which it actually comes to live in a particular environment.

NOMADIC PHASE The period in the activity cycle of an army ant colony during which the colony forages more actively for food and moves frequently from one bivouac site to another (as opposed to the statary phase). At this time, the queen does not lay eggs, and the bulk of the brood is in the larval stage. *See also* Statary phase.

NONDESCENDANT KIN SELECTION The same as collateral kin selection; the propagation of genes identical by descent through kin other than direct offspring.

NUPTIAL FLIGHT The mating flight of the winged queens and males.

NYMPH In general entomology, the young stage of any insect species with hemimetabolous development, in which the young are similar in body structure to the adults. In termites, the term is used in a slightly more restricted sense to designate immature individuals that possess external wing buds and enlarged gonads and are capable of developing into functional reproductives by further molting.

ODOR TRAIL A chemical trace laid down by one insect and followed by another. The odorous material is referred to either as the trail pheromone or as the trail substance.

OLIGOGYNY The occurrence in a single colony of from two to several functional queens. Oligogyny is characterized by worker tolerance to more than one queen, combined with antagonism among the queens, with the result that multiple queens cannot coexist in the same immediate vicinity and must space out.

OMMATIDIUM One of the basic visual units of the insect compound eye. The ommatidia are bounded externally by the facets that together make up the glassy, rounded outer surface of the eye.

OVARIOLE One of the egg tubes that together form the ovary in female insects.

PALPATION Touching with the labial or maxillary palps. The movement can serve as a sensory probe or as a tactile signal to another insect.

PARTHENOGENSIS The production of an organism from an unfertilized egg.

PATROLLING The act of investigating the nest interior and outer nest area. Worker honeybees, for example, are especially active in patrolling and are thereby quick to respond as a group to contingencies when they arise in the nest.

PETIOLE The first segment of the "waist" of aculeate hymenopterans; actually, the second abdominal segment, since the first abdominal segment (propodeum) is fused to the thorax.

PHENOTYPE The observable properties of an organism or superorganism (colony) as they have developed under the combined influences of the genetic constitution of the individuals and the effects of environmental factors.

PHEROMONE A chemical substance or blend of substances, usually glandular secretions, used in communication within a species. One individual releases the material as a signal, and another individual responds after tasting or smelling it. Primer pheromones alter the physiology of individuals and prepare them for new behavioral repertoires. Releaser pheromones evoke responses directly.

PHYLOGENY The evolutionary history of a particular group of organisms; also, the diagram of the "family tree" that shows which species (or groups of species) gave rise to others.

PLEOMETROSIS *See* Polygyny.

POLICING Harassing or killing a nestmate worker or queen, or removing the eggs she lays, especially targeting one that tries to usurp the reproductive role.

POLYANDRY The acquisition by a female of more than one male as a mate. In social insect biology, the mating of a female with more than one male.

POLYGYNY The coexistence in the same colony of two or more egg-laying queens. When multiple queens found a colony together (called pleometrosis) and the queens stay together in the mature colony, the condition is referred to as primary polygyny. When supplementary queens are added after colony foundation, the condition is referred to as secondary polygyny.

POLYMORPHISM In social insects, the coexistence of two or more functionally different castes within the same sex. In ants, polymorphism is defined more precisely as nonisometric relative growth occurring over a sufficient range of size variation within a normal mature colony to produce individuals of distinctly different proportions at the extremes of the size range.

PRESOCIAL Also called pre-eusocial; relating to a species that exhibits social behavior less advanced than eusocial behavior.

PROGRESSIVE PROVISIONING The act of providing the larva with meals at intervals during its development (as opposed to mass provisioning). *See also* Mass provisioning.

PUPA The inactive instar of the holometabolous insects (including the hymenopterans) during which development into the final adult form is completed.

QUEEN A member of the reproductive caste in semisocial or eusocial species. The existence of a queen caste presupposes the existence also of a worker caste at some stage of the colony life cycle. In a functional definition of the reproductive caste, queens may not be morphologically different from workers; such individuals are now called gamergates. Using a morphological criterion, the queen caste is defined by its distinctly different anatomy from the worker caste.

QUEENRIGHT Referring to a colony, especially a honeybee colony, that contains a functional queen.

RECOMBINATION The repeated formation of new combinations of genes through the processes of meiosis and fertilization that occurs in the typical sexual cycle of most kinds of organisms.

RECRUITMENT A special form of assembly by which members of a society are directed to some point in space where work or other collective actions are required.

RECRUITMENT TRAIL An odor trail laid by a single scout worker and used to recruit nestmates to a food find, a desirable new nest site, a breach in the nest wall, or some other place where the assistance of many workers is needed, as in defense of the territory.

RELEASER A sign stimulus used in communication. The term is often used broadly to mean any sign stimulus.

REPLETE An individual ant whose crop is greatly distended with liquid food, to the extent that the abdominal segments are pulled apart and the intersegmental membranes are stretched tight. Repletes usually serve as living reservoirs, who regurgitate food on demand to their nestmates. *See also* Honeypot.

RITUALIZATION The evolutionary modifications of morphological or physiological traits or behavioral patterns that change them into signals used in communication or improve their efficiency as signals.

ROLE A pattern of behavior displayed by certain members of a society that has an effect on other members.

SCLERITE A portion of the body wall bounded by sutures.

SIGN STIMULUS The single stimulus, or one of a very few such crucial stimuli, by which an animal distinguishes key objects, such as enemies, potential mates, and suitable nesting places.

SIGNAL A behavioral action that conveys information to members of the same species or to members of the group. True signals, as opposed to cues, have been shaped or modified by natural selection specifically to serve in communication. *See also* Cues.

SOCIAL FACILITATION The phenomenon in which there is an increase of an activity merely from the sight, smell, or sound (or other form of stimulation) coming from other individuals engaged in the same activity.

SOCIAL INSECT In the strict and usual sense (for "true," or "advanced" social insects), an insect that belongs to a eusocial species—for example, an ant, a termite, or one of the eusocial wasps or bees, beetles, thrips, or aphids. In the broad sense, an insect that lives in a cohesive group, with the group members interacting in a way that binds them together.

SOCIETY A group of individuals belonging to the same species and organized in a cooperative manner. The diagnostic criterion is reciprocal communication of a cooperative nature extending beyond mere sexual activity.

SOCIOBIOLOGY The systematic study of the biological basis of all forms of social behavior.

SOLDIER A member of a worker subcaste specialized for colony defense.

SPECIES The basic lower unit of classification in biological taxonomy, consisting of a population or series of populations of closely related and similar organisms. More narrowly defined, a biological species consists of individuals that are capable of interbreeding freely with one another but not with members of other species under natural conditions.

SPERMATHECA The receptacle in a female insect in which the sperm are stored, also called the sperm pocket.

STATARY PHASE The period in the activity cycle of an army ant colony during which the colony is relatively quiescent and does not move from site to site (as opposed to the nomadic phase). At this time, the queen lays the eggs, and the bulk of the brood is in the egg and pupal stages. *See also* Nomadic phase.

STERNITE A ventral sclerite; a portion of the body wall bounded by sutures and located in a ventral position. *See also* Tergite.

STRIDULATION The production of sound or body vibrations by rubbing one part of the body surface against another. Some insect groups (including grasshoppers, crickets, and many ant species) possess special stridulation devices.

SUBSOCIAL Pertaining to a group in which the adults care for their nymphs or larvae for some period of time.

SUPERORGANISM A society, such as a eusocial insect colony, that possesses features of organization analogous to the physiological properties of single organisms. The eusocial colony, for example, is divided into reproductive castes (analogous to gonads) and worker castes (analogous to somatic tissue); its members may, for example, exchange nutrients and pheromones by trophallaxis and grooming (analogous to the circulatory system). Among the thousands of known social insect species, we can find almost every conceivable grade in the division of labor, from little more than competition among nestmates for reproductive status to highly complex systems of specialized subcastes. The level of this gradient at which the colony can be called a superorganism is subjective; it may be at the origin of eusociality (preferred by E. O. Wilson) or at a higher level, beyond the "point of no return," in which within-colony competition for reproductive status is greatly reduced or absent (preferred by B. Hölldobler).

SYMBIOSIS The intimate, relatively protracted, and dependent relationship of members of one species with those of another. The three principal kinds of symbiosis are commensalism, mutualism, and parasitism.

TANDEM RUNNING A form of communication, used by the workers of certain ant species during exploration or recruitment, in which one individual follows closely behind another, frequently contacting the abdomen of the leader with its antennae. Leader and follower are tightly bound to one another by a continuous exchange of signals.

TARSUS The foot of an insect, the one- to five-segmented appendage attached to the tibia, or lower leg segment.

TERGITE A dorsal sclerite; a portion of the body wall bounded by sutures and located in a dorsal position. *See also* Sternite.

TERRITORY An area occupied more or less exclusively by an animal or group of animals (such as an ant colony) by means of repulsion through overt defense or aggressive advertisement.

TRAIL PHEROMONE A substance laid down in the form of a trail by one animal and followed by another member of the same species.

TROPHALLAXIS In social insects, the exchange of alimentary liquid among colony members and guest organisms, either mutually or unilaterally. In stomodeal (oral) trophallaxis, the material originates from the mouth; in proctodeal (anal) trophallaxis, it originates from the anus.

Trophic egg An egg, usually degenerate in form and inviable, that is fed to other members of the colony.

UNICOLONIAL Pertaining to a population of social insects in which there are no behavioral colony boundaries.

WAGGLE DANCE The dance whereby workers of various species of honeybees (genus *Apis*) communicate the location of food finds and new nest sites. The dance is basically a run through a figure-eight pattern, with the middle, transverse line of the eight containing the information about the direction and distance of the target.

WORKER A member of the nonreproductive, laboring caste in semisocial and eusocial species. The existence of a worker caste presupposes the existence also of royal (reproductive) castes. In Hymenoptera, particularly in ants and bees, the worker caste can be defined morphologically, emphasizing the distinctly different morphology of workers and queens in most species, or functionally, as in some ponerine ant species that lack a morphological queen caste and in several wasp species. In termites, the term is used in a more restricted sense to designate individuals in the family Termitidae, which completely lack wings and have reduced pterothorax, eyes, and genital apparatus.

ZOOLOGY The scientific study of animals.

INDEX

ABOUT THE AUTHORS

BERT HÖLLDOBLER is Foundation Professor of Life Sciences at Arizona State University. Before joining ASU he was the Alexander Agassiz Professor of Zoology at Harvard University (1973–1990), and he held the Chair of Behavioral Physiology and Sociobiology at the University of Würzburg, in Germany (1989–2004). In 2002 he was appointed Andrew D. White Professor at Large at Cornell University.

He is a member of several national and international academies, among them the German Academy of Sciences (Leopoldina), the American Philosophical Society, the American Academy of Arts and Sciences, and the National Academy of Sciences (USA). He is the author of several books, including *The Ants,* which he coauthored with Edward O. Wilson, and for which they received a Pulitzer Prize (1991) for nonfiction writing, and their book *Journey to the Ants,* which was awarded the Phi Beta Kappa Prize. He is the recipient of some of the most prestigious research prizes in Germany, among others the Gottfried Wilhelm Leibniz Prize of the German Science Foundation, the Körber Prize for the European Sciences, and the Alfried Krupp Science Prize. At ASU Bert Hölldobler is a cofounder of the new Center for Social Dynamics and Complexity, and he plays a key role in organizing the new social insect research group at the School of Life Sciences. Along with his wife, he divides his time between Arizona and Germany.

EDWARD O. WILSON was born in Birmingham, Alabama, in 1929 and was drawn to the natural environment from a young age. After studying evolutionary biology at the University of Alabama, he has spent his career focused on scientific research and teaching, including forty-one years on the faculty of Harvard University. His twenty books and more than four hundred mostly technical articles have won him over one hundred awards in science and letters, including two Pulitzer Prizes, for *On Human Nature* (1979) and, with Bert Hölldobler, *The Ants* (1991); the U.S. National Medal of Science; the Crafoord Prize, given by the Royal Swedish Academy of Sciences for fields not covered by the Nobel Prize; Japan's International Prize for Biology; the Presidential Medal and Nonino Prize of Italy; and the Franklin Medal of the American Philosophical Society. For his contributions to conservation biology, he has received the Audubon Medal of the Fund for Nature. Much of his personal and professional life is chronicled in the memoir *Naturalist,* which won the *Los Angeles Times* Book Award in Science in 1995. Still active in field research, writing, and conservation work, Wilson lives with his wife, Irene, in Lexington, Massachusetts.